LIGHT, MOLECULES, REACTION AND HEALTH

LIGHT, MOLECULES, REACTION AND HEALTH

Angelo Albini
Department of Organic Chemistry, University of Pavia, Pavia, Italy

Academic Press is an imprint of Elsevier
125 London Wall, London EC2Y 5AS, United Kingdom
525 B Street, Suite 1650, San Diego, CA 92101, United States
50 Hampshire Street, 5th Floor, Cambridge, MA 02139, United States
The Boulevard, Langford Lane, Kidlington, Oxford OX5 1GB, United Kingdom

British Library Cataloguing-in-Publication Data
A catalogue record for this book is available from the British Library

Library of Congress Cataloging-in-Publication Data
A catalog record for this book is available from the Library of Congress

ISBN: 978-0-12-811659-3

For Information on all Academic Press publications
visit our website at https://www.elsevier.com/books-and-journals

Publisher: Andre Gerhard Wolff
Acquisition Editor: Erin Hill-Parks
Editorial Project Manager: Sara Pianavilla
Production Project Manager: Swapna Srinivasan
Cover Designer: Matthew Limbert

Typeset by MPS Limited, Chennai, India

Dedication

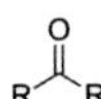

hν

The (selective and quantized) absorption of light generates electronically excited states, in this case by promotion of an electron from the n to the π* orbital

The (triplet) excited states behaves as a highly reactive electrophile

The reaction is atom (H) transfer and yields a pair of highly reactive odd electron species (free radicals)

A variety of reactions, e.g. the one noted below, lead back to the ground state. Excess energy is developed as health

Professor **Ugo Mazzuccato** -
(1929–2017),
to whom this book is dedicated
with gratitude

Contents

7. Emission

8. Conclusion and outlook

Preface

The absorption of a photon by a molecule is a quantized phenomenon and subtracts it to the thermal equilibrated mixture (solvent, solute) promoting (and only it) to an excited state.

$$A + h\nu \rightarrow A*$$

These electronically excited states enjoy but a short lifetime, as one may expect from the high energy involved (typically $h\nu = 40-200$ kcal) and the Boltzman equilibrium, and in fact usually decay even faster. The high energy of such states, however, allows to undergo deep-seated transformations, in a way that may seem too complex, although perfectly rationalizable in the same way as ground state, both qualitatively and by quantitative computations.

One may find in the literature even quite recent studies on the different chemistry of the ground versus excited states for bi- or triatomic molecules, such as O_2, Cl_2, or N_2O. The story becomes more and more complex when proceeding from small, simple molecules to larger ones, where many conformers are present. Biomolecules generally are, in fact, polymers, and quite complex ones. Let us consider as an example a protein with hundreds or thousands of amino acids joined together in a long chain, which is capable of an enormous quantity of different conformations, each of them working as a different molecule in the absorption and involving a photoreaction of its own, from each of the sufficiently long-lived excited states.

Thus there is no reason to be pessimistic about whether it is reasonable to describe on theoretical basis the reactions occurring, but this is surely not applying to my life. On the contrary, chemists are accustomed to extending conclusions that are, although qualitatively, working for polymers, by choosing the role of functional or bulky groups.

The presentation in this book regards the chemical reactions of biomolecules in the cells. These reactions are distinguished from those occurring on excited states in dilute solutions, the preferred medium for mechanistic studies. As a matter of fact, the photochemical reaction occurring in the actual system involves a chemical change on some molecules, heavily controlled by the local structure, then some thermal generally enzymatic transformation in the cell that is the consequence of that reaction, but certainly not identical with it, then, through a couple of further transformations, to a final effect on some organ that undergoes a disease.

Summing up, while the initial reaction is certainly a chemical reaction that initiates a series of thermal steps that, through a series of transformations at the cell level, brings the system to a disease. The mere list above is an obvious sign of the interdisciplinary characteristics of this science, which spans from physics to chemistry, to biology, and to medicine, which rarely works easily and rather requires adapting the method to a specific definition. It seems to be that no book addressing this point has been prepared as yet, while there are excellent texts of say photochemistry, photomedicine, photobiology, etc., the order of presentation is structured with a short initial section that has no pretense of substituting texts of the various sciences listed above, but simply to offer a vocabulary. Then the main reaction involving light as a signal, vision, nonvisual sight, the synthesis of vitamin D from the previtamin in the skin, protection of the internal organs, theranostics, most of which involving concerted photoreactions and possessing robust protecting systems when this is not the case, the use of light for generating active drugs and the combination of emission for guiding therapy (theranostics).

This discipline has certainly not failed to generate the interest of scientists, as is indicated by the Nobel Prize in Biology and Medicine awarded to Nils R. Finsen in 1903 for treating *Lupus vulgaris* (tuberculosis of the skin), with light radiation; to Michael Rosbash, Michael W. Young, and Jeffrey C. Hall, in 2017 for their discoveries of molecular mechanisms controlling the circadian rhythms; to Melvin E. Calvin in 1961 for his discovery of the detailed steps involved in the chlorophyll synthesis[1]; and in 1967 in Chemistry to Ronald Norrish and George Porter.[2]

At least from what I may see, there are no further books on the subject presented here, and I am supported here by the fact that when submitting the proposal to the Publisher, it was readily accepted as something different from a photobiology book. On this point, one may notice, however, that the matter presented here is available not only from publications in (photo)biology and (photo)chemistry, but also in special issues on a number of journals that in part compensate for this need. This is particularly true with reference to the role of free radicals in photobiology, where such special issues have a really important function. Also the amount of science accumulated in these decades is really immense (or, at least, a chemist is not accustomed to such proportions), which forces one to skip most of the material. In particular the choice has been to leave out any patent, although technology advances are quite important in this science, as well as most of the literature not in English or not available through common retrieval means, although again medicine possesses, besides a large international significance also national topics that are significant too. Another important difference is how fast a science changes. I do not think that contemplating the eternal thrushes of mathematics is per se a guarantee of a lively science, but certainly at the moment chemistry takes a fully different approach about every 30 years and biology faster than that. This is obviously a further stumbling block in the project of a book, but one has no choice, and a fast perusal at the reference list immediately makes the result apparent, with all of them, except those of historic values in the present century, and the large majority of them being published within the last lustrum.

I am aware that having a single author to review a field of which he has no experimental experience is a great risk, since it is difficult for him to judge the significance of every experiment, but it seemed to be that putting together such a book in a reasonable time (6 years) was no further possibility. As for the role a practitioner of chemistry may have, I feel that is the best choice although this seems not to be supported by most funding agencies, which rather prefer having physics and biology together with chemistry at most in a menial role for characterizing the material used. If I may recall a personal experience, I happened to participate in a (public) discussion of some research projects, where a participant, illustrious biology practitioner, pointed

1 List of Nobel Prize winners in Physiology or Medicine.

2 List of Nobel Prize winners in Chemistry.

out that what was proposed was not only developing science, but also preparing good scientists who may not only predict how to synthetize the ideal material and instruments for the devised jobs, but also carry out the actual synthesis and assembling of them. To which, the humble undersigned could not avoid noting that such people had been already invented, and they went under the name of chemists. The other point is that while accepting the full responsibility of what is presented in the following, the author acknowledges that this result could not be obtained without the strong help from a number of individuals, in particular Elisa Fasani, PhD; Michela Sturini, PhD; Michele Albini, PhD; Nicola Scuro, Andrea P. Missiroli, MD; Alessandra Bellagente, PhD; Alberto del Ponte, PhD, as well as the continuous support by students who have made the research such a rewarding experience every day, from whom I learned most of the photochemistry, as well as Prof. Ugo Mazuccato, from whom, I feel, I have learned the rest of that science, the best mentor in my remembrance, and to whom I dedicate this worthy book. The same contains a brief account of the senses that communicate with the environment, visual and nonvisual light-sensitive organs, of the skin and how it is affected by sunlight, as well as of light as a healing instrument (PUVA, PDT, and lasers). I attempted to keep the tone of the narrative as light as I was able, but certainly, I do not think this may be the *livre de chevet* of anyone. On the contrary, I attempted to offer a lot of information, perhaps too much, which can be skipped with a good conscience, however.

And now, let us start, I am convinced that the world of light and biomolecules offers much fun, and I concur with Gary Ross, the director of that nice 1998 Hollywood movie "Pleasantville," where it is seen that the whole community finds something worth living when passing from black and white to colors. It goes without saying that I would be grateful for any message about one of the many theoretical or factual mistakes present in the book.

Pavia/Milano, Easter 2017.

CHAPTER

1

Health and light

1.1 The role of light

Light-caused effects consistently involve a reaction of an apolar organic molecule that undergoes some *large change* in the shape. This change allows certain ions to pass through the channel in the membrane suitable for the measure of those cations, the general method chosen by nature in order to register and translate signals. Plants have learned to exploit the large flux of energy impinging on the Planet surface through the complex machinery of chlorophyll photosynthesis (as well as through earlier evolved phototropic systems and to use them for synthesizing a large amount of high energy compounds. However, this has led to large cells enclosed in stiff walls and relatively heavy organisms, not suited for movement. In contrast, animals have learned to move around and chase plants or other animals as they use flexible cells with thin walls fitted with lysosomes and a system for getting energy from chemicals, not the sun, which is much easier (as it were to have a gasoline engine in comparison to a heavy electrical auto fitted with electrical condensers that have to be recharged often). However, the environment where both plants and animals live is the same. Plants have elaborated their chlorophyll system in such a way that no side path potentially damaging the cells is taken competitively when exploiting solar energy. This issue is surely no less serious for animals. As they move, they have to know where they are going and use light for interacting with the environment. Even if they do not use light for their nourishment, they still have the protective part to care for and further have to trust on light for some very important functions, including evolution through mutations (practically only DNA crossing is suitable for this target). As for vertebrates, the very name implies that they have to have a strong basis to which lean against for mechanical work, and their bones are able to answer this requirement is satisfied by the continuous building and redissolving the calcium phosphate, a further function that is regulated by light. This opens up a neighboring ion passage and thus enables a change in the ion internal concentration. Such is the usual vocabulary to which nature makes recourse for messages. When photochemical reactions different from those expected occur, the course of events takes a different path and the function stops or follows a different course. Such alternative paths are likewise typical photoreactions of organic molecules involving the formation or cleavage of covalent bonds. Nature has evolved a large array of repair mechanisms, but these may not be sufficient and light-induced diseases may become apparent.

Light, Molecules, Reaction and Health
DOI: https://doi.org/10.1016/B978-0-12-811659-3.00001-3

More generally, what we call light corresponds to a relatively small range of wavelengths from the electromagnetic spectrum, the visible light and the ultraviolet (UV) light. This is particularly important on two grounds: first, this wavelength range is abundant in nature (it makes about 50% of the electromagnetic energy reaching the Earth surface); second, it happens to involve the correct energy for forming electronically excited states that have energy comparable to that of covalent bonds, and thus are certainly prone to react in some way. The course of photochemical reactions has long puzzled scientists working in the field, because of the qualitative difference from those of ground states. Up to some decades ago, electronic excited states were understood as high energy excited species in a thermostatted bath formed by nonabsorbing molecules that remains unchanged, which was enough to explain the fact that photochemical reactions were independent on temperature, no kinetic could be defined and had no common property with catalytic processes. At any rate, understanding how electron distribution changes upon excitation allows to justify, and to a degree, to predict, the reactions of such states. However, recent research has demonstrated that there are many instances of excited states that react faster than they thermalize, in particular among biomolecules.

The first law of photochemistry [1] states that light has to be absorbed to cause any effect and a look to the absorptions of cells and their components evidences that UV light is strongly absorbed; typical examples are (hetero)aromatic amino acids in proteins, DNA, and coenzymes (Fig. 1.1) [2].

The effect solar light has on plants and the season cycles represent the basis of the chronological organization with the yearly festivals that has been one of the fundamental points in the human history as when every tribe adopted agriculture and settled on a ground of its own. Reports on the action of light, for example, in the decolorizing of various objects, began to appear since at least 2000 years BC. For what is closer to the present topic, the effect of light on biomolecules and health, it may be appropriate to recall that in August 1868 Dr. H. Swete read before the Public Health Section at the annual meeting of the British Medical Association in Oxford a report on his way to obtain "correct comparative observations in estimating the influence of light in health and disease," by using an instrument he had built himself, the "actinograph" (where a strip of albuminized paper dunk in a silver nitrate solution was exposed 10 min to light and then developed) [3].

He hoped to have available soon an improved instrument and to obtain results of interest for those of the medical profession that had made "bettering the condition of the poor their more especial study," as well as to carry out comparative observations that may "throw some light on the value of our health resorts," and contrasted the conditions under which people lived there with those of the crowded city. The appalling death rates reported in the poorest part of the population where not only due to bad air and overcrowding, but also to the want of a due proportion of light. Somewhat roughly, he classed the action of solar light in three groups, the pure visual light, most abundant in the yellow rays (and at noon); the calorific part of light, most abundant in the red (and in the afternoon); and the chemical or actinic effect, which is most intense in the blue (and in the forenoon). In analogy with the fact that plants shoot up rapidly but weakly and died early when exposed to yellow light, he thought that want of actinism in effecting the vital functions of the blood caused the big heads and small bodies of the children he had observed in the dwellings of the poor. Exposure to solar light may thus take a therapeutic significance. Understanding the effect of light on biological processes is not easy, because (1) often heavy skeleton

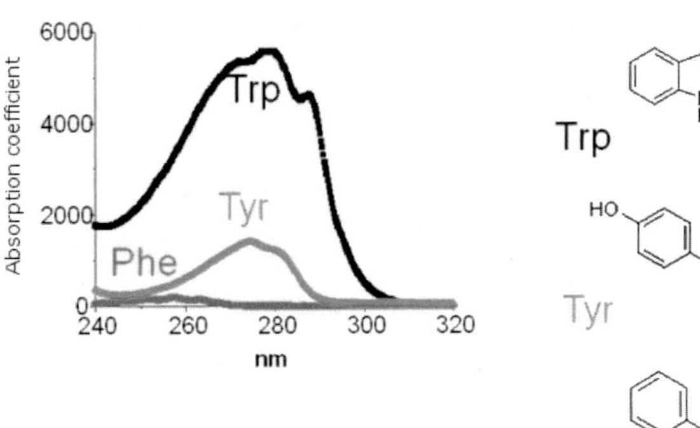

FIGURE 1.1 Absorption spectra of some biomolecules, including amino acids, proteins, and coenzymes.

rearrangements intervene and this has been a stumbling block, which is, however, much better overcome with the now available sophisticated instruments; (2) peculiar structures are extremely effective in some functions, as it is the case for previtamin D precursors, that when purified revealed an unknown potency. Indeed heliotherapy was an old practice, used in ancient times by all known civilizations, and increasingly in the last centuries. Thus Francis Glisson (1697–1777), a professor of medicine at Cambridge, had described this disease associated with want of exposure to light, primarily on the basis of bone deformation observed during dissection, which was known as "wrickden" from the ancient English for "twist" and proposed the name rickets from the Greek rakis, meaning backbone [4].

This disease had been long known, for example, in ancient times Herodotus had reported after visiting the field of a battle between Persians and Egyptians that the skulls of the latter warriors appeared to be much harder than those of the former ones. The skulls of Persians were mostly crushed simply by throwing a pebble to them, and Herodotus had attributed the larger hardness of the skulls of Egyptians to their habit of bearing a shaved head from the young age, thus leaving a free path to the solar light to strengthen the bones. This disease had enormously developed in England starting from the end of the 18th century, in correspondence with the development of the industrial revolution, so that it was said "Morbum Anglorum." The treatment by Folk medicine was combining a diet rich of fish fats, such as cod liver oil, and exposure to solar light. At the beginning of the 20th century the ability of solar light to kill microorganisms was known [5], as was the fact that purulent wounds healed much better when exposed to solar light. From this basis, phototherapy began its course, with the foundation of "sanatories" in mountain sunny resorts. Typically, the treatment involved exposure to the sun of growing portions of the skin or, in order to speed up the healing, also in front of artificial light sources, as exemplified by the then available carbon arcs. Successful application to diseases of large societal impact such as tuberculosis and Lupus vulgaris, made heliotherapy highly considered [6], and gave rise to the period of health tourism, particularly in Switzerland, where, according to a promotional slip, "the medical organization was ensured by 40 medical specialists and more than 3000 beds were available in 80 different clinics which were open all the year around." There were large private establishments and pensions, State sanatoria and clinics reserved for patients of moderate means, Swiss or foreigners. These 80 clinics were divided into two main groups according to the cases to be treated. First there were those specializing in the treatment of the respiratory system and situated mostly in Feydey, the higher part of the village of Leysin (4500–5000 ft), where pulmonary tuberculosis and affections of the bronchial tubes or larynx asthma etc. were treated. In all these establishments, the treatment was based on the rest and fresh air cure, but, in case surgical treatment was necessary, certain clinics possessed all of the necessary equipment for collapse therapy or thoraces surgery, complete radiology installations, analytical laboratories, etc. This medical and surgical equipment was at the disposal of all of the clinics as were also those of the center for cardiology, the center for the examination of respiratory functions and a clinic which was specialized in the treatment of ocular affections. The other group consisted in the heliotherapeutic clinics under the medical direction of Prof. Dr. A. Rollier, situated mostly in and around Leysin village and intended for treating such disorders [7]. The introduction of strong antimicrobials extinguished these practices, but this remains a fascinating chapter in the history of medicine [7–9]. Exposure to solar light was found to exert protection from tuberculosis of the bones, joints, glands, peritoneum, skin, etc., as well as of certain nontuberculosis affections of the bones and joints. Some clinics were equipped to receive patients suffering from tuberculosis of the genital urinary tract. The treatment was based on progressive sun and air cure according to the individual needs and on the orthopedic principles, the immobilizing equipment used here is easily removable, thus permitting the utmost care of the skin and muscles [7].

Understanding in which way light caused such effects was not simple, unfortunately. In the case of vitamin D, it was found that a rat submitted to a rachitic diet would remain healthy if irradiated with UV light. However, "control" experiments led to contrasting observation. Thus positive results were obtained also when the rats were put into a preirradiated jar, or when one of two rats was removed for irradiation and then returned to the jar. The ideas that either air or material objects that had been irradiated continued to convey healthful secondary radiations were investigated but not confirmed. The then commercially important finding that some diets known to promote rachitism would become antirachitic under irradiation marked a step forward. However, this effect did not explain all of the previous findings. Consumption of either small irradiated fecal particles or of feces from irradiated rats was the likely explanation for the recovery of nonirradiated rats, but this was not tested by direct experiment. It was suggested that an alternative possibility, activity of grease from irradiated fur, could be involved [10]. Most of such "mysteries" were solved when the actual UV-activatable compound, dehydrocholesterol, was purified and isolated from samples of cholesterol, where it was present in a proportion 1–200, and its exceptional potency was revealed [11].

About sixty years after Dr. Swede talk, the development of artificial lamps had been considerable, and it was felt that this was the beginning of a new era of artificial lighting, which had a role also in improving health, not only in conquering more illuminated hours. New tungsten filament mercury vapor lamps had been fabricated by using appropriate glasses that allowed to conserve the health-giving properties of short wavelength radiation, while avoiding the attending risk [12].

After World War II some treatments by light entered the general clinical practice, in particular that of the neonatal jaundice, but at the same time the general diffusion of electric lamps for illumination forced citizen to participate as "unwitting subjects in a long-term experiment on the effects of artificial lighting environments on human health." Although, luckily, such experiment caused "no demonstrably baneful effects" [13]. This was clearly not the way to rationalize the sophisticate interaction between health and (artificial) light. The evidence that had been accumulating suggested in fact that the operation of several organs may be seriously affected by the absorption of light and the intervening chemistry began to be explored [14,15].

1.2 Reactions of the excited states

The variety of chemical reactions is impressive, and certainly photochemical reactions are at first sight even more varied and complicated than thermal ones, but they can be discussed in the same way as those not involving light, on the basis of the electronic structure of the reagents. Electronically excited states result from the promotion of an electron from a bonding orbital, often the highest occupied molecular orbital (HOMO) to an antibonding one, often the lowest vacant orbital. A complication may be that, while there is only one ground state (the singlet for organic molecules and many metal complexes, which are closed-shell species, with all of the orbitals doubly occupied or empty, and have only doubly occupied or empty molecular orbitals), electronically excited states necessarily are open-shell species, singlet or triplet, both of them with two semioccupied orbitals, in the first case with the electron spin coupled, in the latter one parallel. As indicated in Scheme 1.1, for every electronic occupancy there are two spin configurations: singlet and triplet. UV-visible spectra are easily registered, at least from 200 nm up, since otherwise oxygen in the air absorbs significantly. Furthermore, one has to take in care that absorption by the components (including the

S_1 T_1 S_0 H C=O H

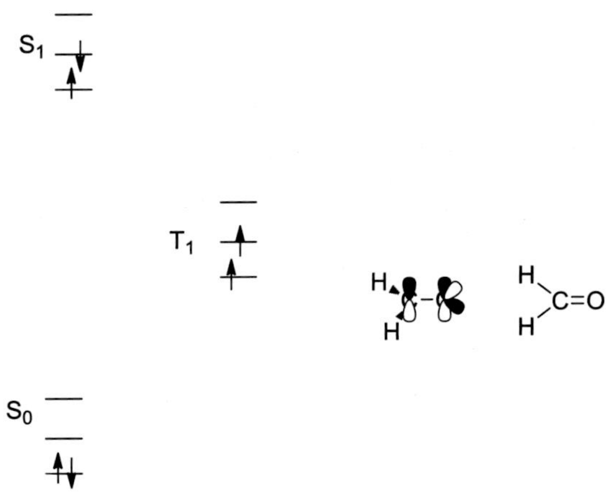

SCHEME 1.1 Contrary to the ground state (S_0), a closed-shell species, electronically excited states are open-shell species and come in pairs, singlet, and triplet for every occupancy of the molecular orbitals. In the example shown, the MOs concerned may be the nonbonding *n* MO sitting on the oxygen atom (in plane) of formaldehyde and the π MO perpendicular to the molecular plane. Thus absorption of a photon by the ground state S_0 $[\pi_2 n_2 \pi^*_0]$ leads to a singlet and triplet for each of the electronic configurations, $^{1,3}[\pi_2 n_1 \pi^*_1]$ and $^{1,3}[\pi_1 n_2 \pi^*_1]$ for the excited state. For any chosen configuration the triplet is always lower in energy than the corresponding singlet.[1]

solvent if the spectrum is registered in solution) and any material placed in between the lamp and the registering photomultiplier does not interfere. Strange as it may seem, this precaution is often neglected. When the material by which the flask is made of ("plastic," usually polymethylmethacrylate in biological laboratories) quite often the interference by the materials is not taken into account (see Fig. 1.2).

In the 1930s various proposals were advanced that there must be a metastable, nonspectroscopic (=not absorbing) state that caused emission at longer wavelength than the spectroscopic states. Therefore mono dimensional diagrams began to be used (Perrin, Jablonski diagrams that indicated the presence of such states).

[1] Electronic states are named by listing all states of a chosen multiplicity and listing them in energy increasing order, S_0, S_1, S_2...T_1, T_2.or by indicating the structure of the two singly occupied MOs, ππ*, σπ*, πσ*, nπ*...and the multiplicity by a superscript figure, usually at the left.$^1\pi\pi^*$,$^3n\pi^*$.

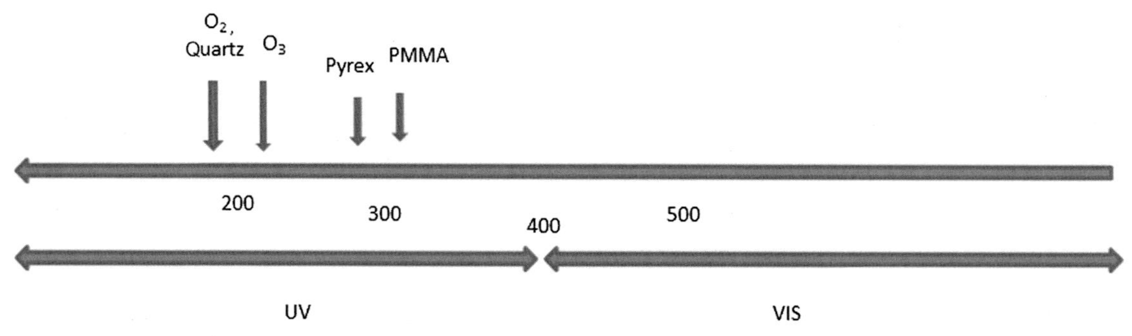

FIGURE 1.2 Range of UV and visible light. The limits to which ozone or oxygen present in air impede measurements are indicated, as well as the corresponding limits for polymethylmethacrylate (PMMA), the "plastic" by which the glassware in biological labs is usually made. Pyrex glass absorbs at about the same wavelength range, "fused," quartz absorbs under 200 nm.

In 1946 G.N. Lewis offered a well-argumented support for such state being the triplet, at the time just introduced by physicists, and developed such a mono dimensional diagram, while conserving for it the name of Jablonski diagram [16,17]. Such a diagram is much more complete and indicates not only the electronic energy of the excited state, but also that of (some of) its vibrational (and rotational) states. Transitions between states are indicated by vertical lines, and are tagged vertical transitions, because in these a large energy change occurs by the absorption or emission of a photon with essentially no change in the molecular geometry (a rule known as the Franck-Condon principle) (Fig. 1.3).

Thus absorption starts from the lowest singlet (S_0) that at room temperature or not much above it has practically no vibrational energy in any of its vibrational modes and leads to a higher singlet S_{nm}, where the subscript letters indicate the *n*th electronic state and the *m*th vibrational state, in one of the vibrational modes ($S_{oo} \rightarrow S_{nm}$). The light-matter interaction is governed by selection rules that determine when such interaction is fruitful and depends on spin and spatial wave functions of starting and arriving electronic states of the molecule. The electronic transitions are fully forbidden when a change of spin multiplicity is involved and they are allowed to various degrees when spin multiplicity is conserved. The degree of permission corresponds to the probability that a transition occurs and determines the intensity of the absorption spectrum. The electronic and vibrational states involved determine the form of the spectra. The likelihood of any vibronic (vibrational + electronic) transition depends on a variety of factors. When the purely electronic transitions are allowed or do not have dominating vibrational contributions, they show a spectrum consisting in an envelope with no feature, while when the purely electronic transitions are forbidden and it is the combination with vibrational quanta that makes the transition (partially) allowed, the spectra directly show the vibrational energy levels. Emission of light is distinguished into two types, the first one is fluorescence, a name chosen because a short-lived emission was first observed on some types of calcium fluorite, known as Bologna stone. The other emission is called phosphorescence, since a long lifetime was first observed on phosphorus samples.[2]

[2] The white phosphorus emission, however, was later recognized as a chemiluminescence, that is, an emission arising from a chemical reaction, not from the absorption of a photon [16].

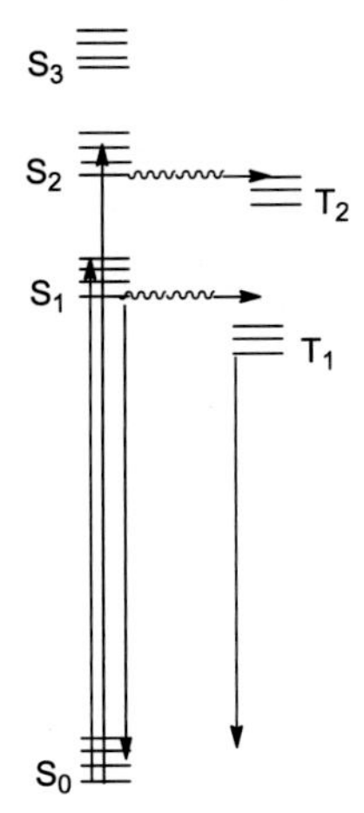

FIGURE 1.3 Schematic representation of vertical transitions (absorption, fluorescence, phosphorescence) and horizontal transitions (internal conversion, intersystem crossing). In vertical transitions quanta of energy are absorbed or emitted while the nuclei do not move, while horizontal transitions involve a change in wave function at a point that is common for both such wave functions.

These transitions, along with absorption, are vertical transitions.

Triplets could not be observed in absorption, because of the strength of the spin prohibition[3] [18]. Furthermore, the concept of photochemical reaction was introduced with the definition of potential energy surfaces (PES), hypersurfaces that represent the electronic energy at various nuclear configurations (and based on the Born-Oppenheimer separation, which recognizes that the wave functions of light electrons can be changed at much faster rate than those of the heavier nuclei). Horizontal transitions involve a conversion of the molecular structure, but no exchange of energy and they are not negligible only at crossing points, where a geometric configuration is actually shared by two wave functions. These are distinguished into internal conversion (IC), when the spin is conserved, and intersystem crossing (ISC), when the spin changes in the process.

The large majority of molecules are in the 0th level of all vibrational modes at 20°C or not much far from room temperature. Thus the vibrational levels detected in the electronic absorption resulting from the combination of electronic and vibrational (=vibronic) transitions are those of the excited states $S_{0,0}\rightarrow S_{1,0}$, $S_{0,0}\rightarrow S_{1,1}$, $S_{0,0}\rightarrow S_{1,2}$, while those of the ground state $S_{1,0}\rightarrow S_{0,0}$, $S_{1,0}\rightarrow S_{0,1}$, $S_{1,0}\rightarrow S_{0,2}$ are present in the emission spectra. The absorption of a photon is by far the most convenient way for generating an electronically excited state, characterized by the high energy (of the same order as that of the chemical bonds) and thus often reacting in some way before losing the electronic energy it has received. Absorption of light is a vertical phenomenon since no change occurs in a vibrational function (the nuclei do not move during the fast, $\sim 10^{-14}$ s, excitation) and the first formed excited state is a snapshot of the ground state, with the same distribution over vibrational and rotational energy as it had originally.

In contrast, it may hardly be that this is the lowest lying conformation of the excited surface, and this leads to hopping through different potential surfaces via horizontal transitions, a phenomenon that is important at crossing points or lines, where the same configuration represents points pertaining to different wave functions (see Figs. 1.4 and 1.5).

All of the processes deactivating the excited states (emission, IC, ISC, chemical reaction, that is, a process leading to a different atomic arrangement) occur competitively. Although emission cuts down severely their lifetime, excited states can react chemically in ms or ns

[3] The likelihood of the transition between two states, using one or more photons can be calculated, thus originating "selection rules." S-T or T-S transitions are strongly prohibited (by a factor of 10^6) so that they are not observed under usual conditions, but also when the spin is conserved a strong prohibition may be observed (up to a factor 10^3), see Ref. [18].

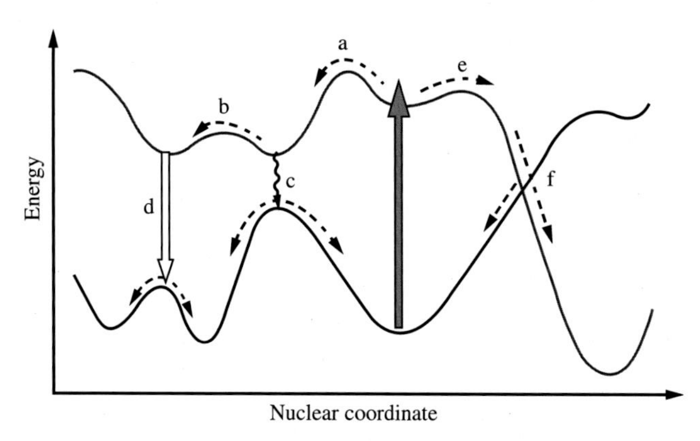

FIGURE 1.4 Schematic representation of the possible reactions from an excited state occurring along a diabatic surface: path c, the system drops from an excited to the ground state surface during the course of the reaction, while the nuclei move. Horizontal transitions occur at crossing points, which pertain at both functions (see, e.g., path f). On the other hand, path b depicts an adiabatic photochemical process, where the excited state of the product is formed, and may emit (a chemiluminescent reaction, path d).

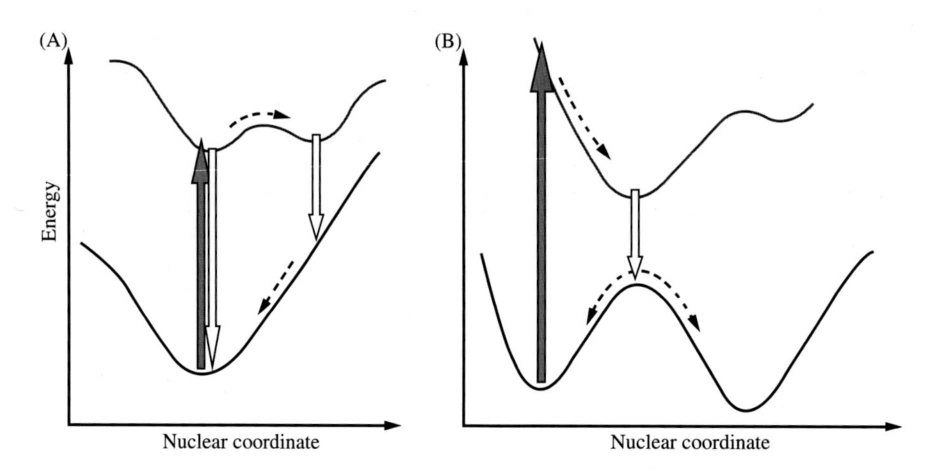

FIGURE 1.5 Electronic excitation often leads to a nonminimum configuration of the excited state.

at room temperature and the reaction may be very competitive. There are at least two chemistries playing a role, the one from the lowest excited singlet and the one from the lowest triplet, besides the one from the ground state (Kasha's three states postulate). A corollary is that irradiation in any band causes the same photochemical and photophysical consequences. Excited states are characterized by their emission (fluorescence from the singlets, phosphorescence from the triplets) and reactions. Further, quenching of the emission give information on the excited state properties, although the long lifetime of triplets made it difficult to observe their emission in solution. Typical examples of generally relevant processes are listed in Schemes 1.2 and 1.3. The generation of excited states has been described by having recourse to the Born-Oppenheimer separation postulate.

Thermal reactions are adiabatic, they remain at every configuration at the lowest possible energy, with no change in the PES, while excited state reactions are by definition diabatic, since they start from an excited state PES and at some configuration have to come down to the ground state configuration, either the starting one (photophysical processes) or a new one (photochemical processes that involve a reaction, that is atoms are differently connected). It is also possible, however, that the product is still formed in the excited state, and

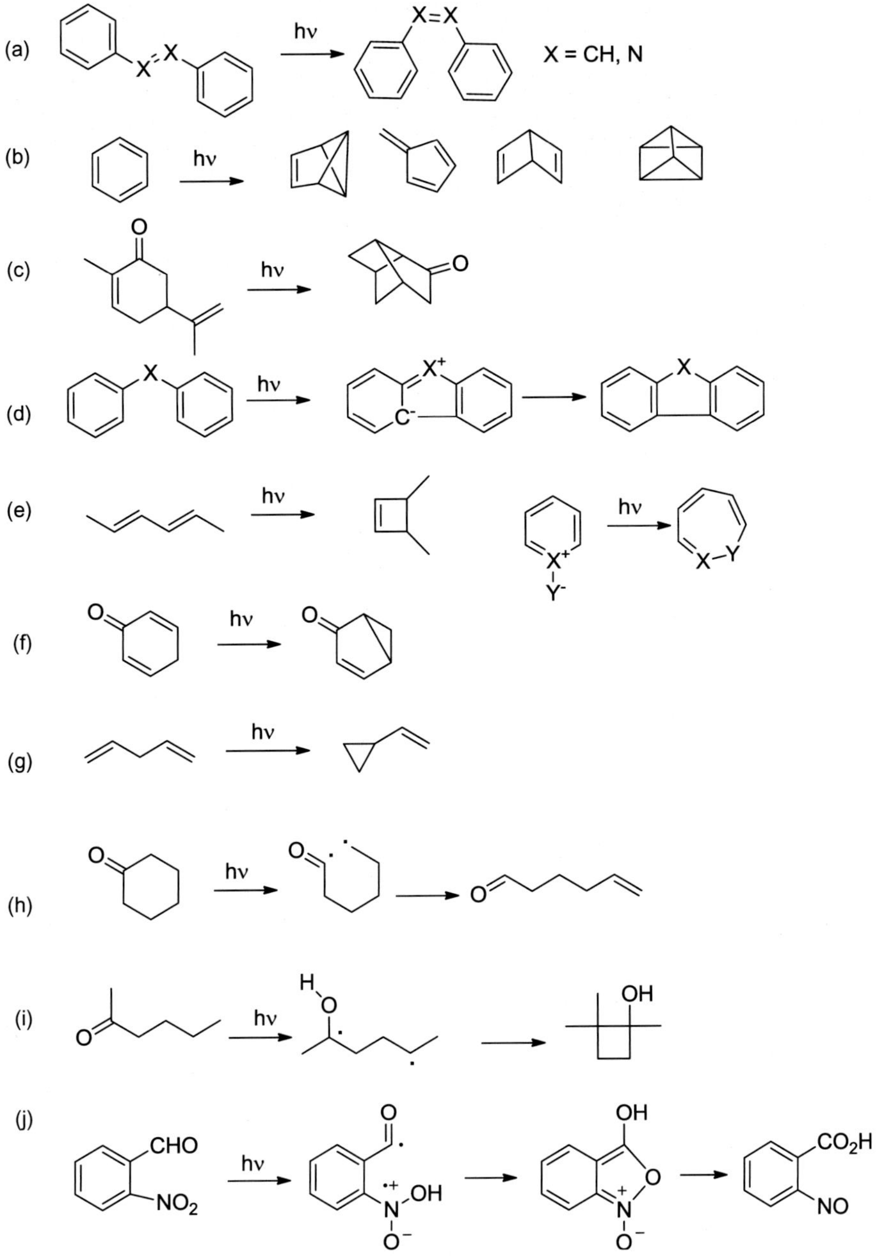

SCHEME 1.2 General photochemical reactions: unimolecular processes.

then decays emitting a luminescence; in this case, the "chemical" bond making and cleaving step is adiabatic.

Electronically excited states are too short-lived (some s down to ms and less for triplets, a few ns down to ps for singlets) and never

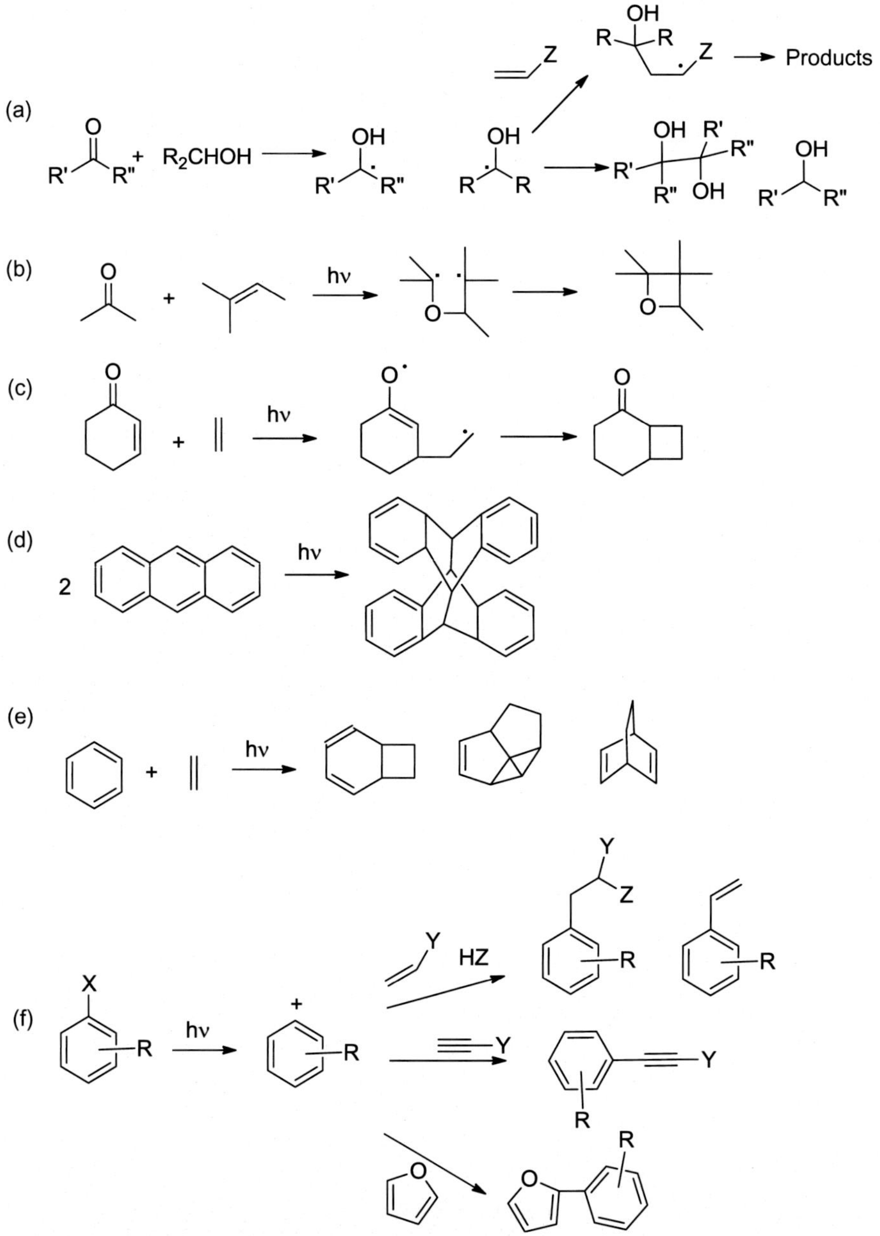

(a) (b) (c) (d) (e) (f)

SCHEME 1.3 General photochemical reactions: bimolecular processes.

accumulate to a large enough steady-state concentration to be detected spectroscopically. In order to reveal such species one has to profit of the fact that their generation is possible with little dependence on conditions. Thus one has to generate as many as possible excited states by a short and intense light-flash, allowing to reveal such short-lived species by some rapid detection. The mechanism of photochemical reactions is quite interesting, even beyond the photochemical proper, since the high starting energy allows to form intermediates of high energy that may have a role in thermally processes as well, typically forming radicals (neutral odd-electron species) or ions.

The first practical embodiment of this idea was developed by R.G.W. Norrish and G. Porter and gained the two scientists the Nobel Prize in chemistry in 1967. This was based on quartz tubes connected to a capacitor that was evacuated [19]. When the pressure was sufficiently low, a discharge started, during 10–20 ms; a small part of the flash was diverted and used for starting a monitoring lamp. The spectrum of the short-lived intermediates formed was registered on a photographic paper at different delay times after passing through a prism in order to accurately monitor the evolution of a species absorbing at that wavelength, or alternatively was measured over a range of λ at predisposed time intervals. An important advancement involved the use of pulsed lasers, rather than light-flash, able to deliver flashes with a lifetime of c.10 ns. In this case, the light-flash was concentrated on a small area of a solution contained in an optical flask, and a probing light and a fast photomultiplier allowed to follow the evolution of triplets, some singlets and the intermediates of most photochemical reactions (radicals, cations, anions). A further extension has been possible by using the pump probe technique, where a sample is hit by a pump flash and this is followed by a probe flash after an adjustable delay. The experiment is averaged over a number of pulses, which allows a resolution in the ps range or below, in order to reveal ultrafast phenomena [20–22].

Flash photolysis has been evolved primarily for absorptions in the UV and visible, due to the large cross sections of molecules at such wavelengths, but the principle has been evolved for any spectroscopic methods [23], while more sensitive instruments were evolved, in particular for infrared (IR) and electronic paramagnetic spectroscopy (EPR) measurements. Again taking advantage of the small dependence on environmental conditions, one may carry out reactions in a matrix, in particular prepared by codistillation with a noble gas at a low temperature, or somewhat less conveniently, in a glassy solvent. Under these conditions, further reactions of primary products and intermediates are mainly hindered, while excited state reactions are not [24].

1.3 Extending the examination to the whole electromagnetic spectrum

Photochemistry refers to the interaction between light (visible, and by extension, UV) and matter, thus to a small portion of the electromagnetic spectrum. This is due to two grounds: (1) this range of wavelengths is largely present in the environment (c.53% of the solar radiation) and (2) this corresponds to the barrier confronted when ground states are promoted to electronically excited states, the protagonists of a rich and valuable (bio)chemistry (Fig. 1.6).

The emission spectrum of the Sunis very similar to that of a black body with a temperature of about 5800K and extends over almost all of the electromagnetic spectrum [26].

The other parts of the electromagnetic field contain radiations both higher and lower in energy than UV–visible radiation, in both cases of common use in medicine (see Table 1.1). The high energy rays found on one extreme are called vacuum UV ($\lambda < 200$ nm), both air and quartz absorb below this limit and instrument must be evacuated to use them) or ionizing radiations since they are able to cause the detachment of core electrons, those closer to the nuclei. Examples are α particles that are helium nuclei (He^{+}), β-particles (electrons), protons,

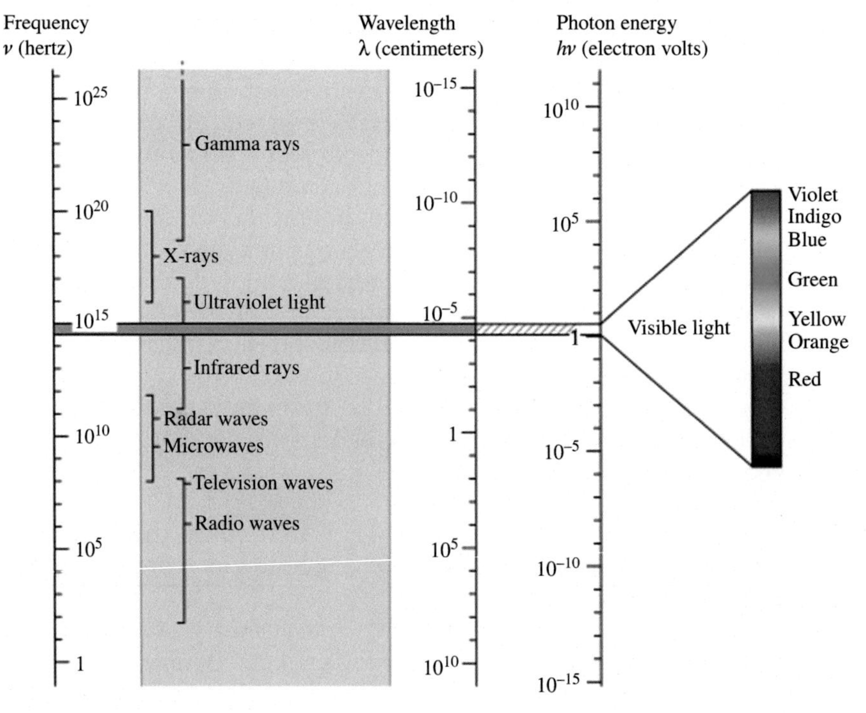

FIGURE 1.6 The electromagnetic spectrum. The narrow range of visible light is shown enlarged at the right. *Encyclopaedia Britannica, Inc. [25], Table 1.*

TABLE 1.1 Electromagnetic spectrum.[a]

Name	Wavelength	Frequency (Hz)	Photonenergy (eV)
Gamma ray	<0.02 nm	>15 EHz	>6.21×10^4 eV
X-ray	0.01 nm–10 nm	30 PHz–30 EHz	124 eV–1.24×10^5 eV
Ultraviolet	**10 nm–400 nm**	**750 THz–30 PHz**	**3.1 eV–124 eV**
Visible light	**390 nm–750 nm**	**400 THz–770 THz**	**1.65 eV–2.89 eV**
Infrared	750 nm–1 mm	300 GHz–400 MHz	1.65 eV–1.24×10^{-3} eV
Microwave	1 mm–1 m	300 MHz–300 GHz	1.24×10^{-6} eV–1.24×10^{-3} eV
Radio	1 mm–100,000 km	3 kHz–300 GHz	1.24×10^{-9} eV–1.24×10^{-3} eV

[a]*The ultraviolet and visible spectra are in bold character.*

γ-rays (emitted by radioactive atoms). For the use in medicine of such particles it is important to know how much they will be able to penetrate in the tissue. This can be calculated on the basis of the linear energy transfer that can be calculated by the Bethe equation [27], the main feature of which is that electrons come close to the light velocity at about 1 meV, while with heavier particles the turning point comes at a considerably higher energy. These aggressive radiations are used in radiotherapy, where they kill cancer cells, for example, in breast cancer [28–35].

In the low energy extreme, IR radiation has energy corresponding to vibrational energies and microwaves to rotational energies. In view of the long wavelength, they penetrate much more in the skin than UV and visible light. In everyday experience one is familiar with microwave ovens and their use for cooking. Actually it has been surmised that microwaves and radiofrequencies may damage biomolecules. In fact, this is true, and some characteristics of these rays are useful, as an example radiofrequencies are known to penetrate into the deeper layers of the skin and produces heath, which in turn produces a tightening of the subdermal layers that is commonly used in therapy. Differences in protein expression have been found in the skin of volunteers exposed to radiofrequency modulated electromagnetic field (mobile phones), suggesting that this may be generally observed (at an energy, however, that is much above that used in microwave ovens, wi-fi transmission, radar, etc.) [36–42].

As it appears from the above discussion, light has differentiated effects on the human body, according to its wavelength and the tissues (Fig. 1.7) [13].

In Table 1.2 the performances of commonly used lamps are compared with skylight, by using the most common measure units (lm,[4] W). It appears that discharge arcs are as efficient as the sun in emitting UV and visible radiations, while incandescent lamps are very weak in that region. Metal halide B arcs and tungsten lamps mainly contribute to the emission in the visible region, but all of the man-made lamps release much heat. Discharge arcs, and in particular metal halide arcs are convenient sources for simulating the sun radiation, quite strong in the visible range.

Many animals (including humans) have a sensitivity range of approximately 400–700 nm. The absorption and scattering by Earth's atmosphere produces illumination that approximates an equal-energy illuminant for most of this range.

A commonly used measure unit for photosynthetic studies is the Einstein $s^{-1}\ m^{-2}$, that is, the quantity of radiant energy in Avogadro's number of photons. As the Einstein is no SI unit, the equivalent mole $s^{-1}\ m^{-2}$ unit is used, referring to the number of photons in a waveband incident per unit time (s) on a unit area (m^2) divided by the Avogadro constant ($6022 \times 10^{23}\ mol^{-1}$) [43,44].

1.4 Looking forward

The efficiency of photochemical reactions is expressed by the quantum yield, that is the ratio between transformed molecules and absorbed light photons. The latter parameter is easily measured in experiment involving transparent solutions, but in an optical turbid sample such as a biological fluid, part of the light impingent is scattered (i.e., rays get deviated due to certain amount in the space) or reflected (i.e., the angle of incidence is equal to the angle of reflection at any given point). Several ways to calculate the fraction of light that is actually absorbed under such conditions have been reported [45,46].

The determining role of light on every aspect of a healthy life is well known. Many physiological functions depend on light. On the other hand, the extremely high energy

[4] The lumen (unit lm) expresses the total luminous flux of a light source and is obtained by multiplying the intensity (in candela) by the angular span over which the light is emitted. The candela measures the luminous intensity in a given direction of a source that emits monochromatic radiation of frequency 540×10^{12} Hz and has a radiant intensity in that same direction of 1/683 wattper unit solid angle (steradian) [43].

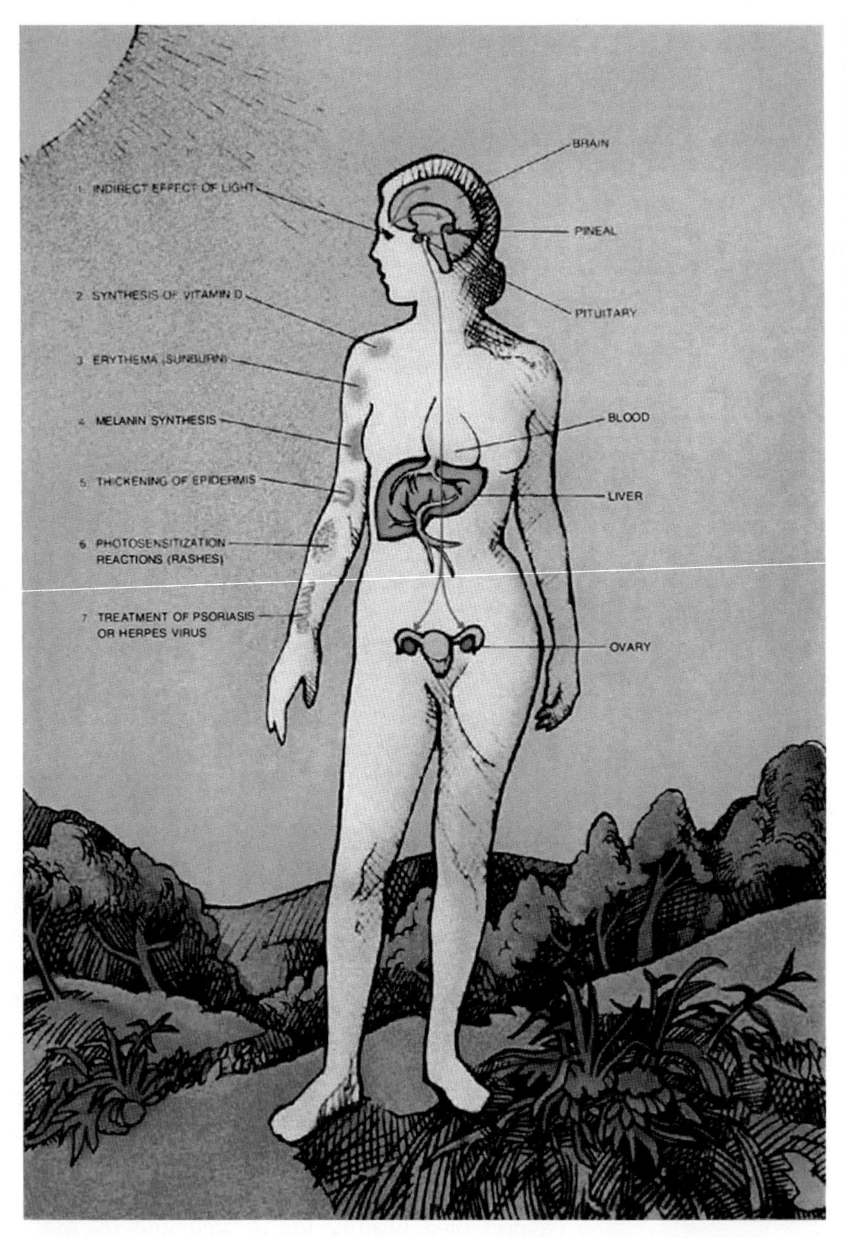

FIGURE 1.7 The (presumed) effect of light on human body on the function of some organ (erythema formation, but also vitamin D synthesis in the skin, melanin synthesis, and thus regulation of circadian rhythms and sleep, as well as effects on internal organs and healing of some diseases). Source: *Reproduced with permission from R.J. Wurtman, The effect of light on human body, Sci. Am. 233 (1) (1975) 69–77.*

photons easily cause serious damage to cells and organs. Living beings, and in particular humans, evolved over geological ages and became adapted to solar light. In present time, artificial light has a paramount role. Man-made chemicals substituted natural compounds in the general use, thus changing the general ecologic equilibrium [47,48].

The changed conditions may result in a dramatic worsening of negative effects of light on health, but also in new perspectives of medicinal applications. This topic is generally

TABLE 1.2 Electrical, photometric, and radiometric characteristics of selected commercial lamps compared to skylight. Modified from Ref. [44].

Parameter	Incandescent lamps (100 W)	Fluorescent cool-white	Discharge (clear mercury)	Discharge (metal halide B)[a]	Discharge (low pressure sodium)	Sun (SKY)[b]
Total lm/W	17	64	48	87	143	124
Radiation per unit of luminous flux (mW/lm)						
250–400 nm	0.08	0.13	1.1	0.33	0.004	1.74
400–700 nm	3.97	2.93	2.6	3.05	1.92	4.56
Thermal	4.8	5.0	8.9	3.9	1.6	0.01
Radiation output (W per lamp)						
250–400 nm	0.14	2.0	23	13	0.13	
400–700 nm	6.9	46	55	122	63	
Thermal	8.4	79	186	156	53	
Distribution of output power as a fraction of electrical input power (mW/W)						
250–400 nm	1.4	8	52	29	0.6	216
400–700 nm	69	188	124	265	276	567
Thermal	84	322	423	339	230	1

[a] *Metal halide B: sodium, scandium, thorium, mercury, lithium iodides.*
[b] *Skylight (SKY): scattered sunlight; does not include emission from the atmosphere.*

regarded as a chapter of biology. While this would be hardly doubted, the chemical aspects are particularly important when electronically excited species are involved, and maintaining a chemical point of view throughout helps to bridge different aspects, from physics to bedside medicine, even when (or perhaps particularly when) the biological effect seems quite far from the molecular change that originates it.

The main tenets of photochemistry were developed during the 1940s, and involved a significant advancement both in the theory and in the experiments [49]. Initially, it was thought that thermalization of the excited states was reached before reaction and this was sufficient to justify Kasha's rule [50] and the fact that these reactions could not be described by means of classic kinetics or in a way similar to catalytic processes. Advanced research demonstrated, however, that many photochemical reactions and in particular those occurring in all important biological phenomena, such as vision and vitamin D synthesis in the skin, could well occur at a faster rate than thermalization. The enormous advancement of spectroscopic techniques that took place in parallel allowed to recognize the reaction pathways, otherwise quite a large stumbling block because of the deep-seated changes in structure typically occurring in photochemistry. A choice of exemplary excited states reactions are listed in Schemes 1.2 and 1.3.

Thus unimolecular reactions include geometric isomerization of (conjugated) alkenes (a) [49–51], π_6 (d), and π_4 (e) electrocyclic reactions [52–54] and related reactions such as the di-π-methane rearrangement (f, g) [55], as well as intramolecular 2 + 2 cycloaddition (c) [56,57], "Norrish type II" intramolecular

γ-hydrogen abstraction often followed by Yang cyclization (i) [58–60], α-cleavage of (cyclic) ketones (h) [61], isomerization of *o*-nitrobenzadehyde and related compounds (j) [62], benzene ring rearrangement to give products with σ bonds between nondirectly bonded carbons (b) [63] (Scheme 1.2).

Correspondingly, generally occurring bimolecular photoreactions include Ciamician's reduction of ketones to pinacols or alcohols (a) and the trapping of the resulting radicals [64], Paternò-Büchicarbonyl-alkene [2 + 2] cycloaddition (b) [65], enone-alkene [2 + 2] cycloaddition (c) [66,67], $\pi_4 + \pi_4$ and$\pi_4 + \pi_2$ aromatic cycloaddition (d) [68], *o*-, *m*- and *p*-benzene alkene cycloaddition (e) [69], conversion of halogenated benzenes and related compound to phenylated alkenes, alkynes, heterocycles (f) [70] (Scheme 1.3).

$$A + h \rightarrow {}^{1}A \rightarrow A + h'$$

$$A + h \rightarrow {}^{1}A \rightarrow {}^{3}A \rightarrow A + h'$$

$${}^{1,3}A + \text{Quencher} \rightarrow A + \text{Quencher}^{*}$$

SCHEME 1.4 Photophysical decays of excited states: fluorescence, phosphorescence, and quenching. Energy transfer to singlets causes quenching of the emission. This is hardly applicable to the triplet due to the long lifetime that causes the complete quenching, in particular in the case of dissolved oxygen.

Furthermore, polymerization processes easily occur under photosensitized conditions [71] and the large structure and richness of detail is ideal for building grafted polymers and similar materials [72]. In order to study the mechanism involved, in particular singlet or triplet processes, quenching studies are useful. Quenching is effective, provided that the quencher possesses a lower-lying excited state, and is particularly efficiently for long-lived triplets (Schemes 1.4 and 1.5).

Steric effects are likewise rationalized in the same way as thermal processes, as indicated in Scheme 1.6 for diastereoselectivity [58] and Scheme 1.7 for enantioselectivity [73]. This is an important characteristic of photochemical reactions, that if the reagent is stabilized by a weak force in various ways in a complex, which is obviously the general case with biomolecules, each conformer can be considered a different species with its own absorption spectrum and covalent bonds are formed based on such photochemical reactions. Thus at least in principle, but in fact for a rather large number of species, a stereospecific photoreaction occurs. The diastereoisomers *syn* and *anti* of the ketones exhibit distinct photochemical reactivities due to conformational preferences; the *anti* isomer undergoes efficient Yang cyclization in 75% – 90% yields with a remarkable

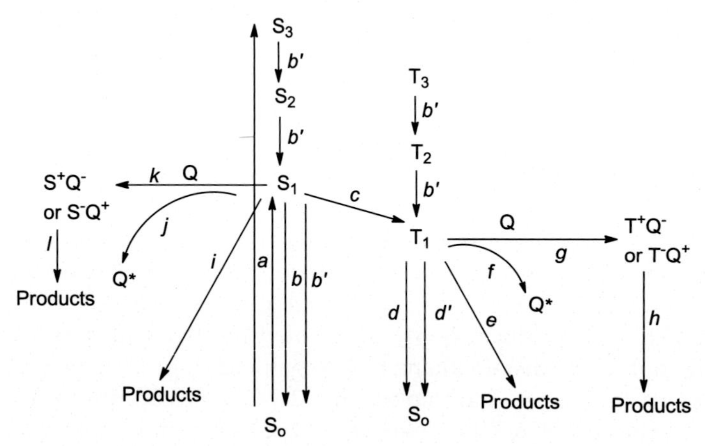

SCHEME 1.5 Overall deactivation processes of excited states: excitation by photon absorption (*a*), fluorescence (*b*), internal conversion (*b'*), intersystem crossing S_1 to T_1 (*c*), phosphorescence (*d*), intersystem crossing T_1 to S_0 (*d'*), quenching by energy (*j*) or electron transfer (*k*) from the singlet and from the triplet (*f* and *g*, respectively), chemical products result directly from excited states (*i* from the singlet, *e* from the triplet) or via redox processes (*l* from the singlet, *h* from the triplet).

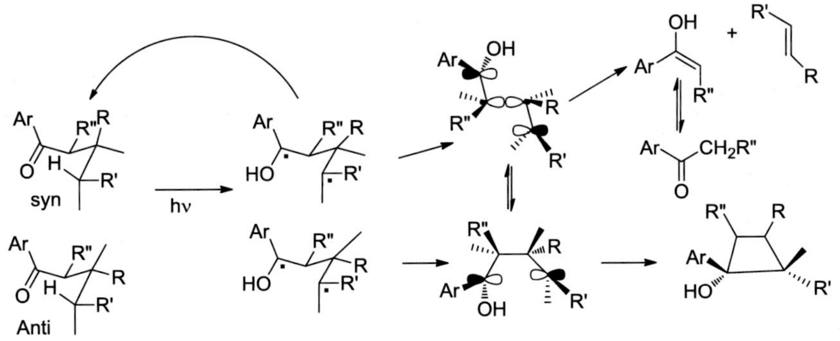

SCHEME 1.6 α,β-substituted butyrrophenone photochemically reacts depending on the *syn* or *anti* arrangement of the substituents.

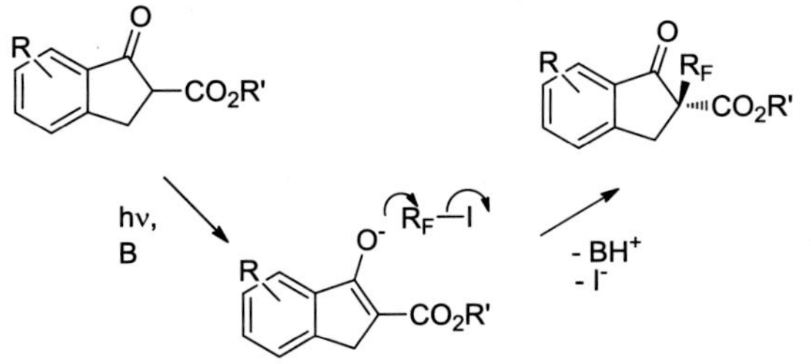

SCHEME 1.7 Stereospecific alkylation of a β-chetoester in the presence of a chyral base.

diastereoselectivity (>90%), the *syn* isomer predominantly undergoes Norrish Type II elimination. The difference in the products from the diastereoisomers are consistent with the formation of precursor diastereoisomeric triplet 1,4-biradicals in which the substituents at α- and β-positions stabilize the geometry (Scheme 1.6). Again by taking advantage from the formation of complexes, the alkylation of ketomalonates occurs enantioselectively when a enantiopure photocatalyst is present, due to the formation of intermediary ions (Scheme 1.7).

A peculiar place in photochemistry (and indeed in chemistry tout court) is played by oxygen. The enormous role oxygen has in chemistry depends on the electronic structure of this molecule. Molecular oxygen O_2, though an even electron-paired species, is nonclosed shell molecule. If the available MOs are filled as usually, after that atomic and bonding σ and π orbitals are filled, two electrons remain that cannot be but antibonding (π*). Therefore they would rather stay separated, and the lowest lying species (the ground state) will have a diradical character that may be represented, though not quite exactly, with structure O–O and will be a triplet (T_0) (see Scheme 1.8).

Being localized on two strongly electronegative atoms, the electrons are less reactive than it would be otherwise, for example, the rate of reaction with organic compounds is negligible at room temperature. However, in the lower excited state (singlet state $^1O_2$), that has a configuration with statistically two electrons on each oxygen atom, oxygen acquires a strong electrophilic character that makes it react with molecules having a character of weak electron donors with rate constants ranging between $0.3 \times 10^9$and $15 \times 10^9\ M^{-1}\ s^{-1}$ at room temperature.

On the other hand, molecular oxygen (ground state, triplet $^3O_2$), interacts with most organic molecules in their excited state through energy transfer. Such a process is an energy transfer quenching and occurs when the accepting state (quencher) has an energy lower than the donating exciting state and there is no spin forbiddance (see Scheme 1.4). This is particularly significant in the case of oxygen, that has a singlet energy lower than that of the triplet of most molecules and the

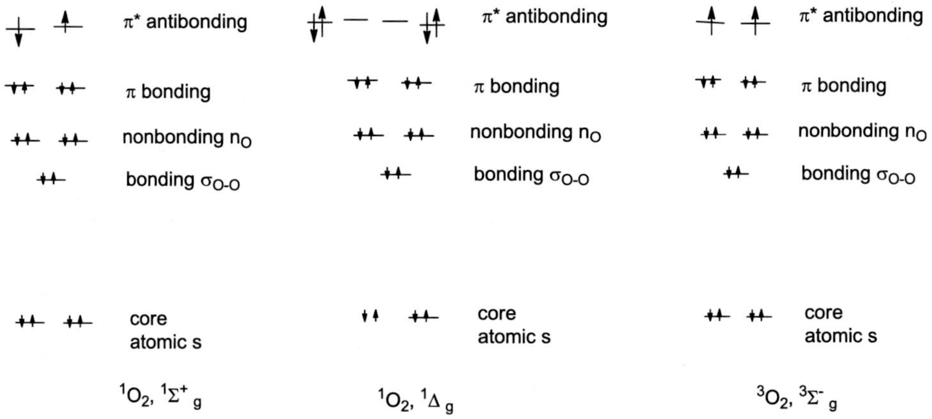

π* antibonding
π bonding
nonbonding n_O
bonding $\sigma_{O\text{-}O}$
core atomic s
$^1O_2$, $^1\Sigma^+_g$

π* antibonding
π bonding
nonbonding n_O
bonding $\sigma_{O\text{-}O}$
core atomic s
$^1O_2$, $^1\Delta_g$

π* antibonding
π bonding
nonbonding n_O
bonding $\sigma_{O\text{-}O}$
core atomic s
$^3O_2$, $^3\Sigma^-_g$

SCHEME 1.8 Low-lying structure of oxygen molecule. All of the states show the same orbital occupancy and differ only for the spin. The lowest state is a triplet $\left({}^3\Sigma_g^-\right)$, the higher states are singlet (doubly degenerate $^1\Delta_g$ and $^1\Sigma_g^+$).

process: $A^{3*} + {}^3O_2 \rightarrow A + {}^1O_2$ is energetically favored and maintains the overall spin[5] [74,75].

This equation corresponds to sensitization of oxygen to give singlet oxygen. Typical sensitizers yielding singlet oxygen are dyes, from eosin derivatives to the "killer red protein" [76].

The efficiency of singlet oxygen [$\Phi(^1O_2)$] formation was determined for a variety of molecules, and found to range from 0.24 to 2.00 [77]. The lowest singlet has a bonding character (the two π orbitals are one occupied and the other one not occupied in the mixed configuration). This enhances the electrophilic character of this state and gives to it a great reactivity with nucleophilic species, including alkenes and DNA bases.

Some general reactions via singlet oxygen are indicated in Scheme 1.9, including (a) pericyclic process, such as [4 + 2] cycloaddition on aromatics; (b) heterocycles (e.g., ozonides from furans); (c) on guanine derivatives; and (d) the ene (Schenck) reaction [78,79].

On the other hand, a diradical species such as triplet oxygen is not radicalic enough to undergo reactions with closed-shell molecules, but is certainly able to trap preformed radicals and to carry out oxygenation reactions. These processes are tagged as sensitization, involving one of the mechanisms showed in Scheme 1.10, often indicated as Type I oxygenation, a chemical activation where it is the organic molecule that is excited, Type II oxygenation, where it is the oxygen molecule O_2 that is "physically" (via energy transfer) activated, and Type III oxygenation that refers to an electron transfer process forming superoxide anion (Scheme 1.10) [80,81].

Excited ketones abstract efficiently hydrogen from alcohols and other substrates. The C-centered radicals are trapped by oxygen in a Type I oxygenation. This reaction competes with quenching of the excited carbonyl and formation of singlet oxygen (Scheme 1.11).

With regard to application in the cell, the most important characteristic is whether an intermediate with some (di)radical character is involved. Processes where this happens, typically triplet processes, have to be avoided. When these processes occur in the presence of oxygen, both physical quenching to form singlet oxygen and chemical trapping of the C-radicals to form highly reactive peroxyl radicals occur that cause heavy damage to the tissues.

5 The overall spin is conserved when two triplets interact.

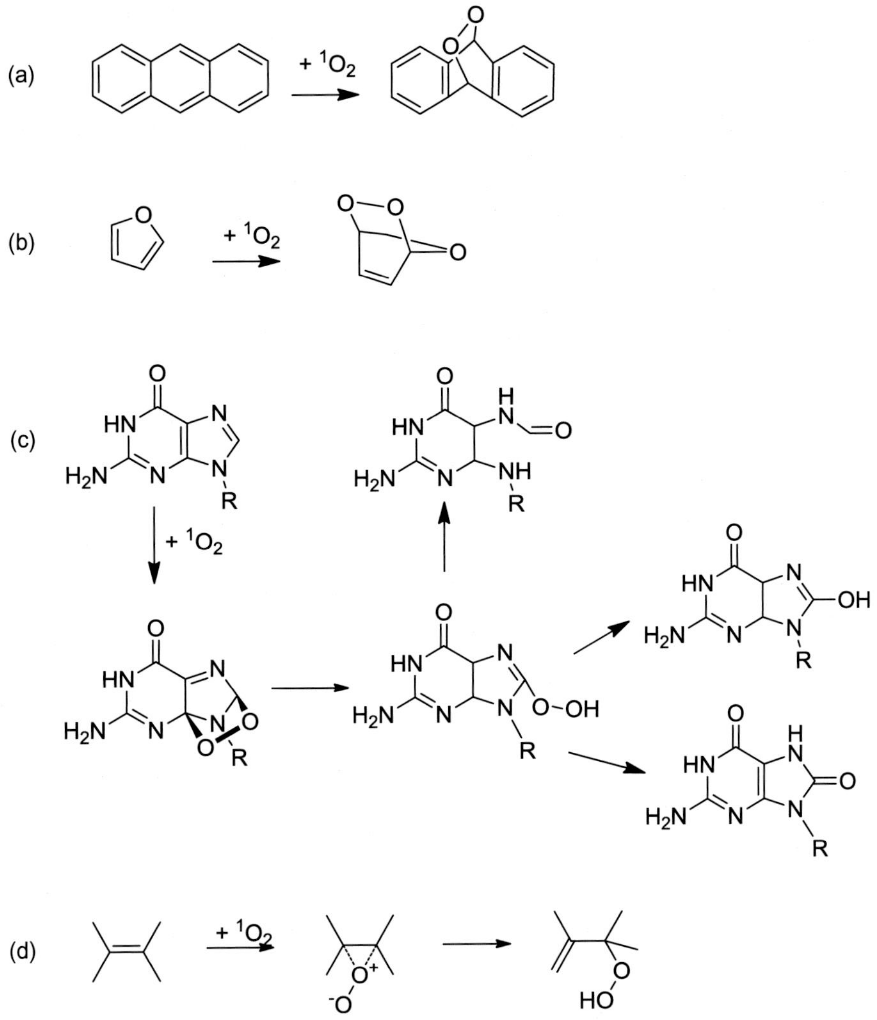

SCHEME 1.9 General photoreactions of biomolecules via singlet oxygen.

1.5 Conclusion

In the last decades, advancing research has found that the postulates expressed in the previous sections are too strict. First, there are a number of photophysical processes and photochemical reactions that exceed the rate of *normal* processes in the sense that glide through barriers at an *ultra* fast rate and should be mentioned here because these include some important reaction of biomolecules [82–84]. Thus the geometric isomerization of (conjugated) alkenes may occur faster than the time required for reaching the thermal equilibration with the environment and IC from the upper excited states decay occurs via a fast leakage mechanism (a "conical intersection"). Thus a coherent nonadiabatic path may be followed and result in a *dynamic* coupling between ground and excited states that is determined by the rate at which the system arrives at the configuration of the conical intersection. Under such conditions, the Kasha's three states postulate is no more valid (Fig. 1.8).

Another all important point involves vertical processes implying two (or more) photons. In fact, while Einstein theory is referred to one-electron processes, it was early established that more electrons can be involved in absorption processes [85]. Since two (or more) photons are implied, such processes are proportional to the second (or higher) power of the absorbed flux (nonlinear absorption) and are important only with a dense light flux. Therefore the experimental verification had to wait the availability of suitable lasers. In this way, a high-lying excited state would be reached by simultaneous absorption of two smaller photons through a process tagged photon upconversion [86]. This is of great theoretical interest, because different selection rules are involved under these conditions, but also because of the useful applications, in particular the extension of the usable fraction of the solar energy, by adding a part of the IR (or at least NIR) irradiation. In this case triplet-triplet annihilation (fusion) leads to a high-lying singlet excited state (photon upconversion). The singlet states then emit (in the visible). The largest application involves the reabsorption of the singlet state emission by a conventional photosolar cell. Triplet fusion ($S_0 + 2\,h\nu \rightarrow T_1 + T_1 \rightarrow S_1 + h\nu'$) is quite convenient for extending the range of usable photons from solar light (see Scheme 1.12) [87]. Notice, however, that since the emitting singlets arise from triplet states, the shape is as usual, but the lifetime is that of the triplet (delayed fluorescence).

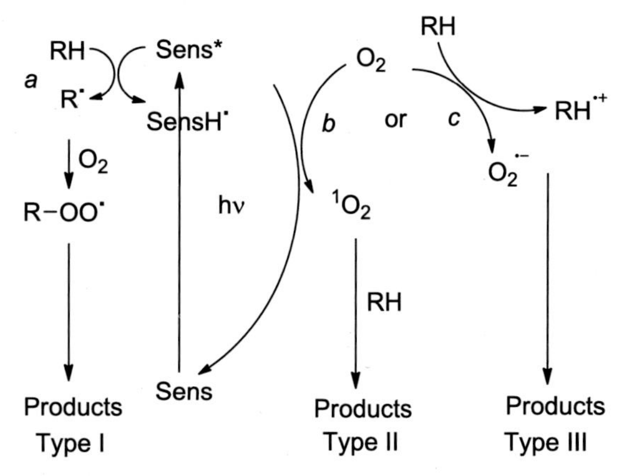

SCHEME 1.10 Photosensitized oxidation, (a) Type I, (b) Type II, (c) Type III.

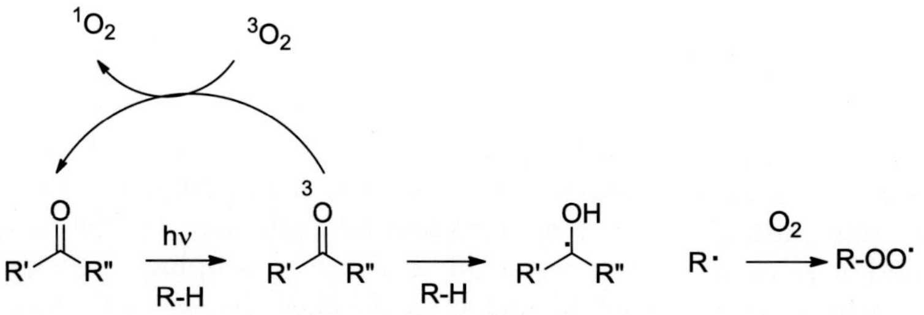

On the contrary, singlet fission is a process in which an organic chromophore in an excited singlet state (S_1) interacts, via various mechanisms, with a neighboring ground state chromophore (S_0') and both are converted into triplet excited states ($S_1 + S_0' + h\nu \rightarrow T_1 + T_1'$). The two chromophores can be of the same kind ("homofission") or of different kinds ("heterofission"). Singlet fission does not occur in single small-molecule chromophores, at least not at the usual excitation energies, and is constrained to multichromophoric systems, because there have to be at least two chromophores to accommodate the two triplet excitations [88,89]. This process allows to

SCHEME 1.11 Competition of different paths in photooxygenation reactions.

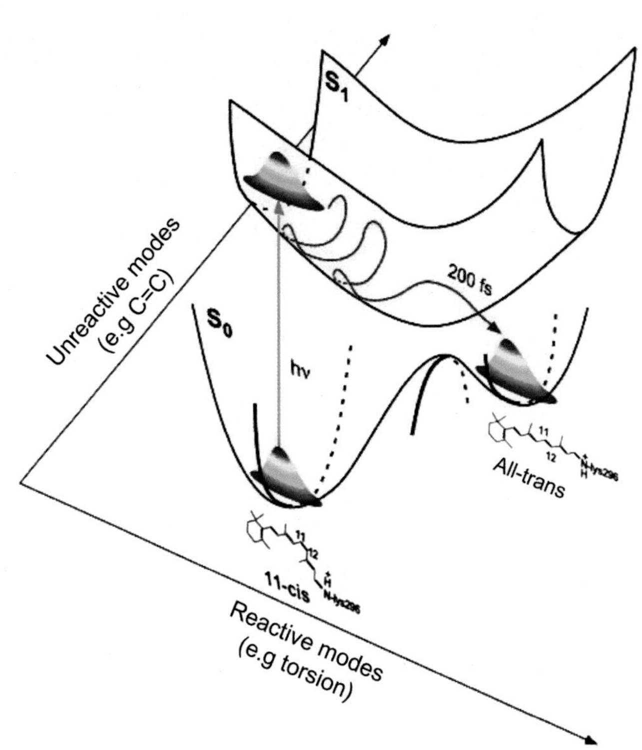

FIGURE 1.8 Model of the excited state potential energy surface in rhodopsin: conical intersection between the S_1 and S_0 potential energy surfaces. In the vicinity of such a surface funnel, extremely strong and localized nonadiabatic coupling can result in rapid and efficient internal conversion prior to excited state equilibration. Excitation with energy greater than 20,000 cm^{-1} (λ < 500 nm) preferentially increases the projection onto *higher* frequency unreactive modes. Thus the wavelength dependence of the quantum yield is best observed with excitation lesser than 20,000 cm^{-1} where the low-frequency torsional modes dominate the Franck-Condon envelope. Source: *Reproduced with permission from J.E. Kim, M.J. Tauber, R.A Mathies, Wavelength dependent cis-trans isomerization in vision, Biochemistry 40 (46) (2001) 13774–13778.*

exploit the bright, easily detected emission in the visible of many organic chromophores. S_0' is often indicated as an "annihilator" and may be chosen among a variety of aromatic molecules or metal complexes (in particular lantanides) [90]. As such, this property is at the base of many microscopic diagnostic methods [91].

In the last decades, highly valuable applications have been found that greatly increased the interest for this topic and fostered the development of theory as well, for example, two photon absorption largely improves exploiting the sun energy, while triplet-triplet annihilation makes it possible to have a high energy excited state, and yet take advantage from the higher penetration of long wavelength absorbing species. It is important to notice that while luminescence in solution does not increase with concentration because of self-absorption, highly efficient emission is observed in the solid state or at any rate under conditions where complexes are the predominant species.

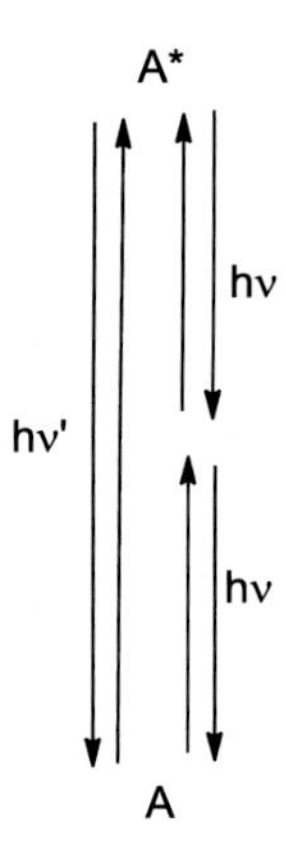

SCHEME 1.12 A singlet excited state A* is reached by simultaneous absorption of two small photons (triplet-triplet annihilation) or, conversely, a high-lying state can reach the ground state by emitting two small photons (singlet fission).

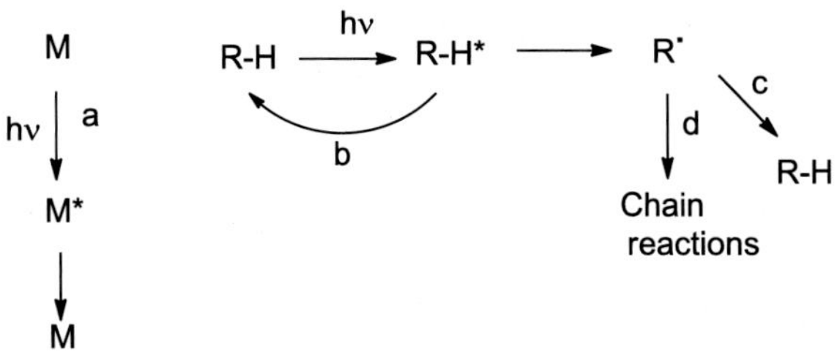

SCHEME 1.13 A photochemical reaction necessarily involves an excited state (path a, via M* from M). However, a large number of such reactions involv a further intermediate,often a radical formed by atom (typically H) transfer (see path b, R· from R. The reaction then pursues via chain or non chain processes.

In conclusion, an overall picture of the interaction between light and biomolecules and cells has been sketched. As it will be discussed in the following sections, highly sophisticated systems are involved. As an example, when maximizing the effect is the target, for example, in vision, the solution adopted is the introduction of an amplifying step. On the other hand, when avoiding damage to the cells is the target, a number of approaches are possible, such as (a) the presence or formation of another molecule (M in Scheme 1.13) that is photostable and absorbs at the same λ as the molecule to be protected, (b) the quenching of the excited state, (c) the trapping of aggressive primary products such as radicals, before they enter a very damaging process, and (d) in particularly chain processes are possible (Scheme 1.13). As an example, melanins exert their protective role by absorbing at the same wavelength as DNA and decay, thus transforming dangerous UV light into much less dangerous vibrational energy or heath. However, if some light is absorbed by melanins and the triplet is formed, a reaction takes place leading to free radicals, namely reactive oxygen species, which are produced in large amounts. Melanin has the capacity to act as a pseudosuperoxide dismutase and can convert superoxide anion radicals to hydrogen peroxide, thus setting a second line of defense [92]. Evolution, however, prefers to put itself on the safest side and uses directly mainly or only photostable molecules. In this context, important results have been obtained by the study of phenols as models for tyrosine and nitrogen-heterocycles pyrrole for heme and indole for tryptophan. Deposition of a UV photon energy into the phenol causes excitation to a strongly absorbing $^1\pi\pi^*$ state. These molecules also possess a weakly absorbing excited electronic state of $^1\sigma\pi^*$character, which intersects both the $^1\pi\pi^*$ state and the electronic ground state (S_0). Stretching along the O−H or N−H bond coordinate leads to homolytic cleavage. Interestingly, this applies to pyrrole and imidazole, but not to the imidazole isomer, pyrazole [93] (Fig. 1.9).

To summarize, photochemical reactions can be classed in a few categories, as indicated in Scheme 1.14.[6] In aromatic and aliphatic hydrocarbons, the excitation causes a very large change in the orbital occupation determining major skeleton rearrangements and, since no particular stabilization is available from the atoms, concerted processes are the only choice (a) [94]. On the other hand, electronically excited states are both much easily reduced [the electron is now transferred from the low-lying lowest unoccupied molecular orbital (LUMO)] and oxidized (the electron is now transferred to the low-lying HOMO), thus electron transfer is a fast process, even if it is easily reversible (b) [95]. In heteroatom-containing compounds, nonbonding orbitals participate in the excitation

[6] $\sigma\sigma^*$ states are not considered because they are quite high in energy and rarely give "clean" processes.

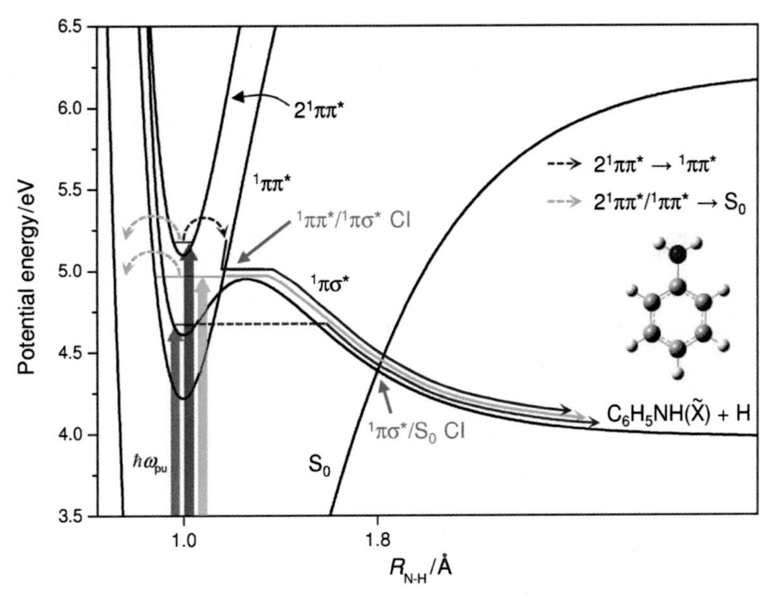

FIGURE 1.9 Energy of excited states of phenol. The surface of a forbidden $^1\sigma\pi^*$ state in such compounds crosses both the $^1\pi\pi^*$ surface and that of the electronic ground state S_0 leading to homolytic cleavage. Source: *Reprinted with permission from G.M. Roberts, V.G. Stavros, Biomolecules photostability and 1πσ* states: linking these with femtochemistry, in: R. de Nalda, L. Bañares (Eds.), Ultrafast Phenomena in Molecular Sciences: Femtosecond Physics and Chemistry, Springer Series in Chemical Physics, Springer International Publishing, Switzerland, 2014, pp. 119-143.*

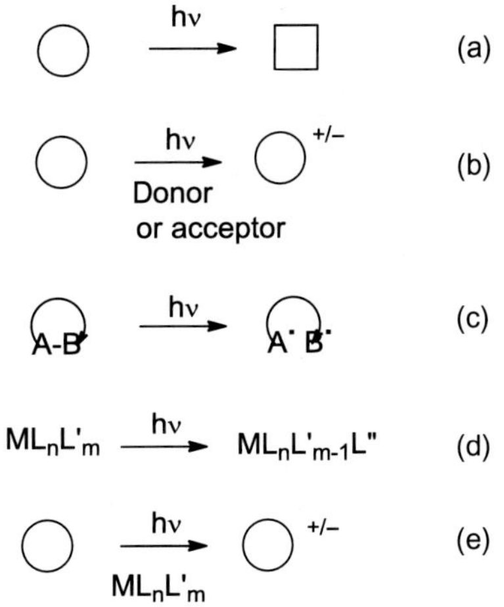

$\bigcirc \xrightarrow{h\nu} \square$ (a)

$\bigcirc \xrightarrow[\text{Donor or acceptor}]{h\nu} \bigcirc^{+/-}$ (b)

A-B $\xrightarrow{h\nu}$ A· B· (c)

$ML_nL'_m \xrightarrow{h\nu} ML_nL'_{m-1}L''$ (d)

$\bigcirc \xrightarrow[ML_nL'_m]{h\nu} \bigcirc^{+/-}$ (e)

SCHEME 1.14 Rough classification of photochemical reactions.

processes and often determine the structure of low-lying excited states nπ* that thus have a diradical character and often undergo intramolecular and inter-molecular reactions dominated by the electron localized on the heteroatom (c) [96]. As for metallic complexes, electronic excitation leads to metal-centered (d–d) transition having ligand centered or metal to ligand or ligand to metal electron transfer character, and accordingly may undergo selective ligand exchange (d) [97] or redox processes (e) [98].

True to the general choice consistently adopted, evolution prefers multiple synergic possibilities, that will be discussed in the following chapters.

References

[1] A. Albini, Some remarks on the first law of photochemistry, Photochem. Photobiol. Sci. 15(3) (2016) 319–324.
[2] F. Solano, Melanins: skin pigments and much more: structural models, biological functions, and formation routes, New J. Sci. (2014) 1–29.

[3] H. Swete, On the importance of obtaining correct comparative observations in estimating the influence of light in health and disease: illustrated by the actinography, Br. Med. J. 2(417) (1868) 656–657.
[4] P.M. Dunn, Francois Glisson (1597-1677) and the 'discovery' of rickets, Arch. Dis. Child Fetal Neonatal Ed. 78(2) (1998) F154–F155.
[5] A. Downs, T.P. Blunt, Research on the effect of light upon bacteria and other organisms, Proc. R. Soc. Lond. 26 (1877) 488–500.
[6] N.R. Finsen, La Phototherapie, Carré et Naud, Paris, 1899.
[7] P. Jarret, R. Scragg, A short history of phototherapy, vitamin D and skin diseases, Photochem. Photobiol. Sci. 16 (2017) 283–290.
[8] H. Bloch, Solar theology, heliotherapy, phototherapy and biological effects, J. Natl. Med. Assoc. 82 (1990) 517–521.
[9] J. Martins e Silva, Breve historia do raquistismo e da scoberta da vitamin D, Acta Reumatol. Port. 32 (3) (2007) 205–229.
[10] N.J. Carpenter, L.I. Zhao, Forgotten mysteries of the early history of vitamin D, Am. Soc. Nutr. Sci. J. 16 (2017) 283–290.
[11] G. Wolff, The discovery of vitamin D: the contribution of Adolf Windaus, Am. Soc. Nutr. Sci. J. 134 (2004) 1299–1302.
[12] M. Luckiesh, Simulating solar light. A new era of artificial lighting, Trans. A.I.E.E. Am. Illuminat. Engin. Soc. 25 (1930) 397–405.
[13] R.J. Wurtman, The effect of light on human body, Sci. Am. 233 (1) (1975) 69–77.
[14] P.H. Hart, M. Norval, V.V. Reeve, The health benefits of UV radiation exposure through vitamin D production or non-vitamin D pathways, Photochem. Photobiol. Sci. 16 (2017) 281–282.
[15] R.C. Häder, R. Worrest, Effect of enhanced solar ultraviolet radiation on aquatic ecosystems, Photochem. Photobiol 53 (1991) 717–725.
[16] G.N. Lewis, M. Kasha, Phosphorescence and the triplet state, J. Am. Chem. Soc. 66 (1944) 2100–2116.
[17] G.N. Lewis, M. Calvin, Paramagnetism of the phosphorescent state, J. Am. Chem. Soc. 67 (1945) 1232–1233.
[18] L. Stanton, Selection rules for pure rotation and vibration-rotation hyper-Raman spectra, J. Raman Spectrosc. 1 (1973) 53–57.
[19] G. Porter, Flash photolysis and some of its applications, Science 160 (3834) (1968) 1299–1307. and In Porter, G. (1972). Nobel Lecture-Chemistry. Elsevier, Amsterdam, 1963–1970.
[20] J.C. Scaiano, Laser flash photolysis, a tool for physical organic chemistry, in: R.A. Moss, M.S. Platz, M. Jr Jones (Eds.), Reactive Intermediates in Organic Chemistry, John Wiley& Sons, Hoboken, NJ, 2004 (Chapter 18).
[21] A. Beeby, Pump – probe laser spectroscopy, in: D.L. Andrews, A.A. Demidov (Eds.), Introduction to Laser Spectroscopy, second ed., Springer, NY, US, 2002, pp. 105–137.
[22] Y. Chiang, M. Gaplovsky, A.J. Kresge, K.H. Leung, C.M. Ley, et al., Photoreactions of 3-diazo-3H-benzofuran-2-one; dimerization and hydrolysis of its primary photoproduct, a quinonoid cumulenone: a study by time-resolved optical and infrared spectroscopy, J. Am. Chem. Soc. 125 (2003) 12872–12880.
[23] B. Thrush, Flash photolysis and the study of free radicals in the infrared, J. Chem. Soc. Faraday Trans. 2 (1986) 2125–2128.
[24] V.E. Bondybey, M. Rasanen, A. Lammers, Rare gas matrices, their photochemistry and dynamics: recent advancement in selected areas, Annu. Rep. Prog. Chem., Sect. C Phys. Chem. 95 (1999) 331–372.
[25] https://www.britannica.com/science/electromagnetic-radiation/Continuous-spectra-of-electromagnetic-radiation
[26] G. Kopp, J. Lean, A new, lower value of total solar irradiance: evidence and climate significance, Geophys. Res. Lett. 38 (2011) L01706.
[27] D.R. Grimes, D.R. Warren, M. Partridge, An approximate analytical solution of the Bethe equation for charged particles in the radiotherapeutic energy range, Sci. Rep. 7 (2017) 9781.
[28] Early Breast Cancer Trialists' Collaborative Group (EBCTCG), Effect of radiotherapy after breast-conserving surgery on 10-year recurrence and 15-year breast cancer death: meta-analysis of individual patient data for 10 801 women in 17 randomised trial, Lancet 378 (2011) 1707–1716.
[29] E.R. Brown, D.T. Woolard, T. Gelmont, A. Samuels, Remote detection of biomolecules in the THz region, Proc. Int. Symp. Microwave (2002). 7321452, 378.
[30] T.A. Bucholz, Radiotherapy and breast cancer survival, Lancet 378 (9804) (2011) 1680–1682.
[31] T.J. Whelan, J.P. Pignol, M.N. Levine, J.A. Julian, R. MacKenzie, S. Parpia, et al., Long-term results of hypofractionated radiation therapy for breast cancer, N. Engl. J. Med. 362 (2010) 513–520.
[32] A. Goldhirsch, W.C. Wood, S. Coates, R.D. Gelber, B. Thürlimann, H.J. Senn, Strategies for subtypes--dealing with the diversity of breast cancer: highlights of the St. Gallen International Expert Consensus on the Primary Therapy of Early Breast Cancer, Ann. Oncol. 22 (2011) 1736–1747.
[33] A. Wickberg, L. Holmberg, H.O. Adami, A. Magnuson, K. Villman, G. Liljegren, Sector resection

with or without postoperative radiotherapy for stage I breast cancer: 20-year results of a randomized trial, J. Clin. Oncol. 32 (2014) 791–797.

[34] R.T. Hitchcok, Radio frequency radiation, in: D. Harbison, M.M. Bourgeois, G.T. Johnson (Eds.), Hamilton & Hardy's Industrial Toxicology, sixth ed., John Wiley & Sons, Hoboken, NJ, 2015.

[35] P. Sigmund, Particle penetration and radiation effects, Springer Series in Solid State Sciences, Springer-Verlag, Berlin Heidelberg, 2006, p. 151.

[36] P. Boscol, M.B. Di Sciascio, S. D'Ostilio, A. Del Signore, M. Reale, P. Conti, et al., Effects of electromagnetic fields produced by radiotelevision broadcasting stations on the immune system of women, Sci. Total Environ. 273 (2001) 1–10.

[37] N.R. Kumar, S. Sangwan, P. Badotra, Exposure to cell phone radiations produces biochemical changes in worker honey bees, Toxicol. Int. 18 (2011) 70–72.

[38] P. Vecchia, R. Mattes, D.L. Ziegelberge, R. Sauners, A. Swerdlow, Exposure to high frequency electromagnetic fields, biological effects and health consequences (100kHz–300GHz), International Commission of Non Ionizing Protection, Health Phys. 102 (2012) 305–321.

[39] J.R. McNamee, V. Chauhan, Radiofrequency radiation and gene/protein expression: a review, Radiat. Res. 172 (2009) 265–287.

[40] A. Anu Karinen, S. Heinävaara, R. Nylund, D. Leszczynski, Mobile phone radiation might alter protein expression in human skin, BMC Genomics 9 (1998) 7.

[41] A.B. Copty, N.O. Yair, B. Itai, M. Golosovsky, D. Davidov, Evidence for a specific microwave radiation effect on the green fluorescent protein, Biophys. J. 91 (2006). 1413–1423.

[42] M. Havas, When theory and observation collide: can non ionizing radiation cause cancer? Environ. Pollut. 221 (2017) 501–505.

[43] D.L. DiLaura, K.W. Houser, R.G. Mistrick, G.R. Steffy (Eds.), Lighting Handbook, Illuminating Engineering Society (IES), NY, 2011.

[44] R.W. Thimijan, R.D. Heins, Photometric, radiometric and quantum light units of measure: a review of procedures for interconvertion, Hortscience 18 (1983) 818–822.

[45] R.M.P. Doornbos, R. Lang, C. Aalders, C.E. Cross, H.J. C.M. Sterenborg, The determination of *in vivo* human tissue optical properties and absolute chromophore concentrations using spatially resolved steady-state diffuse reflectance spectroscopy, Phys. Med. Biol. 44 (1999) 967–981.

[46] B. Chance, M. Cope, E.N. Gratton, N. Ramanujan, B. Tromberg, Phase measurement of light absorption and scatter in human tissue, Rev. Scient. Instrum. 69 (1998) 3457–3481.

[47] T. Longcore, C. Rich, Ecological light pollution, Front. Ecol. Environ. 2 (2004) 191–201.

[48] S. Sauvé, S. Bernard, P. Sloan, Environmental sciences, sustainable development and circular economy: alternative concepts for trans-disciplinary research, Environ. Development 17 (2016) 48–56.

[49] G.S. Hammond, J. Saltiel, A.A. Lamola, N.J. Turro, J.S. Bradshow, D.O. Cowan, et al., Mechanism of photochemical reactions. XXII. Photochemical *cis-trans* isomerization, J. Am. Chem. Soc. 82 (1964) 3197–3217.

[50] M. Kasha, Phosphorescence and the role of the triplet state in the electronic exitation of complex molecules, Chem. Rev. 41 (1947) 401–419.

[51] R.S.H. Liu, G.S. Hammond, Photochemical reactivity of polyenes: from dienes to rhodopsin, from microseconds to femtoseconds, Photochem. Photobiol. Sci. 2 (2003) 835–844.

[52] A. Staykov, A. Areephong, B. Jetsuda, B. Wesley, Y. Kazunari, B.L. Feringa, Electrochemical and photochemical cyclization and cycloreversion of diarylethenes and diarylethene-capped sexithiophene wires, ACS Nano 5 (2011) 1165–1178.

[53] E. Havinga, J.L.M.A. Schlatman, Remarks on the specificity of the thermal and photochemical transformations in the field of vitamin D, Tetrahedron 16 (1961) 146152.

[54] M. Garavelli, C.S. Page, P. Celani, M. Olivucci, M.E. Schmid, S.A. Trushin, et al., Reaction path of a sub-200 fs photochemical electrocyclic reaction, J. Phys. Chem. A 105 (2001) 4458–4469.

[55] H.E. Zimmerman, D. Armesto, Synthetic aspects of the di-p-methane rearrangement, Chem. Rev. 96 (1996) 3065–3112.

[56] J. Jiannan Zhao, J.L. Brosmer, Q. Tang, Z. Yang, K.N. Houk, P.L. Diaconescu, et al., Intramolecular 2 + 2 photocycloaddition through visible light induced energy transfer, J. Am. Chem. Soc. 139 (2017) 9807–9810.

[57] R.M. Coates, P.D. Senter, W.R. Baker, Annelative ring expansion via intramolecular [2 + 2] photocycloaddition of alpha, beta-unsaturated gamma-lactones and reductive cleavage: synthesis of hydrocyclopentacyclooctene-5-carboxylates, J. Org. Chem. 47 (1982) 3597–3607.

[58] N. Singhal, A.L. Koner, P. Mal, P. Venugopalan, W.M. Nau, J.M. Moorthy, Diastereomer-differentiating photochemistry of β-arylbutyrophenones: Yang cyclization versus Type II elimination, J. Am. Chem. Soc. 127 (2005) 14375–14382.

[59] J.N. Moorthy, A.L. Koner, S. Samanta, N. Singhal, W. M. Nau, R.G. Weiss, Diastereomeric discrimination in the lifetimes of Norrish Type II triplet 1,4-biradicals and stereocontrolled partitioning of their reactivity (Yang cyclization versus Type II fragmentation), Chem. Eur. J. 12 (2006) 8744–8749.

[60] P.J. Wagner, A.E. Kempaininen, Type II processes in phenyl ketones. Triplet state reactions as dependence of gamma and delta-substitution, J. Am. Chem. Soc. 94 (1972) 7495–7499.
[61] F.D. Lewis, C.H. Hoyle, H. Magyar, H.G. Heine, W. Hartmann, Substituent effects on the photochemical a-cleavage of deoxybenzoin, J. Org. Chem. 63 (1998) 488–492.
[62] M.V. George, J.C. Scaiano, Photochemistry of ortho-nitrobenzaldehyde and related studies, J. Phys. Chem. 84 (1980) 492–495.
[63] R. Papadakis, S. Ottosson, The excited state antiaromatic benzene ring: a molecular Mr.Hyde? Chem. Soc. Rev. 44 (2015) 6472–6493.
[64] A.S. Muhammed, M.A.S. Al-Amoudi, J.M. Vernon, Photochemical addition of secondary alcohols to maleimides, J. Chem. Soc. Perkin Trans. 2 (1999) 2667–2670.
[65] T. Bach, Stereoselective intermolecular [2 + 2]-photocycloaddition reactions and their application in synthesis, Synthesis 5 (1998) 683–703.
[66] A. Tröster, A.R. Bauer, T. Bach, Enantioselective intermolecular [2 + 2] photocycloaddition reactions of 2H-1-quinolones, J. Am. Chem. Soc. 138 (2016) 7808–7811.
[67] H. Sakuragi, K. Tokumaru, K. Ito, K. Terakawa, K. Kicuchi, R.C. Caldwell, et al., Conformations and competitive [2 + 2 + 1 cycloadditions of intramolecular exciplexes of some anisylalkenyl 9-phenanthrenecarboxylates, J. Am. Chem. Soc. 104 (1982) 6796–6797.
[68] N.C. Yang, J. Libman, L. Barrett, M.H. Hui, R.L. Loeschen, Photochemical addition of acyclic 1,3-dienes to 9-cyanoanthracene and 9-anthraldehyde, a 4 + 2 cycloaddition, J. Am. Chem. Soc. 94 (1972) 1406–1408.
[69] R.U. Remy, C.G. Bochet, The arene–alkene photocycloaddition, Chem. Rev. 116 (2016) 9816–9849.
[70] A. Albini, M. Fagnoni, Arylation reactions: the photo-S_N1 path via phenyl cation as an alternative to metal catalysis, Acc. Chem. Reac. 38 (2005) 713–721.
[71] O. Tatsuro, M. Tomimoto, T. Mizutani, S. Furusawa, Cyclic acetal-photosensitized polymerization 1. Photooligomerization of 2-substituted 1,3-diphenylimidazolizine compounds, Polym. Bull. 6 (1982) 631–638.
[72] J.H. Choi, R. Ganesan, D.K. Kim, C.H. Jung, J.M. Hwang, T.C. Nho, et al., Immobilization of biomolecules by using iron irradiation-induced graft polymerization, J. Polymer Chem. 47 (2009) 6124–6134.
[73] Ł. Woźniak, J.J. Murphy, P. Melchiorre, Photo-organocatalytic enantioselective perfluoroalkylation of β-ketoesters, J. Am. Chem. Soc. 137 (2015) 5678–5681.
[74] C. Grewer, H.D. Brauer, Mechanism of the triplet-state quenching by molecular oxygen in solution, J. Phys. Chem. 98 (1994) 4230–4235.
[75] P.B. Merkel, D.R. Kearns, Radiationless decay of singlet molecular oxygen in solution. An experimental and theoretical study of electronic-to-vibrational energy transfer, J. Am. Chem. Soc. 94 (1972) 7244–7256.
[76] R.B. Vegh, K.M. Solntsev, M.K. Kuimova, S. Cho, L. Liang, L. Bernard, et al., Reactive oxygen species in photochemistry of the red fluorescent protein "Killer Red", Chem. Commun. 47 (2016) 4887–4889.
[77] S. Kanfer, N.J. Turro, Reactive forms of oxygen, in: D. L. Gilbert (Ed.), Oxygen and Living Properties. An Interdisciplinary Approach, Springer, New York, 1981, pp. 47–59.
[78] K. Gollnick, A. Griesbeck, Singlet oxygen photooxygenation of furans. Isolation and reactions of (4 + 2)-cycloaddition products (unsaturated sec.-ozonides), Tetrahedron 41 (1985) 2057–2068.
[79] S. Jain, A. Kushwah, P.K. Paliwal, G.N. Babu, pH Dependent photooxygenation of guanine by singlet oxygen in the presence of rose Bengal, Int. J. Chem. Sci. 8 (2010) 763–768.
[80] K. Yamaguchi, S. Yabushita, T. Fueno, K.N. Houk, On the mechanism of photooxygenation reactions. Computational evidence against the diradical mechanism of singlet oxygen ene reactions, J. Am. Chem. Soc. 103 (1981) 5043–5046.
[81] W. Adam, S. Bosio, A. Bartoschek, A.G. Griesbeck, Photooxygenation of 1,3-dienes, in: W. Horspool, F. Lenci (Eds.), Handbook of Organic Photochemistry and Photobiology, 25, CRC Press, Boca Raton, FL, 2004, pp. 1–19.
[82] J.E. Kim, M.J. Tauber, R.A. Mathies, Wavelength dependent cis-trans isomerization in vision, Biochemistry 40 (2001) 13774–13778.
[83] E. Vauthey, Investigations of bimolecular photoinduced electron transfer reactions in polar solvents using ultrafast spectroscopy, J. Photochem. Photobiol. A Chem. 179 (2006) 1–12.
[84] P. Rosspeintner, B. Lang, E. Vauthey, Ultrafast photochemistry in liquids, Ann. Rev. Phys. Chem. 64 (2013) 247–271.
[85] M. Goeppert-Mayer, Über Elementarakte mit zwei Quantensprüngen, Ann. Phys. 9 (1931) 273–295.
[86] N.S. Makarov, M. Drobizhev, G. Wicks, E.A. Makarova, E.A. Lukyanets, A. Rebane, Alternative selection rules-for one- and two-photon transitions in tribenzotetraazachlorin: quasi-centrosymmetrical-conjugation pathway of formally non-centrosymmetrical molecule, J. Chem. Phys. 138 (2013) 214314.

[87] V. Jankus, E.W. Snedden, D.B. Brigh, V.L. Whittle, J.A.G. Williams, A. Monkman, Energy upconversion via triplet fusion in super yellow PPV films doped with palladium tetraphenyltetrabenzoporphyrin: a comprehensive investigation of exciton dynamics, Adv. Funct. Mat. 23 (2013) 384–393.

[88] M.S. Smith, J. Michl, Singlet fission, Chem. Rev. 110 (2010) 6891–6936.

[89] I.O. Koshevoy, Y.-C. Lin, Y.-C. Chen, A.J. Karttunen, M. Haukka, P.-T. Chou, et al., Rational reductive fusion of two heterometallic clusters: formation of a highly stable, intensely phosphorescent Au-Ag aggregate and application in two-photon imaging in human mesenchymal stem cells, Chem. Commun. 46 (2010) 1440–1442.

[90] J. Zhu, Q. Liu, W. Feng, Y. Sun, F. Li, Upconvertion luminescent materials: advances and applications, Chem. Rev. 115 (2015) 395–465.

[91] G. Zonios, A. Dimou, I. Bassukas, D. Galaris, A. Tsolakidis, E. Kaxiras, Melanin absorption spectroscopy: new method for noninvasive skin investigation and melanoma detection, J. Biomed. Opt. 13 (2008) 01401.

[92] M.M. Brenner, V.J. Hearing, The protective role of melanin against UV damage in human skin, Photochem. Photobiol. 84 (2008) 539–549.

[93] G.M. Roberts, V.G. Stavros, Biomolecules photostability and $^1\pi\sigma^*$ states: linking these with femtochemistry, in: R. de Nalda, L. Bañares (Eds.), Ultrafast Phenomena in Molecular Sciences: Femtosecond Physics and Chemistry. Springer Series in Chemical Physics, Springer International Publishing, Switzerland, 2014, pp. 119–143.

[94] F. Bernardi, M. Olivucci, M.A. Robb, G. Tonachini, Can a photochemical reaction be concerted? A theoretical study of the photochemical sigmatropic rearrangement of but-1-ene, J. Am. Chem. Soc. 114 (1992) 5805–5812.

[95] A. Weller, Photoinduced electron transfer in solution: exciplex and radical ion pair formation free enthalpies and their solvent dependence, Z. Phys. Chem. (NF) 133 (1982) 93–98.

[96] H.E. Zimmerman, A new approach to mechanistic organic photochemistry, in: W.A. Noyes, G.S. Hammond, J.N. Pitts (Eds.), *Advances in Photochemistry*, 1, John Wiley & Sons, Hoboken, NJ, 1963, pp. 183–208.

[97] B.A. Albani, C.B. Durr, C. Turro, Selective photoinduced ligand exchange in a new tris-heterolepticRu(II) complex, J. Phys. Chem. A 117 (2013) 13885–13892.

[98] V. Balzani, Electron Transfer in Chemistry, Wiley-VCH Verlag, Weinheim, Germany, 2001.

CHAPTER

2

Interaction with the environment: Skin

2.1 Light in the environment

The skin is the largest organ of human body and its function is that of separating from the environment and protecting the body and its organs. Naked skin exposed to solar light receives a fraction of solar light hitting the planet, that is the portion of the light emitted by the sun that arrives to the earth surface, less the part that is absorbed or scattered by the atmosphere and depending on weather conditions as well as on societal--and thus clothing--habits. The overall solar radiation hitting the earth has a virtual intensity of 1400 W (J/s)/m^2 or 8.4 J/cm^2/min. About 98% of the emission is in the wavelength range 200–2700 nm (Fig. 2.1A). In particular, 40%–45% involves the visible (by humans) light range (200-700 nm), less than 5% the short wavelength range [ultraviolet (UV), 100–400 nm] and the rest to the long wavelength component [infrared (IR), 700–4500 nm].

The absorptions due to the gasses present in the atmosphere appear as holes in the black body radiation emitted by the sun as it reaches the earth surface, overall, about 50% of the original radiation [3]. The dose arriving at the earth surface in moderate climates ranges between 1 and 6 J/cm^2 for UV-B and 40 and 140 J/cm^2 for UV-A per day, according to the season, and thus depends primarily on the angle formed by the rays from the sun with the planet surface at the considered latitude.

As for the absorption of the light, the measure unit used by physics practitioners is based on a general concept, cross section: if the particles of a gas can be treated as hard spheres in contact, then the probability of all encounters is proportional to the sphere surface, $\sigma = \pi(2r)^2$, and when an exploring ray of flux ϕ_0 of n particles travels through a thin layer of a material of thickness dz, then the flux would be attenuated by the amount $d\phi/dz = -n\sigma\phi_0$, and in integral form $\phi/\phi_0 = e^{-snz}$ with an exponential dependence on concentration. In chemistry, a slightly different approach is used for arriving at the same relation [4]. Chemistry practitioners are accustomed to use molecular absorptivity (ε), that is the coefficient characteristic of every substance that enters the Lambert-Beer law, $A = \varepsilon bc$, where b is the path traversed by the probing ray, c is the molar, and A is the absorbance, the reciprocal of the logarithm of the transmittance [$A = \log(1/T)$] that is thus conveniently proportional to the molar concentration. The law is valid only for dilute solutions and for molecules with defined molecular weight (MW). Most biomolecules are polymers and in the case of melanin not homopolymers but heteropolymers (made of different monomers), so that one may have at most an

Light, Molecules, Reaction and Health
DOI: https://doi.org/10.1016/B978-0-12-811659-3.00002-5

average composition and MW and no precise structure. More importantly, molecules are strictly adherent one to another, in something that is more akin to the solid state than to an ideal solution, so that the transmission is ill defined and measured. The attenuation transmitted light by a sample can be determined by having recourse to total reflectance instrument with an integrating sphere, and where the material is not suitable for this method by calculating it from different, such as temperature bursts or γ-rays [5].

Light hitting the skin, as any other object, is in part scattered, in part reflected, in part absorbed, and in part further transmitted. The diffuse reflectance of surfaces can be measured by means of an integrating sphere that affords an average of the beam intensity over all angles of illumination and observation. The behavior of the beam depends on the characteristic of the medium. Raylegh scattering is the mainly elastic scattering involving particles much smaller than the light wavelength and corresponds to the oscillations of the electric field of the particles with the same frequency of the electric field. It is directed toward every direction and is four times stronger in the violet than in the red. This is what makes the sky appear blue, as the beams of this color are for a large fraction scattered by the gases of the atmosphere (and essentially for the same reason, lakes and oceans appear blue to our eyes, although water is colorless, and the same holds for glaciers, where a large number of small air bubbles are present, compressed by the weight of ice). On the contrary, light arriving directly from the sun (see Fig. 2.1A) [6] appears yellow (and red at the sunset, when the oblique rays have to cross a longer path through the atmosphere) because they have lost a large part of the blue.

A different behavior results when an incident beam with a wavelength comparable to the dimensions of the particles is operating. Here a different scattering, known as Mie scattering, takes over, although partially overlapping with Raylegh scattering. This is much less dependent on the wavelength and predominantly is directed in the same direction of the beam.

Furthermore, the body fluids are turbid (think of particles of fat in milk, or of droplets of water in clouds), and contain molecules with dimensions just below 1 mm. The most extensive penetration of light into tissues occurs in the near-IR region of 700–900 nm. Here blood absorption (due to highly absorbing hemoglobin and deoxyhemoglobin, corresponding at the 5%–10% of the tissue volume occupied by blood) rapidly decreases, while that of tissue water increases and scatter predominates by two order of magnitude. Light in the range of 630-670 nm where the eye sensitivity and penetration depth are well matched, will likely provide the best chance of visual stimulation[1,7].

In order to determine the absorption of a component of a solution in a biological fluid, it is appropriate to take into account not only the absorption coefficients (μ_a) due to it, but also the scattering coefficient (μ_s) that determines the phase of the emerging light ray. As for the skin, the impinging ray has an intensity of c.1 J/cm^2 for the UV-A and c.900 J/cm^2 for the visible light [7].

A parallel to tanning of biological tissues is the yellowing of still water under irradiation, observed in a light intensity dependent manner via the formation of a variety of organic matters.

A phenomenon in a sense parallel to ozone depletion is the browning of superficial waters due to the increased presence of dissolved organic matters (DOMs) that may lead to a dramatic decrease of the transparence, for example, diminish the penetration of UV light from 10 to 2 m[8].

Reports by an (UN sponsored) International Committee on environmental effects of ozone depletion and its interactions with climate change are regularly published [9] (Fig. 2.2).

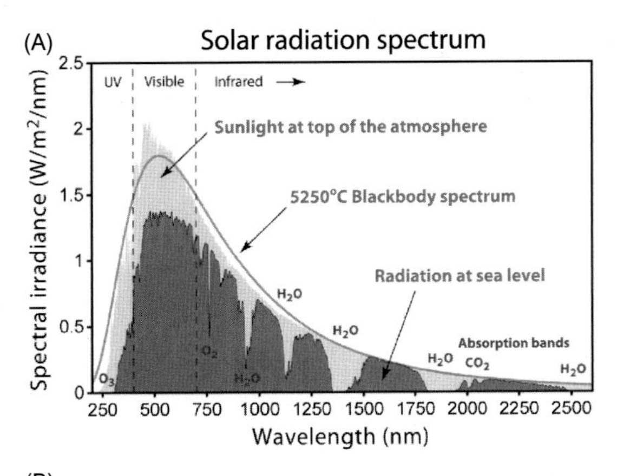

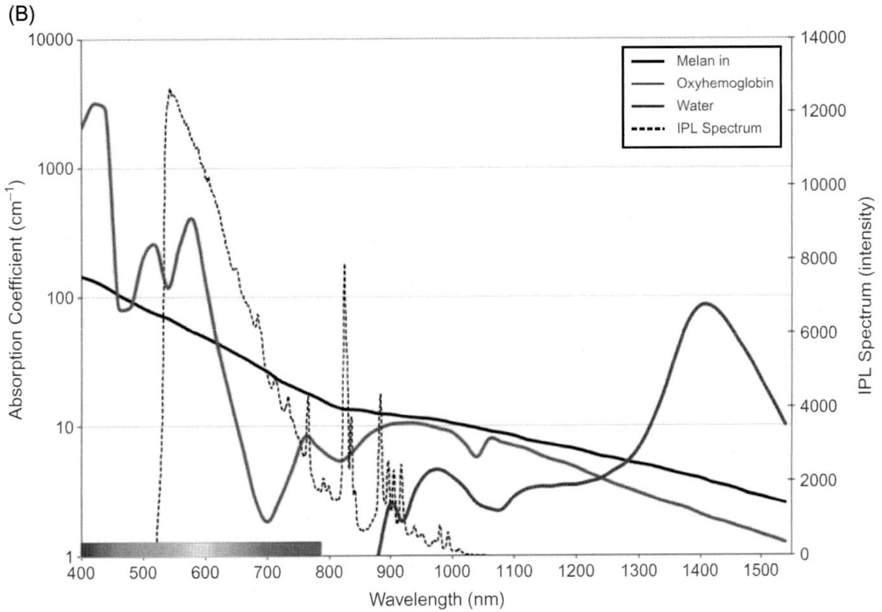

FIGURE 2.1 (A) Spectrum of solar radiation above the atmosphere and sea level [1]. (B) The near-infrared therapeutic window of 600–1000 nm, where tissue absorption is small compared to scattering is apparent. Source: *(A) Reproduced with permission from R. Kortum, E. Sevick, Quantitative optical spectroscopy for tissue diagnosis, Ann. Rev. Phys. Chem. 47 (1996) 555–606. (B) Ash, C, Dubec, M, Donne, K, Bashford, T (2017) Effect of wavelength and beam width on penetration in light-tissue interaction using computational methods. Lasers Med Sci. 2017; 32(8): 1909–1918 [2].*

2.2 Microanatomy of the skin

The skin is the largest organ in the body, with a surface of ccirca .1.8 m^2 and is composed of four functionally different layers of keratinocytes (KCs) at different stages of differentiation (Figs. 2.2).

Stratosphere, and smoke resulting from wildfires and the burning of biomass, attenuate sunlight before it reaches terrestrial and aquatic ecosystems (Fig. 2.3). DOM, although derived primarily from terrestrial ecosystems, attenuates sunlight primarily in aquatic ecosystems (Fig. 2.4). Attenuation of UV radiation

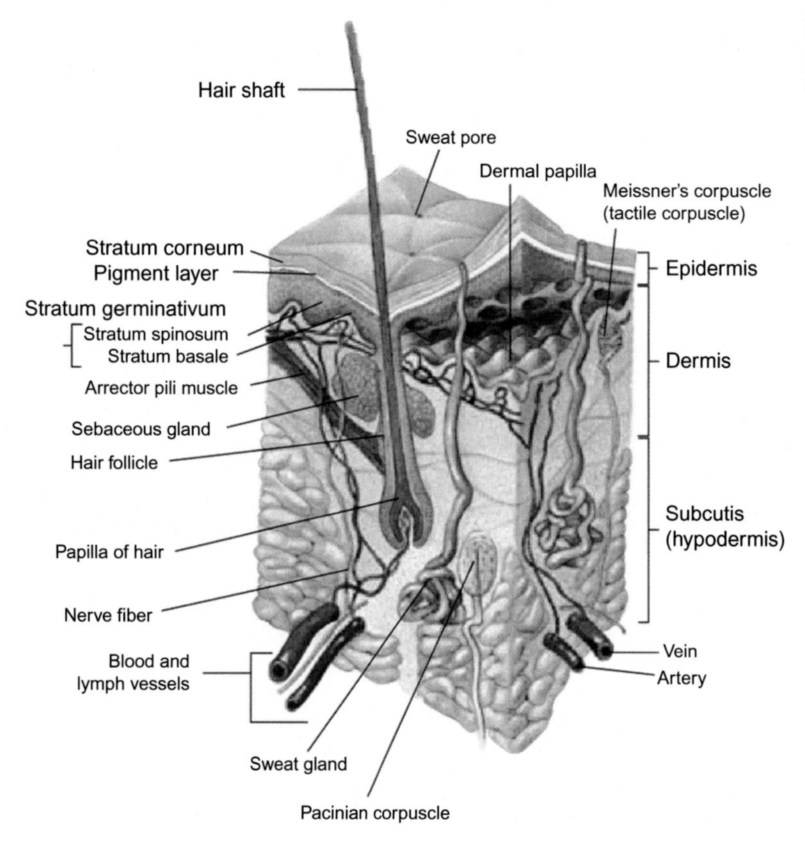

FIGURE 2.2 Microanatomy of the skin. Reprinted with permission from [10].

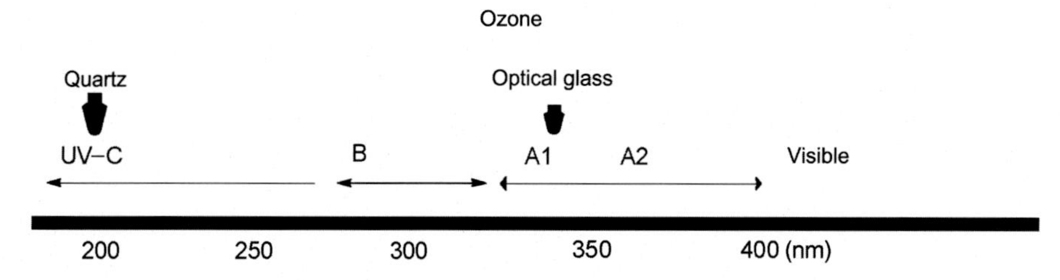

FIGURE 2.3 Borders of UV and visible light, and absorption by quartz, glass and ozone light.

(UVR) by ozone is much more restricted to the shorter wavelengths of UV-B, whereas both DOM and especially smoke attenuate UV-A radiation and visible light in addition to UV-B. These three ubiquitous environmental substances (ozone, DOM, and smoke) are therefore critical in regulating UV-to-visible light ratios in the environment, which influence important processes ranging from DNA damage and vitamin D production (both of which show maximum response to very short wavelengths of UV-B) to the direction of migration of zooplankton in the water column. For example, the common freshwater crustacean Daphniais attracted to longer wavelength visible light but avoids UVR. Wavelength-specific

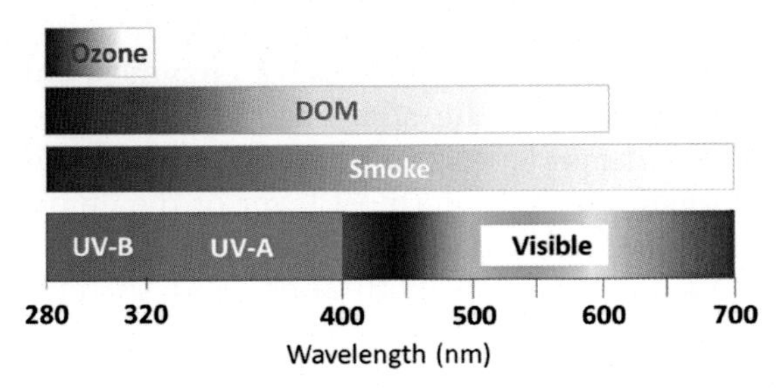

FIGURE 2.4 Atmospheric ozone, 90% of which is located in the lower stratosphere absorbs the largest part of the UVC, causing, along with smoke, the greenhouse effect [12]. The absorption by dissolved organic matter (DOM) similarly causes absorption of a part of UV in surface waters.

attenuation by particulates is not shown here as it depends on size and chemical composition of the particulate material [12] (Fig. 2.5).

The thickness of these layers varies in different sites, different species, and under different conditions. From the inner to the external layers, these are the basal layer (stratum basale) that is placed on the basal lamina separating the dermis and epidermis. This layer contains the epidermal stem cells that proliferate, providing the cells for the upper differentiating layers. They are large, columnar cells that form intercellular attachments with adjacent cells through desmosomes and adherent

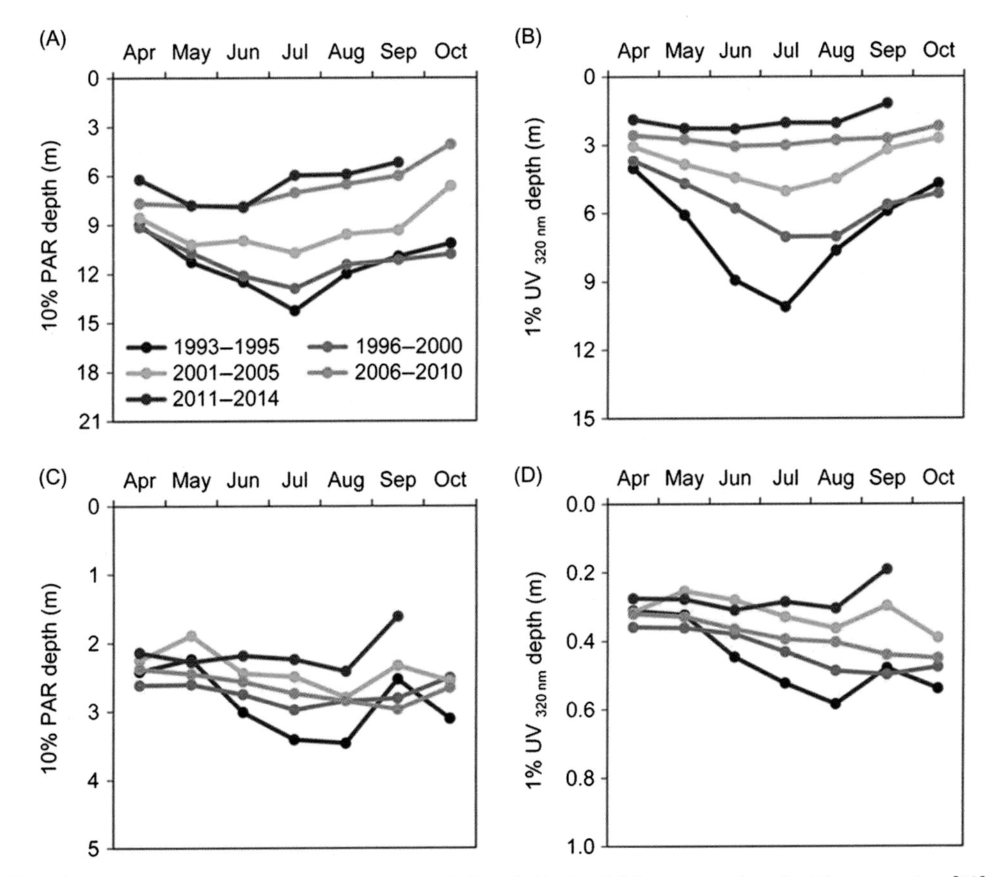

FIGURE 2.5 Yellowing due to the UV penetration in North England lakes , reproduced with permission [13]. (A and C) PAR depth, photosynthetically active radiation (PAR) deptj of penetration (m) and penetration of UV light (B and D) during the summer season in various years.

junctions. Desmogleins (Dsg) 2 and 3 and desmocollins (Dsc) 2 and 3 are the main proteins of cadherins group, which take care of adhesion, in the desmosomes forming the basal layers, whereas both P- and E-cadherins participate in the adherents junctions of these cells. However, as the KCs move out of the basal layer and begin to differentiate, Dsg 1 and Dsc 1 become the dominant desmosomal cadherins, whereas P-cadherin is no longer produced and it is rather E-cadherin that becomes the main cadherin in adherent junctions. At this level, the expression of proteins (claudins and occludins) that form the tight junctions is likewise initiated. An asymmetric distribution of integrins on their lateral and basal surface makes their attachment to the basal lamina and adjacent cells possible and helps to regulate their proliferation and subsequent differentiation. In the first phase, the fibrous proteins keratins K5 and K14 are produced in KCs, while later a switch to producing keratins K1 and K10 in the stratum spinosum occurs. Furthermore, cells of this layer also begin to produce involucrin, a component of the cornified envelope, and the enzyme transglutaminase-K, which is responsible for the ε-(γ-glutamyl)lysine crosslinking of involucrin and other substrates into the insoluble cornified envelope. The stratum granulosum, the uppermost nucleated layer, which is above the stratum spinosum, owns its name for the presence of electron-dense granules of keratohyalin. The larger of the two granule types contains profilaggrin, precursor of filaggrin, which appears to function as a bundling protein for the keratin filaments, although more recently it is thought to play a role in maintaining the water content of the epidermis, because it is degraded into smaller peptides that change the osmotic properties [14]. The smaller granules contain loricrin, which becomes a major component of the cornified envelope. The stratum granulosum also contains lamellar bodies and lipid-filled vesicles. These are responsible for secreting their lipid content (and lipid-processing enzymes) into the junction of the stratum granulosum and the stratum corneum. Finally, the stratum corneum is the outermost layer of the epidermis. In this layer, the external cornified envelope undergo full maturation, and surround the bundled keratin filaments that are attached to the lipid envelope. In this way, they contribute to the processing of the lipids secreted by the lamellar bodies. The net result is a resilient impermeable structure protecting the viable layers that are underneath. Enhancing the pH by treatment with strong bases such as 1,1,3,3-tetramethylguanidine or 1,8-diazabicyclo[5,4,0]undec-7-ene in ethylene glycol-ethanol mixed solvent causes the rapid activation of a serine protease, which in turn catalyzes the rapid degradation of corneodesmosomes. Western blotting revealed degradation of Dsg1, a key corneodesmosome structural protein, in parallel with loss of corneodesmosomes [15].

All of the above processes are regulated, both in vivo and in vitro, by calcium ions. A calcium gradient within the epidermis (with the lowest concentration in the stratum basale and increasing concentrations until the outer stratum granulosum, where it reaches its maximum) promotes the sequential differentiation of KCs. In the meantime, they migrate through the different layers of the epidermis to form the permeability barrier of the stratum corneum. Calcium ions likewise promote differentiation by both outside–in and inside–out signaling, including the formation of desmosomes adherent junctions and tight junctions. This maintains the cell-to-cell adhesion, while they promote the activation of various kinases and phospholipases that produce second messengers. In turn these regulate intracellular free calcium and PKC activity, which are required for the differentiation process. The calcium receptor plays a central role by initiating the intracellular signaling events that drive differentiation in response to extracellular calcium. Calcium concentration in the stratum corneum is very high in part because those relatively dry cells are not able to dissolve the ions. An increase in extracellular calcium concentrations induces an increase in intracellular free calcium concentrations in KCs. In epidermis are also found melanocytes, responsible for the production of melanin pigments, Langerhans cells (LCs) as antigen presenting cells, as well as Merkel cells interacting with nerve endings. The dermis is the area of supportive connective tissue between the epidermis and the underlying subcutis. It is a fibrous and elastic tissue that gives the skin its flexibility and strength. Appendances are sweat glands and hair roots and also blood and lymph vessels. The dermis is made up of fibroblasts, which produce extracellular matrix proteins like collagens, elastin, and structural proteoglycans, and also includes immune cells such as mast cells (MCs) and macrophages. The subcutis is the layer of loose connective tissue and fat beneath the dermis.

2.3 Interaction between light and skin

As it will be discussed below, the UV-B part of the spectrum causes severe damages to the skin while UV-C can be disregarded, because this radiation does not reach the earth surface. This is because of the large amount of ozone present in the stratosphere (12-50 km) around the planet absorbs completely this wavelength range (ozone absorbs strongly in the 290–320 nm range, UV-B, and fully the 200–290 nm range, UV-C). In the stratosphere, ozone is formed photochemically by combination of oxygen atoms (in turn resulting from the photolysis of oxygen molecules into atoms) and molecules. The thus formed oxygen atoms then attack oxygen molecules reforming ozone, or react with ozone molecules to reform oxygen molecules. The cycle goes on, with the participation of further small molecules, if present. This causes the formation of toxic and unpleasant compounds ("photochemical smog"), but not the change in the ozone concentration (Scheme 2.1).

$$O_3 \xrightarrow{h\nu} O_2 + O \qquad O \xrightarrow{O_2} O_3 \qquad O \xrightarrow{O_3} 2\,O_2$$

SCHEME 2.1 Ozone formation in the gas phase.

Thus O_3 formation and destruction are in a steady state in the stratosphere and this is essential for maintaining the role of this gas as UV filter. Ozone is also destroyed by further odd electron species, in particular nitrogen oxides (NOs) and hydroxy radicals. Such reactions are again in balance, however, with the rates of formation. Further photochemically formed radicals have a role, in particular for the quality of the air (photochemical smog, arising from combustion gases under appropriate wind conditions causes serious respiratory troubles and diffuse irritations), but not a thinning of the ozone layer.

Different is the role of chlorofluorocarbons (freons), chemically stable and highly volatile compounds that have been largely used in the past years. Large industrial applications are refrigeration, as aerosol propellants in making plastic foams, as cleaners for electronic parts, and as coolants in automobile air conditioners exactly because of their stability in the ground state. However, these are photochemically active and interfere in the ozone cycle. Typically, the least energetic bond (C–Cl in fluorochlorohydrocarbons, freons) is cleaved upon irradiation and chlorine atoms are formed, and these in turn destroy ozone through a (catalytic) photochemical process giving rise to the phenomenon of ozone depletion and ozone hole, which has been recognized internationally (Scheme 2.2) [16–22].

$$CClFCl_3 \xrightarrow{h\nu} CFCl_2^{\cdot} + Cl^{\cdot}$$

$$\left.\begin{array}{l} Cl^{\cdot} + O_3 \rightarrow ClO^{\cdot} + O_2 \\ ClO^{\cdot} \quad O^{\cdot} \rightarrow Cl^{\cdot} + O_2 \end{array}\right\} \text{net result} \quad O_3 + O^{\cdot} \rightarrow 2\,O_2$$

SCHEME 2.2 Consumption of ozone under irradiation in the presence of chlorofluorocarbons.

$$CF_nCl_m + h\nu \rightarrow CF_nCl_{m-1}\cdot + Cl\cdot$$
$$O_3 + Cl\cdot \rightarrow ClO + O_2$$
$$2ClO \rightarrow 2Cl\cdot + O_2 \text{ (chain mechanism)}$$

The chemical stability of freons makes their migration up to the stratosphere (not in the troposphere, where the concentration of the reacting species is not sufficient) and while shifting over the continents. Furthermore, these compounds are highly persistent (many decades). The realization of this situation led to the ban of the production of such fluids (Montreal Protocol, 1987 [9]), although it was apparent that ozone consumption was going to remain a significant process for many years to come. Between the 1970s and the 2000s, average stratospheric ozone decreased by approximately 4%, contributing an approximate 4% increase to the average UV intensity at the earth's surface. However, much larger changes have taken place in some locations, especially in the southern hemisphere because of the seasonal "ozone hole" caused by the reaction of halogenated hydrocarbons [17].

A similar situation is found in bromine derivatives, such as halons (e.g., $CFBr_3$) used in fire extinguishers and methyl bromide used as soil sterilant (in the case of $CFBr_3$, photochemical cleavage gives $Br\cdot$ and $CFBr_2$, both of them much stable than the chloro derivatives) [23]. The situation is consistently monitored and in fact sign of recovering is observed [24], although volcano eruptions introduce a further variable [25].

As concentrations of ozone depleting substances decrease over the next decades, greenhouse gases will become the main cause of change in the atmosphere composition. However, in the last two decades, changes in solar UVR in northern mid-latitudes have been mainly controlled by clouds and aerosols rather than by ozone [26,27].

Absorption of light may cause a reaction through the excited states. The efficiency of photochemical reactions is expressed by the quantum yield, that is the fraction of the excited molecules (that thus have absorbed a photon) that undergo a specific chemical reaction, $\Phi_{reac} = n_{reac}/n_{abs}$. As it is usual for chemists this relation is referred to moles of particles, $\Phi_{reac} = M_{reac}/M_{abs}$. The molar unit for photons is called Einstein and correspond to an Avogadro number of photons, and was introduced by M. Bodenstein [28]. The name of photon for a quantum of light is due to GN Lewis [29]. Einstein referred[1][30–32] the quantum yield (Φ) to the first stage of any photochemical reaction, that is the excited state, and in this sense it was clear that $\Phi = 1$ [31]. However, if the excited state converted to the final product only in part, then the *product* had $\Phi < 1$, in many cases $\Phi << 1$. On the contrary, excited states have a quite high energy, and starting from a high point it was all but impossible that the process would lead to species that were still (thermally) reactive, and enter in a chain process, so that $\Phi > 1$, or $\Phi >> 1$ may well result for the final product formation. The question was a main point of discussion in the 1920; it requires some decades to arrive at a general consensum, despite the fact that an international meeting was organized by the Royal Chemical Society exactly to this purpose [33] (Figs. 2.6 and 2.7).

The complex consequences of exposure to solar light involve erythema, tanning, urticaria,

[1] The Einstein's equivalent law was adopted as the second law of photochemistry, causing some disagreement in the scientific community [28–30].

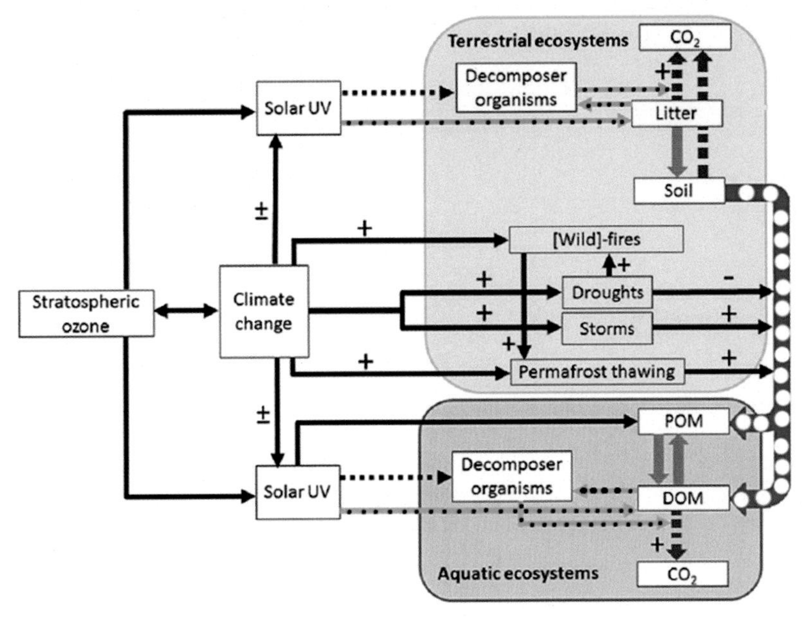

FIGURE 2.6 Biological mechanism of tanning and the role of melanin. Interactive effects of solar (ultra violet) UV light and various surfaces radiation and climate change on processes flows within and between terrestrial and aquatic ecosystem. Key to symbols: Black arrows indicate linkages between environmental actors; + shows increase in a process of flows; - a decrease in a process of flow. Dashed arrows indicate direct effects of solar UV radiations on decomposing organism. Gray arrows indicate a flow of carbon within ecosystem. White dots arrows indicate a flow of carbon from terrestrial to aquatic ecosystems. Black small dots arrows refer to the process of "priming." Dashed arrows indicate a production of carbon dioxide in terrestrial and aquatic ecosystems. POM and dissolved organic matter (DOM) stand for particulate and dissolved organic matter, respectively [22].

or even skin cancers, according to the intensity and wavelength involved, as well as the skin type [35].

When a photochemical reaction is undesired, for example, because it causes toxic phenomena in the organism, one may intervene in three ways, exactly analogously to what is done in reinforcing the light fastness of a dye or of a material. First, a filter may be introduced that absorbs competitively with the compound considered. Second, a quencher of the excited states is added that react fast with excited state leading back to the ground state. Third, rather than contrasting the photochemical reaction, an additive may be chosen to inhibit the high reactivity of the (primary) photoproducts; the typical application is using compounds that in some way trap radicals that are generated during exposure to light, thus inhibiting chain reactions that are otherwise quite common with such intermediates (Scheme 2.3).

UV light causes both direct and indirect effects on health. The best-defined direct benefit is the synthesis of previtamin D in the skin, while direct adverse effects include skin cancers, cataracts, and reactivation of some viral infections. Indirect effects include those resulting from changes in food quality and disinfection of surface waters used for drinking. It is likely that for any individual there is an optimal level of exposure, but this is highly variable and difficult to define. As warmer temperatures predominate in the future, time

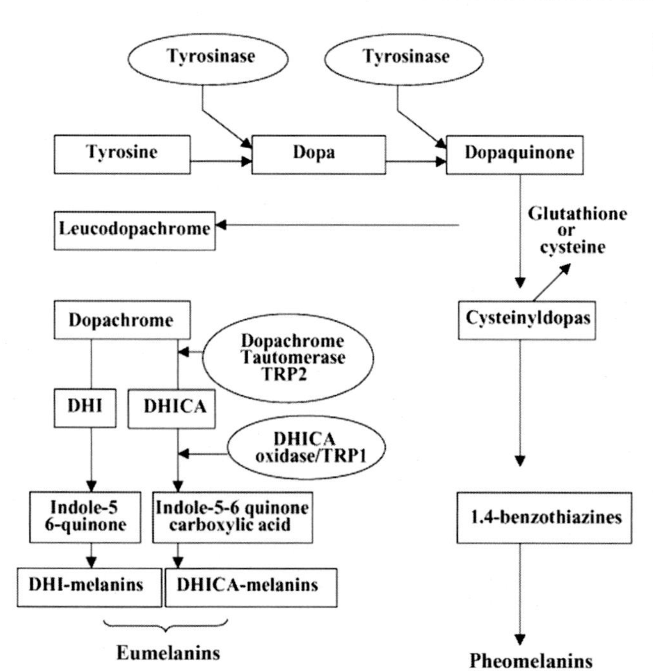

FIGURE 2.7 Enzymatic reaction connected with tanning [34].

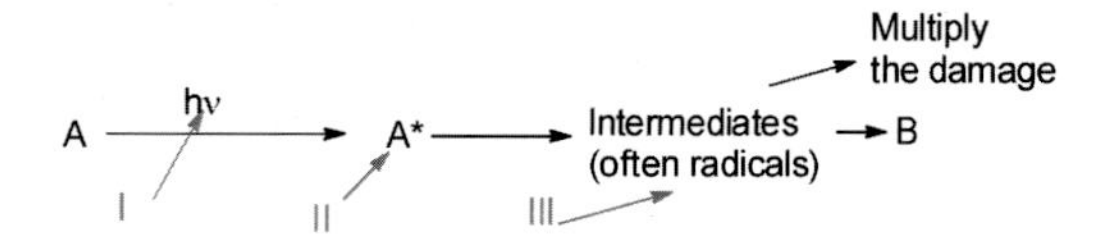

SCHEME 2.3 Ways to inhibit undesired photochemical reactions.

spent outdoors will increase and thus exposure to both UV-A and UV-B radiation. Research has firmly established that UV-B radiation is the cause of DNA damage and skin cancers, but new evidence suggesting that UV-A irradiation inhibits the repair of DNA damage through a number of pathways and fosters local invasion of tumor cells (metastasis) has been documented [36]. Furthermore, both UV-A and UV-B radiation cause an immune suppression and it is likely that the development and spread of skin cancers will increase during the course of the 21st century [37]. The overall incidence of cutaneous malignant melanoma and nonmelanoma skin cancer continues to increase in most countries for which data are available, but is decreasing in several countries in younger age groups [38,39].

Scientific data show that sunscreen provides protection from DNA damage and from sunburn following exposure to UVR, and thus reasonably enjoying a protection against skin cancer. Regular use of sunscreen was reported to be associated with a lower number of nevi (moles), a marker of melanoma risk, in both children and adults. In many countries there is a high level of knowledge of the risk connected with exposure to UV light and a positive attitude toward sun protection, but compliance remains low, and a risky sun exposure behavior and a preference for tanning are common. As for teenagers, proximity to tanning facilities and geographic characteristics (living in one or another part of a country, in the presence of smog or other air contamination or not, and attending a rural high school) are associated with intentional tanning [40].

2.4 Natural skin protection and tanning: melanogenesis

In order to withstand light flux, pigments are present in the skin, mainly of the family of melanin (Scheme 2.4) [41].

Melanins are located within the top layer of the skin (epidermis), and contribute, along with hemoglobin, which is found in the bottom layer (i.e., the vascular network of the dermis) to the pinkish color which we perceive for skin (this corresponds to c. 50% back scattering from within the dermis in Caucasian subjects) [42,43].

UV and visible light have some effect on skin and internal organs. UV-C, 100–290 nm, is absent from the impinging rays, because it is completely absorbed by ozone present in the atmosphere; UV-B, 290–315 nm, is absorbed by DNA and various biomolecules and causes a reddening and a thickening of the skin and rashes (erythema) and, after some hours, tanning, but also the photochemical synthesis of previtamin D from the dehydrocholesterol obtained from food. UV-A, often further subdivided into A1, 315–340 nm, A2, 340–400 nm (visible light, 400–700 nm), also cause erythema. Artificial light is used therapeutically (either visible light for photodynamic therapy, via oxygen activation, or UV light for photochemotherapy) [44,45].

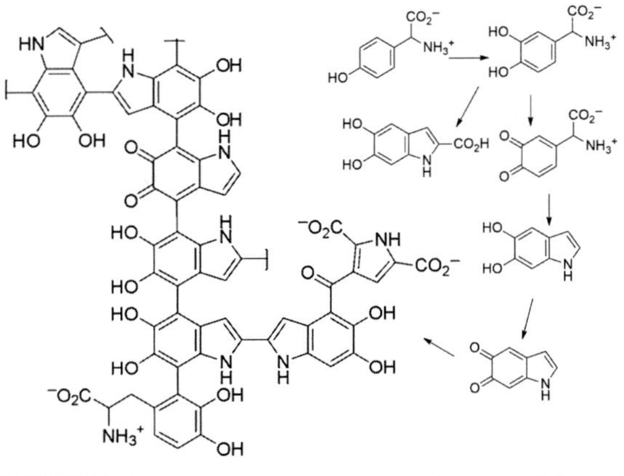

SCHEME 2.4 Melanogenesis, chemical aspects. Most of the compounds indicated have been isolated and characterized. The oxidation of tyrosine to dopamine and to dopaquinone and dopachrome is followed by a polycondensation leading to pheomelanin. The reaction of indolecarboxylic acid occurs in the same way but is much slower. The sections of the polymeric structure shown pertain due to the red pigment pheomelanin. This, as well as the other polymelanins, is extensively conjugated polymer that therefore absorbs intensively over the UV and visible region of the solar spectrum.

Two main types of melanins are present in the skin, eumelanin and pheomelanin. The latter is a black, insoluble polymers formed by polymerizations of dihydroxyindoles and dioxoindolines and ultimately by oxidation of tryptophan, which occurs as an enzyme-controlled reaction in the first phase and then via an uncontrolled reaction, where quinoides dioxoindolines have a major role. Eumelanin is an alkali soluble, red–brown polymer, based on benzothiazine monomers. The type and amounts present give rise to the different skin types that dermatologists distinguish (as well as to hair and eye color) (see Table 2.1) [46].

Notice that classification in one of the above types is to be used only within the original conditions. Thus the minimal erythemal dose (MED) determined under different conditions (as an example in the presence of a sensitizer) correlates only weakly with the individual skin phototype according to Table 2.1 [47].

Both polymerizations begin from the oxidation of tyrosine to dopamine, in the case of eumelanin via dopaquinone and dopachrome and then polycondensation of this quinone under the action of Cu(II) enzymes. As a side path, conjugate addition of cisteine via the sulfhydryl group onto the quinone moiety leads to 2- and 5-cysteindopa and from the latter one to a benzothiazine ring. A third pigment (neuromelanin) is found in the brain and arises from the polymerization of dopamine with a minor amount of cysteine (Scheme 2.5). As it appears from Scheme 2.5, no small regular constituent is apparent in the structure of

TABLE 2.1 Skin color types (originally proposed by T. B. Fitzpatrick) [43].

Skin type	Typical features	Tanning ability
I	Pale white skin, blue/green eyes, blond/red hair	Always burns, does not tan
II	Fair skin, blue eyes	Burns easily, tans poorly
III	Darker white skin	Tans after initial burn
IV	Light brown skin	Burns minimally, tans easily
V	Brown skin	Rarely burns, tans darkly easily
VI	Dark brown or black skin	Never burns, always tans darkly

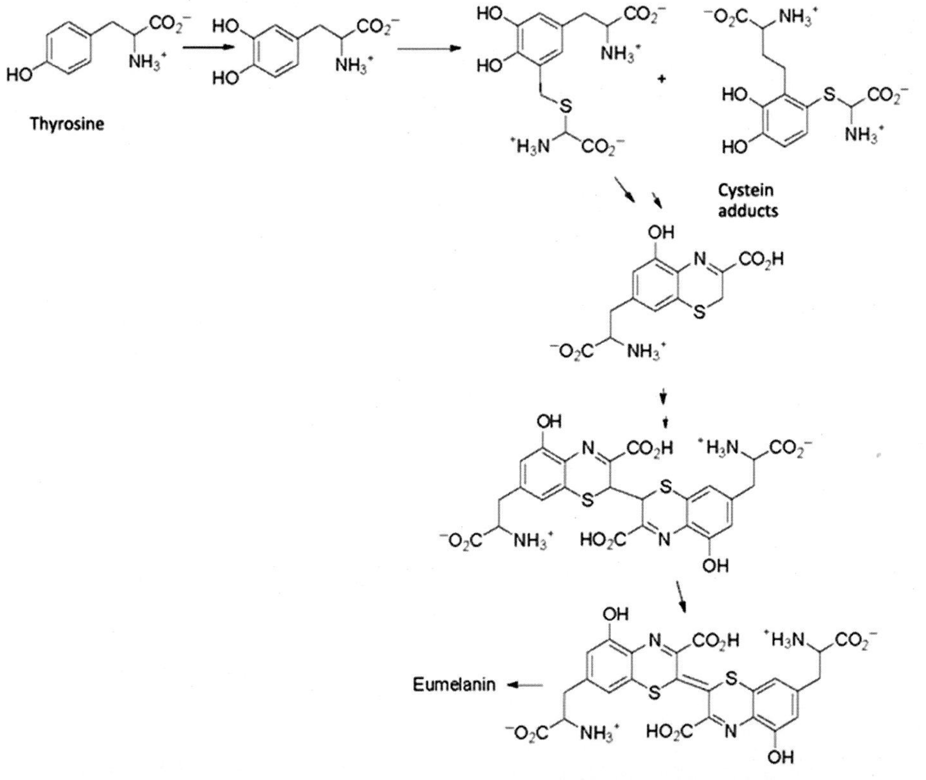

Thyrosine

Cystein adducts

Eumelanin

SCHEME 2.5 The highly favored addition of cystein gives thioethers and the sulfur containing, water soluble eumelanin.

melanins, that are thus to be considered heteropolymers [45,46].

The first step in eumelanogenesis after the oxidation by phenol oxidase to dopamine and in competition with decarboxylation that leads to dopaquinone, which then undergoes spontaneous cyclization to produce cyclodopa, and this then rapidly undergoes a redox exchange with another quinone molecule to produce one molecule each of DOPA chrome and DOPA. DOPA chrome is then spontaneously decomposed by decarboxylation at neutral pH to form 5,6-dihydroxyindole (DHI). Indolecarboxylic acid reacts much slower than the corresponding decarboxylate compound, as one would expect since this is a much poorer electron donor, and cyclizes and polymerizes under the control of dopamine conversion enzyme to melanin. In the meantime, the strongly colored pigments migrate toward the keratinocytes. In melanocytes, the late melanosomes bind to microtubules and undergo actin-dependent transport to the cell periphery, and then are transported to KCs.

The first effect, involving an increased production of the dark pigment melanin by melanocytes results in the tanning of the skin. The degree of tanning depends primarily on a genetic, indeed ancestral, mechanism and is related to the skin type of each individual, and thus on the geographical origin. Darkest skin types are found in tropical regions, particularly in grass lands, where nothing hinders solar rays, while in populations originating from predominating forested land the skin easily have a less dark tone. Furthermore, darkest skin types are frequent among people from Artic areas, be it due to exposure to light reflected by snow or to the fat-rich diet. Abundant exposure to solar light may thus have serious adverse effects, but boosts the immune system, protecting from several serious diseases and promotes previtamin D formation through a photochemical reaction in the skin.

Melanocytes are melanin-producing neural-crest–derived cells located in the bottom layer of the epidermis, of which there are about 1200 units/mm^2 of the skin independently of the human race, but depending on the location on the body. This transfer melanosomes, the organelles that synthesize, store, and transport melanins to KCs, the outermost layer of epithelial cells [48–52].

In the last cells, melanin is localized above the nucleus in the form of a cap-like structure to protect the cellular DNA. Mechanistically, as shown in Fig. 2.8 [54–58], melanogenesis goes through various phases, starting with the formation of relatively amorphous and spherical vescicles from the endoplasmatic reticulum (Stage I). The vescicles are then trafficked to Stage Imelanosomes and transform them into elongated, fibrillar organelles (Stage II melanosomes). Continuation of the melanin polymerization and uniform deposit of the pigment on the internal fibrils (Stage III) until all of the structural details are blurred (Stage IV) completes the synthesis of melanosomes. Early formed melanosomes derive from endoplasmic reticulum and various subcellular components, as well as a number of specific proteins that guarantee the correct functioning of this structure, in particular with regard to tyrosinase, tyrosinase-related protein 1, and dopachrome tautomerase. UV-B light stimulates the production of melanine that is controlled by the shock protein gp96.

More specifically, KCs modulate the activity of intracellular trafficking machineries required for efficient pigment transfer [59].

Pigment transfer from the melanosomes to the keratinocytes occurs by a shedding vescicle cargo mechanism. The pigment globules are connected to the filopodia of melanocyte dendrites and are released from various areas of the dendrites of normal human melanocytes derived from darkly pigmented skin. The globules are then captured by the microvilli of normal human KCs, also derived from darkly pigmented skin, which incorporated them in a protease-activated receptor-2-dependent manner. After the pigment globules have been ingested by the KCs, the membrane that surrounded each melanosome cluster is gradually degraded, and the individual melanosomes are then spread into the cytosol and distributed primarily in the perinuclear area of each KC. This suggest that a melanosome transfer

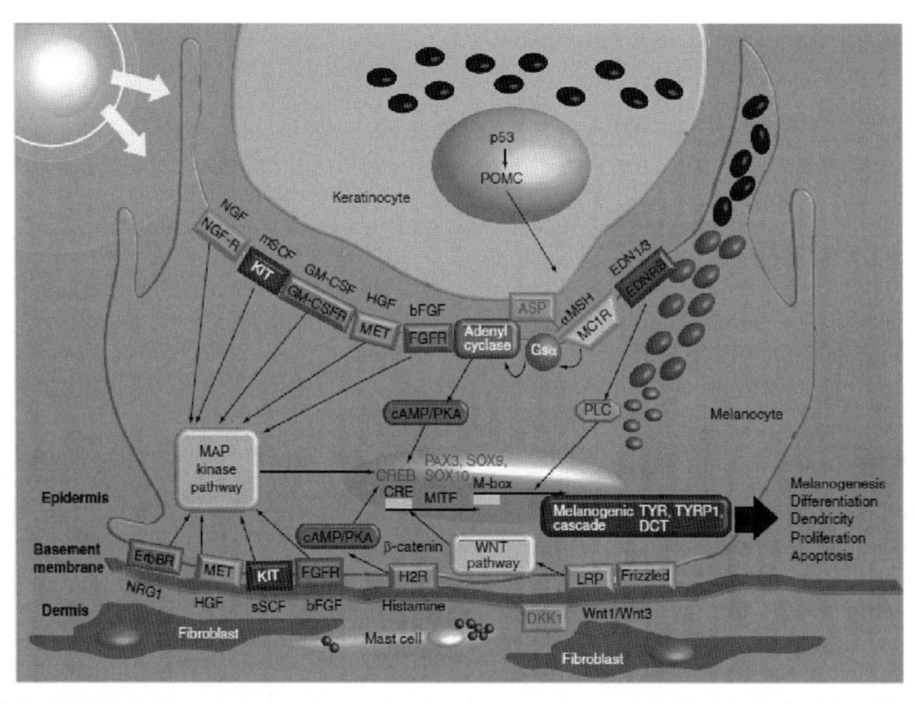

FIGURE 2.8 Schematic representation of receptors, ligands, and other factors that regulate pigmentation of human skin under UV radiation. *LRO*, Lysosome-related organelle; *ACTH*, adrenocorticotropic hormone; *ASP*, Agouti signal protein; *DCT*, DOPAchrome tautomerase; *DKK*, Dickkopf; *DHI*, 5,6-dihydroxyindole; *DHICA*, DHI-2-carboxylic acid; *DOPA*, 3,4-dihydroxyphenylalanine; *EMI*, epithelial–mesenchymal interactions; *EMT*, epithelial-mesenchymal transitions; *bFGF*, basic fibroblast growth factor; *HGF*, hepatocyte growth factor; *HOX*, homeobox; *MC1R*, melanocortin 1 receptor; *MITF*, microphthalmia transcription factor; *αMSH*, α-melanocyte-stimulating hormone; NGF, nerve growth factor; *POMC*, proopiomelanocortin; *SCF*, stem cell factor; *TYR*, tyrosinase, *UV*, ultraviolet [57]. Source: *Reproduced from T. Kondo, V.J. Hearing, Update on the regulation of mammalian melanocyte function and skin pigmentation, Expert Rev. Dermatol. 6 (2011) 97–108.*

pathway is involved wherein, melanosomes are transferred from melanocytes to KCs via the shedding vescicle system. This packaging system generates pigment particles, normal human melanosomes are distributed individually in the cytosol with no surrounding membranes, and are not the type of exosome enclosed by intracellular multivescicular bodies. Rather melanosomes bind directly from the plasma membrane into the extracellular space. The pigment globules are large (2–7 mm), larger than shedding vescicles (1 mm) and are released into the extracellular space enclosed by plasma membrane and they transfer intracellular organelles and components to neighboring cells in a manner distinct from exocytosis. Such pigment globules are generated not only from the tips but also from various areas of the dendrites of normal human melanocytes. In this way, the process is much faster than transferring individual melanosomes separately.

Melanosomes are packed in globules enclosed by the melanocyte plasma membrane, released into the extracellular space from various areas of the melanocyte dendrites, phagocytosed by KCs, and are then dispersed around the perinuclear area.

Microphthalmia-associated transcription factor (MITF) regulates mammalian pigmentation by responding to environmental factors, including UV, as well as to factors secreted from KCs, fibroblasts, and other cells. MITF controls not only melanogenesis, but also differentiation, dendricity, proliferation, and apoptosis [60].

UVR induces immediate pigment darkening by chemical modification of melanin, and transfer of melanosomes to KCs. UV-B exposure also leads to delayed tanning by further synthesis of melanin over several days after UV exposure and persists for weeks. UV-induced pigmentation plays a protective role by preventing DNA damage and accumulation of mutations. Given the importance of melanin and skin pigmentation in providing protection from solar radiation, the evolutionary aspects of human skin pigmentation have received much attention [39,53,55–61]. Fig. 2.9 has shown the cooperation between KCs and melanocytes [62].

UV light is strongly absorbed by biological molecules (see Fig. 2.1B) that therefore react and are thus is often indicated as UVR. The time required for the development of an erythema is about 8 h, but changes depending on the wavelength and the portion of skin chosen (the trunk is the best choice for a comparison) [63–73]. A quantitative comparison requires at any rate more elaborated instruments, of which portable versions are on the market, however, as well as by the use of the appropriate mathematics [73]. The immediate skin darkening caused by UV-B and UV-A fades within 24 h, but UV-A also causes a more persistent darkening (up to 10 days).

The effect of solar light depends on many factors, such as solar elevation angle (SEA), the scattering in the atmosphere, the reflection on various surfaces, such as water, sand, and snow. Reflection may increase the fraction of light received, depending on the position and this may lead to a radiation received greater than 100% of the direct light. The classification of light types is not universally accepted (as an example, if the wavelength taken as the UV-A/UV-B boundary is at 320 nm, as defined by the International Commission on Illumination, UV-B is 4.34% of the irradiation from the sun; if this boundary is taken at 315 nm, however, it is 2.11%, in both cases as measured in June

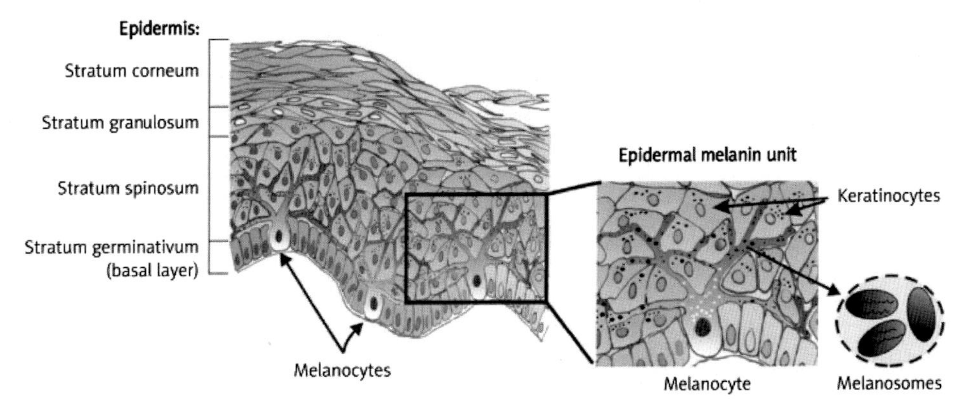

FIGURE 2.9 Schematic presentation of the keratocytes-melanocytes cooperation. Proliferation, differentiation, and melanogenesis are under control of factors secreted by surrounding keratinocytes (nervous growth factor receptor, NGFR, prostaglandins receptors, EP1/EP3/FP, melanocortin receptor, MC1R, ethydium bromide, ETBR, and 2(α), melanocyte stimulating hormone, MSH, α-MSH, endothelin-1, ET, granulocyte macrophage colony stimulating factor, GM-CSF, fibroblast growth factor, FBGF1/2, fibroblast growth factor, SGF, receptor tyrosine kinase, c-kit). Signal transduction then leads to cAMP response element-binding protein to microphtalmia associated transcription factor, MITF-M, and tyrosinase-related TYR, TYRP [62].

in Copenhagen) [74]. The impinging light is stronger and richer of short wavelength radiations as far as the measuring is done closer to the equator. As for the irradiation received by each individual, this varies greatly depending not only on the above physical factors, but also on societal, religious, and cultural factors. As an example, English and Scandinavian people tend to be exposed more than people arriving in those northern lands from more sunny countries, because of the habit the latter have of looking for shade. As for the effect on each individual, differences in keratocytes are present at birth and are not modified by extrinsic factors, such as UVR. Thus in dark skinned people they are more numerous, larger, and elongated. As a consequence, degradation is delayed and visible pigmentation more marked. The action of eumelanin is the most important one for individuals with dark skin and hair, whereas pheomelanin is found in individuals with red hair and Type I and II skin (see Table 2.1 and Fig. 2.10) [75,76].

2.5 Artificial skin protection

Evolution chose to protect sensitive biomolecules by filtering away UV rays, and an increased melanin production takes place in the irradiated skin. This can be artificially imitated by placing topically on the skin compounds that absorb light themselves, thus avoiding excitation of biomolecules, either sunscreens arising from plants or man-made sunblockers.

In order to be effective, quite important is the way the sunscreen (cream or lotion) is applied [77]. Thus the creams usually contain 5%–10% of the active material, and the amount applied must ensure that the layer applied is thick enough that it operates evenly on all of the body. Furthermore, cream must be applied quite often (typically advised before exposing oneselves to sun irradiation and then after 30 min exposure).

Natural filters are based, as indicated above, on polyphenols, and they have to be consumed irreversibly to give oxidized photoproducts (quinones) if they have to be active. In contrast, a requisite of the man-made filters (see typical structures gathered in Scheme 2.6 and spectra in Fig. 2.11) is to have them reacting reversibly and this because it is undesirable to put a large amount of not natural compounds (and of their photoproducts) on the skin.

In practice, however, it should be taken into account that people tend to apply much lower amount of such compounds and that these are easily washed away, whatever it may be reported on the bottle about the resistance to sweet and sea water. The active ingredients are either: "physical" or "chemical" [78–80]. The physical active ingredients (that remain unchanged because they *reflect* light), such as titanium dioxide or zinc oxide, are expected to scatter UV light. Nanoparticles of these oxides and of quantum dots (semiconductor particles of the same dimension as light wavelengths that have special optical properties) have also been used but have been found to penetrate to some extent in the body, and in view of the well-known phototoxic properties of these oxides, may not be considered safe [81]. With "physical" sun screens the results depends on the dimensions of the particles; the smaller the best rule is acceptable for the cosmetic point of view, but a shift to short wavelengths may also occur, and thus protection may be insufficient. A cause of the often observed poor reproducibility of in vitro texts has been the unavailability of suitable skin substitutes. Transpore tapes are largely used for this function, but the results are poor reproducible because the pores on the surface can be unequally filled [82]. Furthermore, recent research has evidenced that activity may extend to blue visible light. Some photosensitivity protection factors were developed

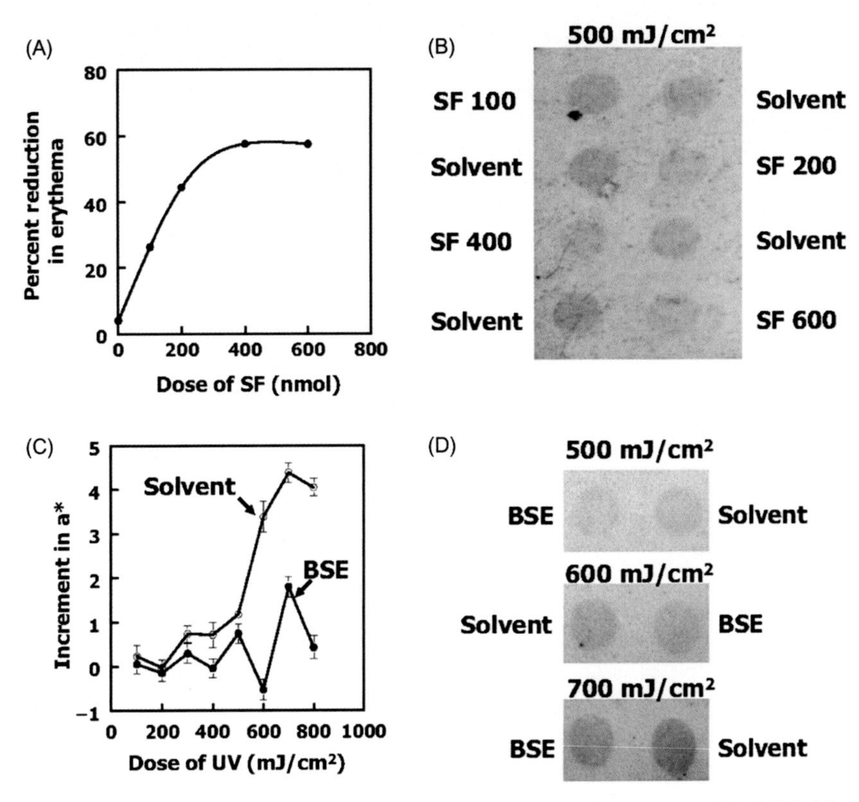

FIGURE 2.10 SF-rich BSEs protect human skin against erythema caused by 311-nm UVR. (A) Inhibition of skin erythema development by topical treatment of a male volunteer with a range of SF doses. The circular 2.0-cm-diameter spots received 100, 200, 400, or 600 nmol SF as BSE in 25 μL of 80% acetone/20% water on 3 days at 24-h intervals. Control spots received 25 μL of solvent only. The erythema values (a*) were measured 4 days before radiation with 500 mJ/cm^2 of UVR and 24 h after radiation. The 4-day mean a* value for the solvent-treated areas before radiation was 6.70 ± 1.16. Inhibition of erythema formation (%) was calculated from [a* (untreated) − a* (treated)/a* (untreated)] × 100. The untreated values (zero dose) were calculated from the increment of two areas that received 25 μL of BSE in 80% acetone/ 20% water containing 400 nmol of unhydrolyzed glucoraphanin (the inactive glucosinolate precursor of SF). (B) Photograph of four pairs of spots of individuals (described in A) who received 100, 200, 400, or 600 nmol doses of SF (as BSE) or solvent only. (C) Effect of topical treatment with SF-containing BSE on erythema response to a range of doses of UVR. With the use of 16-window template, horizontally adjacent pairs of spots were treated with either 200 nmol of SF in 25 μL of 80% acetone/20% water or solvent alone on 3 successive days at 24-h intervals and 24 h later were radiated with 100–800 mJ/cm^2 of UVR. The increments in a* values for each spot after UVR with respect to their 4-day means before UVR are plotted as a function of UV dose. The visually determined minimum erythema dose was 600 mJ/cm^2. (D) Photographs of pairs of BSE- and solvent-treated spots that received 500, 600, or 700 mJ/cm^2 of UVR. The complete set of percent reduction values for this subject are shown in Table 2.1 (subject 2) [60].

by using light at 420 ± 30 nm that were suitable for people sensitive to visible (blue) light [83,84]. Creams are usually formulated by using pigmentary titanium dioxide on an inert base, such as a nonionic surfactant (as BPC cetomacrogolcream base) with the addition of iron oxides and/or burnt sugar to give a range of cosmetically acceptable tints. Despite the somewhat amateurish approach, all of the marketed creams appear to have no

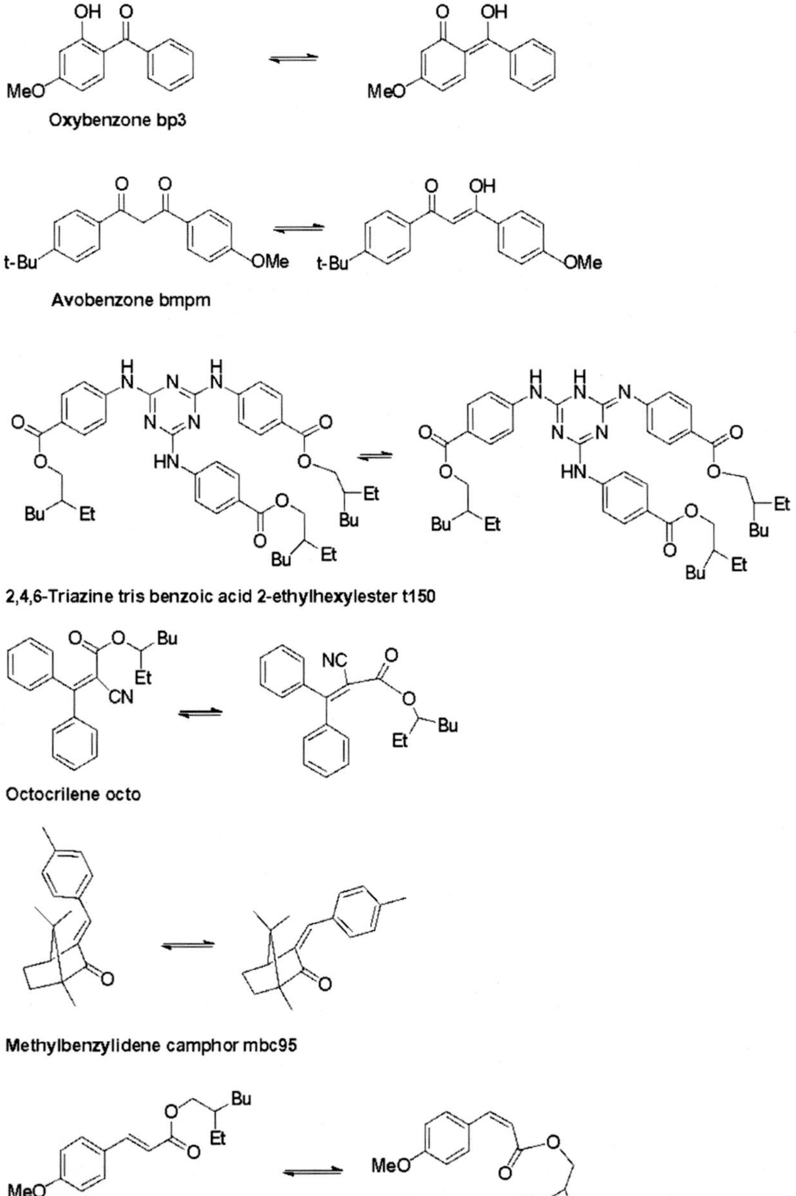

Oxybenzone bp3

Avobenzone bmpm

2,4,6-Triazine tris benzoic acid 2-ethylhexylester t150

Octocrilene octo

Methylbenzylidene camphor mbc95

Octylmethoxycinnamate m40

SCHEME 2.6 Structure of largely used sunscreens. As it is apparent all of the reactions are reversible, based on the cheto–enolic equilibrium (the upper three) or on the cis–trans equilibrium.

significant toxicity [84]. In general, natural sunscreens operate by H donations. Artificial sunscreens, on the contrary, are reversible dyes because the idea is to put as low as possible amount of organic material on the skin (Scheme 2.7).

"Chemical" filters (i.e., organic conjugated compounds, such as stilbene derivatives,

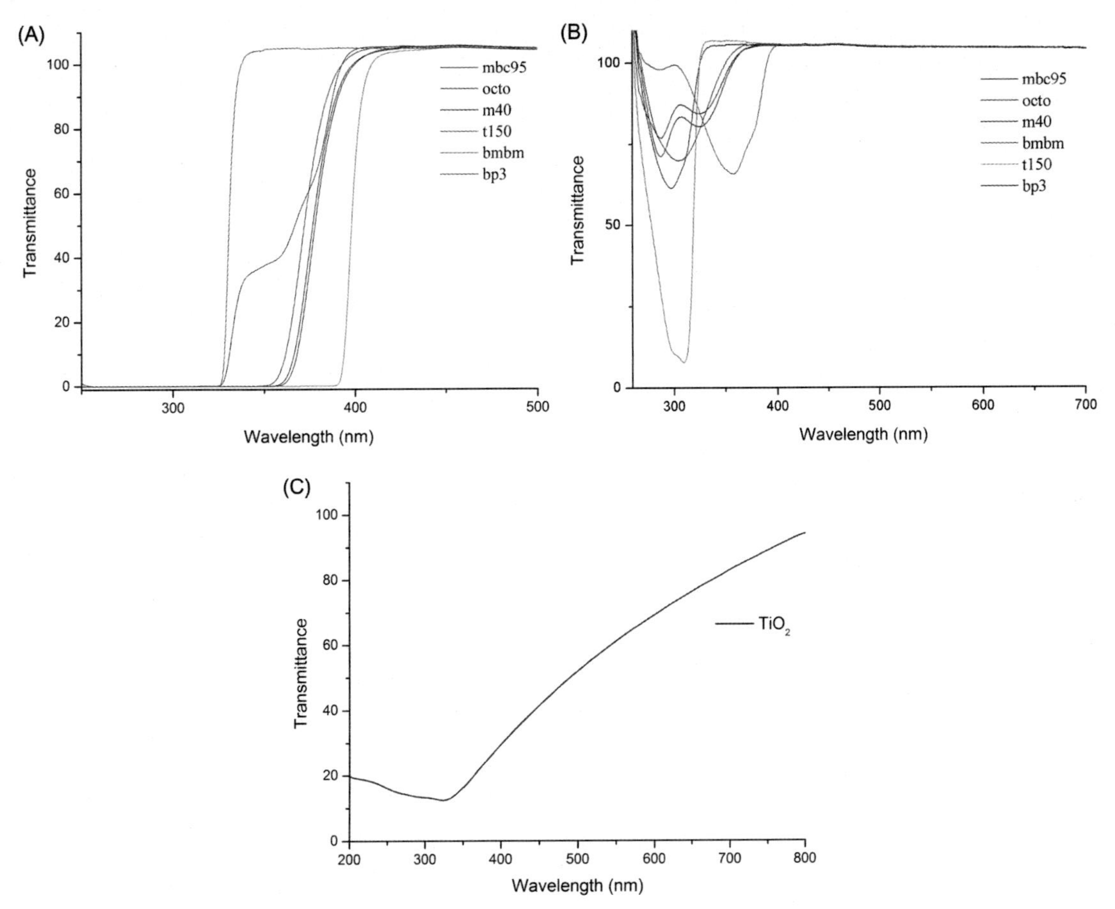

FIGURE 2.11 (A) Spectra of transmittance of commonly used optical filters (2 mg cm^3 solutions; bp3, oxybenzone; bmbm, avobenzone; t150, 2,4,6-triazine *tris*-benzoic acid 2-ethylhexyl ester; m40, octylmethoxycinnamate; octo, octocrylene; mbc95, 4-methylbenzylylidene camphor); (B) as in (A), after 1 to 100 dilution, in order to evidence the different UV-B or UV-A1 protection; (C) reflectance spectrum of a TiO_2 suspension (2 mg/cm^3). *Courtesy: Prof. D. Dondi, Pavia.*

β-dibenzoylmethanes, cynnamates, and others compounds that *absorb* light) are a versatile alternative and are all based on a photochemically efficient, but thermally 100% reversible, isomerization. No compound is known that has such high photostability as this application would require, however, the above-mentioned classes of molecules that undergo a thermally reversible photoreaction are sufficiently suited for this job and are commonly used. These "ultrafast" (occurring in the subpicosecond time interval) isomerization and relatively slow back reactions have been well documented by fast spectroscopy (see an example for a 2-hydroxybenzophenone in Scheme 2.8) [85].

Naturally occurring compounds that have a photoprotective role in vegetables, in particular flavonoids (the basic structure is shown in Scheme 2.9), are largely used in (biological) cosmetic preparations [86]. Both natural polyphenol (the flavanoid morelloflavone, several glycosides of which are present in the pericarp

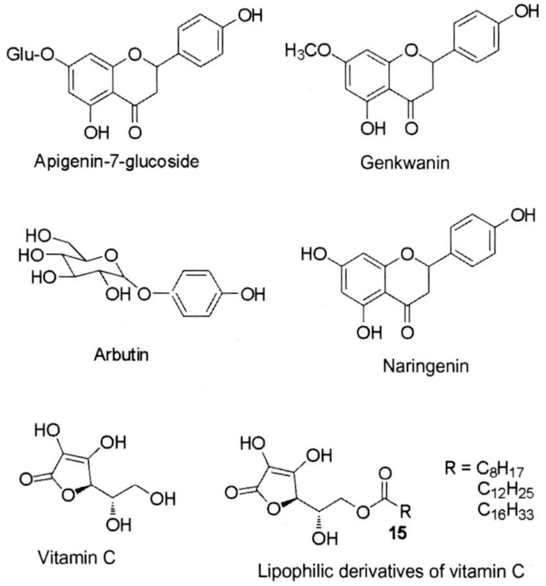

Apigenin-7-glucoside Genkwanin

Arbutin Naringenin

Vitamin C Lipophilic derivatives of vitamin C

R = C_8H_{17}, $C_{12}H_{25}$, $C_{16}H_{33}$ **15**

SCHEME 2.7 Largely used sunscreens with acetal, phenol, or sugar structure.

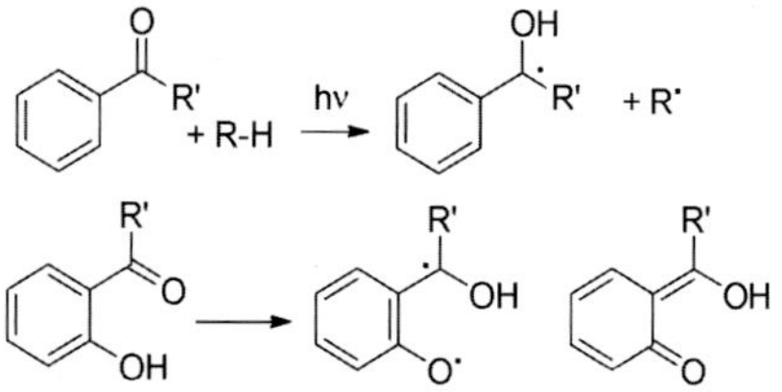

SCHEME 2.8 Inter- and intramolecular H transfer to triplet aromatic carbonyls.

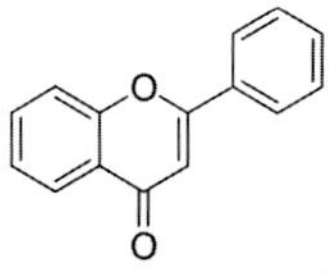

Flavone

SCHEME 2.9 Basic structure of flavones.

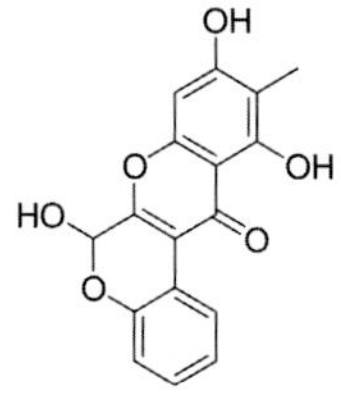

SCHEME 2.10 Boeravinone B.

extract of various species of *Garcinia brasiliensis*) and synthetic phenols, such as Ronacare are used [87–89] (Scheme 2.10).

As for chemical sunscreens, it is difficult to find fully satisfactory filters that absorb intensively over all of UV-B and UV-A1 and A2 and are of as low (irreversible) photochemical reactivity as this can be neglected. In practice only, a few structural types are used, and all of them react efficiently, but the product isomerized back with 100% efficiency. An interesting case are aromatic carbonyls that absorb strongly and feature a relatively long-lived, highly reacting lowest-lying triplet. This has a radicalic character and easily transfer a hydrogen atom from a nearby molecule (the solvent, e.g., an alcohol) leading to a pair of radicals and to the fast destruction of the compound. When, however, an easily transferred H atom is present in the *same* molecule, in an easily accessed site, as it is the case for 2-hydroxybenzophenone, *intramolecular* H transfer completely takes over and leads to an overall even-electron intermediate that has no radicalic character. This intermediate is rather considered as the enolic form of the ketone, to which it rapidly reverts back in the ground state, thus achieving the conditions for reversibility. The course of the photochemical reaction has been followed in detail in some cases and it has been evidenced that an ultrafast decay (in the femtosecond range) is involved. Thus the benzophenone derivative is excited into the second excited singlet, with no significant atom movement, according to the Frank–Condon principle and decays to S_1 in 100 fs, while the H atom of the OH group begins to bond to the ketone moiety. In a slightly longer time (400 fs) some pyramidalization at the (formerly) ketone moiety occurs and is completed in the vibrationally hot S_0

state (5–8ps). The last intermediates rearrange in part directly, in part via the enol form, to the starting molecule (Scheme 2.11) [85].

The use of sunscreen has become common practice, but this has coincided with a rapid increase of skin cancer. One of the possible explanations is that many of the compounds used as sunscreens are able to penetrate in the body by crossing the skin barrier. In this way, such compounds that are mostly good oxygen sensitizers would continue to sensitize oxygen deeper within the dermis and make damage more serious. A solution proposed has been using sunscreens that are less active as oxygen sensitizer or less skin-penetrating. As an example, high melting organic compounds, known as photosensitizers, can be converted into solid emulsions from the melt and then crystallized to have uniform nanocrystals [90]. Further reversible systems largely used exploit either again thermally reversible hydrogen transfer to give an enolic form, as in the case of dibenzoylmethanes, or geometric E/Z isomerization, as in the case of cynnamates (Scheme 2.12). With the cynnamates a small amount of the Z isomer builds up, but this hardly change the absorption of the mixture.

As mentioned above, it is difficult to obtain a single sunscreen that is equally satisfactory for the absorption in both UV-A1/2 and B regions. Thus combinations of a number of filters are most often used, in such a way that all of the significant wavelength range is absorbed. Unfortunately, this not rarely leads to a new photochemical reaction, not possible when each of the components is irradiated alone. A representative case is that of dibenzoylmethanes and cinnamates, both of which are, as indicated above, fully reversible in their photoreaction. When a mixture of the two sunscreens is irradiated the short-lived enolic form of the dibenzoylmethane adds irreversibly to the cinnamate through a cross-cycloaddition (de Mayo reaction) to form a cyclobutane that finally opens up via a retroaldolic reaction and gives an open-chain

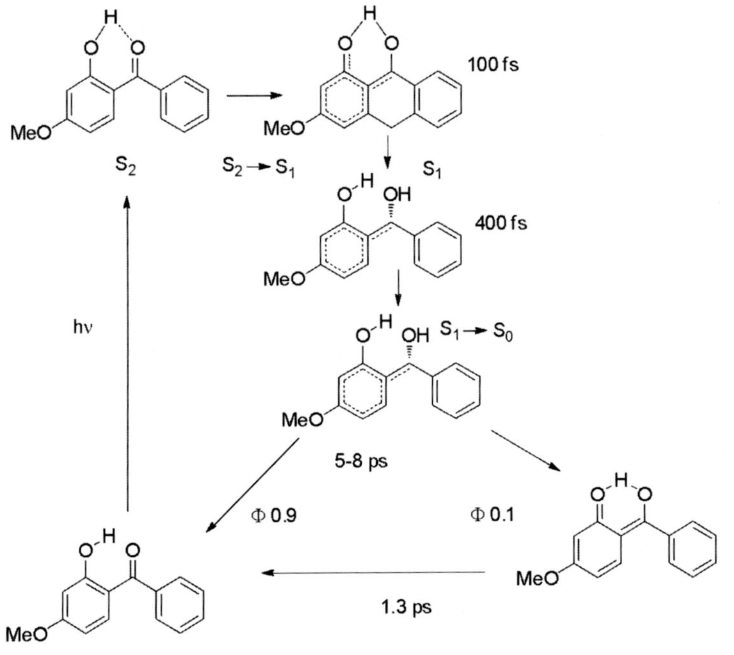

SCHEME 2.11 Reversible H transfer in 2-hydroxybenzophenone derivative.

dicarbonyl, so that both compounds are rapidly and irreversibly consumed (Scheme 2.13) [91].

Furthermore, research is being revealing new structures that may be useful for new sunscreens, for example, the mycosporine-like amino acids as well as scytonemin elaborated by cyanobacteria (Scheme 2.14) [92–95].

There are also substances that are able to operate both as filter and sensitizers. An example is offered by a natural anthraquinone, parietin (Scheme 2.15), which in a microcrystalline state acts as a filter, while it operates as an oxygen sensitizer in solution [96].

The effect of natural versus artificial light and the specific effect of UV, visible, and IR irradiations and erythema formation and its persistence have been determined [51].

The action spectrum for UV-induced tanning and *erythema* are almost identical, but UV-A is more efficient in inducing tanning whereas UV-B is more efficient in causing erythema. In vitro studies evidenced that *melanin* can react with DNA and *act* as a photosensitizer-producing reactive oxygen species (ROS) after UV-A radiation [97].

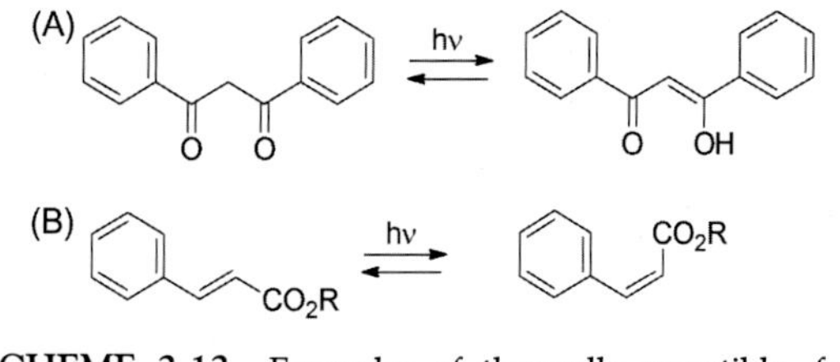

(A) O O hν OH
(B) CO_2R hν CO_2R

SCHEME 2.12 Examples of thermally revertible, fast photoisomerizations: H-transfer in dibenzoylmethanes (A), E/Z isomerization of cynnamates (B), the two main classes of organic sunscreens.

Suntan, which nature developed as a natural defense mechanism against sun light, is generally considered as desirable, at least in the western hemisphere. Increased tanning may be obtained by using sun beds, and thus increasing the light flux, a practice that is not encouraged because it causes skin aging (or even cancer). It is possible to apply specific compounds, ranging from bronzes that are simply make ups that can be taken away by washing with soap, to some compounds that have a chemical role, such as a triose, dihydroxyacetone, that condenses with the amino group in amino acids (the Maillard reaction) present in the cells (actually the dead cells on the outer layer of the epidermis) to form brownish compounds called melanoidins [98].

For a quantitative evaluation of tanning, mathematical models have been developed that are able to account for the extreme heterogeinicity of the samples and suggested that the transmittance through the epidermis is c.20%, relatively independent on its thickness. Melanosomes, four times as much in black epidermide than in Caucasian one, acted both as a filter and as scattering centers [61,62].

Artificial sunscreens are characterized by a sun protection factor (SPF). This value expresses the fact that UV-B causes an frythema, and the minimal dose at which a perceptible erythema is formed (minimal erithemal dose) can be experimentally determined on a portion of the skin of healthy volunteers (usually the inner part of the elbow or the mid-back). This area is covered with a uniform layer of sunscreen and the time required to give the same reddening on the protected versus an untreated area is measured.

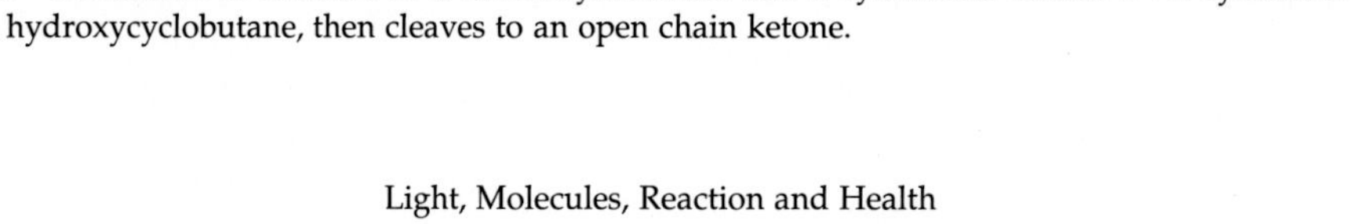

O OH + CO_2R hν Ph HO COPh Ph CO_2R → Ph Ph O Ph O CO_2R

SCHEME 2.13 Irradiation of mixture of a dibenzoylmethane and a cynnamate causes 2 + 2 cycloaddition. The first intermediate, a hydroxycyclobutane, then cleaves to an open chain ketone.

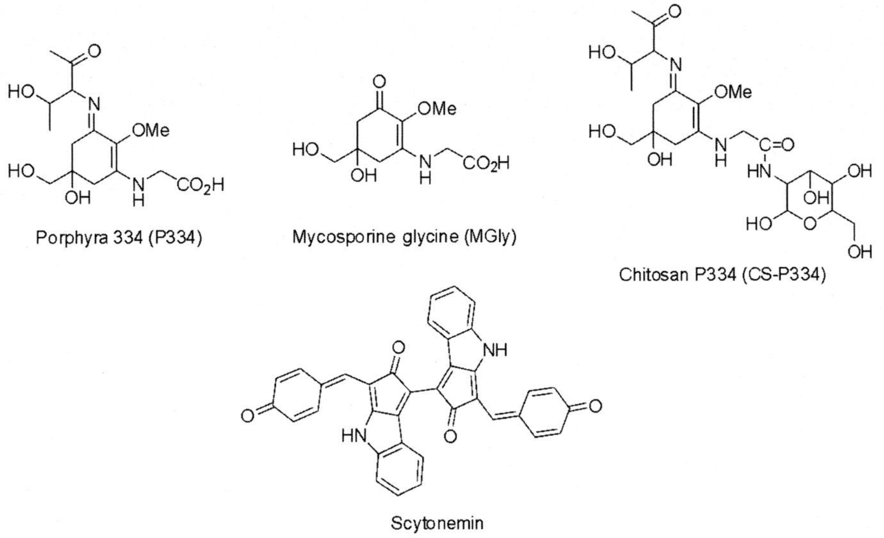

Porphyra 334 (P334)

Mycosporine glycine (MGly)

Chitosan P334 (CS-P334)

Scytonemin

SCHEME 2.14 Examples of new sunscreens.

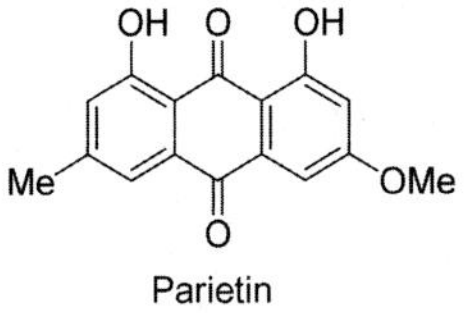

Parietin

SCHEME 2.15 Parietin.

The absorption data of organic molecules are usually reported as the wavelength of maximal absorption (λ_{max}) and the absorptivity at that wavelength (ε_{max}). This is useful, because the absorbance of a solution (A), of molar concentration (c), the logarithm of the reverse of the transmittance (T), can be expressed through the Beer law as proportional to the molar concentration of the substance and the depth of the path traversed by the measuring ray (a)

$$A = \log\log 1/T = a\varepsilon c$$

The sun protecting factor SPF is defined as the ratio below

$$\mathrm{SPF} = \frac{\int A(\lambda)E(\lambda)}{\int A(\lambda)E(\lambda)\mathrm{MPF}} \approx \frac{1}{T}$$

where $A(\lambda)$ and $E(\lambda)$ represent respectively the absorption of the filter and the erythematic efficiency of solar light by each wavelength, MPF is the monochromatic protection factor calculated by comparing the amount of time needed to produce a sunburn on sunscreen-protected skin to the amount of time needed to cause a sunburn on unprotected skin (roughly proportional to the transmittance of that filter at that λ), and integrals refer to the wavelength range explored. A study often referred to show that when applying a cream, most people do not use it evenly, but leave some part of the body unprotected. Even when the cream is applied evenly, the amount used is 0.5 to 1 mg/cm^2 because 90% of the people do not even know which is the correct amount (2 mg/cm^2, applied 1–30 min before exposure and reapplied after any exercise that may take it away). This amount corresponds to the values reported on commercial products and indeed this is the value recommended by FDA and actually proved to be active. In the practice, however, people tend to use much lower amounts, which strongly impairs the performance of the creams. When applied following instructions, the result is satisfactory [71,99].

Notice also that, in view of the logarithmic relation between absorption and transmittance the SPF factor grows only by a small degree when either the concentration of the molecule or the absorbance at that wavelength grow by a large factor. Thus by equating SPF and 1/T, one has that SPF = 10 corresponds to $T = 10\%$, with SPF = 20, $T = 5\%$, with SPF = 30,$T = 3\%$ and obviously SPF values of greater than 50 make no sense, because at this point any further additions only minimally decreases the residual 2% transmittance.

A simplified classification has been more recently adopted for commercial sunscreens, and refers to four categories, low protection (SPF below 15, marked as either 6 or 10, when the exact factor was measured between 6 and 9.9 or 10 to 14.9), medium protection (SPF 15 and over, marked 15, 20, or 25, when 15 to 19.9, 20 to 24.9, or 25 to 29.9, respectively), high protection, SPF 30 and over, marked either 30 or 50 (measured 30 to 49.9 or 50 to 59.9, respectively), and very high protection, SPF over 50 and marked 50 + (measured >60). It should be noted that the action of UV filters used in sunscreens is exactly the same as that operating for additives protecting from light compounds used for car parts or other sun-exposed surfaces, mostly indeed the same chemicals are used as sunscreens. The classification of these compounds as light-absorbing sensitizers as if this were their only action is inappropriate, as most of these compounds also interfere with subsequent steps of the damaging process of the skin, for example, by trapping reactive intermediates involved in the process, as radicals [100].

Among the most important advancements in the field of UV filters, one should mention the issue of a regularly checked ISO methodology [71]. Thus the precise evaluation of the amount of sunscreen applied is not always easy, because part of the cream remains on the fingers, and the layer is not uniform (recommended value, 2 mg/cm^2, or a film thickness of 0.0002 cm). A detailed examination of the issue has shown that the result depends on the chemical and physical nature of the sunscreen and its light absorption (whether it forms a homogeneous phase with the vehicle, whether a uniformly thick film is formed from the evaporation of the vehicle, etc., and how the experimentally measured data compare with the computed ones), on the lamp and the vehicle used and on the roughness of the substrate. Thus SPF was calculated in silico for common UV filters for OW-C, oil-in-water cream; OW-S, oil-in-water spray; WO, water-in-oil, and CAS, clear lipo-alcoholic spray. These formulations show a different SPFs in vitro and different thickness distribution in films. A good agreement has been found between the results obtained in vitro and in silico. The accordance was good and their thickness frequency distribution was decreasing with the thickness in the order WO > GEL > OW-C >OW-S > CAS (notice that OW-S and CAS have the lowest viscosity among the screens examined). The results indicate that the variations in SPF between formulations are determined mainly by their film-forming properties. Very small film thicknesses have a determining role as well as emulsion type and viscosity.

Extended examination demonstrates that the use of the film thickness frequency distribution with a suitable computational method provides accurate predictions of sun protection efficacy [101].

Another, largely used parameter is the UV index that is no experimental measure but rather compares literature values of the sensitivity with the UV-visible spectrum of the compound considered and can be determined by a computer program, available also as for use on an inexpensive portable instrument. This indicates the risk of getting a sunburn at a given time and location. The UV index calculated on this parameters is based on a linear scale, for example, doubling the index means that a sunburn will be formed in a half the time than in an unprotected skin under those conditions, originally defined with values up to 10, corresponding to a sunny noon in Toronto, but then used in an open-end scale to adapt it to more sunny countries [102]. A convenient option is using DURHAM Erythema Tester, an all-in-one device that contains both a UV source and a template that delivers 10graded irradiances increasing in 26% intervals in a single exposure, without opening or closing the instrument [103]. In using these parameters it should be taken into account that these of course change according to the time of day (peak between 10 a.m. and 4 p.m.), sky conditions (although a significant percentage of the sun's damaging UVR can pass through clouds), closeness to reflective surfaces, such as water, sand, concrete, snow, and ice. Furthermore, season of the year (late spring, early summer), altitude (an increase of c.4% should be added forever 300 m shift in altitude), closeness to the equator are all factors that make getting sunburn easier. In general, noninvasive, in vitro computational systems based on the UV spectrum of each filter and the literature erythemal power are largely used and predict the SPF dependably. In the idea of establishing a strong relation between irradiation and damage caused, various sources have been used, from intense pulsed light (IPL) treatment to treatment with low-energy lamps [104]. The mean light intensity for maintaining a healthy skin until 80 years of age was found to be to be 2.54 min (0.14 MED) for unprotected skin and 127 min with the use of a sunscreen with SPF of 50 [68].

Despite the difference of each individual, a level of exposure that can be tolerated with minimal risk has been determined by using apposite solar simulators. These are built in such a way as to deliver a "nonextreme" irradiation, well suited for studying the effect of light on the skin [80]. Considerable effort has been given to the determination of the exposure to solar light that can be allowed with no serious risk of causing against clinical, cellular, and molecular effects induced by daily UV exposure (DUV-R) through in vivo and in vitro studies and can be adapted to every situation. An efficient model is the UV daylight spectrum representing a nonextreme sun exposure, with an efficient evaluation of the zenithal time of the sun (SEA lower than 45 degrees, for latitudes from 60 degrees south to 60 degrees north, during all the months of the year), with a UV-A/UV-B ratio of 27. In everyday outdoor activities, this type of exposure does not induce any visible short-term effect but may lead to long-term UV-induced deleterious consequences.

Using xenon arc solar simulators with appropriate filters, it is possible to obtain a "UV-solar simulated radiation" (UV-SSR) including UV-A and UV-B wavelengths, with a UV-A/UV-B ratio close to 10. This UV-SSR spectrum is very useful for the photobiology studies and reproduces summer zenithal sunlight with a high UV-B erythemogenic spectral portion. Although the irradiance of such solar simulators is in most cases larger than solar light to which we are exposed in the environment, the MED, for a given skin color phototype, have been shown to be comparable to those found in such outdoor situations [79].

The change in color of the skin when exposed to solar light has been determined by reference to the variation of both luminance (ΔL^*) and skin pigmentation (ΔE) (see Figs. 2.12 and 2.13). As it is apparent, large effects were limited to the first days and

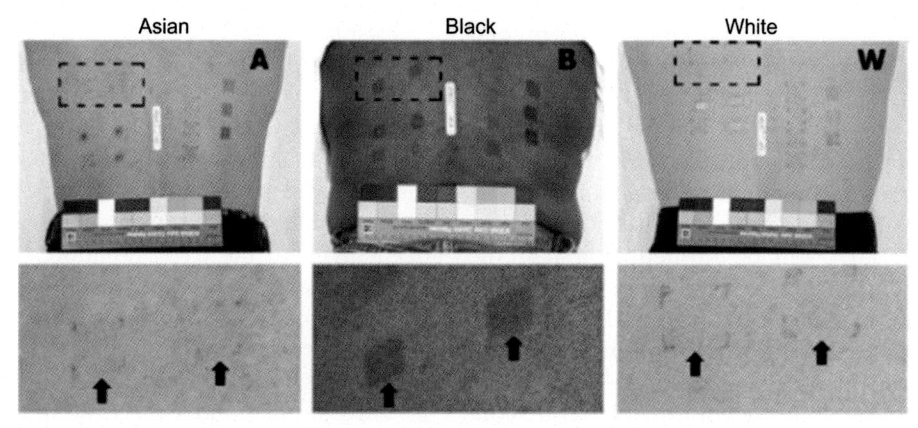

FIGURE 2.12 Tanning in different pigmented skin, reproduced with permission [56].

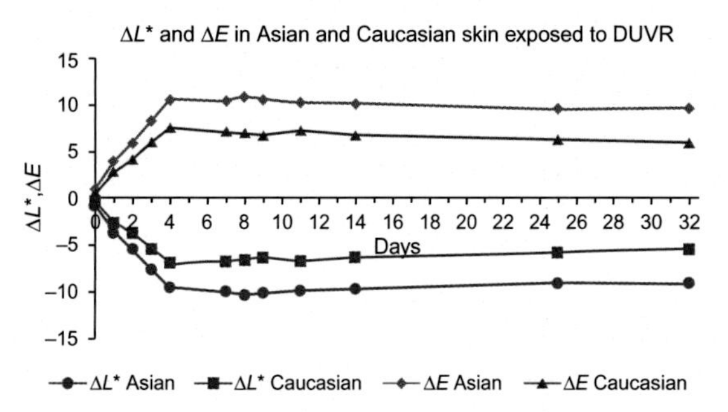

FIGURE 2.13 Variation of tanning according to entries, reproduced with permission [80].

occurred in contrary directions when Asian and Caucasian skin were exposed [56].

Such daily sun exposure, simulated in the laboratory by the DUV-R spectrum, induced in vivo significant clinical effects such as disturbed hydration, altered biochemical properties and alteration of the microtopography of skin, and increased pigmentation. Biological alterations and damage were also observed, including an increase in the epidermal thickness, a decrease in number of LCs together with an increase of their size, urocanic acid (UCA) isomerization, an increase in number and size of melanocytes and melanin deposition, an increase in KC proliferation as well as SBC formation and p53 accumulation. The dermis was also affected with the induction of tenascin, a decrease in fibrillin and procollagen I, and a reduction of glycosaminoglycan deposition. Oxidative stress involving both dermis and epidermis occurred under these conditions, and the alteration of genes involved was also evidenced. Efficient daily UVR protection, including UV-B and UV-A absorption, is required to avoid the suberythemal cumulative effects of such sun exposure. In vitro and in vivo photoprotection studies showed that, in addition to UV-B protection, a sufficient UV-A protection is also essential to reach a significant prevention efficacy against DUV-R-induced damage. Thus the SPF value is not by itself sufficient to express the efficacy of protection against clinical, cellular, and molecular effects induced by daily UV exposure[80].

2.6 Light effect on hair and teeth

Closely related to the effect on skin is that on hair.

IPL sources have been successfully used for hair removal. The destruction of human hair follicles and surrounding tissues following IPL treatment offers an opportunity of a deep study of the bulge and the bulb regions. Each hair is composed of three lineages, forming concentric layers of terminally differentiated, dead hair cells (Fig. 2.14).

During the resting phase (telogen), bulge hair follicle stem cells (HFSCs) are kept inactive by bone morphogenetic protein (BMP) and fibroblast growth factor (FGF) 18 signals from the neighboring K6 + bulge and from nearby fibroblasts and adipocytes. BMP inhibitors and proactivating FGFs from the dermal papilla (DP) overcome the inhibitory cues, leading to entry into anagen. At the base of the bulge, some hair germ (HG) cells become wingless/integrated. This multipotent progenitor, which express sonic hedgehog (SHH). SHH triggers bulge HFSCs to divide symmetrically and grow the outer root sheath (ORS), which then pushes the signaling center away, returning the bulge to quiescence, and then progressively returning the ORS cells to quiescence in a cascade as it grows downward. Progenitors that maintain contact with the DP continue to produce SHH, which fuels the DP to elevate signaling and expand the multipotent progenitor pool. While the hair bulb grows, it envelops the DP. Interactions at the elaborated interface between the DP and hair bulb establish microniches, each of which contains unipotent progenitors that produce the seven concentric differentiating layers of the hair shaft and its channel, or inner root sheath. These progenitors exhaust their proliferative capacity and an apoptotic phase (catagen) ensues, during which the follicle is degenerated and restored back to its resting size as it enters the next telogen. Note that some cells within the upper/mid ORS are spared, and these become the new bulge and new HG for the next hair cycle.

The human hair is composed by a high- and a low-sulfur proteins and the former type is about 40% of the hair proteins. Hair develops from multipotent clonogenic KCs, which are similar to multipotent stem cells [106]. Hair follicle progenitor cells are maintained in an

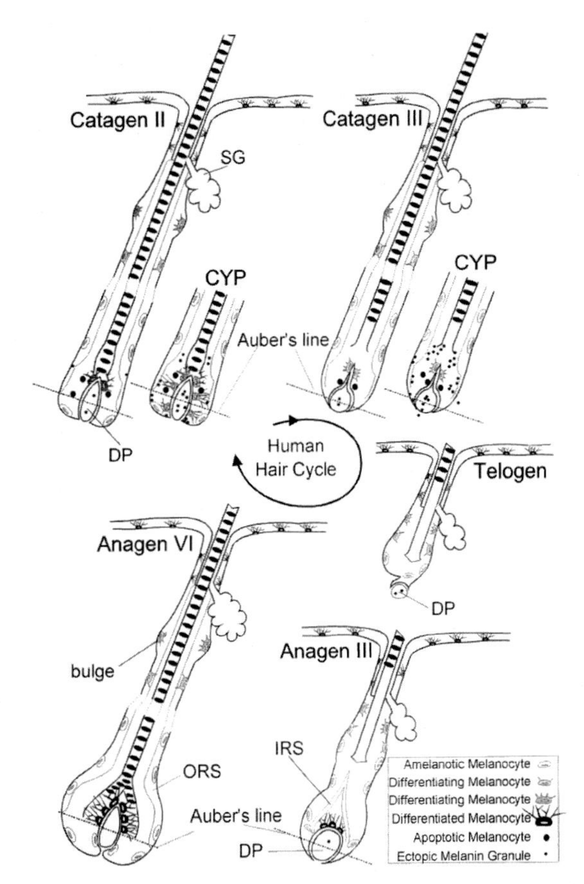

FIGURE 2.14 Schematic representation of the status of the "hair follicle melanin unit" during the hair growth cycle in normal human scalp and after cyclophosphamide treatment in C57BL/6 mouse back skin. The ®gure depicts ®ve selected hair cycle stages, showing early regressing follicles (catagen II and III), a resting HF (telogen), a HF in early phase of active growth (anagen III), and a mature anagen VI HF. CYP-induced catagen in C57BL/6 mice are also depicted in early catagen. Note the distribution of melanocytes in the HF; mature melanotic melanocytes are located primarily in the human epidermis, upper infundibulum, and hair bulb. Melanocytes in less mature stages of differentiation are also located in the ORS and perhaps also in the matrix. Source: *Reproduced with permission from T.Desmond, B. Vladimir, P.Ralf. The Fate of Hair Follicle Melanocytes During the Hair Growth Cycle. The journal of investigative dermatology. Symposium proceedings / the Society for Investigative Dermatology, Inc. [and] European Society for Dermatological Research. 2000, 4. 323-32P [105].*

undifferentiated state by the transcription factor Lhx2 [107]. The role of stem cells makes this a scientifically highly relevant topic. In principle, removal of hair by light can be achieved by three mechanisms, viz photothermal destruction through local heating, photomechanical destruction through the generation of shock waves, or photochemical destruction through the creation of toxic mediators such as singlet oxygen or free radicals. Light sources that destroy hair photothermally include the long-pulsed ruby (694 nm), long-pulsed alexandrite (755 nm), long-pulsed diode (810 nm), and long-pulsed Nd:YAG (1064 nm). An intense-pulsed noncoherent light as well as radiofrequency also injures hair photothermally. A Q-switched Nd:YAG laser (1064 nm), with or without the addition of a topical carbon suspension, destroys hair mechanically. This treatment was found to operate a safe and effective epilation. In some cases, however, symptoms similar to heavy metal toxicity have been reported [108–113].

The reliability of hair regrowth by the effect of light has been proposed but not definitively confirmed. It resulted that low light laser therapy may be a promising treatment option for patients who do not respond to either finasteride or minoxidil, and who do not want to undergo hair transplantation. This technology is increasingly used for the treatment of alopecia, although at present it appears to work better for some people than for others [114]. Using 655-nm red light and 780-nm IR light once a day for 10 min gave satisfactory performances. Such treatments have been confronted with medical treatments available [115,116]. The other age-related change is, of course, that of color. This is due to melanin again and is attributed to melanocytes stem cells apoptosys. When the stem cells are lost the color disappears [117]. As long as it is present, melanin also has a significant role of protection on the skin [98].

As discussed above, tanning determined by the production of melanin pigments in neural crest-derived melanocytes, and its transfer to epithelial cells, has an obvious aim, that is protecting the human skin by screening harmful UVR. Related phenomena occur in the hair, but this has primarily an important role in social/sexual communication, and further allows the rapid excretion of heavy metals, chemicals, and toxins from the body by their selective binding to melanin [118].

Follicular melanogenesis differs from the roughly parallel epidermal melanogenesis by its cyclic nature, as opposed to the continuous course of the latter. Thus the cycles involve periods of melanocyte proliferation (during early anagen), maturation (mid-to-late anagen), and melanocyte death via apoptosis (during early catagen) and each time results in the reconstruction of an intact hair follicle pigmentary unit. After about 10 cycles, however, gray and white hairs appear. This suggests that an age-related, genetically regulated, end of the ability to generate pigments of each of the hair follicles. The chemical basis of this effect is the accumulation with age of mutations following oxidations of nuclear and mitochondrial DNA as well as the dysregulation of antioxidant mechanisms or pro/antiapoptotic factors within the cells. A reduction of the tyrosinase activity of hair bulbar melanocytes and the insufficient migration of melanocytes from a reservoir in the upper ORS to the pigment-permitting microenvironment close to the DP of the hair bulb leads to gray hairs, in which the pigment is diluted, as well as to white (not pigmented at all) hair, although the main cause of the "gray" appearance of the hair is mainly due to the formation of a mixture of pigmented and white hair [119].

IPL epilation principally targets pigmented structures. A recent study documented that a collateral damage depleted stem cells, and damage at the DP was observed only with high-energy treatment modalities. Extrapolated to frequently treated hairs, these observations explain why some hairs grow back after a single IPL treatment [104].

As for teeth, hydrogen peroxide and carbamide (a 1-to-1 complex of urea and hydrogen peroxide 35% that decomposes spontaneously developing the components and, in contact with teeth, gaseous oxygen and OH • radicals) are largely used as whitening agents for cosmetic purposes [120].

$$O=C(NH_2)_2 \cdot H_2O_2 \rightarrow O_2 + 2\,NH_3 + CO_2$$

Some concern has been expressed over the safety of their use. A recent study has documented that light activation offers no benefits for the amount of whitening achieved, persistence of the whitening treatment, or avoidance of tooth sensitivity from the whitening treatment. General conclusions are that

- Home-based bleaching (following manufacturers instructions) results in less tooth sensitivity than in-office bleaching.
- The optimal regimen to obtain persistence of tooth whitening is to follow an in-office treatment with monthly home-based touch-up treatments using OTC products.
- Aggressive bleaching with high concentrations of hydrogen peroxide office-based products causes enamel softening, surface roughness, and an increase in the susceptibility of the tooth to demineralization, based upon in vitrofindings.
- Dental restorations are susceptible to unacceptable color change even when using the home-based OTC systems.
- In-office bleaching of restored teeth using a 35% hydrogen peroxide product caused tooth sensitivity in all cases. Teeth with restorations have a significantly greater chance of becoming sensitive and result in a greater degree of pain when exposed to whitening regimens [121]. The efficiency of the treatment has been tested by using different reference systems (shade guide, shade visual unit, recorded as shade guide units detachement from the baseline, thus through an essentially binary system, and CIE La*b* scale using a chromometer [122–124]). The agreement between the different modes of measure was good, and the whitening obtained significant, although the maximum result was obtained at different times, in a concentration dependent manner. In another study, it was shown that the enamel of teeth was considerably demineralized due to a reduced crystal size and crystallinity and led to obvious changes in the surface morphology of enamel surface [125,126]. The effect of drugs and light on the teeth enamel has been reviewed [127] A nonphysician supervised application is liable to cause unexpected problems of some seriousness, however.

2.7 Role of previtamin D

UV-B makes formation of previtamin D competitive with damage to the skin [128].

A feeling of well-being is by most people associated with exposure to solar light. This is in part due to pleasant warming of the skin, in part to a variety of chemical reactions, such as the enhanced production of β-endorphin, a 31-amino-acid peptide derived from a protein (propiomelanocortin) that is an effective endogenous opioid compound with a strong pain-relieving effect, as well the liberation of nitric oxide, a known vasodilator. A chronic excessive exposure to sunlight has to be avoided, since it increases the risk of nonmelanoma skin cancer, but the avoidance of exposure to the sun increases the risk of vitamin D deficiency. The synthesis of previtamin D in the skin and the likelihood of cancer force two targets to compete. More light absorbed enhances synthesis of previtamin D and busts the immunological activity, but cancer formation too. The balance between the two actions is not easily reached in solution [129]. Sun exposure (5–10 min for arms and legs or the hands, arms, and face, 2 or 3 times per week) and increased dietary and supplemental vitamin D intakes guarantees vitamin D sufficiency [130].

There is no doubt that solar UV exposure is the most important environmental risk factor for the development of nonmelanoma skin cancer and sun protection is of particular importance to prevent these malignancies, especially in risk groups. An important link that improved our understanding of these new findings was the discovery that the biologically active vitamin D metabolite, 1,25-dihydroxyvitamin D ($1,25(OH)_2D$) is not exclusively produced in the kidney, but in many other tissues such as prostate, colon, skin, and osteoblasts. Extrarenally produced $1,25(OH)_2D$ is now considered to be an autocrine or paracrine hormone, regulating various cellular functions including cell growth. It has been shown that strict sun protection causes vitamin D deficiency, particularly in risk groups, such as babies and individuals of Asian and Black origin, or nursing home residents or patients under immunosuppressive therapy and surely the detection and monitoring of vitamin D deficiency in sun-deprived risk groups is of high importance and must be adequately treated [131]. However, 90% of all requisite vitamin D has to be formed in the skin through the action of the sun [132,133] and this is a significant problem, for a connection between vitamin D deficiency and a broad variety of serious diseases including various types of cancer, bone diseases, autoimmune diseases, hypertension, and cardiovascular disease (CVD) has now been clearly indicated in a large number of epidemiologic and laboratory studies. On the other hand, it is at present unclear whether the vitamin D concentration in the blood has any causative effect on the onset or progression of myopia, it may well be that this is only a marker of the less time spent outdoor, which surely favors myopia [134].

The discovery of vitamin D and the demonstration that it is formed through a photochemical reaction in the skin certainly is one of the most important accomplishments in photobiology. However, the newest research has accumulated more and more evidence that on one hand exposure to solar radiation has important effects on health not involving vitamin D, and on the other that vitamin D has a much larger function in nature than governing the calcium metabolism.

2.8 Vitamin D synthesis in the skin: friend or foe

Only a fraction of the required vitamin D is under normal conditions, taken from food, while the largest part need to be photochemically synthesized in the skin through the action of solar or artificial UV-B radiation. The skin is a key organ of the human bodys vitamin D endocrine system and represents both the site of vitamin D synthesis and a target tissue for biologically active vitamin D metabolites. In humans, vitamin D is generated in KCs [135].

In the crowded UV-B absorption region, another group of strongly absorbing compounds are previtamins D. The so-called vitamins D (more appropriately considered as prehormones, not vitamins, since the active forms are not present in food, although some precursors are) are a group of *seco*-steroid hormones (see Scheme 2.16) that are photochemically synthesized in the skin.

The photochemical reaction (see Scheme 2.17) is the isomerization of a cyclohexadiene (7-didehydrocholesterol) to give an open-chain triene (previtamin D), a reaction subjected to strict steric control (only a disrotatory opening, that is, the extremes of the triene has to rotate either both clockwise or both counterclockwise, respectively). This is possible only with the stereochemistry indicated above, otherwise the intermediate would be the energetically

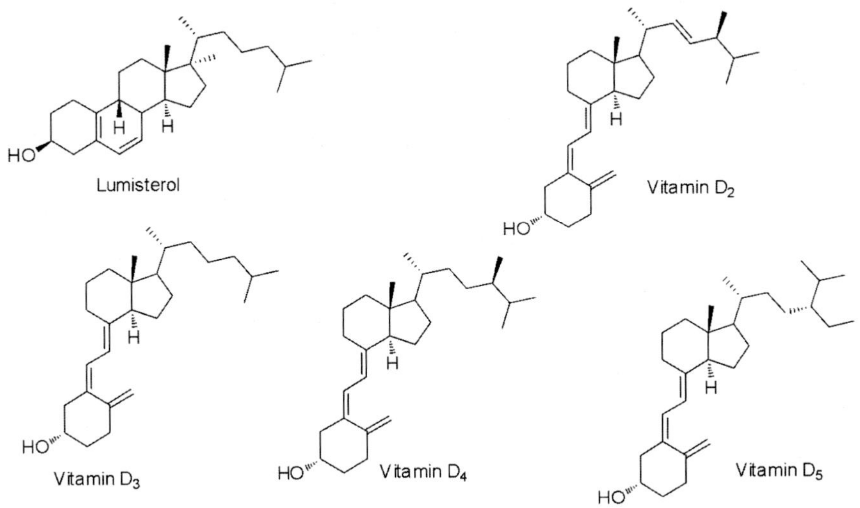

SCHEME 2.16 Vitamins D chemical structures: Vitamin D_1, a 1-to-1 molecular complex of vitamin D_2 and lumisterol; vitamin D_2 (ergocalciferol), present in some vegetables, organisms; and vitamin D_3 (cholecalciferol), typical of animal tissues and vitamins D_4 and D_5.

unachievable *trans* cyclohexene. This is followed by thermal reactions (a sigmatropic isomerization, with a C–H bond migrating along the π system) and rotation along the single bond. Two hydroxylation steps are required in order to obtain the active compound, and involve the tertiary position in the chain (a reaction occurring in the liver) and the cyclohexane ring (in the kidney), respectively.

The identification of all the products involved has represented a historic high point of photochemical research, the course of the reaction is essentially the same also in solution, and the optimization of the industrial synthesis of this compound in solution has been continuosly ameliorated and is probably susceptible of further improvement [132]. As indicated in Scheme 2.17, a series of thermal and photochemical reactions have to occur. This means that conditions are to be controlled accurately and at any rate, the temperature cannot be lowered, because the isomerization of the side chain has to be fast enough to accumulate the isomerized triene. All of the reactions are reversible and consecutive, so that in solution the synthesis of the vitamin D_3has to be carried out up to where the highest percentage of the desired compound is obtained, not to the highest conversion. As for the location in the cell, human KCs that contain the enzymatic machinery (cytochrome P450, family 27, subfamily B, CYP27B1) for the synthesis of the biologically most active natural vitamin D metabolite, $1{,}25(OH)_2D_3$.

This is an autonomous vitamin D_3 pathway. Production of $1{,}25(OH)_2D_3$ in the skin may mediate intracrine, autocrine, and paracrine effects on KCs and on neighboring cells. KCs, sebocytes, fibroblasts, melanocytes, macrophages, and other skin immune cells express the vitamin D receptor, that is a member of the superfamily of trans-acting transcriptional regulatory factors, which also contains the steroid and thyroid hormone receptors, the retinoid-X receptors and retinoic acid receptors. As many as 500–1000 genes may be controlled by

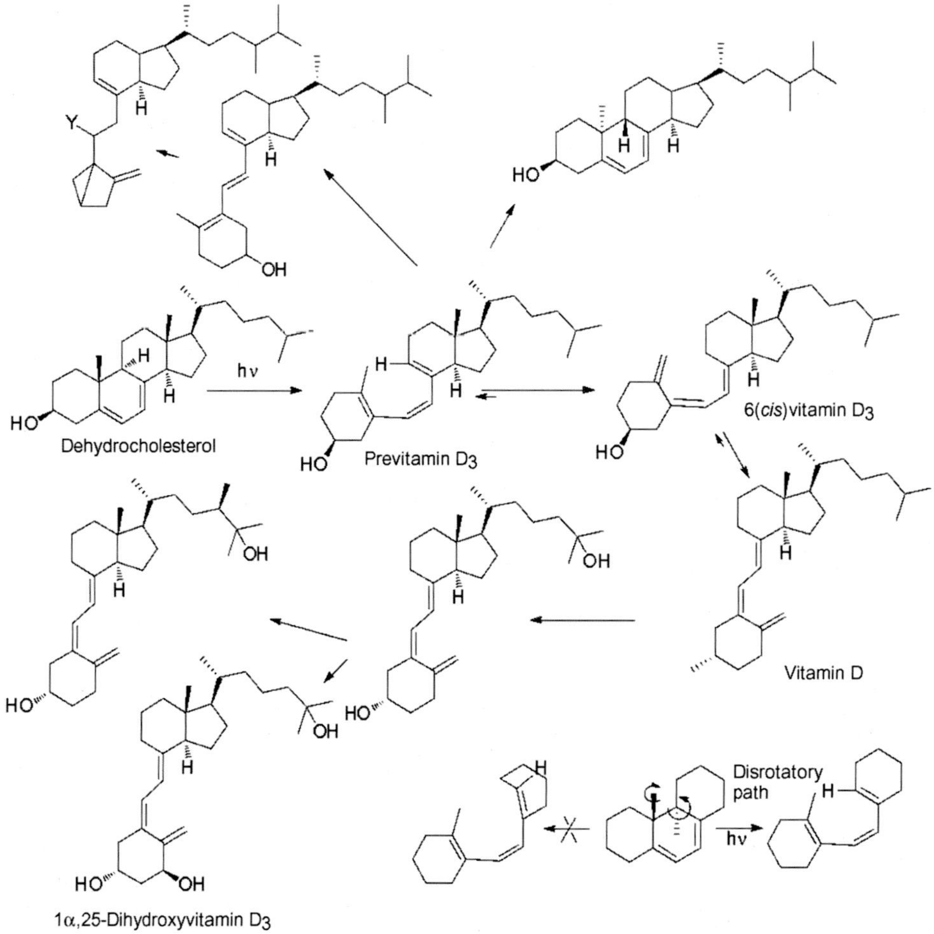

SCHEME 2.17 Thermal and photochemical steps in the conversion of dehydrocholesterol into vitamin D via previtamin D_3 and the ensuing further thermal hydroxylations. 1,25-Dihdroxy vitamin D_3 is the active form of vitamin D_3 and usually indicated simply as vitamin D.

vitamin D receptors ligands that regulate a broad variety of cellular functions including growth, differentiation, and apoptosis [135].

Vitamins D have two physiological functions. First they govern the concentration of calcium and phosphate in serum, although at a concentration that is supersaturating with respect to bone mineral. In the body, these hormones stimulate active intestinal absorption of both calcium and phosphate and mobilize calcium when required. A strong and persistent low level of 25-hydroxyvitamin D (serum level, <10 nM/L) causes hypocalcemia, secondary hyperparatiroidism, secondary hypophosphatemia, and osteomalacia. The last syndrome is a disorder of the skeletal metabolism, corresponding to infant rickettsia, and regards populations such as elderly people that are forced to stay home, or people that for religious or traditional reasons wear clothings that almost wholly hinder exposure to solar light, as well as people for whom the intake of calcium and vitamin D is reduced, for example, due to resections or gastroinestinal bypasses, celiac disease, hereditary deficit, or resistance to vitamin D absorbance from food [136]. An oncogenic hypophosphatemia is also known. Osteomalacia may be asyntomatic or cause localized or diffused osteoarticular pain, muscular weakness, trouble in getting up and moving around, rarely up to spasms, cramps, and tetania. The most significant results are fractures, involving in particular vertebras (biconcave deformity). The demineralization is also indicated by the pseudofractures (Looser-Milkman streaks, weak lines, often bilateral or symmetric and perpendicular to the cortical surface, generally at the media part of the femoral diaphysis or to the ischiopubic ramus of the pelvis). Calcium is absorbed primarily from food in the intestine, in particular the duodenum, via the epithelial cell's brush border membrane and is bound to one of the calbindin, a family of vitamin D–dependent proteins. This complex brings calcium ions to the body, via the basement membrane, a selectively permeable membrane. The concentration of calcium and phosphate ions in the blood is strictly monitored by the parathyroid glands and the parafollicular cells. A high level of calcium in the serum activates parafollicular "C" cells and the calcitonin level leading to a fixation of Ca^{2+} in the bones by inhibiting the action of osteoclasts and promotes expulsion of this ion, but more strongly of HPO_4^{2-}, in the urine. On the contrary, too low level of Ca^{2+} in serum acts on the parathyroid glands and the hormone they produce. This causes an increase of vitamin D_3 in the kidneys and indirectly activate osteoclasts, resulting in a release of calcium from bones (that thus functions as a reservoir of this ion) and further from kidneys and gut (Scheme 2.18).

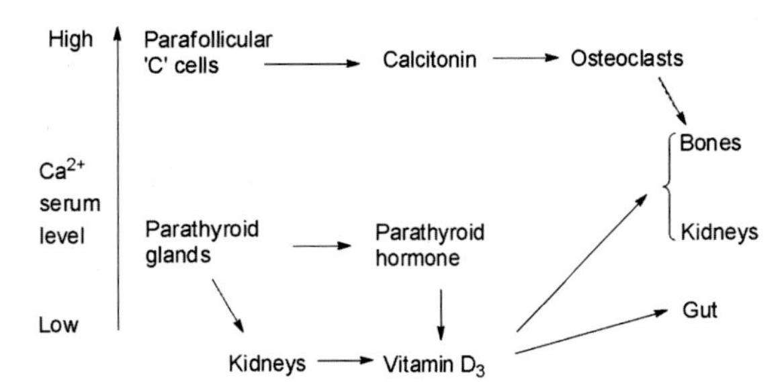

SCHEME 2.18 Overview of general physiologic features and functions of vitamin D [76]. A high level of calcium in the serum activates parafollicular "C" cells and enhances the calcitonin level leading to a fixation of Ca^{2+} in the bones by inhibiting the action of osteoclasts and promotes expulsion of this ion, but more strongly, of HPO_4^{2-}, in the urine. On the contrary, too low level of Ca^{2+} in serum acts on the parathyroid glands and the hormone they produce.

The synthesis of vitamin D_3 in the skin in a single exposure cannot overcome 10%–20% of the amount of 7-dehydrocholesterol (DHC) originally present in epidermis, while additional UV-B light transforms previtamin D_3 into inactive metabolites, tachysterol, and lumisterol (see Scheme 2.16).

Vitamin D_2 is plant derived, produced exogenously by irradiation of ergosterol. Vitamin D_2, like vitamin D_3, is available from foods (e.g., vitamin D_3 is found in cod liver oil), and can enter the circulation through gastrointestinal absorption. Studies have shown that dietary supplementation with vitamin D_2 produces similar effects [137].

An important issue is whether an insufficiency of vitamin D during pregnancy may have an effect on the newborns. A recent study demonstrated that a low level or vitamin D may cause an insufficient lungs development from the initial phase (embrional) of the development and this may have a long-lasting consequence, although it was disregarded whether this resulted from light itself or via vitamin D [138]. The localization of the synthesis of vitamin D and the entire vitamin D endocrine system are schematized in Fig. 2.15.

DCs are able to synthesize $1,25(OH)_2D_3$ in vitro as a consequence of increased 1α-hydroxylase

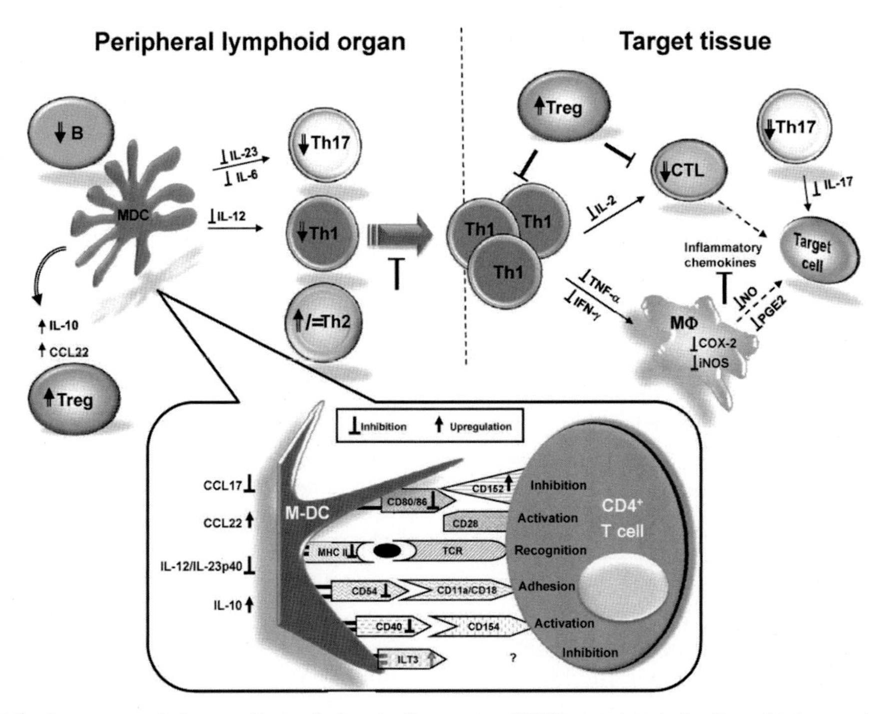

FIGURE 2.15 Immunomodulatory effects of vitamin D receptor (VDR) agonists: induction of tolerogenic myeloid dendritic cells (MDCs) promoting the development of $CD4^+CD25^+Foxp3^+$Treg cells. VDR agonists inhibit in MDCs, but not in plasmacytoid DCs, expression of surface costimulatory molecules, for example, CD40, CD80, CD86, as well as MHC Class II and CD54 molecules. Cytokines affecting differentiation into Th1 and Th17 cells, interleukin (IL)-12, and IL-23, respectively, are also inhibited in MDCs. Conversely, expression of surface inhibitory molecules such as ILT3 and of secreted inhibitory cytokines such as IL-10 are markedly upregulated. Chemokines potentially able to recruit $CCR4^+$ regulatory T cells like CCL22 are also upregulated, whereas the CCR4 ligand CCL17 is downregulated. Upon interaction with MDCs, $CD4^+$ T cells upregulate expression of the inhibitory molecule CD152 (CTLA-4). DCs expressing low levels of costimulatory molecules, secreting IL-10, and expressing high levels of inhibitory molecules (e.g., ILT3) favor the induction and/or the enhancement of $CD4^+$ $CD25^+$ $Foxp3^+$ Treg cells. VDR agonists can modulate the inflammatory response via several mechanisms in secondary lymphoid organs and in target tissues. In secondary lymphoid organs, VDR agonists inhibit IL-12 and IL-23 production and downregulate costimulatory molecule expression (CD40, CD80, CD86) expressed by MDCs while upregulating their IL-10 production and ILT3 expression. MDC modulation by VDR agonists inhibits development of Th1 and Th17 cells while inducing $CD4^+CD25^+Foxp3^+$ regulatory T cells and, under certain conditions, Th2 cells. VDR agonists can also inhibit the migration of Th1 cells, and they upregulate CCL22 production by M-DC, enhancing the recruitment of $CD4^+CD25^+$ regulatory T cells and of Th2 cells. In addition, VDR agonists exert direct effects on T cells by inhibiting IL-2 and IFN-γ production. In target tissues, pathogenic Th1 cells, which can damage target cells via induction of cytotoxic T cells (CTL) and activated macrophages (MΦ), are reduced in number, and their activity is further inhibited by $CD4^+CD25^+$ Treg cells and by Th2 cells induced by VDR agonists. IL-17 production by Th17 cells is also inhibited. In MΦ, important inflammatory molecules such as cyclooxygenase-2 (COX-2) and inducible nitric oxide synthase (iNOS) are inhibited by VDR agonists, leading to decreased production of nitric oxide (NO) and prostaglandin E_2 (PGE_2). MΦs, as well as DCs and T cells, can synthesize $1,25(OH)_2D_3$ and this may also contribute to the regulation of the local immune response. Blunt arrows indicate inhibition; broken arrows indicate cytotoxicity.

expression, and this could also contribute to promote regulatory T-cell induction. It is also possible that 1,25$(OH)_2D_3$ may contribute to the physiologic control of immune responses, and possibly also be involved in maintaining tolerance to self-antigens, as suggested by the enlarged lymph nodes containing a higher frequency of mature DCs in vitamin D receptor (VDR)-deficient mice. This appealing concept has been recently highlighted by the observation that vitamin D_3 induced by sunlight in the skin is hydroxylated by local DCs into the active hormone, which in turn upregulates on activated T cells expression of the epidermiotropic chemokine receptor CCR10, a primary VDR-responsive gene, enabling them to migrate in response to the epidermal chemokine CCL27. Thus the autocrine production of 1,25$(OH)_2D_3$ by DCs can program the homing of skin-associated T cells, which could include regulatory T cells able to counteract the proinflammatory effects induced in the skin by sun exposure. Interestingly, B cells can also synthesize 1,25$(OH)_2D_3$and can preferentially expand Foxp3$^+$ Treg cells, suggesting that the tolerogenic potential of B cells could perhaps be associated with their capacity to produce 1,25 $(OH)_2D_3$ [128].

Photechemical electrocyclization of dehydrocholesterol is followed by cytochrome P450 oxidase (CYP27) regulated further hydroxylation.

When photochemically synthesized, vitamin D is bound to carrier proteins and transported to the liver, where it is hydroxylated at the 25 positions by CYP2R1 and CYP27A1. This is the main circulating form and the easiest analyzed in the serum. Then this compound is transported to the kidneys and other tissues where it undergoes a further hydroxylation at the 1-α position by CYP27B1. The trihydroxy compound has an important effect on the kidneys and further is transported by DBP to many other vitamin D receptor-positive target tissues, such as bones, intestine, and the parathyroid gland, where they exert both a genomic and a nongenomic effect [139,140].

The large versatility of complexation and reactions regulated by 1,25$(OH)_2D$ opens up a large selection of targets for clinical application, with the provision that functional selectivity can be achieved to match what the cell-specific genomic selectivity would seems to promise. This realization has spawned great interest in developing 1,25$(OH)_2D$ analogs to do just that [141]. UV-B irradiation of cultured human KCs induced the conversion of 7-DHC significantly in the presence of calcitriol, suggesting a synergic effect that may be potentially useful in the treatment of skin diseases, such as psoriasis [142,143].

Contemporary research has evidenced that vitamin D does not play only the known role in calcium homeostasis, but has a much larger action spectrum, involving also diseases with a variety of symptoms induced by the mutation of a single gene (pleiotropic action) [144]. An important step forward has been the recognition that these compounds do not circulate as the free hormones, but bound to specific proteins. As an example, a recent study determined the binding of vitamin D throughout the human genome identified 2776 genomic positions occupied by the binding sites and 229 genes with significant changes in expression in response to vitamin D [145]. Binding sites were significantly enriched near autoimmune and cancer-associated genes [146–148]. Notably, genes with vitamin D binding included the interferon regulatory factor 8, associated with multiple sclerosis, and tyrosine-protein phosphatase nonreceptor type 2, associated with Crohn's disease and type 1 diabetes mellitus (T1DM). A variety of tissues (e.g., brain, skin, breast, colon, cardiac muscle, immune cells) are able to introduce the third hydroxy group and activate the vitamin [149]. The positive action of vitamin D, and the relation with disorders resulting from its scarceness or inefficient synthesis, has been demonstrated in diseases ranging from respiratory infections in nursing homes, to breast cancer survival, schizophrenia, bipolar disorder, multiple sclerosis, hypertension, and cardiovascular mortality [150–152].

25-OH vitamin D bound to specific DBP/GC-binding proteins (globulins) that circulate in the blood, but it may become trapped by droplets of fats, from which they are liberated only when there are reductions in adiposity. In this way this metabolite is protected from degradation in the liver. In accordance with this observation, exercise, not just staying outdoor, is usually associated with better vitamin D status [153,154]. Remaining many hours exposed to the sun, on the other hand, causes the formation of the "overirradiation" products, such as lumisterol and suprasterols, which have little of the "classic" vitamin D activity, but which may contribute to protection from UV-induced DNA damage in the skin [155]. The concentration of the vitamin is variable, apparently because the operation of different enzymes, in particular of 7-DHCreductase, which converts 7-DHC, the precursor of vitamin D, to cholesterol, and the enzyme CYP2R1 of the P450 family, the main 25-hydroxylase of vitamin D [139].

Complexes with vitamin D response elements present in the promoter regions of the target genes are formed and result in up- or downregulation of transcription. There is now considerable evidence to suggest that 1,25-$(OH)_2D$ can also utilize other signal transduction mechanisms in order to generate rapid, nongenomic responses. A rapid-acting, nongenomic pathway for 1,25-$(OH)_2D$ action has been proposed to generate a variety of cell-specific responses within seconds to minutes.

This involves the hormone binding to a membrane receptor, the identity of which is still not clear [139,152,153].

2.9 Vitamin D and the immunological system

The other function of vitamin D is immunological. During evolution, the ability of many organisms to synthesize vitamin D photochemically represented, and still represents, a major driving factor for the development of life on the earth. One of the results of the shortened exposure to sun light in contemporary sedentary lifestyle is the predisposition to vitamin D deficiency, which results in a rising rate of immune and inflammatory diseases reported in many parts of the world. In particular, in developed countries, almost a half of the population is affected by an allergic disease at some stage of the life, usually as food allergy or eczema in the first year of life, then progressing to allergic rhinitis and asthma in subsequent years. Although an estimated 90%–95% of the human vitamin D requirement can be afforded by photochemical (UV-B) synthesis in the skin, the dietary source is quite important [130,135]. Several studies have demonstrated that 1,25$(OH)_2D_3$ efficiently promotes the terminal differentiation and decreases the proliferation of cultured human KCs in a dose-dependent way. Furthermore, vitamin D mediates effects on many cell types present in the skin that are involved in immunological reactions, in particular the antimicrobial innate immune system is under its direct control. Vitamin D, together with lipopolysaccharides synergically induces cathelicidin antimicrobial peptide (AMP) expression in many cell types [152,155].

It appears to be generally accepted that a level of 400 IU/day (an IU corresponds to 0.025 mcg) of vitamin D was sufficient for reaching 50 nmol/L required in newborn children, while avoiding hypercalcaemia (notice, however, that the vitamin D concentration in breast milk is quite low and solar light absorption remains a must). The allergic effect of vitamin D seems not to have been unambiguously established; it appears that children who have low cord blood 25-hydroxy vitamin D levels have a larger probability of developing eczema and wheeze (but not asthma) at a later age. The changes of the characteristics of prenatal and infant optical properties of skin have an important diagnostic role [156–158].

A study of the effect of UV light on the systemic immune functions showed that after two weeks of narrowband UV exposure, a significant increase on the amount of both circulating vitamin D_3 and circulating regulatory T cells took place. Proliferative and IL-10 responses to anti-CD3/CD28 were reduced independent of the vitamin D level, suggesting that light and vitamin D_3 levels may affect particular immune function independently [159]. One of the most important effects of vitamin D involves multiple sclerosis. Initial observations showed an inverse relationship between sunlight exposure and the incidence of sclerosis [160,161] and were followed by studies on animals that suggested that in the animal model of multiple sclerosis, experimental autoimmune encephalomyelitis was suppressed by vitamin D_3, or by the hypercalcemia that was produced under such conditions. Recent studies supported, however, that there was a narrow band of light (300–315 nm) that prevented the encephalitis, while not affecting vitamin D_3 in the serum [162]. Multiple sclerosis is characterized by Th1 and Th17 expression, and UVR has a possible beneficial role for Th1-mediated autoimmune diseases (as well as T1DM and rheumatoid arthritis) [163]. The three largest causes of perinatal deaths are prematurity, infection, and asphyxia due to fetal growth restriction. Vitamin D_3 deficiency is an inflammatory condition, which in turn is overrepresented in intrauterine fetal death. Vitamin D_3 sufficiency has an immunomodulatory, antiinflammatory effect and produces antibacterial proteins that might be involved in the etiology of both prematurity and infections. Both striated and coronary muscle strength are dependent on adequate vitamin D_3 levels. If the fetus is vitamin D deficient, the fetal heart might be more vulnerable to hypoxic stress, as supported by recent data demonstrating that women with vitamin D-lack in early pregnancy were more likely to be delivered by emergency cesarean. The results need, however, to be confirmed by other groups and further work has to be done in order to reduce the risk of bias. If so, those with high risk for coronary heart disease, stroke, venous thromboembolism, or diabetes should be recommended safe sun exposure habits if living in a low UV intense region [164].

Another explored relation is that with obesity that is increasing in many parts of the world. A recent study does support that the development of obesity and two of its metabolic comorbidities, type-2 diabetes and metabolic syndrome, are affected by either vitamin D or nitric oxide, although it was felt that more epidemiological and clinical research was required that focuses on measuring the direct associations and effects of exposure to UVR in humans [165].

2.10 Ultraviolet light effect involving molecules different from vitamin D

Peropsin, an opsin member initially detected in human ocular tissue, is, however, present in various tissues, including the skin. Peropsin expression in human skin has been determined and its potential role in KCs proposed [166–168]. Numerous peroxin proteins serve several functions including the recognition of cytoplasmic proteins that contain peroxisomal targeting. Peropsin is located suprabasally layer in the skin where KCs are well differentiated. A key concept developed in recent years is that absorption of light by a small number of G proteins is transformed into a robust downstream signal [169] (Fig. 2.16).

Furthermore, the photopigment rhodopsin is expressed in human melanocytes and is involved in UV phototransduction which induces early melanin synthesis. Rhodopsin is expressed and localized on the plasma

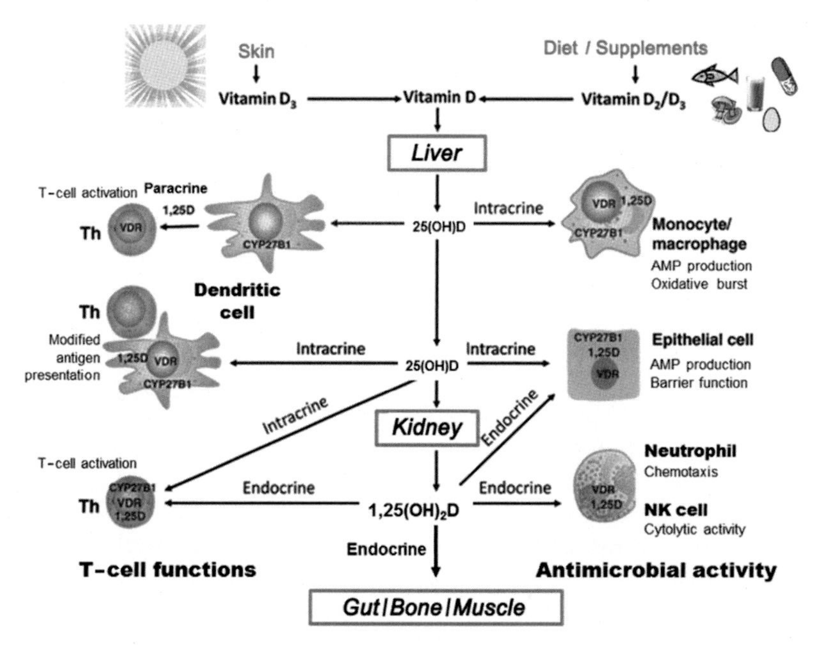

FIGURE 2.16 Mechanisms for innate and adaptive immune responses to vitamin D. Ergocalciferol (vitamin D_2) from the diet and cholecalciferol (vitamin D_3) from the diet or produced from the action of UV-B on the skin are metabolized in the liver to form 25-hydroxyvitamin D (25(OH)D), the main circulating form of vitamin D. Target cells such as monocytes, macrophages, and dendritic cells expressing the mitochondrial vitamin Dactivating enzyme 1-α hydroxylase (CYP27B1) and the cytoplasmic vitamin D receptor (VDR) can then utilize 25(OH)D for intracrine responses via localized conversion to 1,25-dihydroxy-vitamin D (1,25(OH)2D; calcitriol, shown in the figure as 1,25D for intracellular locations). In monocytes and macrophages this promotes antibacterial responses to infection. In dendritic cells, intracrine synthesis of 1,25(OH)2D inhibits dendritic cell maturation, thereby modulating helper T-helper (Th) cell function. Th cell responses to 25(OH)D may also be mediated in a paracrine manner, via the actions of dendritic cell-generated 1,25(OH)2D. Intracrine immune effects of 25(OH)D also occur in epithelial cells expressing the VDR and the 1-α hydroxylase (CYP27B1). However, other leukocytes such as neutrophils and natural killer (NK) cells do not appear to express CYP27B1 and are therefore likely to be directly affected by circulating levels of 1,25(OH)2D synthesized by the kidneys or locally produced in and secreted from tissue macrophages and dendritic cells. VDR-expressing Th cells are also potential targets for systemic 1,25(OH)2D, although intracrine mechanisms have also been proposed. In a similar manner, epithelial cells can respond in an intracrine manner to 25(OH)D, but may also respond to systemic 1,25(OH)2D to promote antibacterial responses [169].

membrane in epidermal keratinocytes (NHEKs), and only violet light among several wavelengths within the visible range significantly increased the expression of rhodopsin mRNA. Moreover, rhodopsin overexpression decreased the mRNA expression levels of KC differentiation markers, such as keratin-1 and keratin-10, and these decreased expression levels were recovered by a rhodopsin-directed siRNA. In addition, it was demonstrated that violet light significantly decreased the phosphorylation levels of cAMP responsive element-binding protein and that it more effectively decreased the phosphorylation of such protein when rhodopsin was overexpressed. Subsequent experiments showed that 380 nm light elicited the strongest Ca^{2+} response among the wavelengths tested. Importantly, light-induced Ca^{2+} transients occurred only when rhodopsin was expressed and localized on the plasma membrane in NHEKs, and only violet light among several wavelengths within the visible range significantly increased the expression of rhodopsin mRNA.

Decreased the mRNA expression levels of KC differentiation markers, such as keratin-1 and keratin-10. Violet light also decreased the mRNA expression levels of KC differentiation markers and these decreased expression levels were recovered by a rhodopsin-directed siRNA. Moreover, it was further demonstrated that violet light significantly decreased the phosphorylation levels of cAMP responsive element-binding protein and that it more effectively decreased the phosphorylation of such protein. In addition, it was found that pertussis toxin, a Gαi protein inhibitor, restored the rhodopsin induced decrease in the differentiation markers in NHEKs [167].

It appears to have been clearly established that supplementation of vitamin D per os has no effect on blood pressure. Furthermore, a genetic study showed that a reduced genetic vitamin D production correlated linearly with an increased and all-cause mortality, but no increase in cardiovascular mortality, despite the fact that vitamin D level inversely correlate with cardiovascular health [170]. A way out from this dilemma was that a different molecule was involved, and recent studies suggest that this is NO (Fig. 2.17), as it was experimentally demostrated [168].

A recent investigation of the habit in sun exposure versus death, concerning a 15-year time span showed that, compared with highest sun exposure group, the subdistributional hazard ratio (sHRs) of CVD mortality among sun exposure avoidance and moderate avoidance people were 2.3, for 95% confidence interval (CI) 1.8–3.1, and 1.5 for 95% CI 1.2–1.8, respectively. The corresponding sHRs for non-cancer–non-CVD death were 2.1, 95% CI 1.7–2.8, and 1.57, 95% CI 1.3–1.9, and for cancer 1.4, 95% CI 1.04–1.6 and 1.1, 95% CI 0.9–1.4, respectively (Fig. 2.10). Low life expectancy is traditionally related to overweight, inactivity, and smoking, but the data above document that the major life style risk factors related to all-cause mortality. A low sun exposure habit was found to be a risk factor in the same range as smoking with a 0.6 to 2.1 years shorter mean life expectancy rate, over the 15 years of the survey, as well as the evolutionary advantage in being less pigmented when living far from the equator.

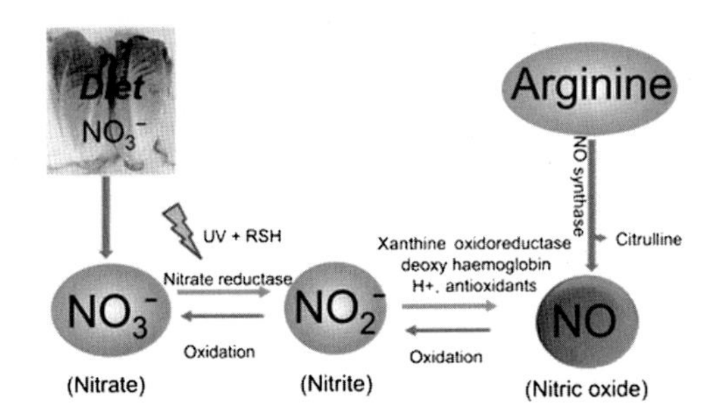

FIGURE 2.17 Summary of nitric oxide production pathways involving the skin [168].

This study documents the public health implications, with an apparent cancer (and CVD) related life-expectancy decrease, not found when attempting to establish a relation with vitamin D level. In summary, people with coronary heart disease, stroke, venous thromboembolism, or diabetes should be recommended safe sun exposure habits if living in a low UV intensity region. Notice, however, that summing together all the cancer cases contrasts with the fact that cutaneous malignant melanoma is the skin cancer mainly related to increased mortality and is related to episodic overexposure to UVR. There seems to be a relationship between higher sun exposure and malignant melanoma incidence but an inverse relationship to prognosis. High UV exposure increases the incidence, while low sun exposure habits/vitamin D levels have been linked to thicker, more aggressive melanomas with shorter survival times.

The new findings indicate that there is a need for modification of guidelines regarding sun exposure. They may also add to our knowledge regarding the increasing incidence of diabetes mellitus and increased mortality among non-Caucasians in western countries. According to the present knowledge, in a low solar intensity region we should aim for sound and safe sun exposure habits, especially for those at increased risk of CVD or noncancer/ non-CVD.

2.11 Skin diseases caused by light

Acute consequences of exposure to sunlight are sunburn and tanning, chronic ones photoaging and photocarcinogenesis, including the formation of malignant tumors after repeated exposure to solar radiation. As an example, the first sign of photoaging in Japanese subjects exposed to solar light is the appearance of freckles around 20 years of age. Fine wrinkles then appear after 30 years of age, and benign skin tumors, seborrheic keratoses, can occur after 35 years of age [68]. The skin and eyes are most sensitive to damage by UV light at $\lambda < 290$ nm wavelength, which is in the highest energy UV-C band, and is almost never encountered except from artificial sources like welding arcs or "germicidal" lamps. Most sunburns are caused by longer wavelengths, simply because these are more prevalent in the sunlight at ground level. In a previous section, it has been shown that the ozone present in troposphere and stratosphere filters away all of the UV-C, but only a part of the longer wavelengths (however, it must be taken into account that in recent decades, the incidence and severity of sunburns has increased worldwide, partly because of chemical damage to the atmosphere's ozone layer).

The reactions of the skin occurs through two paths, either photoallergy or phototoxicity. Acute phototoxic reactions involve erythema, edema, and rash, and later hyperpigmentation. Long-term phototoxicity involves chronic sun-caused damage and may lead to the different variety of skin cancer. Cancer formation, particularly when a contact with exogenous chemicals has intervened (including psoralens, coal tar, several drugs, such as tetracyclines, sulfamides an many others). Photoallergy involves either an immediate antibody formation, as in solar urticaria, an immunoglobulin E-mediated hypersensitivity, or a delayed cell-mediated reaction, as typically when caused by exogenous chemicals (including smoke) [171,172]. There has been some confusion in the nomenclature, which has sometimes hindered recognizing diseases. Recent classification, as accepted by international bodies, differentiates primary and secondary dermatoses. Primary photodermatoses are caused by a photosensitizing substance and are referred as idiopathic when the etiology is not known. They include polymorphous light eruption, solar urticaria, hydrozoa vacciniforme, actinic prurigo, and chronic actinic dermatitis. The first one is the most common, at least in Europe and the States, and the key symptoms are severe pruritic skin and lesions. Macular, papular, papulovesicular, urticarial, multiforme-, and plaque-like variants are differentiated morphologically, although generally maintained for each individual, hence the name polymorphous attributed to this disease. A good UV sunscreen protects the skin and at any rate the skin lesions resolve spontaneously within several days of ceasing sun exposure and do not leave behind any traces. Many patients develop tolerance over the course of the sunny period of the year, meaning that lately even prolonged sunbathing can be tolerated later in the season, in accord with the fact that the symptoms develop in March-June, not in full summer when solar light is strongest. Solar urticaria is a rare (0.08% of all urticaria cases) but severe disorder [173]. Urticarial skin lesions appear a few minutes after exposure and an anaphylactic shock may occur after whole-body exposure. The action spectrum ranges over the entire UV spectrum up to visible light. Most patients react to both

UV-A and visible light. In vitro preirradiation of the patient's plasma or serum is also an option because some patients develop an urticarial reaction at the site of injection of the irradiated plasma. As for the papular or papulovesicular type, differential diagnosis indicates photoallergic eczema, ictus, or prurigo simplex. The underlying cause of plaque-type lesions may be delayed onset solar urticaria, erythema multiforme, and even lupus erythematosus (LE), particularly the tumid type. However, a latency of 1–3 weeks after sun exposure and a long healing time is typical for LE. A dermatological maxim states that "a patient gets polymorphic light eruption (PMLE) on holiday but brings LE home." Although PMLE is the most important differential diagnosis for cutaneous LE, it virtually never changes into LE, even if nonspecific positive antinuclear antibodies are present. Further variation are hydrozoa vacciniforme, a very rare disorder (prevalence: 0.34 per 100,000) with acute onset characterized by numerous hemorrhagic vesicles on the face and the hands, which heal with varioliform scarring, and chronic actinic dermatitis, a term that is used as an umbrella term, as it combines different medical conditions, such as persistent light reaction, actinic reticuloid, and photosensitive eczema. The hypothesis that a photoallergy leads to chronification of the inflammatory lesions, which remain even after elimination or avoidance of the photoallergen, via persistence of the sensitizer was presented but not proven as yet. Actinic prurigo is a rare photodermatosis that develops in childhood and has a chronic-persistent course. Photoallergic reactions are less frequent than phototoxic reactions and, unlike the latter ones, occur only if a specific sensitization has been acquired. Sensitization may involves contact or oral ingestion and important topical photoallergens include halogenated salicylanilide, fenticlor, hexachlorophene, bithionol, and in rare cases also sunscreens.

IR light may be lower energy, but is still of significant value (typically corresponding to molecular vibrational energies). In view of the long wavelength, they penetrate further low in the skin. In everyday experience one is familiar with microwave ovens and their use for cooking food. Therefore, it is not unreasonable to expect that microwaves and radiofrequencies may damage biomolecules. In fact, such an effect operates, and indeed is exploited for therapeutic application, as is the case for radiofrequencies that are known to penetrate into the deeper layers of the skin and produces heat, with a tightening of the subdermal layers. Differences in protein expression have been found in the skin of volunteers exposed to radiofrequency-modulated electromagnetic field (mobile field), suggesting that this may be generally observed (at an energy, however, that is much above that used in microwave owens, wi-fi transmission nets, radar, and all of the electromagnetic fields in which we are immersed). As an example, bees changed somewhat their behavior when exposed to cell phone radiation and their secretion increased their nutritional content up to a certain irradiation time, then it dropped somewhat. On the contrary, radiofrequencies appeared to reduce cytotoxic activity in the peripheral blood of women, although with no dose–response effect [174–176].

2.12 Skin photoaging and the role of free radicals

Sunburns are more frequent with fair skin people, but certainly not uncommon for dark skin people. A control of the symptoms may be obtained also by administrating substances that are strong reducing agents or radical traps. The use of sun beds leads to an effect similar to solar light (skin aging). Photoaging of the skin depends on an amount of melanin in the skin. In addition to direct or indirect DNA damage, UV-A activates cell surface receptors of KCs and fibroblasts in the skin, which leads to a breakdown of collagen in the extracellular matrix and a shutdown of new collagen synthesis. Presumably, the collagen is only imperfectly repaired and the structural integrity of the skin is lost with formation of a solar scar, and ultimately clinically visible skin atrophy and wrinkles. Oxidation of cellular biomolecules by exposure to UV light could be prevented by previous antioxidant treatment [177], it is important that protective antioxidants are present when required. Thus vitamin E has excellent properties as antioxidant, but is so sensitive to sun light that it is not easily available when required. Transdermal delivery is an efficient solution [178]. The appearance of skin reflects a combination of one's general health, ethnicity, life style, diet, and age and has been long known to be strongly increased by exposure to solar light. These features determine the color, texture, firmness, and smoothness of the skin. Intrinsic aging is a naturally occurring process that relates closely to chronological age. At a microscopic level, chronologically aged skin can be characterized by an atrophic epidermis with flattening of the dermal-epidermal junction and loss of the downward projections of epidermis lying between upwardly directed DP (Rete pegs), as well as a decrease in the number of fibroblasts and collagen, resulting in a much thinner dermis than that observed in young individuals. The mechanism of UV light–causing photoaging involves activation of growth factors and cytokine receptors and dermal cells. Downstream signal transcription involves activation of nitrogen-activated protein kinase pathways (extracellular signal-regulated kinase, c-jun *N*-terminal protein kinase, and p38). These pathways converge in the nucleus of cells and form an activated complex of transcription factor activator protein 1 (cFoscJun), which induces matrix metalloproteinases that finally degrade skin connective tissue [179]. The other way around, research has identified parameters that characterize a youth skin, although this may be unchivalrous. Thus unbiased clustering revealed gene expression signatures characteristic of older women with skin youthfulness (SY) compared with older women without skin youthfulness, after accounting for gene expression changes associated with chronological age alone. Gene set analysis was performed using Genomica open-access software. A study identified a novel set of candidate SY genes demonstrating differences between the SY and the non-SY group and showed that immunologic gene sets are the most significantly altered in SY, suggesting that the immune system plays an important role in SY [180].

Furthermore, UV irradiation generates ROS in human skin in vivo; some of the oxysterols result from free radical oxidation, other from enzymic oxidation catalyzed by cytochrome P450. A study of the effect of isoflavone genistein, that has an antioxidant activity, and *N*-acetylcysteine (NAC), which can be converted into the endogenous antioxidant glutathione, showed that UV irradiation caused a decrease of hydrogen peroxide in human skin in vivo to impair responses to UV light that eventuate in photoaging in human skin. Although both genistein and NAC inhibited stress-activated mitogen-activated protein kinases (MAPKs), genistein had a broader effect on both extracellular signal-regulated kinases (ERKs)and c-Jun N-terminal kinases (JNKs), the kinases that bind and phosphorylate Ser-63 and Ser-73, whereas NAC inhibited only the former, and not JNK. This is consistent with the fact that geninstein was unable to block the phosphorylation of the epidermal growth factor receptor (EGF-R). As the extracellular ERK pathway is most closely associated with activation of EGF-R, the block of UV-induced EGF-R phosphorylation by genistein is consistent with its inhibition of ERK activation. As for NAC, the fact that, despite its inability to block EGF-R photophosphorylation by UV, it significantly prevented ERK activation indicates the presence of a redox-sensitive step in ERK activation by UV irradiation, as has been reported for ERK activation by growth factors and hormones [181].

NAC can significantly block UV induction of hydrogen peroxide and genistein is also known to possess an antioxidant activity, suggesting that ERK may have been mediated in part through a similar ROS quenching mechanism that is independent of its ability to block EGF-R activation.

Genistein similarly prevented JNK activation by UV, suggesting that in human skin in vivo the EGF-R signaling cascade directly, or perhaps indirectly, via ROS it generates, leads to JNK phosphorylation in vivo [179,181,182].

IR-A (700–1400 nm) accounts for 30% of total solar radiation and has the capacity to penetrate deeper into the skin, reaching subcutaneous tissues (Fig. 2.1B). Nowadays, there is compelling evidence that associates solar IR-A radiation with premature aging and the progression of malignancies. The others range are IR-B, 1400–3000 nm and IR-C, 3000 nm^{-1} mm. Interestingly, for many years IR-light–based therapies have been used clinically to promote wound healing, protect muscles from stress, and reduce proinflammatory cytokine and chemokine production. The apparent dichotomy of IR-A and its effect on the skin (good or bad? Friend or foe?) is explained by the capacity to control the intensity, time of exposure, and heat production during clinical exposure to IR-A [183].

IR-A regulates as much as 600 genes in human skin that are involved in extracellular matrix homeostasis, apoptosis, cell growth, and stress responses. This results in an increase of the mitochondrial production of ROS and a decrease of adenosine triphosphate (ATP) synthesis, and in turn activation of MAPKs and caspases (leading to apoptosis). The kinases play a critical role in controlling the expression of metalloproteinase-1, called also interstitial collagenase and in this way extracellular matrix destruction. Moreover, studies using different experimental models showed that IR radiation enhanced the deposit of elastotic material in the dermis while decreasing collagen [184]. Epidermal hyperplasia/thickening, increased senescent marker expression (i.e., telomerase expression and activity), angiogenesis (by increasing vascular endothelial growth factor production and CD31 positive cells), erythema, and swelling are also characteristics present in IR-radiated skin. Finally, IR-A also triggers a significant decrease in antioxidant capacity in the skin (specifically by destroying carotenoids such as β-carotenes and lycopene), as well as the activation-recruited mast cells (MC_{TC}), enhance oxidative stress and inflammation, promoting premature aging [170] (compare Fig. 2.18).

The skin is the only organ chronically exposed to the environment, and the resulting interaction with environmental factors can strongly influence skin physiology, as it is the case with eyes, leading to extrinsic aging. By far the most studied source of extrinsic skin is downward projections of epidermis lying between upwardly directed DP. The term "photoaging" describes the effect of chronic UV-light exposure on the skin, estimated to account for up to 90% of visible skin extrinsic aging, and is characterized by dryness, a rough texture, increased skin laxity, irregular pigmentation, lesions of small vessels (telangiectasia) or small bowel bleeding (angioectasia), a yellowish color, thickening, deep creases, fine wrinkles, solar elastosis, and an extensive decrease of fibrillar collagen (types I and III) due to the decrease in transforming growth factor TGF-β levels and the activation of activator protein-1 (AP-1). This emerging information reveals that chronological aging and photoaging share fundamental molecular

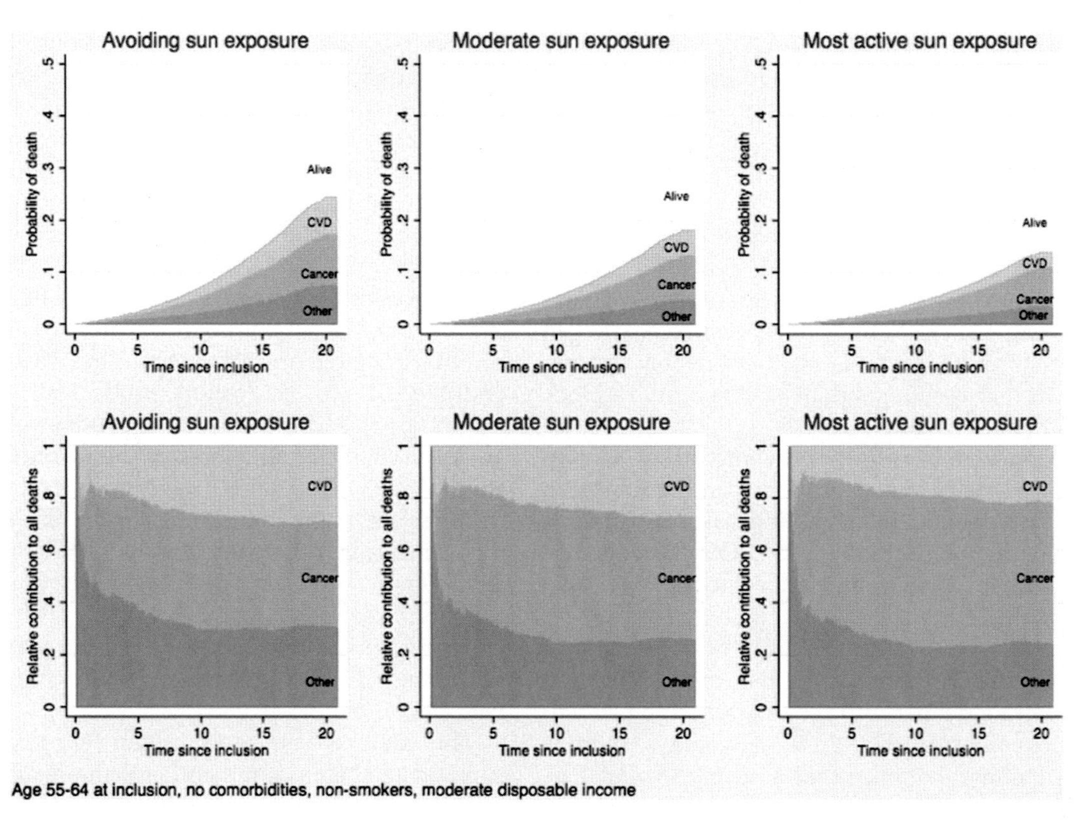

FIGURE 2.18 Probability of death by sun exposure habits in a competing risk scenario. Upper three graphs show death categorized into cardiovascular disease (CVD), cancer, and other (noncancer/non-CVD) according to time in years since study inclusion. Bottom three graphs show relative contribution to death by sun exposure habits. As compared to highest sun exposure group, the subdistributional hazard ratio (sHRs) of CVD mortality among sun exposure avoidance and moderate exposure were sHR = 2.3, 95% CI 1.8–3.1, and 1.5, 95% CI 1.2–1.8, respectively. The corresponding sHRs for noncancer–non-CVD death were 2.1, 95% CI 1.7–2.8, and 1.57, 95% CI 1.3–1.9, and for cancer 1.4, 95% CI 1.04–1.6, and 1.1, 95% CI 0.9–1.4, respectively [170].

pathways and skin damage by UV irradiation and by the passage of time. Thus chronological aging and photoaging share fundamental molecular pathways. Indeed, new insights regarding convergence of the molecular basis of chronological aging provides new opportunities for the development of new antiaging therapies [185].

Two main processes that induce skin aging are conveniently distinguished: intrinsic and extrinsic. A stochastic process that implies random cell damage as a result of mutations during metabolic processes due to the production of free radicals is also implicated. Intrinsic aging reflects the genetic background and depends on time. Extrinsic aging is caused by environmental factors such as sun exposure, air pollution, smoking, alcohol abuse, and poor nutrition. Extrinsically aged skin is characterized by photo damage as wrinkles, pigmented lesions, patchy hypopigmentations, and actinic keratoses. Timely protection including physical and chemical sunscreens, as well as avoiding exposure to intense UV irradiation, is most important. A network of antioxidants such as vitamins E and C, coenzyme Q10, alpha-lipoic acid, glutathione, and others can reduce signs of aging. Further anti-aging products are three generations of retinoids, among which the first generation is broadly accepted. A diet with lot of fruits and vegetables containing antioxidants is recommended as well as exercise two or three times a week [186].

On the other hand, a photoaged epidermis shows an excessive volume increase (hyperplasia), an increased number of melanocytes and melanomas, as well as the presence of atypical KCs and persistence of these specialized cells as they rise into the stratum corneum (parakeratosis) of which they determine the thickness. Again, it has been demonstrated that this transformation is not determined by UV light alone, but also by visible and IR radiation. The mechanism involves light absorption by melanin and an energy transfer to form singlet oxygen (Scheme 2.19) that is known to attack lipids, proteins, and DNA. Such results should alarm public health authorities because the general belief that UV filters are able to withstand the unpleasant consequences of exposure easily leads to remain longer to the sun [13,14].

Spending long periods in the sun and being fully unprotected from light may result in an irreparable damage to the skin, skin photoaging, and possibly the formation of tumors. The old recipe of staying at most for small periods of time in the sun allows to combine obtaining the positive effects and avoid serious risks. Actually, it has been demonstrated that c.50% of the radicals formed in the skin during irradiation result from the visible and IR regions of the sun emission and a free radical concentration threshold can be determined (c.4×10^{12} radicals for mg). Under this limit, primarily formed oxygen-centered radicals (ROS) predominate, while above it secondary carbon-centered lipid radicals (LOS) predominate and skin photoaging and cancer may be induced [187] (Fig. 2.19) [188].

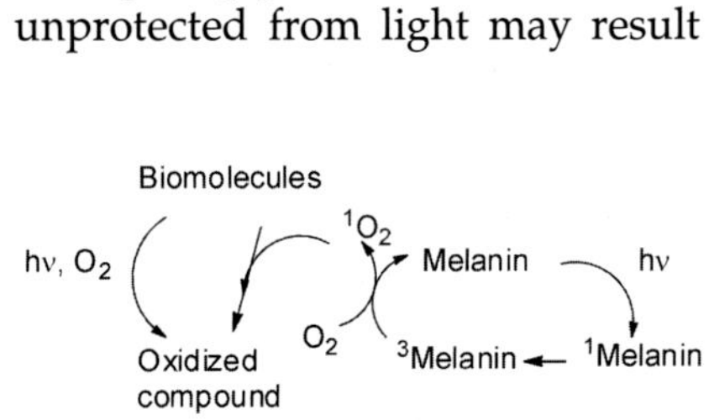

Biomolecules
hν, O_2
$^1O_2$
Melanin
hν
Oxidized compound
O_2
^{3}Melanin
^{1}Melanin

SCHEME 2.19 Oxidation of molecules by direct irradiation or by sensitized reaction via energy transfer. The triplet is a virtual state that cannot be reached from the ground singlet S_0, but is formed through intersystem crossing. This applies, for example, to the oxidation of skin biomolecules either by direct absorption (UV-B) or by sensitization by melanin (UV-A and visible light) that leads to indirect oxidation.

Molecular oxygen is the second component of the atmosphere (roughly 21% at sea level) and its amount varies according to history (top value has been around 30% roughly 2.5 million years ago, no oxygen in primeval atmosphere until cyanobacteria began to afford it), altitude and temperature [189,190]. For complex eukaryotic organisms and all currently living aerobic species, molecular oxygen is a central molecule in cellular respiration. However, some oxygen containing compounds are highly reactive and toxic to cells, and it has become customary to design them as ROS, and analogously reactive nitrogen species (RNS) and reactive sulfur species (RSS) or, more generally, reactive species or reactive oxygen intermediates (ROI). These species include both even electron species, such as oxygen itself, its excited (singlet) state, ozone, hydrogen peroxide, (hydro)peroxy radical derivatives of hypohalogenous acids (HClO, HBrO, HIO) and odd-electron species, such as superoxide radical anion, hydroxyl and hydro(alkoxy)peroxy radicals, carbonate radicals, etc. RNS include nitric oxide radical (NO or NO•), nitrogen dioxide radical (NO_2•), nitrite (O = N-O•), and peroxynitrite (O = N-OO•). RSS include sulfides, polysulfides, thiosulfate, disulfides, as well as species formed upon

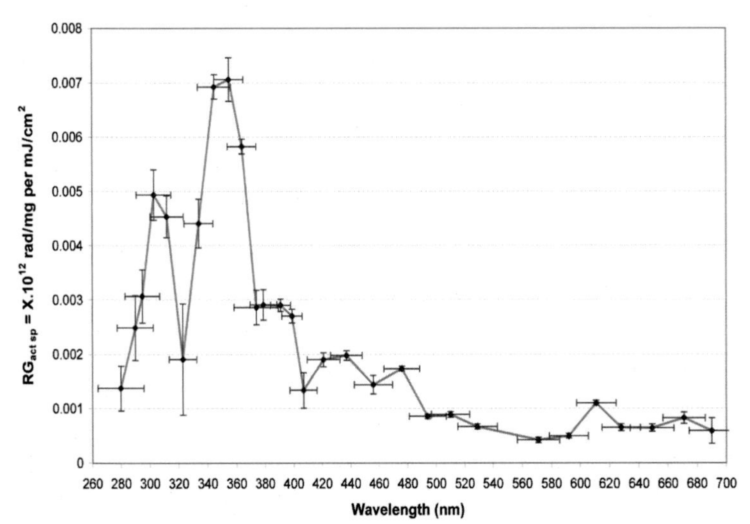

FIGURE 2.19 Action spectrum of the formation of free radicals [188].

incomplete reduction of molecular oxygen, namely superoxide radical anion ($O_2\bullet^-$), hydrogen peroxide (H_2O_2), and hydroxyl radicals (OH•), while ROS includes both ROI, ozone (O_3) and singlet oxygen ($^1O_2$). Further oxygen-centered radicals as peroxyl (ROO•), alkoxyl (RO•), semiquinone ($SQ\bullet^-$), and carbonate ($CO_3\bullet^-$) radicals as well as related even-electron species, such as organic hydroperoxides (ROOH) are also frequently enclosed in the definition of ROS. A somewhat more encompassing definition includes within ROS also even electrons compounds such as hypochlorous (HOCl), hypobromous (HOBr), and hypoiodous acids (HOI), as well as charged species such as radical ions (R^+, R^-).

ROS, in particular, hydroxyl and peroxyl radicals, hydrogen peroxide, and superoxide radical anion, have long been studied as the active intermediate for mild oxidation processes [191,192].

Elapsing time produces a characteristic pattern of changes, indicated as skin photoaging that might, however, also be caused much before the chronological age due to (essentially oxidative) stress, mainly involving radicals.

The predominant processes from primary radicals involve σ-bond fragmentation (a, a′),

a) $R^\bullet + X{-}H \longrightarrow R{-}H + X^\bullet$

a′) $R{-}A^\bullet \longrightarrow R^\bullet + A$

b) $R^\bullet$ + =\\Z ⟶ R\\•/\\Z

c) $R^\bullet + M^{n+} \longrightarrow R^+ + M^{(n-1)+}$

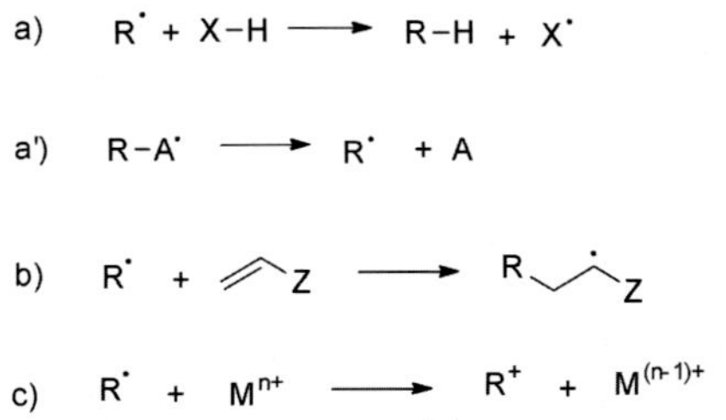

SCHEME 2.20 Double-strand DNA breaks. Source: *Reproduced from V. Manova, D. Gruszka, DNA damage and repair in plants – from models to crops, Front. Plant Sci. 6 (2015) 885 [193].*

addition to a π-bond (b) and single electron transfer with metal ions (c) (Scheme 2.20).

The reactions of these odd-electron species are very fast (k 10^3 to $10^8\ M^{-1}\ s^{-1}$), and may result, roughly, in unimolecular fragmentation (of an even electron fragment) (Scheme 2.21), as well as isomerization and addition reactions.

Photochemistry is an expedient way for generating radicals and study their properties under very mild condition, and further have an important role in many biological reactions.

$$R\text{-}CO_2^- \xrightarrow{-e} R\text{-}CO_2^{\bullet} \longrightarrow R^{\bullet} + CO_2$$

$$R\text{-}N_2^+ \longrightarrow R\text{-}N_2^{\bullet} \longrightarrow R^{\bullet} + N_2$$

SCHEME 2.21 Cleavage of an even-electron fragment from a radical.

$$\text{–C–H} \rightarrow \text{–C}^{\bullet} \xrightarrow{O_2} \text{–C–O–O}^{\bullet} \rightarrow \text{–C–O}^{\bullet}$$

$$\text{–C–O–OH}$$

$$\text{=C–C–H} \xrightarrow{^1O_2} \text{C–O–OH}$$

SCHEME 2.22 Lipid oxidation models, showing formation of hydroperoxides via radicals and via singlet oxygen.

2.13 Enzymatic antioxidants

The term "antioxidant" refers to any molecule capable of stabilizing or deactivating free radicals before they attack cells. Humans have evolved highly complex antioxidant systems (enzymic and nonenzymic) [191], which work synergistically, and in combination with each other to protect the cells and organ systems of the body against free radical damage. Antioxidants contain a relatively weak covalent bond that can be extracted by radical reagents (e.g., S-H, PhO-H), or an easily oxidized moiety (as in polyphenols), or a trap radicals via formation of adducts (e.g., by addition onto conjugated alkenes). The antioxidants can be of endogenous or exogenous origin, for example, as a part of a diet or as dietary supplements. Some dietary compounds that are not able to quench free radicals, but enhance endogenous activity may also be classified as antioxidants, that is of molecules (or physical agents) able to hinder the oxidation of other molecules, a fact of obvious importance when a chain process is involved, as typically via radicals [194]. Generation of C-centered radicals by thermal homolysis is followed by oxygen trapping to yield peroxyl radicals and the strong electrophilic oxyl radicals. At room temperature singlet oxygen reacts with allyl derivatives to yield allyl hydroperoxydes [193] (Scheme 2.22).

An ideal antioxidant should be readily absorbed from the food and quench free radicals, and chelate redox metals at physiologically relevant levels. It should also work in both aqueous and/or membrane domains and effect gene expression in a positive way. Endogenous antioxidants play a crucial role in maintaining optimal cellular functions and thus systemic health and well-being. However, under conditions which promote oxidative stress, endogenous antioxidants may not be sufficient and dietary antioxidants may be required to maintain optimal cellular functions. The most efficient enzymatic antioxidants involve glutathione peroxidase, catalase, and superoxide dismutase (SOD). Glutathione peroxidase is present in two forms, one of which is selenium (Se)-dependent and the other, Se-independent. The differences are due to the number of subunits, catalytic mechanism, and the bonding of Se at the active center, and glutathione metabolism is one of the most important antioxidative defense mechanisms present in the cells [195] (Fig. 2.20).

There are four different Se-dependent glutathione peroxidases present in humans, and these are known to add two electrons to reduce peroxides by forming selenoles (R-Se-OH). Therefore the antioxidant properties of these selenoenzymes allow them to eliminate peroxides as potential substrates for the Fenton reaction. Se-dependent glutathione peroxidase acts in association with the polyamide glutathione (GSH), which is present in high concentrations in cells and catalyzes the conversion of hydrogen peroxide or organic peroxide into water or alcohol while simultaneously oxidizing GSH. It also competes with catalase for hydrogen peroxide as a substrate and is the major source of protection against low levels of oxidative stress. Catalase is a heme enzyme present

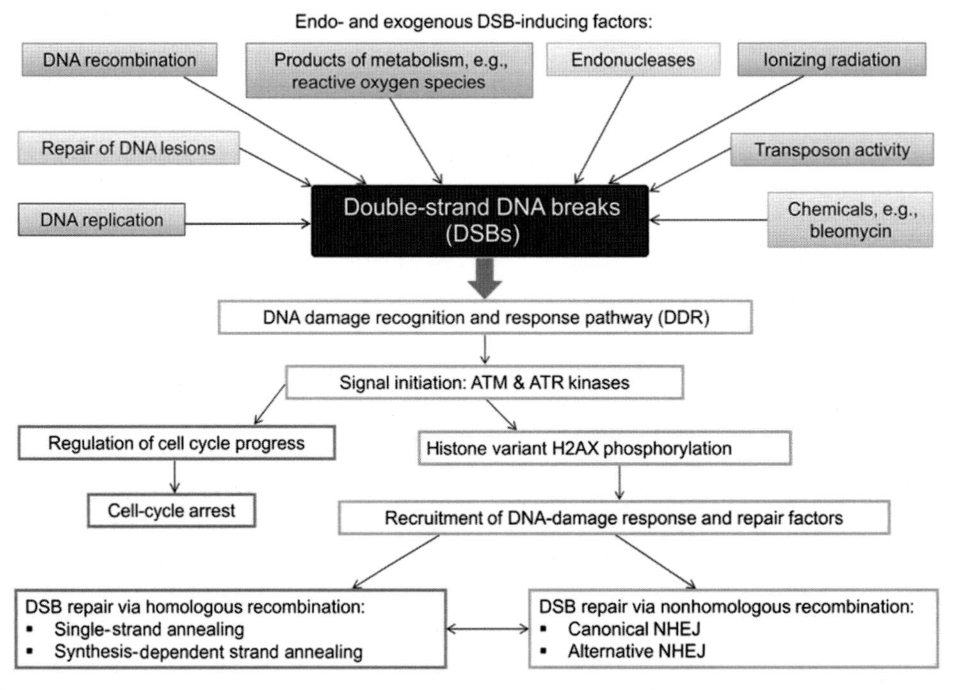

FIGURE 2.20 Summary of nitric oxide production pathways involving the skin [193].

in the peroxisome of aerobic cells and is very efficient in promoting the conversion of hydrogen peroxide to water and molecular oxygen. Catalase has one of the highest turnover rates for all enzymes: one molecule of catalase can convert approximately 6 million molecules of hydrogen peroxide to water and oxygen each minute. SOD in turn catalyzes the conversion of superoxide anions to dioxygen and hydrogen peroxide. SOD exists in several isoforms, which differ in the nature of active metal center, amino acid composition, cofactors, and other features. Three forms of SOD are present in humans: cytosolic Cu, Zn-SOD, mitochondrial Mn-SOD, and extra cellular-SOD. SOD neutralizes superoxide ions by going through successive oxidative and reductive cycles of transition metal ions at its active site. Cu, Zn-SOD has two identical subunits with a molecular weight of 32 kDa and each of the subunit contains as the active site, a dinuclear metal cluster constituted by copper and zinc ions, and it specifically catalyzes the dismutation of the superoxide anion to oxygen and water. The mitochondrial Mn-SOD is a homotetramer with a MW of 96 kDa and contains one manganese atom per subunit, and it cycles from Mn(III) to Mn(II), and back to Mn (III) during the two-step dismutation of superoxide [196–200].

Extracellular SOD contains copper and zinc, and is a tetrameric secretary glycoprotein having a high affinity for certain glycosaminoglycans such as heparin and heparin sulfate, however, its regulation in mammalian tissues occurs primarily in a manner coordinated by cytokines, rather than as a response to oxidative stress [201].

2.14 Nonenzymatic antioxidants

Nonenzymatic antioxidants include vitamins E and C, thiol antioxidants [glutathione, thioredoxin (TRX), and lipoic acid], melatonin, carotenoids, natural flavonoids, and other compounds. Some antioxidants can interact with other antioxidants regenerating their original properties; this mechanism is often referred to as the "antioxidant network." There is growing evidence to support a link between increased levels of ROS and disturbed activities of enzymatic and nonenzymatic antioxidants in diseases associated with aging [202,203]. Some example of common antioxidants are shown in Scheme 2.23.

2.14.1 Thiols

The major thiol antioxidant is glutathione (GSH), a triamide that results from the condensation of three amino acids, glutamate, cysteine, glycine, not a tripeptide, however, because the glutamate is linked at the γ amino group, not to the usual α. This is a multifunctional intracellular antioxidant and is considered to be the major thiol-disulfide redox buffer of the cell. It is abundant in cytosol, nuclei, and mitochondria, and is the major soluble antioxidant in these cell compartments. Glutathione has also been shown to play a role in cell senescence since studies involving human fibroblasts have shown that the intracellular glutathione level has a strong influence on the induction of a postmitotic phenotype, and that depletion of glutathione may play a significant role in the cellular aging in human skin. The reduced form of glutathione is GSH, glutathione, while the oxidized form is GSSG, glutathione disulfide. The antioxidant capacity of thiol compounds is due to the sulfur atom,

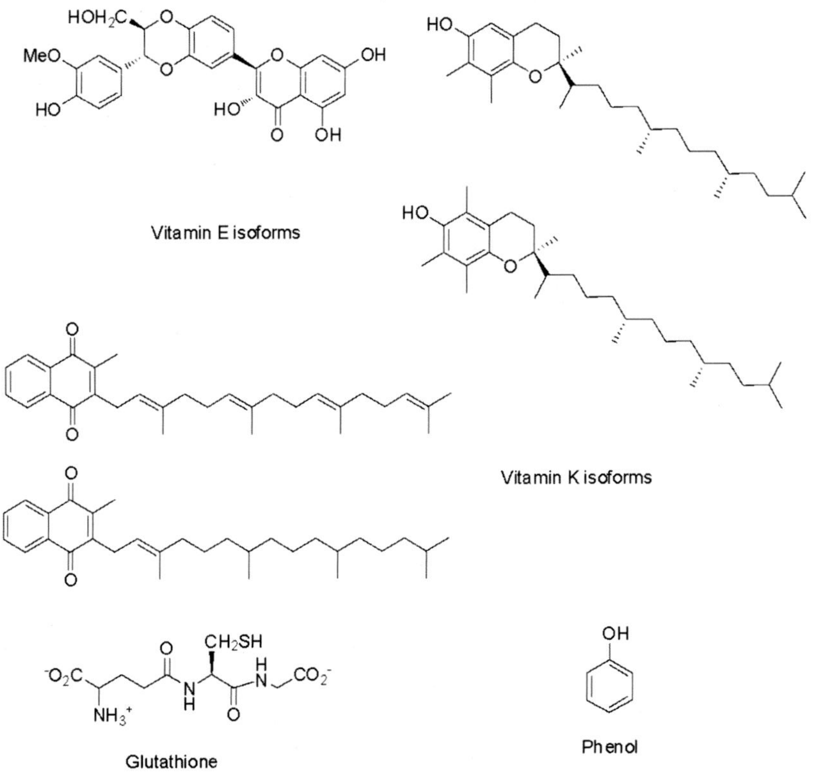

Vitamin E isoforms

Vitamin K isoforms

Glutathione

Phenol

SCHEME 2.23 Molecules known for their antioxidant action: phenols, vitamins E, vitamins K, and glutathione.

which can easily accommodate the loss of a single electron. Oxidized glutathione (GSSG) is accumulated inside the cells, and the ratio of GSH/GSSG is a good measure of oxidative stress of an organism [204].

The main protective roles of glutathione against oxidative stress are that it can act as a cofactor for several detoxifying enzymes, participate in amino acid transport across plasma membrane, scavenge hydroxyl radical and singlet oxygen directly (an overall spin allowed process), and regenerate vitamins C and E back to their active forms [205].

Another thiol antioxidant is the TRX system; this is a group of small proteins with oxidoreductase activity (RSH/RSSR) and are ubiquitous in both mammalian and prokaryotic cells. It also contains a disulfide and possesses two redox-active cysteins within a conserved active site (Cys-Gly-Pro-Cys). TRX contains two adjacent −SH groups in its reduced form that are converted to a disulfide unit in oxidized TRX when it undergoes redox reactions with multiple proteins [206].

TRX levels are much lower than those of GSH; however, TRX and GSH may have overlapping as well as compartmentalized functions in the activation and regulation of transcription factors.

The third important thiol antioxidant is the natural compound α-Lipoic acid (ALA), which is a disulfide derivative of octanoic acid and is sometimes referred to as thiothic acid. It is both water and fat-soluble, and therefore, is widely distributed in both cellular membranes and the cytosol of eukaryotic and prokaryotic cells. ALA is readily absorbed from the diet and is converted rapidly to its reduced form, dihydrolipoic acid (DHLA). Both ALA and DHLA are powerful antioxidants and they exert their effects by scavenging free radicals, metal ion chelation and antioxidant recycling, and repairing protein damage due to oxidative stress either in the cytosol or hydrophobic domains. DHLA is a stronger antioxidant than lipoic acid and can act synergistically with other antioxidants such as glutathione, ascorbate, and tocopherol. However, it can also exert prooxidant properties both by its iron-reducing ability and by its ability to generate sulfur-containing radicals that can damage proteins [207,208].

2.14.2 Vitamins E (α-tocopherol) and C (ascorbic acid)

Vitamin E is a fat-soluble vitamin existing in eight different forms. In humans, α-tocopherol is the most active form, and is the major powerful membrane bound antioxidant employed by the cell. The main function of vitamin E is to protect against lipid peroxidation (LPO), and there is also evidence to suggest that α-tocopherol and ascorbic acid function together in a cyclic-type of process. During the antioxidant reaction, α-tocopherol is converted to an α-tocopherol radical by the donation of a labile hydrogen to a lipid or lipid peroxyl radical, and the α-tocopherol radical can therefore be reduced to the original α-tocopherol form by ascorbic acid [209–212]. Long-term UV exposure is known to induce cell damage which may cause skin carcinogenesis. Vitamin E a natural biological antioxidant, has been shown to protect against UV light-induced cell damage, to treat skin diseases and several types of cancer or to decrease oxidative stress. For this reason, vitamin E–based formulations, may be employed as protective agents. To obtain an effective photoprotection, a significant amount of antioxidant needs to arrive at the site, but skin barrier properties and vitamin E poor stability under direct exposure to UV light limit its use [175]. Vitamin E has also been shown to inhibit low density lipoproteins (LDL) oxidation in vitro, and also increases LDL oxidative resistance, decreases agonist-induced platelet aggregation, and preserves agonist-induced vasodilation ex vivo [213,214].

Long-term supplementation with vitamin E in hypercholesterolemic patients and/or chronic smokers has shown to increase levels of autoantibodies against oxidized LDL and it has recently been shown to prevent ischemic heart disease. However, evidence from other clinical trials is controversial and confusing since some studies have failed to show a link between dietary supplementation of α-tocopherol and LDL resistance to oxidation. This may be due to the fact that vitamin E not only acts as an antioxidant but can also interact with enzymes and modulates genes involved in atherosclerosis [215,216].

Vitamin C is an important and powerful water-soluble antioxidant and thus works in aqueous environments of the body [217]. Its primary antioxidant partners are vitamin E and the carotenoids and further it works along with the antioxidant enzymes. Vitamin C cooperates with vitamin E to regenerate α-tocopherol from α-tocopherol radicals in membranes and lipoproteins, and also raises intracellular glutathione levels thus playing an important role in protein thiol group protection against oxidation [191,218]. Vitamin C supplementation in healthy humans has shown recently that its intake results in significant reduction of oxidative stress and inflammation as shown by a reduction in the concentration of F2-isoprostanes, prostaglandin E2 (PGE2), and monocyte chemotactic protein-1 [219].

2.14.3 Carothenoids

These are mainly colored pigments present in plants and microorganisms and epidemiological studies have revealed that an increased consumption of a diet rich in carotenoids is correlated with a lower risk of age-related diseases. Carotenoids contain conjugated double bonds and their antioxidant activity arises due to the ability of these to delocalize unpaired electrons. An example of carotenoid is β-carothene (Scheme 2.24).

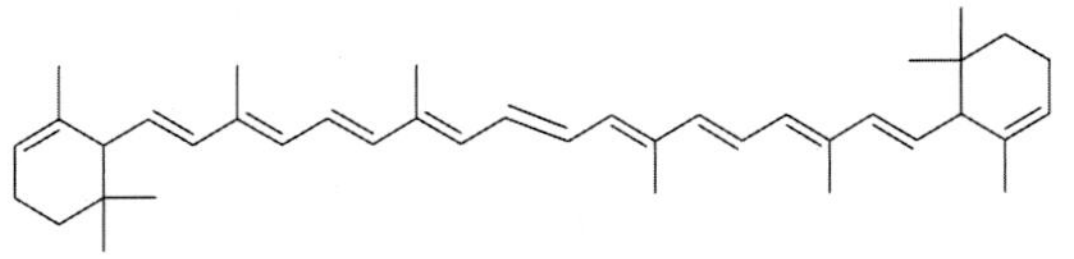

SCHEME 2.24 β-carothene.

The conjugated structure is also responsible for the ability of carotenoids to physically quench singlet oxygen without degradation and for the chemical reactivity of carotenoids with free radicals. The efficacy of carotenoids for physical quenching is related to the number of conjugated double bonds present in the molecule, which determines their lowest triplet energy level. They can also scavenge peroxy radicals thus preventing damage in lipophilic compartments, however, the carotenoid β-carotene can also act as a prooxidant causing an increase in LPO. The concentrations of carotenoids and the partial pressure of oxygen are also important factors in their effectiveness as antioxidants. Carotenoids, in particular β-carotene, exhibit antioxidant properties at low oxygen partial pressure but become prooxidants at high pressures of oxygen and similarly, at high carotenoid concentrations, prooxidant behavior is displayed [191].

2.14.4 Flavonoids

These are a broad class of low molecular ubiquitous groups of plant metabolites and are an integral part of the human diet. Flavonoids are benzo-γ-pyrone derivatives consisting of phenolic and pyrane rings and during metabolism hydroxyl groups are added, methylated, sulfated, or glucuronidated. There is intense interest in flavonoids due to their antioxidant and chelating properties and their possible role in the prevention of chronic and age-related disease [220].

Flavonoids are present in food mainly as glycosides and polymers and these comprise a substantial fraction of dietary flavonoids. The biological properties of flavonoids are determined by the extent, nature, and position of the substituents and the number of hydroxyl groups. These factors also determine whether a flavonoid will act as an antioxidant or as a modulator of enzyme activity, or whether it possesses antimutagenic or cytotoxic properties. The most reported activity of flavonoids is their protection against oxidative stress. Thus flavonoids can scavenge peroxyl radicals, and are effective inhibitors of LPO, and can chelate redox-active metals, and thus prevent catalytic breakdown of hydrogen peroxide (Fenton chemistry). However, under certain conditions, flavonoids can also display prooxidant activity and this is thought to be directly proportional to the total number of hydroxyl groups, which have been further reported to modulate cell signaling [191].

A protective role flavonoids in the diet of humans has been indicated in some large prospective studies. In vitro inhibition of LDL oxidation by flavonoids has also been demonstrated while total antioxidant capacity is increased and LDL oxidizability is reduced after consumption of several natural products that are rich in flavonoids. A high flavonoid intake is also associated with a lower mortality rate from coronary heart disease and lower incidence of myocardial infarction in older men, and a reduced risk of coronary heart disease in postmenopausal women has been observed [221–226]. The Zutphen Elderly study (a systematic survey of health and diet on a cohort of elderly people) also demonstrated an inverse relationship between consumption of catechin, a predominant flavonoid in tea and ischemic heart disease mortality in a cohort of 806 men [222]. In support of this, black tea consumption has shown a decrease in markers of oxidative stress and inflammation in patients with coronary artery disease. Carotenoids have shown to increase LDL oxidative resistance in ex vivo studies, however, in another study involving elderly healthy subjects supplementation with a carotene mixture or lycopene had no effect on oxidative modification of LDL in vitro, despite significant increase in plasma and LDL concentrations of lycopene, α-carotene, and β-carotene. In contrast, in another clinical trial a significant decrease in serum LDL cholesterol was observed which was in parallel with an increase in serum lycopene. In addition, in patients with diabetes mellitus increased susceptibility to LDL oxidation was normalized by natural β-carotene or lycopene dietary supplements [223–225].

The relationship between antioxidants and gene expression has also been investigated and evidence is emerging that antioxidants may prevent CVD influencing gene expression directly or via gene promoters, via control of regulatory signals, and via posttranscriptional pathways [227,228]. The role of phytochemicals in the inhibition of cancer and inflammation has also been extensively studied and it is now clear that these exert their action by modulating phase I and phase II enzymes and by modulating the cell-signaling pathways involved in inflammation. In future the role of proteomics and nutrigenomics will be important in determining the diet-gene relationship. Since free radicals are implicated in the pathogenesis of neurodegenerative diseases, the role of antioxidants in their prevention has been gaining popularity. It has been reported that the concentration of antioxidant varies within the different regions of the brain and some enzymatic antioxidants, such as catalase, are found in lower concentrations in the brain when compared to other tissues [191].

2.15 Antioxidants action against age-related diseases

The human body has a host of mechanisms such as the DNA-repair systems to deal with free radical induced damage and depending

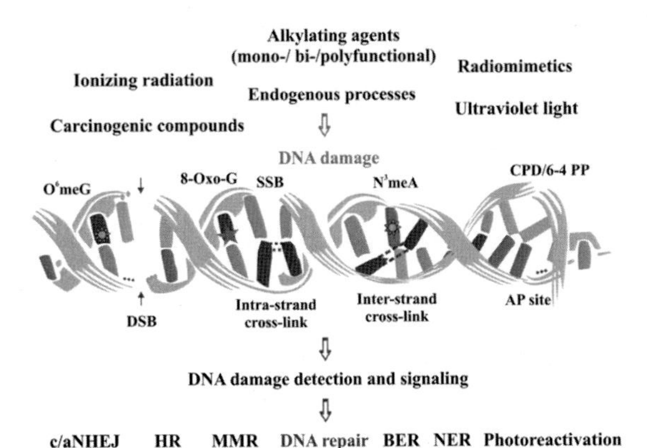

FIGURE 2.21 **Schematic representation of the major DNA lesions induced by various external and endogenous factors, and the types of DNA repair mechanisms employed to remove them from the eukaryotic genome** [193].

on the circumstances, environmental and genetic factors can either increase or decrease the incidence of diseases associated with old age (Fig. 2.21).

Epidemiological studies have demonstrated that diet plays a crucial role in the prevention of age-related chronic diseases especially if combined with regular physical activity and abstaining from smoking. Free radicals and oxidative stress are recognized as important factors in the biology of aging and of many age-related diseases [229]. One mechanism to slow down the aging process and the decline in the vital body functions is to modulate oxidative stress by calorie restriction, however, this is difficult to achieve. Hence dietary components with antioxidant activity have received particular attention because of their potential role in modulating oxidative stress associated with aging and chronic conditions. Several studies have indicated potential roles for dietary antioxidants in the reduction of age-related diseases. This is supported by the fact that in elderly subjects a higher daily intake of fruits and vegetables is associated with an improved antioxidant status compared to subjects consuming diets poor in fruits and vegetables. Therefore the use of antioxidants by this group may lower the prevalence of diseases associated with old age; evidence supporting this is briefly outlined below [191].

Mediterranean diet, which is rich in fruits and vegetables, has been shown to reduce the incidence of CVD. Evidence is now emerging that some dietary antioxidants besides displaying traditional antioxidant potential can influence signaling pathways and gene expression relevant in atherosclerosis by mechanisms other than antioxidative ones. Vitamin C has been shown to inhibit LDL oxidation in vitro primarily by scavenging free radicals and another ROS, thereby preventing them from interacting with LDL [189]. The observational data in humans suggest that vitamin C ingestion is associated with reduced CVD; however, the results of randomized controlled trials have been mainly disappointing. A randomized double-blind crossover trial has shown a positive correlation of plasma vitamin C with LDL resistance to oxidation. In contrast to this, in another recent study no correlation between vitamin C and LDL resistance to oxidation has been reported [230].

The effect of antioxidants has also been investigated on the vascular endothelium since it plays a key role in the regulation of vascular tone and its dysfunction correlates with CVD. Garlic, which is high in antioxidants, inhibits the ability of platelets to aggregate, increases antioxidants levels and also inhibits LDL oxidation. It also increases intracellular glutathione (GSH) levels in vascular endothelial cells by modulation of the GSH redox cycle specifically increasing GSSG activity and SOD activity [231]. The role of antioxidants in preventing platelet aggregation is still a matter of controversy and contradictory results have been obtained with the antioxidant vitamins C and E [231]. Although the intake of dietary flavonoids is inversely correlated with the risk of mortality from coronary artery disease. Its role in platelet function has provided contrasting evidence. For example, cocoa supplementation in healthy subjects significantly increased flavonoids levels and decreased platelet aggregation, in support of this in another study chocolate consumption also decreased platelet aggregation [232,233]. In contrast, some studies have shown no effect of flavonoids on platelet aggregation despite an increase in its plasma concentration [234]. The effectiveness of carotenoids on platelet function is scarcely by data [188]. A variety of antioxidants have been investigated for the reduction of oxidative stress associated with Alzheimer's disease (AD). It has also been reported that the concentrations of vitamins A, C, E, and β-carotene in plasma, serum, or cerebrospinal fluid are lower in AD patients than in controls and supplementation with these vitamins is useful in the prevention of AD [212,235–237]. Vitamin E is also reported to slow the rate of motor dysfunction in Huntington's disease (HD) [238,239]. In contrast, no effects of these antioxidants on AD have also been reported. It has been suggested that flavonoids may have neuroprotective effects both in vitro and in vivo possibly by their abilities to scavenge ROS. This is supported by the fact that polyphenols found in blueberry have been shown to reverse age-related declines in neuronal signal transduction as well as cognitive and motor deficits and increase hippocampal plasticity. In addition, Concord grape juice reverses the course of neuronal and behavioral aging possibly through a multiplicity of direct and indirect effects that can affect a variety of neuronal parameters and curcumin, a powerful antioxidant from the curry spice turmeric reduces oxidative damage and amyloid pathology associated with AD. Garlic is also reported to protect against age-related maculopathy and cataract formation in the elderly. The role of garlic in preventing cerebral aging and dementia is also supported by other studies, which indicate that phytochemicals displaying antioxidant properties can improve neurological dysfunctions [231,240].

Ginkgo extract has also been investigated in the prevention of neurodegenerative diseases and has a beneficial effect in the treatment of AD patients. In contrast no efficacy of Gingko extract on AD subjects has also been noted. Hence larger studies are needed to clarify the therapeutics effects of Gingko extract in AD subjects [241].

Melatonin is a potent free radical scavenger (Scheme 2.25). Its levels decline with age, and patients with neurodegenerative diseases have significant reductions of this substance. It also displays neuroprotective and antioxidant properties against amyloid β-protein–mediated oxidative damage, and displays immunomodulatory properties, and thus can play a role in healthy aging [242].

An early biochemical change in Parkinson's disease (PD) patients is a reduction in total glutathione levels. Infusion of GSH in PD patients has been demonstrated to improve the symptoms but the therapeutic effects only lasted between 2 and 4 months after GSH treatment was stopped. However, the role of induction of endogenous antioxidants in the prevention of neurodegenerative diseases needs further investigation [243].

There is evidence to indicate that ROS are involved in cancer initiation and promotion and malondialdehyde (MDA) concentration is increased in patients with neoplasms [191]. Consumption of potent dietary antioxidants can lower the effects of oxidative DNA damage in the aged besides lowering the overall risk of cancer. A recent study has indicated

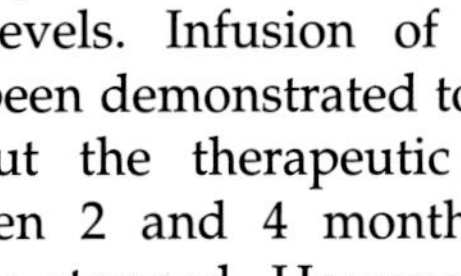

SCHEME 2.25 Melatonin.

that a combination of antioxidants is a powerful adjunctive preventive treatment for cancer since the total activity of antioxidant enzymes such as SOD and catalase is reduced in certain types of cancers [244,245]. Many studies have shown that vitamin C protects against cell death triggered by various stimuli and this protection is associated with its antioxidant property. Vitamin C supplementation studies have shown a reduction in markers of oxidative DNA, lipid, and protein damage, and in support, vitamin C has been shown to regulate factors that can influence gene expression, apoptosis, and other cellular functions [246]. Intervention with vitamin E supplementation has shown a reduction in the risk of colorectal adenomas and prostate cancer. However, controversy surrounds the effectiveness of vitamins in reducing cancer and negative effects of vitamins C and E have also been reported. A study has revealed that vitamin E at doses of 400 IU or more can actually increase the risk of death [211].

However, no risk was reported when vitamin E was used at 200 IU or less. There is also an association between cancer incidence and various disorders of GSH-related enzyme functions especially the alterations of glutathione S-transferases.

Carotenoids also display antiproliferative properties when tested in various cancer cell lines. Increased intake of lycopene has been reported to attenuate alcohol-induced apoptosis in 2E1 cells, and reduces the risk of prostate, lung, and digestive cancers. This cancer preventive property of lycopene is associated with its antioxidant property and its ability to induce and stimulate intercellular communication via gap junctions which are known to play a role in the regulation of cell growth, differentiation, and apoptosis [247].

In this context the redox state of the cell is also important, as there is evidence to show that redox balance is impaired in cancer cells compared with normal cells, which may be related to oncogenic stimulation. Antioxidants may prevent cancer by inducing phase II detoxifying enzymes and activating transcription factors and endogenous antioxidant enzymes such as glutathione peroxidase and catalase. It is known that altered levels of antioxidant enzymes and nonenzymatic antioxidants as well as changes in the related signal pathways are evident in many human cancers. Evidence is also emerging that flavonoids such as garlic, green tea, silibinin, and curcumin have cancer preventive properties. Many of these dietary compounds appear to act on multiple target signaling pathways, which include downregulation of COX-2 and downregulation of the transcription activators, AP-1 and NF-κB known to be extremely important in tumor promoter-induced cell transformation and tumor promotion, and both are influenced differentially by the MAPK pathways. Further support for the role of flavonoids in preventing cancer has come from grape seed proanthocyanidins, which have been reported to inhibit UVR-induced oxidative stress and activation of MAPK and NF-κB signaling in human epidermal KCs. Dietary antioxidants may also prevent cancer by potentially suppressing angiogenesis by inhibiting IL-8 production and the cell junction molecule VE-cadherin. These studies concur with the epidemiologic, clinical, and animals' studies that consumption of antioxidants is associated with a reduced risk of cancer among the elderly [234,248–250]. Oxidative stress is increased with aging and is a contributing factor for the initiation and progression of complications in diabetes mellitus such as lens cataracts, nephropathy, and neuropathy. The use of antioxidants in preventing and treating diabetes has been investigated over the last decade. Dietary supplementation with a combination of antioxidants and vitamins C and E has been reported to cause a reduction in oxidative stress markers in patients with type 2 diabetes. In contrast, clinical trials involving vitamin E supplementation on diabetic complications have shown conflicting data. Vitamin E has a greater effect in protecting LDL oxidation in type 2 diabetics who are at a greater risk of CVD. Similarly, conflicting data have also been obtained with vitamin C supplementation and decreased fasting plasma insulin levels and improved insulin action have been reported, whilst another study has reported no effect. Further conflicting data have been observed in diabetic postmenopausal women who display a higher level of oxidative stress and supplementation with high vitamin C in this group has been reported to have an increased risk of mortality from CVD. However, the invasive potential of cancer cells is also increased in the presence of abnormally high levels of Mn-SOD. Flavonoids also have a role to play in the treatment of diabetes as these have shown to protect against hyperglycemic and alloxan-induced oxidative stress in experimental animal models. In support, in clinical trials flavonoids have shown to offer protection against type 2 diabetes in a large cohort of women. Melatonin has been shown to reduce diabetic nephropathy and neuropathy in experimental animal models but more work is required to assess its efficacy in humans [191,225,251].

Finally there is evidence to support the use of lipoic acid in treating type 2 diabetes, since it displays strong antioxidant properties and increases glucose uptake through recruitment of the glucose transporter-4 to plasma membranes, a mechanism that is shared with insulin-stimulated glucose uptake. Lipoic acid is also reported to improve neural blood flow, endoneural glucose uptake, metabolism, and nerve conduction. It probably exerts its effect in diabetic patients by reducing lipid accumulation in adipose and nonadipose tissue, by increasing glucose uptake and by activating pyruvate dehydrogenase complex, which is known to play a major role in the oxidation of glucose-derived pyruvate. It is clear that more human clinical trials are required in order to establish the role of antioxidants in the prevention and treatment of diabetes [223,225,242,252].

Various essential metals play an important role in controlling oxidative reactions in biological tissues by activating the enzymes. For example, copper (Cu) is an essential cofactor in a number of critical enzymes including cytochrome C oxidase and copper, zinc-SOD (CuZn-SOD). Although unregulated Cu is also a well-known prooxidant, it can exert its antioxidative effects through the action of transporter proteins such as metallothionein and ceruplasmin. A Cu deficiency–induced decrease in the activity of CuZn-SOD in humans and animals has been reported. Cu deficiency also decreases the activity of ceruloplasmin, which requires Cu for its ferroxidase function, and it can also lead to a reduction in enzymes of the oxidant defense system such as Se-dependent glutathione peroxidase (Se-GPX) and catalase. Furthermore, a deficiency in Cu can also alter other ROS scavengers including metallothionein (a Cu and Zn containing protein) and the nonprotein thiol, glutathione. Copper and zinc are also essential cofactors for enzymes involved in the synthesis of various bone matrix constituents and could be important in the elderly since they may play an important role in reducing bone loss in osteoporosis [253]. Iron (Fe) is an essential constituent of catalase enzymes, hemoglobin, and myoglobin, but is also a prooxidant (via Fenton reactions) when it is present in excess. In the presence of lipids iron creates oxidative stress and it has been suggested that subjects with high levels of lipids and serum iron are at an increased risk of cancer. Thus iron chelators such as albumin, haptoglobin, lactoferrin transferring and urate also have an important role to play in preventing oxidative stress-related diseases.

Se is another important cofactor and epidemiological findings that have linked a lowered Se status to neurodegenerative and CVDs as well as to an increased risk of cancer. There is an association between Se reduction and DNA damage, oxidative stress and some evidence that Se may affect not only cancer risk but also progression and metastasis. Se intervention in subjects with a lower Se status has shown some benefits in reducing the incidence and mortality in all cancers but more specifically in liver, prostate, colon-rectal, and lung cancers. Its protective effects appear to be associated with its presence in the multiform of glutathione peroxidases, which are known to protect DNA and other cells that maybe damage from oxidative stress.

The element manganese (Mn) is another cofactor involved in antioxidant defense mechanisms and is a vital component of Mn-SOD enzyme, which plays a crucial role in protecting mitochondria from free radical attack. Zinc (Zn) another component of SOD is also involved in antioxidant defense systems and protects the vascular and immunological systems from the damaging effects of free radical species, and it is also a key constituent or cofactor of over 300 mammalian proteins which may have a role in the prevention of initiation and progression of cancer. Evidence supports the fact that Zn deficiency can impair the host protective mechanisms designed to protect against DNA damage, thus increasing the risk of cancer, and it also plays an important role as an antioxidant and/or as a cofactor in keeping the skin healthy, thus it can play an important role in healthy aging [191].

Coenzyme Q10 (CoQ10) (ubiquinone) is a fat-soluble quinone that transfers electrons from complexes I and II to complex III in mitochondrial respiratory chain, this process being coupled to ATP production. In its reduced form, CoQ10 also inhibits LPO and can protect mitochondrial inner-membrane proteins and DNA from oxidative damage, and is the most widely used cofactor supplement in the treatment of mitochondrial disorders [254]. CoQ10 is commonly used for treatment of cardiomyopathy, and neurological disorders such as fibromyalgia, CVD, neurodegenerative diseases, cancer, diabetes mellitus, male infertility, and periodontal disease, and can thus prevent age-related mitochondrial dysfunction. Most promising results have been obtained in the treatment of neurological disorders while its use in the treatment of CVD and diabetes has produced contradictory data [255].

Likewise, it appears to be some relation between PD and malignant melanoma, although no precise etiology has been established [256]. Riboflavin (vitamin B_2) formula is another cofactor, which is converted to flavin dinucleotide, which serves as a coenzyme for glutathione reductase and other enzymes. Low intakes of riboflavin have been associated with different diseases including cancer and CVDs and there is some evidence that treatment with riboflavin can provide some benefit against diseases associated with oxidative stress [257]. In Scheme 2.26 are shown the chemical structure of some vitamins.

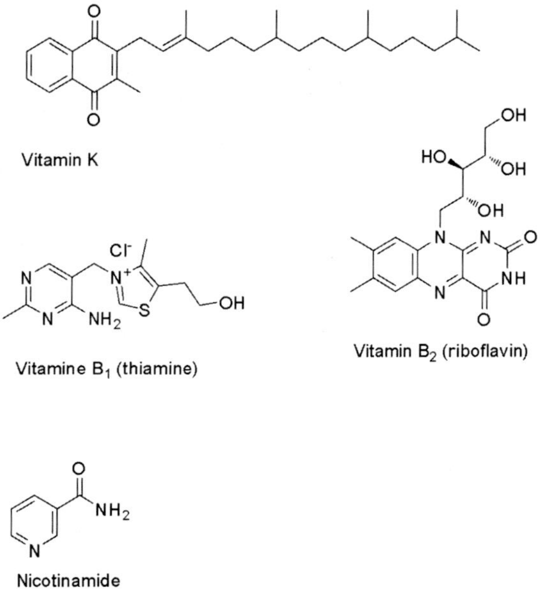

Vitamin K

Vitamine B_1 (thiamine)

Vitamin B_2 (riboflavin)

Nicotinamide

SCHEME 2.26 Vitamins K, B_1, B_2, and PP.

Thiamine (vitamin B_1) is another cofactor that has been investigated for its role in the treatment of oxidative stress–related diseases. Thiamine diphosphate is the active form of thiamine and it serves as a cofactor for several enzymes, which are important in the biosynthesis of reducing equivalents used in oxidant stress defenses. Thiamine deficiency has been linked to the promotion of neurodegeneration and an increase in oxidative stress; however, its effectiveness in treating diseases associated with free radicals is still unclear [258,259]. Nicotinamide, is a precursor for both nicotinamide adenine dinucleotide (NAD/NADH), and nicotinamide adenine dinucleotide phosphate (NADP). It plays an important role in energy metabolism, signal transduction, cellular injury, aging, and shows significant inhibition of oxidative damage induced by ROS [191].

Carnitine, which transfers long-chain fatty acids across the mitochondrial membrane, has also been investigated for its property to scavenge free radicals. It has been shown to protect mitochondrial membrane damage during the aging process in an experimental model, and to display antioxidant properties in the prevention of acetic acid–induced colitis. It has also been reported to inhibit hepatocarcinogenesis via improvement of mitochondrial dysfunction in an experimental model, and it has been reported to improve fatigue symptoms in cancer patients. A complication is cancer cachexia [260]. The biochemical interaction between free radicals, antioxidants, and cofactors needs to be considered further. Results from long-term trials are needed to evaluate the safety and beneficial role of these cofactors in the prevention and treatment of diseases associated with free radicals. There is now universal agreement that free radicals are involved in the physical, biochemical, and pathological changes associated with aging. Oxidative damage to proteins, lipids, and DNA accumulates and increases with age. This is associated with age-related diseases such as CVD, neurodegenerative diseases, cancer, and diabetes. The human body deals with the pathological effects of ROS by utilizing the endogenous antioxidant system (e.g., enzymes such as SOD), and by the ingestion of exogenous antioxidants in the diet (e.g., flavonoids). The presence of cofactors is also important for the antioxidants to exert optimum effects. If the oxidative stress exceeds the protection afforded by antioxidants the aging process and some of the diseases associated with it can accelerate.

This change in the world population has been accompanied by rise in living standards leading to lifestyle and behavior changes that are having an adverse impact on population health. This increase in older people is likely to place greater financial burden on the health services and high social costs for individuals and society if not managed properly.

As indicated above, mitochondria are the main site of oxidation in cells. These organelles are surrounded by two membranes the main components of which are phosphatidylcholine and phosphatidylethanolamine (PE) with a well-defined lipid composition; most of these lipids are synthesized in the endoplasmic reticulum, although de novo lipid synthesis and remodeling of mitochondrial lipids are important for maintaining the structural integrity and function of mitochondria. Cardiolipins are mainly located in the inner membrane. Radical formed by light can cause oxidations finally giving 4-hydroxynonenal (4-HNE), a highly reactive unsaturated aldehyde that forms a variety of polymeric adducts that in turn have an important role in the development of many cancer forms (Scheme 2.27). This is because of the strong electrophilic character that leads to the formation of covalent adducts with nucleophilic functional groups in macromolecules such as proteins (particularly through the SH group), DNA, and lipids or lipid-containing substances.

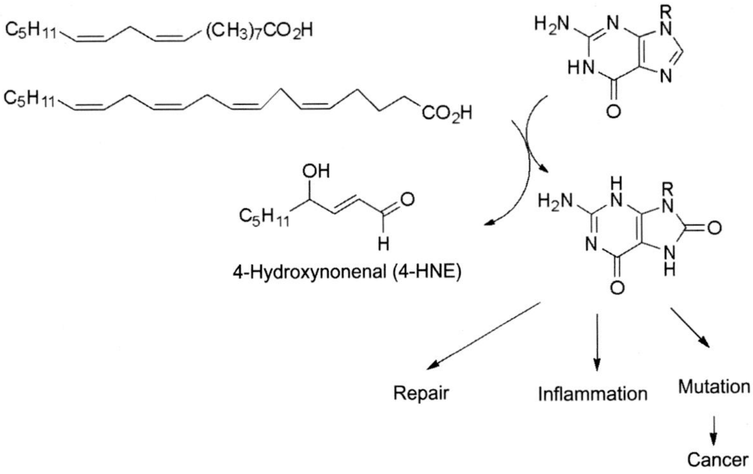

SCHEME 2.27 A schematic view of the formation of 4-hydroxynonenal in cancer cell.

In Scheme 2.27 it has been shown how UV induces inflammation going through C- and O- centered radicals, and aldehydes such as hydroxynonenal like proceeds analogously. This conjugated aldehyde is produced through ROS-induced LPO (at the highly reactive *bis*-allyl position) of mitochondrial and plasma membranes and its adducts with biological macromolecules are involved in cancer initiation, progression, metabolic reprograming, and cell death (Scheme 2.28).

Biological consequences of 4-HNE addition are reducing membrane integrity, affecting protein function in cytosol, causing nuclear and mitochondrial DNA damage, inhibiting electron transfer chain activity, activating mitochondrial uncoupling proteins activity, reducing tricarboxylic acid cycle activity, and inhibiting aldodehydrogenase 2 (an aldodehydrogenase, that is, found only in this location and catalyzes the oxidation to 4-hydroxynonenoic acid) [261].

Healthy aging involves the interaction between genes, the environment, and life styles, and in order for the elderly to live independently and relatively disease and disability free requires that healthy life styles are promoted throughout life. The most modifiable lifestyle factors are physical activity and diet, and the elderly population should be encouraged to take up physical activity since it has a positive effect on decreasing the risk of many diseases associated with old age. The elderly should also be encouraged to consume a diet rich in antioxidants as there is evidence that such a diet especially in combination with a healthy life style can lower the rate of all-causes and cause-specific mortality by greater than 50% in the 70–90 years old. Although some of the evidence that certain dietary antioxidants and some cofactors can reduce free radical mediated damage and promote healthy aging is controversial, the elderly should be encouraged to take exogenous antioxidants and cofactors, which have shown efficacy in scientific studies. However, more controlled studies are needed in order to investigate the efficacy and safety of antioxidants and cofactors, and their mode of action especially in the elderly in order to compensate the low absorption [262].

A variety of defense mechanisms have been evolved by cells, in particular in mithocondria, to transform the toxic aldehyde into less reactive molecules or to limit the toxicity of 4-HNE. 4-HNE macromolecule adducts in

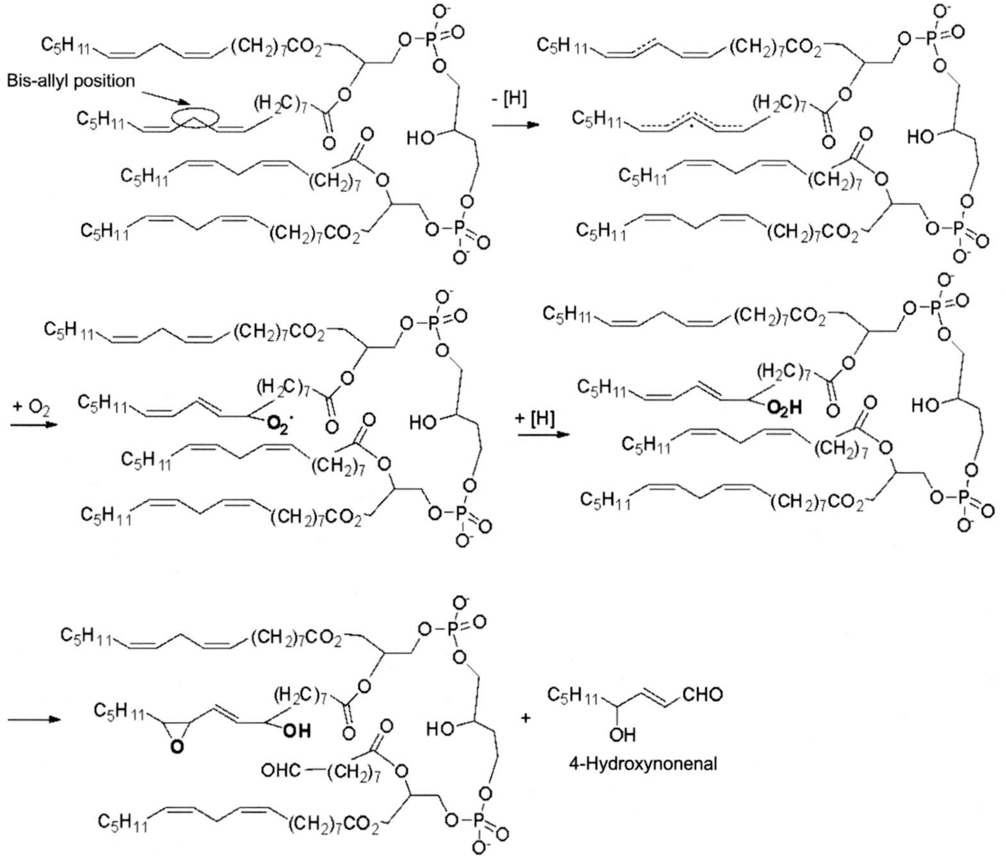

SCHEME 2.28 Reactive oxygen species interaction with lipids mainly involves hydrogen abstraction from the activated bis-allylic position. The thus formed pentadienyl radical adds oxygen to form an hydroperoxide. The arrangement of this intermediate through intrachain oxygen transfer leads to 4-hydroxynonenal. (Participating atoms are shown in bold character).

mitochondria are involved in cancer initiation and progression by modulating mitochondrial function and metabolic reprograming. 4-HNE protein adducts have been widely studied but the DNA modification by lipid electrophiles has yet to emerge. The biological consequence of PE modification remains to be defined, especially in the context of cancer, and also manipulation of mitochondrial ROS generation, LPO, and production of lipid electrophilic C–centered radicals have to be considered [261].

Vitamin E is the major lipid-soluble antioxidant and plays a vital role in protecting membranes from oxidative damage through a nonenzymatic process. Its primary activity is to trap peroxy radicals in cellular membranes. Actually this term refers to a group of substances that include both tocopherols and tocotrienols. The highly unsatured γ-tocopherol, present in corn and soybean oil and in margarine and dressing, is the most common form of the vitamin, at least in the North American diet. The most biologically active form of vitamin E, and the second-most common through a nonenzymatic process of vitamin E in the diet, is α-tocopherol, which is most abundant

$$O_2^{\bullet-} + O_2^{\bullet-} \xrightarrow{SOD} O_2 + H_2O_2$$

SCHEME 2.29 Enzymatic (superoxide dismutase, SOD) dismutation of superoxide anion into water and hydrogen peroxide.

Uric acid

Bilirubin

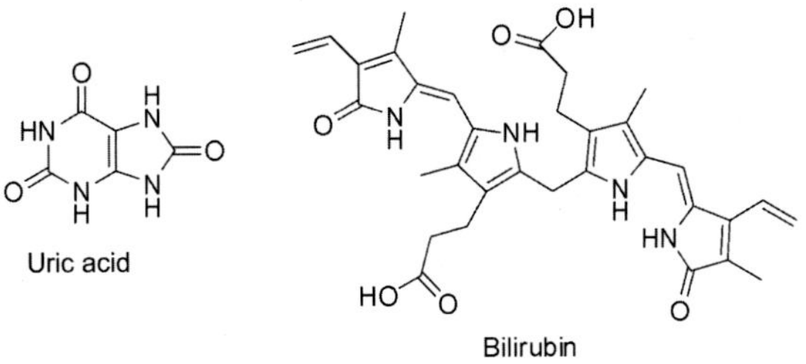

SCHEME 2.30 Uric acid and the tetrapyrrole pigment, bilirubin.

in wheat germ oil, sunflower, and safflower oils. As a fat-soluble antioxidant, this compound interrupts the propagation of ROS that diffuse through biological membranes or through a fat, while the lipids it contains are oxidized by reacting with more-reactive lipid radicals and form more stable products. Attention should be given to the fact that to high an amount of tocopherols in the diet (>1 g per day) may cause hyper-vitaminosis E, with an associated risk of vitamin K deficiency and the related bleeding problems. The activity of vitamin E is, as mentioned, nonenzymatic and thus the vitamin is consumed and needs to be reduced back. This is accomplished by SOD enzymes that catalyze (acceleration factor c.10,000 times with respect to the noncatalyzed reaction) the conversion of two superoxides into hydrogen peroxide and oxygen (Scheme 2.29) [263,264]. The electron transfer process has been accurately characterized both in aprotic and protic conditions.

SODs are metal-containing enzymes that depend on a bound metal for their antioxidant activity. In mammals, the manganese-containing enzyme is most abundant in mitochondria, while the zinc or copper forms predominant in cytoplasm. SODs are inducible enzymes--that is, exposure of bacteria or vertebrate cells to higher concentrations of oxygen results in rapid increases in the concentration of SODs.

Vitamin C is a water-soluble antioxidant that can reduce radicals from a variety of sources. It also appears to participate in recycling vitamin E radicals. Oxidized vitamin C can be reduced back by glutathione. Compelling evidence for antioxidant protection of lipids by vitamin C in biological fluids, animals, and humans, both with and without iron cosupplementation has been reported, although it is conceivable that metal ions could contribute to oxidative damage through the production of hydroxyl and alkoxyl radicals and thus vitamin C also may function as a prooxidant under certain circumstances [265]. When an overload of free radicals cannot gradually be destroyed, their accumulation in the body generates a phenomenon called oxidative stress. This process plays a major part in the development of chronic and degenerative illness such as cancer, autoimmune disorders, aging, cataract, rheumatoid arthritis, cardiovascular, and neurodegenerative diseases [191].

Glutathione is the most important intracellular defense against damage by ROS. As mentioned above it results from the condensation of three amino acids (glutamyl-cysteinyl-glycine). The cysteine features an exposed free sulfydryl group (SH) that is very reactive, providing an abundant target for radical attack. Reaction with radicals oxidizes glutathione, but the reduced form is regenerated in a redox cycle involving glutathione reductase and the electron acceptor NADPH (Scheme 2.23) [266]. In addition to these diversely active compounds, there are numerous small molecules that function as antioxidants. Examples include bilirubin, uric acid, flavonoids, and carotenoids [267,268] (Scheme 2.30).

2.16 Skin cancer

Carcinogenesis is a complicated, multistage process in which healthy cells are transformed into abnormal cells as a result of a series of mutations and changes in the patterns of gene expression. Histologically, cancer lesion can be distinguished through a series of tests. Factors predisposing to malignancy include inherited traits, environmental agents, diet, and the risk of cancer increases with age. Cancer development can be described by three stages: initiation, promotion, and progression, and ROS can act in all these stages of carcinogenesis. It is also well established that free radicals are known to react with all components of DNA, thus damaging both the bases and the deoxyribose backbone causing mutations in crucial genes, which ultimately may lead to cancer (Fig. 2.22) [269] (Fig. 2.23).

The main photochemical reaction in melanoma cancer, however, involves a "concerted" $\pi^2 + \pi^2$cycloaddition; this occurs only between facing thymine bases and gives the cis cyclobutane dimer, apparently through the singlet manifold [271]. Ab initiomolecular dynamics simulations show that dimerization of thymine to give a cyclobutane is an ultrafast process requiring the population of the singlet doubly excited state, ^{1}DE (Fig. 2.24).

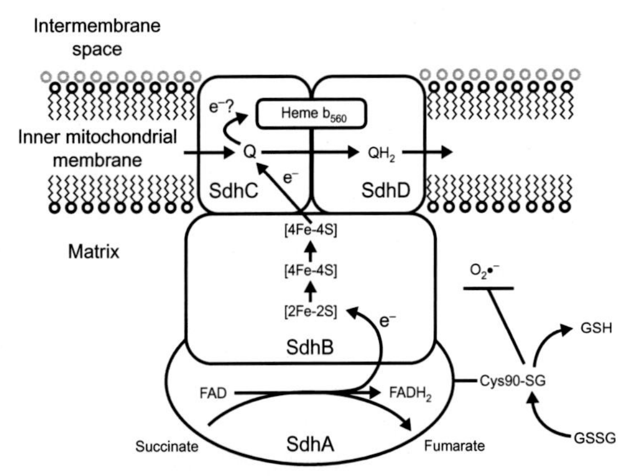

FIGURE 2.22 Modulation of Sdh by S-glutathionylation. Representative diagram illustrating electron flow from succinate to ubiquinone in Sdh with subsequent production of fumarate. The S-glutathionylation of Cys90 on SdhA subunit is required to maintain succinate oxidation activity and diminish O_2• − production. Deregulation of SdhA S-glutathionylation results in diminished activity and increased mitochondrial ROS production which is associated with ischemic damage of cardiac tissue [269].

In fact, time-resolved measurements and calculations have shown that the process is a nice manifestation of the Woodward-Hoffmann rules, that prescribes that π^2s + π^2s cycloaddition cannot occur from the ground state, although the suprafacial-suprafacial addition is possible when the DE state is involved.

The triplet state has no role in the reaction, since it cannot be formed neither by intersystem crossing nor by singlet fission. Electronic effects and conformational control are essential to promote cyclobutane formation. The DE lies at a higher energy than the commonly employed laser wavelengths or the UV spectrum of the sun at the earth's surface, therefore it cannot be populated by direct excitation. This state can be reached by internal conversion from the bright singly excited state through overcoming a small energy barrier. This fact explains the low yield of the reaction.

The other reaction leads to (6-4) adducts that are formed on the millisecond time scale through a four-membered ring intermediate, an oxetane, resulting from the photocycloaddition 2 + 2 of the C = O bond of a thymine and the $C_5 = C_6$ of another one. By absorbing UV light (325 nm, that is, present in significant amount in solar light) the (6-4) adducts are converted into their Dewar isomers (Scheme 2.31). In the crystal structures dT(6-4)T photoproduct adopted a closed circular structure by forming a covalent bond between the C_6 atom of the 5′-base and C_4 atom of the 3′-base. The 5′-side thymine base of the T(6-4)T segment was in a half-chair conformation while the 3′-side pyrimidine base was in a planar conformation. Both base planes were nearly

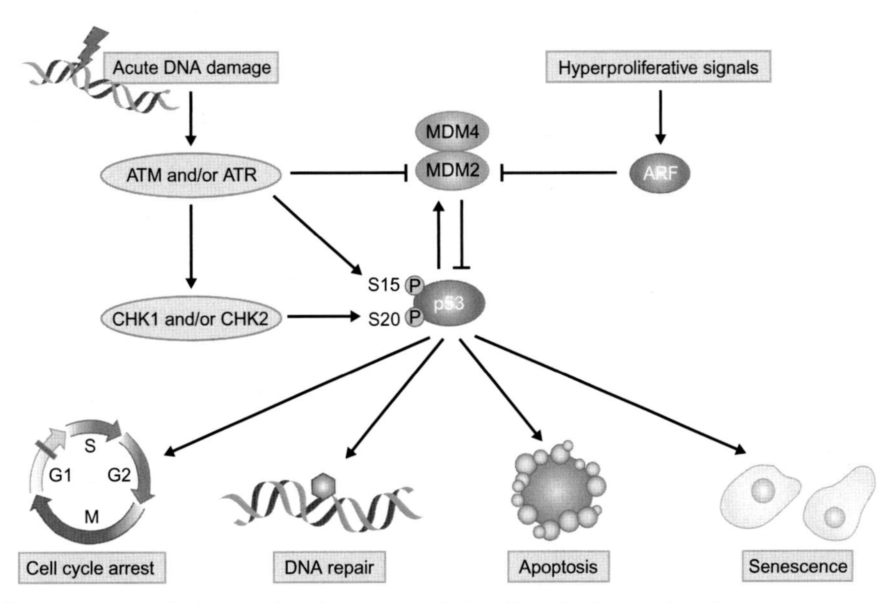

FIGURE 2.23 The most well-elaborated molecular models for p53 activation are those in response to acute DNA damage signals and hyperproliferative signals. p53 induction by acute DNA damage begins when DNA double-strand breaks trigger activation of ataxia–telangiectasia mutated (ATM)—a kinase that phosphorylates the CHK2 kinase—or when stalled or collapsed DNA replication forks recruit ataxia telangiectasia and RAD3-related (ATR), which phosphorylates CHK1. p53 is a substrate for both the ATM and ATR kinases, as well as for CHK1 and CHK2, which coordinately phosphorylate (P) p53 to promote its stabilization. Phosphorylation of p53 occurs at several sites, particularly at the amino-terminus, such as at serines 15 and 20. These phosphorylation events are important for p53 stabilization, as some of the modifications disrupt the interaction between p53 and its negative regulators MDM2 and MDM4. MDM2 and MDM4 bind to the transcriptional activation domains of p53, thereby inhibiting p53 transactivation function, and MDM2 has additional activity as an E3 ubiquitin ligase that causes proteasome-mediated degradation of p53. Phosphorylation also allows the interaction of p53 with transcriptional cofactors, which is ultimately important for activation of target genes and for responses such as cell cycle arrest, DNA repair, apoptosis, and senescence. Hyperproliferative signals similarly activate p53 through perturbation of the MDM2–p53 interaction. These signals can function by liberating the E2F transcription factor, which can stimulate transcription of the ARF tumor suppressor. ARF in turn inhibits MDM2 by antagonizing MDM2 ubiquitin ligase activity, and/or sequestering MDM2 to nucleoli. As a consequence, ARF activation enhances p53 stability and activity, promoting p53 responses such as apoptosis or cellular senescence [270].

perpendicular to each other. In the solution structure of the DNA duplex-decamer, which contained dT(6-4)T/dAA nucleotide pairs, all nucleotides, except for the 3′-side pyrimidine of the dT(6-4)T segment, retained hydrogen bonds with complementary bases, and the DNA duplex flanking the T(6-4)T segment was kinked by 44 degrees [272,273].

The permanent modification of genetic material induced by free radicals represents the first step involved in mutagenesis, carcinogenesis, and aging. In support of this free radical-mediated damage to DNA has been found in various cancer tissues, and there is also a direct link between the size of benign tumors and the amount of DNA oxidized product, 8-hydroxyguanine (8-OH-G) adduct formation; indicating that the level of 8-OH-G may be important in the transformation of benign to malignant tumor [191]. This damage to the DNA can result either in arrest or induction of transcription, induction of signal transduction pathways, replication errors, and genomic instability, all of which are associated

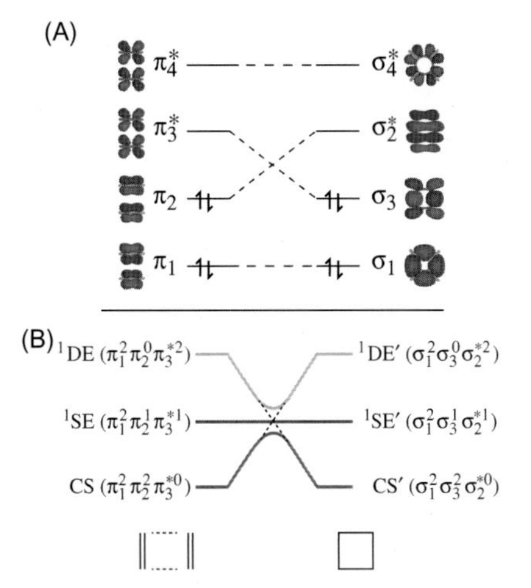

FIGURE 2.24 Woodward–Hoffmann orbital (A) and state (B) correlation diagrams for the [2 + 2] cycloaddition of two ethylene molecules on singlet surfaces. Doubly excited (DE), singly excited (SE), and closed-shell (CS) states are indicated by cyan, blue, and green, respectively [271].

with carcinogenesis. A high level of oxidative stress can induce apoptosis or even necrosis; however, a low level of oxidative stress can stimulate cell division and thus promote tumor growth. ROS probably enhance the final irreversible stage of carcinogenesis, which is characterized by accumulation of additional genetic damage, leading to the transition of the cell from benign to malignant. This includes basal cell carcinoma (BCC), a slowly growing malignant epithelial skin tumor predominantly affecting middle-aged and fair-skinned individuals.

The biochemical and physiological damage induced due to free radical–mediated oxidative stress can be counteracted by antioxidants, as indicated below, and accumulation of the damaged material minimized.

The way by which skin cancer develops is quite varied. Thus squamous cell carcinoma (SCC) and lip cancers develop proportionally

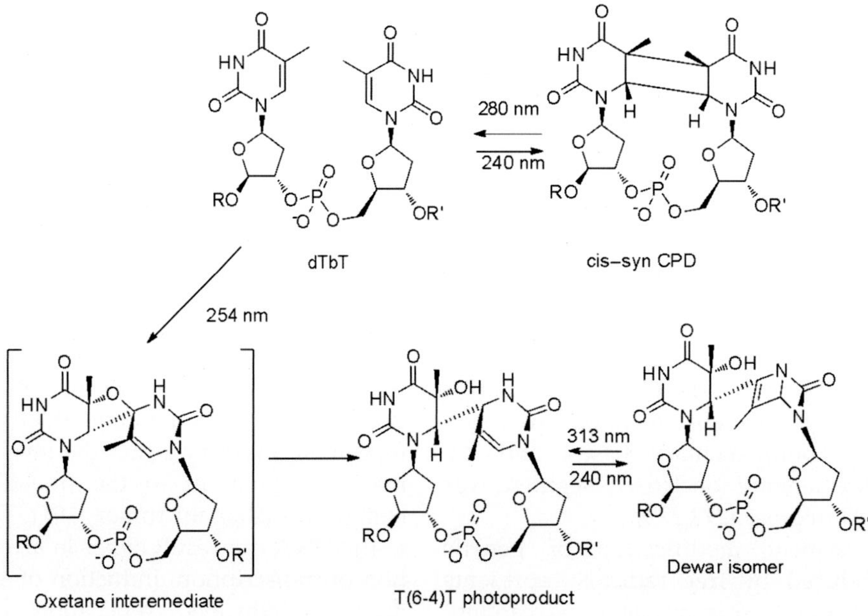

SCHEME 2.31 DNA photoproducts formed by ultraviolet radiation.

to the exposure to UV light, while BCC and malignant melanoma incidence appears to be mostly related to intermittent exposure to solar light during childhood and adolescence, although this statement is certainly oversimplified and is modified by the increased use of artificial UV sources in tanning salons. The mutagenic effect of both UV-B and UV-A components in p53 genes of skin along with the observation of UV-B-induced cytosine to thymine (C→T) transitions and CC→TT mutations at bipyrimidine sites tumors has documented this point [274,275].

It has been also shown that C→T transitions are generated in human skin by UVR within the 310–340 nm range at 5-methylcytosine containing bipyrimidine sites in cytosine phosphate-guanine (CpG)-rich sequences. Histologically, epithelial skin carcinomas are classified as follows. BCC is a slowly growing malignant epithelial skin tumor predominantly affecting middle-aged and fair-skinned individuals. Its incidence is increasing worldwide and also affects younger age groups. BCC can appear clinically as nodular, superficial, morpheiform, or pigmented lesions. The absence of a pigmented network and the presence of at least one positive feature, which includes ulceration (not associated with a recent history of trauma), multiple blue/gray globules, leaf-like areas, large blue/gray ovoid nests, spoke-wheel areas and arborizing (tree-like) telangiectasias, multiple brown to black dots/globules, blue/white veil-like structures and vascular features such as dotted, linear irregular, hairpin, and comma vessels. These features linearly increase with the rate of pigmentation of the lesion, thus presenting diagnostic difficulties for heavily pigmented. Even a pigment network, the negative dermoscopic feature, of which detection should exclude a diagnosis of BCC, can be observed in a minority of cases. Moreover, BCC may display in some rare cases of nonclassic BCC patterns, including short fine superficial telangiectasia, multiple small erosions, concentric structures, and multiple in-focus blue/gray dots (not pepper-like). These new structures could represent early features in BCC, helping to diagnose those lesions lacking classical dermoscopical BCC patterns. Besides arborizing vessels and superficial fine telangectasias, BCCs may display, in a smaller percentage, hairpin vessels, comma vessels, dot vessels, and polymorphous vascular structures. Particularly for this last one, the recognition of other BCC-specific criteria is crucial for a correct diagnostic differentiation with melanoma.

Similarly, to BCC, in SCC the vascular pattern represents the main dermoscopic feature. Conversely, SCC is less rich in dermoscopic structures and a polymorphous vascular pattern is the hallmark of the lesion. It consists of irregularly shaped and distributed hairpin vessels, irregularly distributed dotted, glomerular, and linear vessels ulceration, keratin crust and surface scales may be present. The distribution of hairpin vessels is extremely relevant because this kind of vessel represents the typical vascular structure observed in seborrheic keratosis where, differently from SCC, vessels are regularly distributed throughout the tumor.

Hairpin vessels constitute the main vascular structure also in keratoacanthoma. They are distributed peripherally and surround a mass of keratin localized in the central part of the lesion, ranging in color from yellowish to brown. As for SCC, linear irregular and glomerular vessels may sometimes be observed [276].

Eukaryotic life forms evolved largely because of the ability to extract some 18 times more energy from food sources via oxygen-dependent complete oxidation in mitochondria than is possible from glycolysis. The very process of mitochondrial respiration, however, actually generates more reactive forms of oxygen, including the superoxide anion radical ($O_2{\bullet}^-$), hydrogen peroxide (H_2O_2), and even the highly oxidizing hydroxyl radical (•OH). Thus oxygen represents something of a double-edged sword for aerobic organisms; it is an absolute requirement for life, yet it threatens the very life it supports (about 500,000 errors occur per cell per day). All organic molecules, very much including DNA, are susceptible to oxidative damage from a wide variety of oxygen- or nitroxygen-based reactive species. Cells, organs, and organisms utilize a wide variety of antioxidant compounds (largely from fruits and vegetables) such as vitamins E and C, and a plethora of polyphenols, antioxidant enzymes such as SODs, glutathione peroxidases, peroxiredoxins (Prxs), and glutaredoxins and reductive enzyme cofactors such as glutathione and TRX to attempt to minimize the amount of oxidatively generated damage that occurs to DNA, proteins, and lipids. Nevertheless, substantial damage does occur on a daily basis (about 5000–10,000 per human cell per day) [277]. Tumorigenesis in the skin involves a complex phenomenon that causes alterations of the oncogenic, tumor-suppressive, and cell-cycle control signaling pathways. These pathways include (1) mutated patched tumor suppressor gene (PTCH) (in the mitogenic SHH pathway) and mutated p53 tumor-suppressor gene in BCCs; (2) an activated mitogenic RAS pathway and mutated p53 in SCCs; and (3) an activated RAS pathway, inactive p16, and p53 tumor suppressors in melanoma. It may be stated that all cells have the potential of becoming mutated, but, as long as the tumor suppressing agents are present, this path is blocked. Several tumor suppressing genes are involved in photocarcinogenesis, in particular p53 (SCCs and BCCs), p16 (melanoma), and PTCH (BCCs and possibly SCCs) genes. UVR disturbs the genetic pathways governed by these molecules. In comparison to melanomas, p53 mutations in nonmelanoma skin cancers are much more frequent (~10%–90% and ~1%–20% in nonmelanoma and melanomas, respectively). Furthermore, while mutations in the nonmelanomas skin cancers generally occur as cytosine to thymine transitions at the pyrimidine sites (C:T → T:A and C:C → T:T transition), melanomas usually show cytosine-guanine base pair substitution by thymine-adenine pairs. This, in turn, suggests (1) the absence of intense selective pressure of UVR on the p53 gene in melanomas and (2) although mutations of p53 are likely to be involved in the development of melanomas, they do not play as large of a role as in SCCs and BCCs and p53 mutations are early events in nonmelanomas' skin cancer. The most frequent alterations include point mutations resulting in amino acid substitutions and deletions. These alterations lead to abrogation of p53-dependent pathways involved in cell cycle [278].

The development of novel preventive and therapeutic strategies depends on their understanding of the molecular mechanism behind the well-known clinical phenotype of chronic photodamage of the skin. In that sense it has been shown recently that topically applied retinoids not only are useful in the repair of photoaged skin but also interfere with the UV-initiated induction of matrix-degenerating metalloproteases (MMPs) and thus, if applied prior to exposure might at least help in the prevention of photoaging. Repair enzymes, antioxidants, iron chelators, and inducers of the heat shock response are other promising candidates for the prevention and treatment of the adverse effects of UV on the skin and evaluation of their clinical efficacy is awaited [279].

2.17 DNA photochemistry

As it has been recently shown, the photochemistry of models of DNA and DNA itself can be summarized by Scheme 2.32. The initial photochemical reactions that then develop in skin cancer are cyclobutane and oxetane formation by 2 + 2 cycloaddition of thymine as

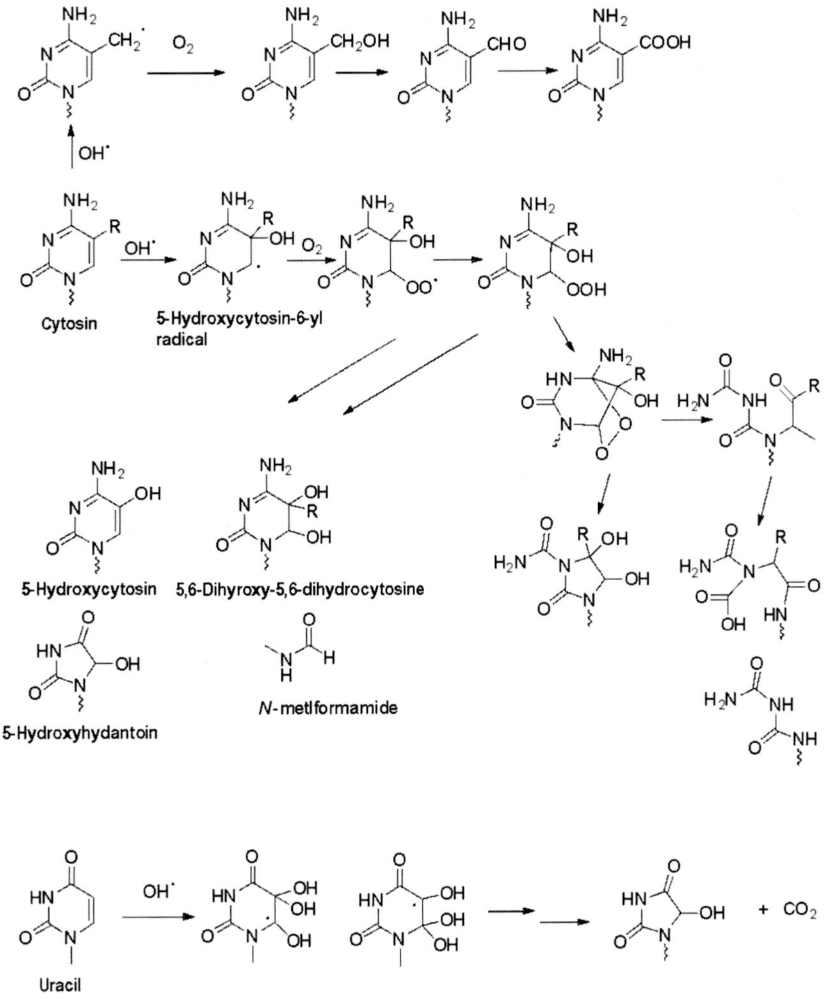

SCHEME 2.32 Radical intermediates and stable final products of cytosine and uracil.

well as formation of glycols by water addition to thymine, uracil, cytosine, and guanine [280–285].

DNA modification in mammalian is due to a small number of chemical changes, although these have grown from the single example of 5-methylcytosine to its oxidized derivatives involving the methyl group to give 5-hydroxymetylcytosine, 5-formylcytosine, and 5-carboxycytosine (Scheme 2.33) [281]. The oxidation occurs stepwise and is promoted by the 10–11 translocation enzyme family, in turn regulated by the oxidized methylcytosine amount (Fig. 2.25). Guanine is the easiest oxidized of the four DNA bases and is known to form quadruplex structures. The final oxidation products differ according to whether they are formed from single stranded (ss), double stranded (ds), or quadruplex DNA. In a study on a dsDNA d(TGGGGT)/(ACCCCA) the guanine oxidation products were identified.

Oxidized polypeptides are removed from cells before aggregation and crossed-link. Alternatively, heavily oxidized proteins and

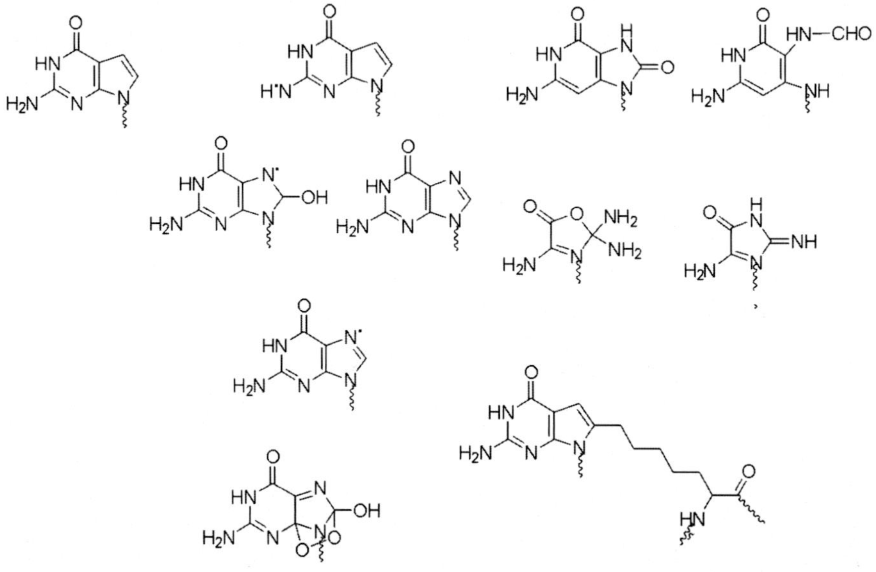

SCHEME 2.33 Guanine oxidation pathways in single-stranded, double-stranded, and quadruplex DNA [285].

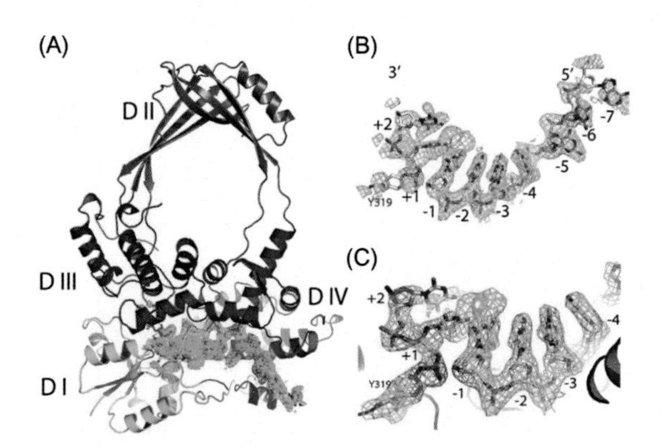

FIGURE 2.25 Structure of the topoisomerase I and DNA covalent complex. (A) Ribbon representation of the overall structure with four domains shown in different colors. The bound DNA fragments are shown with electron density map (green, 2fo-fc map at 1.5σ cutoff). (B) Annealed difference map (fo-fc, 1.5σ cutoff) around the DNA fragments. The difference map was calculated in the absence of bound DNA. (C) Closer view of the active site of the covalent intermediate. The continuous density (final 2fo-fc map) from Y319 to the 3′ dinucleotide (TT) is shown at 1.0σ cutoff. A clear density break of scissile bond is shown [286].

lipids, often in cross-linked mixed-form aggregates are engulfed by phagocytic cells and degraded [280].

The different mechanisms involved in the oxidative degradation of DNA bases are summarized in Figs. 2.26 and 2.27 [287].

2.18 DNA repair mechanism

Several overlapping paths are available to DNA for repairing (it has been estimated that about 200 enzymes take care of this job), reflecting the evolutive importance of this intervention.

The key feature of cancer is the development of genomic instability involving the accumulation of numerous mutations. Several paths lead to genetic instability. These include epigenetic alterations such as DNA methylation or remodeling of chromatin via histone protein modifications, recombination, base excision repair (BER), mismatch repair, and nucleotide excision repair (NER). The last one

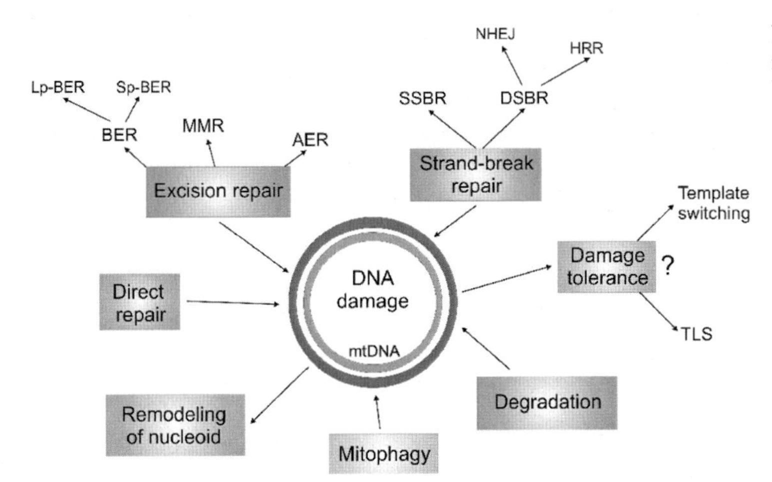

FIGURE 2.26 Different mechanism involved in the oxidative degradation of DNA [287].

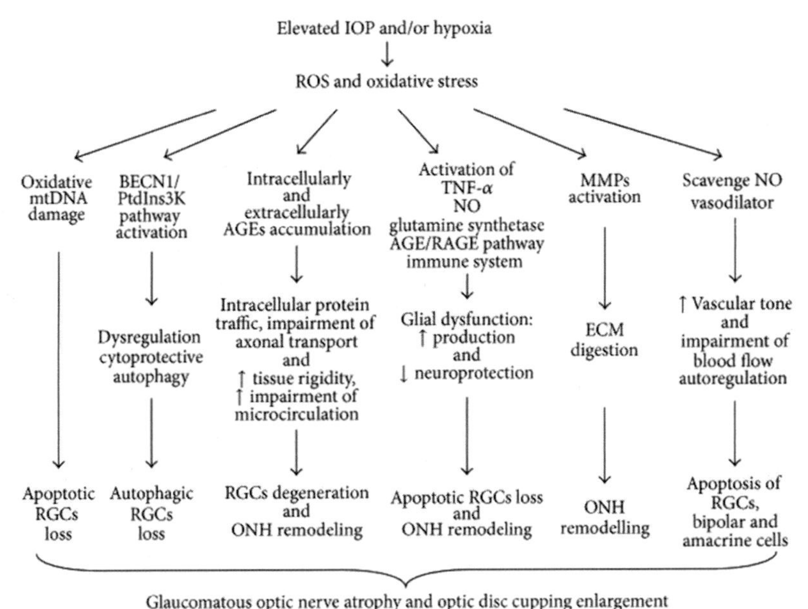

FIGURE 2.27 Schematic overview of the influence of ROS and the oxidative stress on the retina and the optic nerve head changes in the course of glaucomatous neurodegeneration. *IOP*, Intraocular pressure; *ROS*, reactive oxygen species; *mtDNA*, mitochondrial deoxyribonucleic acid; *RGCs*, retinal ganglion cells; *BECN1/PtdIns3K*, Beclin 1/phosphatidylinositol 3-kinase; *AGEs*, advanced glycation end products; *ONH*, optic nerve head; *TNF-α*, tumor necrosis factor alpha; *NO*, nitric oxide; *AGE/RAGE*, advanced glycation end product/receptor for advanced glycation end product; *MMPs*, matrix metalloproteinases; *ECM*, extracellular matrix [288].

is highly versatile and regards a wide class of helix distorting lesions including exogenous sources such as the photoproducts formed by sunlight, a major carcinogen in natural environment (Fig. 2.28).

The majority of the mutations, UV signature transitions, are C→T or tandem CC→TT translocations, occurring at bipyrimidine sequences, the specific targets of UV-induced lesions (Fig. 2.29). Among the protooncogenes that have been found altered in skin tumors are the RAS gene family, N-RAS, Ki-RAS, and Ha-RAS. The reported frequencies of RAS gene modifications (point mutations, amplification, rearrangement) in skin cancers from the general population varies from study to study ranging

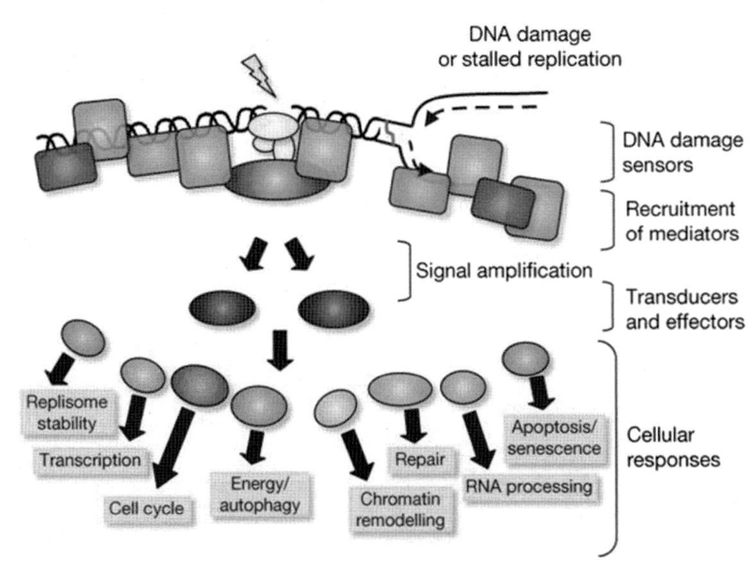

FIGURE 2.28 The presence of a lesion in the DNA, which can lead to replication stalling, is recognized by various sensor proteins. These sensors initiate signaling pathways that have an impact on a wide variety of cellular processes [289].

from under 5% up to 40% reflecting heterogeneity in samples and the type of analyses. The majority of mutations were located at codon 12 and found in all RAS genes with a preponderance for N-RAS alterations. All the mutations involve opposite bipyrimidine sequences, that are strong involved in UV-induced lesions, supporting the general role of unrepaired lesions in skin carcinogenesis. Interestingly, in xeroderma pigmentosum (XP) tumors the high RAS mutation frequency observed was accompanied by high levels of amplification and rearrangement of Ha-RAS and c-myc, a mitogenic oncogene [290]. Oxidation paths involving an electron transfer mechanism have an important role, as indicated in Figs. 2.29 and 2.30.

Evolution has generated a very large array of enzymes (>100) able to repair damaged DNA through a BER mechanism. BER is initiated by DNA glycosylases, which recognize and remove specific damaged or inappropriate bases, forming abasic sites that are then cleaved by an abasic endonuclease. The resulting single-strand break can then be processed by either of the two mechanisms, short-patch BER (where a single nucleotide is replaced) or long-patch BER (where 2–10 new nucleotides are synthesized) (Fig. 2.31).

DNA glycolyases (NEIL) that have a kinetic preference for oxidatively modified guanines and are able to remove hydantoins, whether from single- or double-stranded DNA or from G-quadruplex DNA. These may be grouped into three classes. Type 1 readily removes piroiminodihydrohydantoins, 5-guanidinohydantoins, and 5-carboxyamido-5-formamido-2-iminohydantoins from single-and double-stranded DNA, via a replication coupled repair, while Type 2 participate in the hydantoins removal from single-stranded DNA via a transcription-coupled repair and Type 3 liberates hydantoins from single-stranded DNA and G4, but very slowly, and appears to behave rather as a regulatory protein [293,294].

Most studies have suggested that in mammalians the nucleotide excision pathway (NER) removes helix-distorting bulky excision DNA lesions, while the BER is involved in small, nonbulky lesions. However, evidence has accumulated that there is some cross-talk

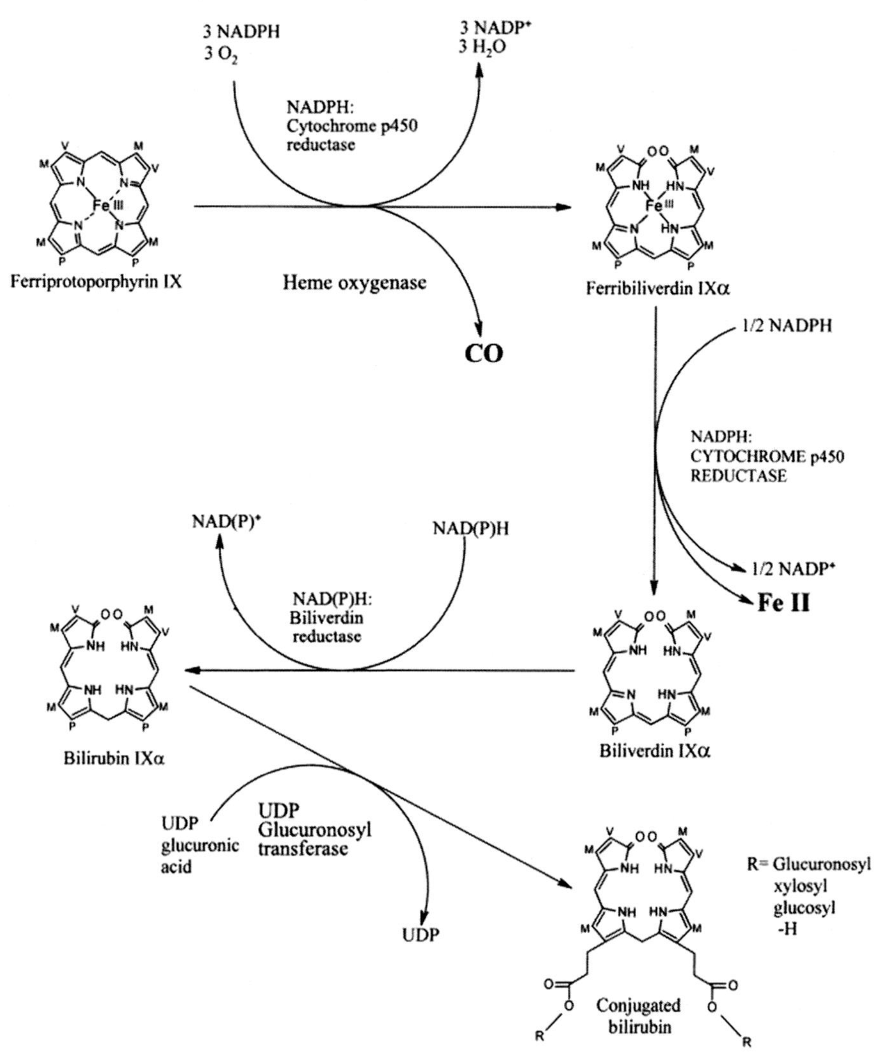

FIGURE 2.29 Heme degradation. Microsomal heme oxygenase (E.C. 1:14:99:3, heme-hydrogen donor:oxygenoxidoreductase) catalyzes the rate-limiting step in heme metabolism. Both heme oxygenase isozymes (HO-1 and HO-2) oxidize heme (ferriprotoporphyrin IX) to the bile pigment biliverdin-IXα (BV), in a reaction requiring 3 moles of molecular oxygen. NADPH:cytochrome p-450 reductase, reduces the ferric heme iron as a prerequisite for each cycle of oxygen binding and oxygen activation. The cleavage of the heme ring frees the α-methene bridge carbon as carbon monoxide, and generates the biliverdin-iron complex (BV-Fe-III). An additional NADPH-dependent reduction releases Fe-II from BV. The BV is reduced to BR by NAD(P)H:biliverdin reductase. Heme side chains are marked: *M*, methyl; *V*, vinyl; *P*, propionate [289].

between the two repair mechanisms. Thus intrastrand cross-linked DNA lesions guanine (C8)−thymine (N3) are substrates of both BER and NER, while 5,8-cyclopurines are substrates only of NER, while structure factors determine NER or BER susceptibility in other cases (for spiroimidodihydantoins, 5-hydantoins, and G*−T* lesions) [295].

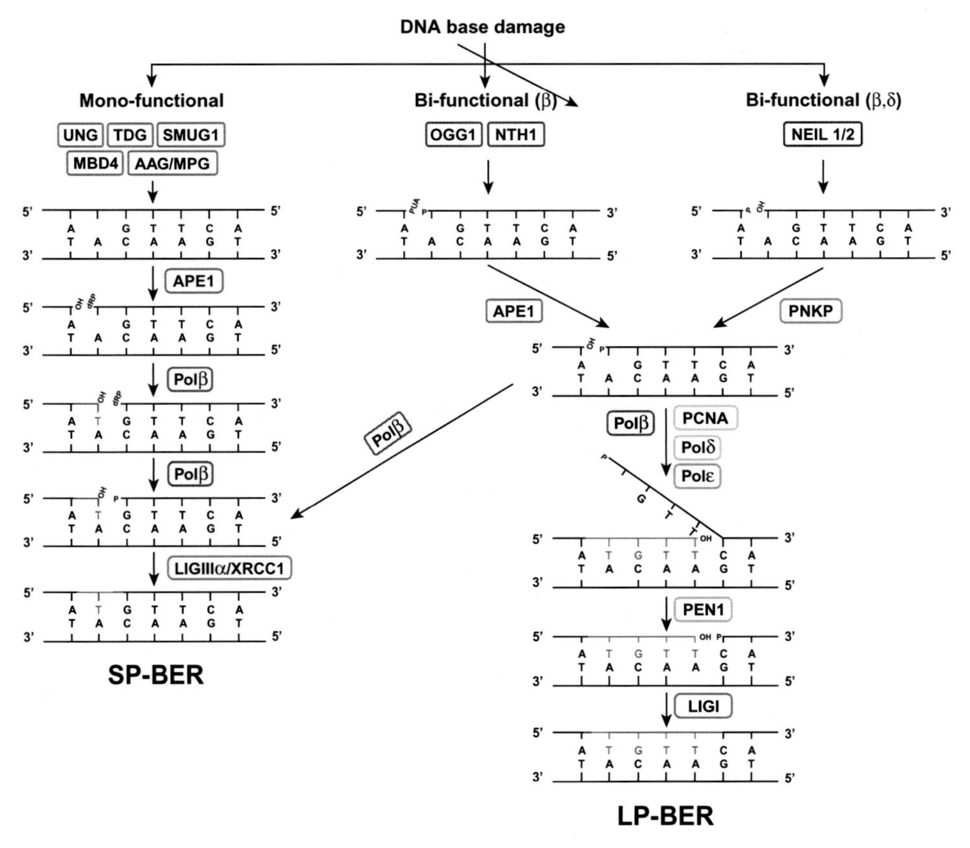

FIGURE 2.30 Schematic illustration of base excision repair (BER) [291].

A systematic examination of the toxicity of nanomaterials is necessarily carried out specifically, since the high surface-to-volume ratio may involve a high toxicity. It was found that neither cerium or titanium dioxide had any significant toxicity, while ZnO and AgNMs were cytotoxic and caused cell death, as demonstrated by comet assay [292]. Exocyclic adducts to DNA bases are formed as a consequence of exposure to certain environmental carcinogens, such as vinyl chloride ethylene oxide, as well as polychlorobiphenyls often present in ground water and endogenously as a consequence of the inflammation and LPO. Complex family of LPO products gives rise to various DNA adducts, which can be grouped into two classes: (1) small etheno-type adducts with strong mutagenic potential, and (2) bulky, propano-type adducts, which block replication and transcription. Etheno-DNA adducts are efficiently removed from the DNA by BER, while nucleotide incision repair (NIR) agents bind rapidly to these lesions, but removes them very slowly [295] (Scheme 2.34).

In contrast, an important role is served by DNA glycosylases that intervene in repair mechanisms such as BER, the AlkB protein and NIR enzymes. Small etheno-type DNA adducts are repaired by these mechanisms, while substituted propano-type and substituted etheno-type lesions are repaired by NER and homologous recombination. The mechanism of excision is outlined in Scheme 2.35.

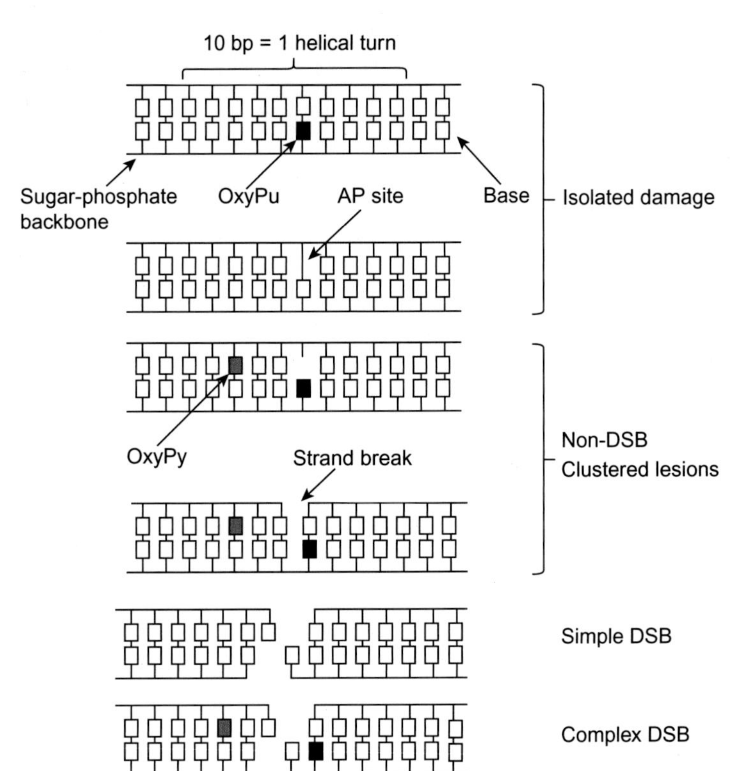

FIGURE 2.31 Scheme of oxidatively generated DNA damage induced by ionizing radiation. Non-DSB clustered lesions comprise 2, 3, or more oxidatively generated base damage (noted oxyPu, oxyPy, U), abasic site, SSB within one or two helix turns of DNA, produced by a single radiation track, the complexity of which increases with increasing LET. Complex DSBs are DSB associated with several oxidized bases and abasic sites [292].

Changes of the level and activity of several enzymes removing exocyclic adducts from the DNA was reported during carcinogenesis. Also several beyond repair functions of these enzymes, which participate in regulation of cell proliferation and growth, as well as RNA processing were recently described. In addition, adducts of LPO products to proteins was reported during aging and age-related diseases [296].

Solar energy acts in two ways, viz the small fraction of the UV-B present in the UV solar (5%) spectrum is directly absorbed by DNA and causes directly a lesion, while the remaining part is UV-A and visible and operates via photosensitization and the production of the various ROS [282–285].

Ionizing radiations can cause clustered DNA lesions that result from the combination of different lesions such as strand breaks, oxidatively generated base damage, abasic sites within one or two DNA helix turns. DSB-clustered lesions include DSB and several base damage and close lying abasic sites (complex DSB). Non-DSB clustered lesions include single strand break, base damage, and abasic sites. At radiation such as X-rays or γ-rays, clustered DNA lesions are three to four times more abundant than DSB. Complex DSBs are repaired by slow kinetics or left unrepaired and cause cell death or mitosis. In surviving cells, large deletions, translocations, and chromosomal aberrations are observed [292] (Fig. 2.32).

Detaching and repairing the affected proteins is a delicate matter, since the genome integrity must be preserved. It could be imagined that each damaged base would have to have a specific glycosylase that recognized it, requiring an enormous number of functionally distinct enzymes. This has proven not to be the case,

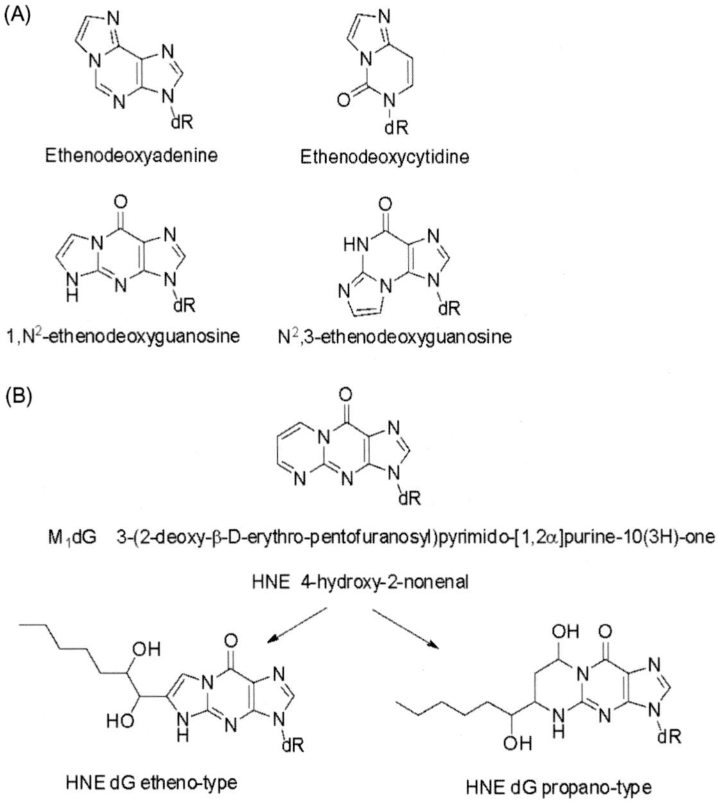

(A)

Ethenodeoxyadenine Ethenodeoxycytidine

1,N^2-ethenodeoxyguanosine N^2,3-ethenodeoxyguanosine

(B)

M_1dG 3-(2-deoxy-β-D-erythro-pentofuranosyl)pyrimido-[1,2α]purine-10(3H)-one

HNE 4-hydroxy-2-nonenal

HNE dG etheno-type HNE dG propano-type

SCHEME 2.34 Various etheno (by two carbon atoms) and propano-type (by three atoms) bridged nucleotides. (A) Small etheno bridged and (B) bulky exocyclic adducts.

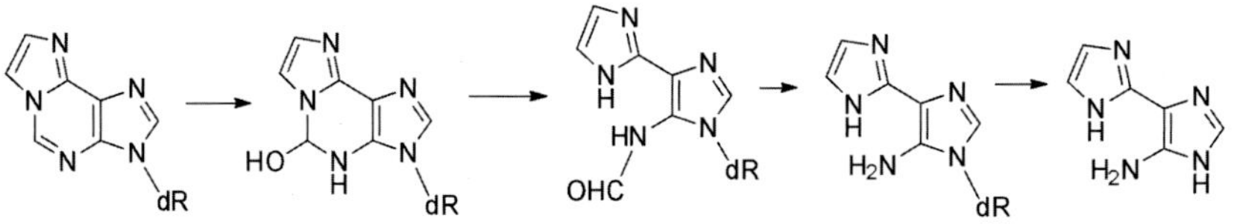

SCHEME 2.35 Mechanism of the degradation of ultraviolet-damaged nucleotides.

however, and rather each glycosylase has broad substrate specificity and a relatively small number of homologous enzymes are able to protect genomes that range from virus to human. Thus endonuclease III (Nth), first discovered in *Escherichia coli*, recognizes and removes a broad spectrum of oxidized pyrimidines as well as 2,6-diamino-4-hydroxy-5-N-methylformamidopyrimidine. Formamidopyrimidine DNA glycosylase (Fpg) likewise removes 8-oxoguanine (8-oxoG) and the further oxidation products of 8-oxoG, such as spiroiminodihydantoin (Sp) and guanidinohydantoin (Gh). The mammalian lineage also removes formamidopyrimidines but not Sp or Gh. Endonuclease VIII (Nei) has significant substrate overlap with *N*th and some substrate overlap with Fpg. The last compounds recognize a broad spectrum of pyrimidines, formamidopyrimidines, and Sp and Gh. An intriguing question is how lyase choose the

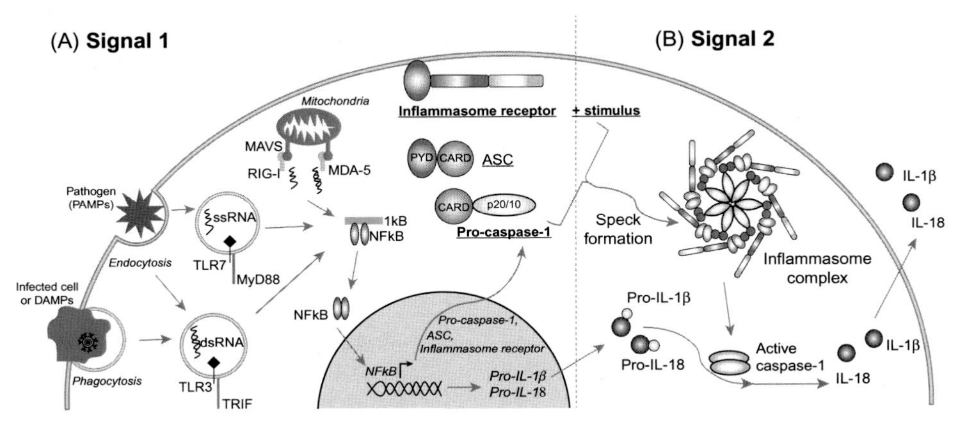

FIGURE 2.32 Overall mechanism of activation of the inflammasome by pathogens. (A) Signal 1: Pathogens which contain PAMPs infect the cell, or infected cell components which contain DAMPs are phagocytosed. PRRs (such as TLR3, TLR7, RIG-I, and MDA-5) detect the PAMPs/DAMPs and stimulate the production of NF-κB, leading to increased cytosolic expression of procaspase-1, ASC, inflammasome receptor (NLRP1, NLRP3, NLRC4, or AIM2) and proforms of IL-1β and IL-18. (B) Signal 2: The inflammasome receptor is activated by a secondary stimulus detected within the cytosol (such as a PAMP or DAMP), causing oligomerization of the inflammasome receptor, ASC and autoproteolytic cleavage of procaspase-1. Active caspase-1 then cleaves the proforms of IL-1β and IL-18, which is secreted from the cell [297].

correct point where to operate, among so many intact bonds. Binding of the enzyme to DNA occurs in the minor groove, causing a bend of the DNA backbone and the extrusion of the lesion from the major groove into the enzyme's active pocket. Glycosylases individuate the damages and periodically insert a wedge residue into the DNA stack through a redundant process [296].

A dramatic advance in understanding the mechanism involved in DNA repair was obtained from studies on XP, a disorder that is characterized by defect in the NER, and results in a dramatic loss of neurons in multiple regions of the brain and the spinal cord, not in the brain's white matter, as observed in the Cockayne syndrome. Scientists exposed to UV light patient cells observed a correlation between the poor cell survival after the exposure to light and the severity of the neurological damage caused. On the other hand, UV light cannot penetrate the skull and thus the hypothesis was presented that some type of endogenous DNA lesions exist in cells that require NER (but not other repair pathways) for its removal. In the absence of NER, this hypothetical damage would be expected to accumulate, ultimately blocking transcription of essential genes and leading to cell death. Further work supported that oxygen-centered radicals were able to form plasmids that were substrates for NER, and cyclic DNA derivatives were explicitly considered [298].

Base lesions in cellular DNA caused by ROS, can be removed by the BER mechanism. Typical lesions result from intramolecular crosslinking between the C-8 position of adenine or guanine and the 5′ position of 2-deoxyribose. Plasmid DNA-containing oxidized bases, such as purine cyclodeoxynucleosides, are stable and accumulate in the DNA of nonregenerating cells. The accumulation of this form of DNA damage well accounts for the progressive neurodegeneration observed in XPA individuals [299].

Cyclopurines were found to accumulate in old tissues due to oxidative stress in a way dependent on the organ [300] (Scheme 2.36).

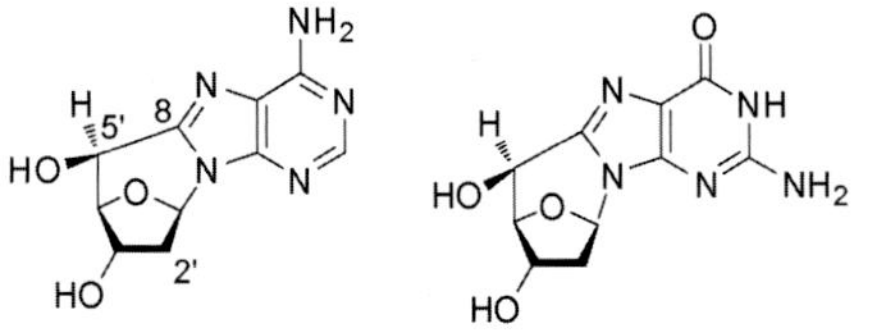

5',8-Cyclo-2'-deoxyadenosine 5',8-Cyclo-2'-deoxyguanosine

SCHEME 2.36 Structure of cyclopurines.

Mitochondria, the sites of oxidative phosphorylation, have defense and repair pathways to cope with oxidative damage. For mitochondrial DNA, an essential pathway is BER, which acts on a variety of small lesions. There are instances, however, in which attempted DNA repair results in more damage, such as the formation of a DNA-protein crosslink trapping the repair enzyme on the DNA. That is the case for mitochondrial DNA polymerase γ acting on abasic sites oxidized at 1-carbon of 2-deoxyribose. Such DNA-protein crosslinks presumably must be removed to restore function. In nuclear DNA, ubiquitylation of the crosslinked protein and digestion by the proteasome are essential first processing steps, DNA nucleobase modifications must be rapidly eliminated if the genome has to be preserved [301]. This is accomplished by a range of DNA glycosylases that are able to exploring and initiating the repair of specific types of base modifications in DNA by hydrolyzing the *N*-glycosidic bond between the damaged base and the 2′-deoxyribose sugar, affording abasic (apurinic/apyrimidinic, AP) sites. Monofunctionalglycosylases pass the AP site to a 5′ AP endonuclease that hydrolyzes the phosphodiester DNA backbone at the AP site. Further trimming of the remaining sugar fragments allows for gap-filling by a DNA polymerase to replace the excised nucleotide. Lastly, DNA ligase provides the finishing touches of repair by sealing the phosphodiester backbone. 8-Oxo-7,8-dihydroguanine is one of the most important oxidation products. The trouble is that shape of the oxoguanine is not suited for the normal DNA configuration and replicative polymerases often misincorporate adenine in front of oxoguanine. If this mispair persists within DNA without being repaired, a G:C to T:A transversion mutation permanently incorporates into the genome, and actually G:C to T:A mutations have been shown to be the predominant type of somatic mutation in protein kinase genes in lung, breast, and colorectal cancers [302]. Variants of MUT proteins, MUTYH, have an essential role for achieving a mismatch specific adenine glycosylase that initiates the BER path by facilitating the way over two carbenium/like intermediates.

A revised mechanism has been proposed for MutY involving two nucleophilic displacement steps similar to the accepted mechanism of "retaining" *O*-glycosidases (Scheme 2.37).

After the location of 8-oxoG:A base pair and the adenine base insertion into the MutY active site, protonation of N7 by Glu^{43} assists cleavage of the *N*-glycosidic bond via an oxacarbenium ion intermediate in an SN1-like way. During the glycosidic bond cleavage the stabilization of the positive charge is ensured by Asp^{144} that then rapidly reacts with the oxacarbenium ion forming the acetal covalent intermediate. This causes adenine vacation of the active site and a nucleophilic water molecule can diffuse into the active site. In addition, the second hydrolysis step appears to proceed via an SN1-like mechanism leading to a second highly reactive transient oxacarbenium ion. Glu^{43} is the active site assisting the water molecule in the nucleophilic attack at the $C_{1'}$ carbon of the acetal intermediate forming the β-anomer of the final abasic site product.

Mitochondrial DNA is distinguished from the nuclear counterpart, but likewise is sensitive to oxidative stress, as an example by ROS. Mechanisms of defense to protect mitochondrial genome integrity include SODs, the

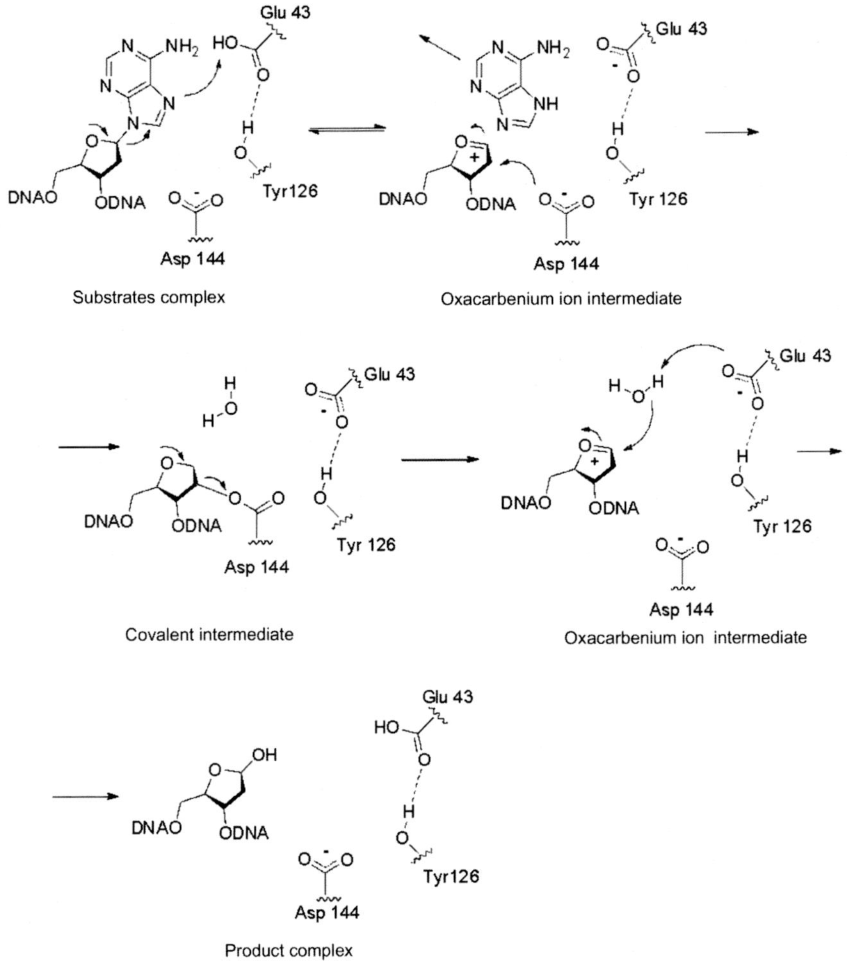

SCHEME 2.37 Revised mechanism for MutY-catalyzed adenine excision. The recently proposed double displacement mechanism for MutY enzymes based on methanolysis studies revealed retention of configuration at the anomeric carbon. This result suggests the involvement of a covalent intermediate, most likely involving Asp^{144}, in analogy to similar work with retaining *O*-glycosidases [302].

glutathione (GSH) peroxidase, TRX, and Prx pathways [289]. Manganese SOD, for example, is present in the mitochondrial matrix and converts superoxide ($O_2^{\bullet -}$) to H_2O_2 (that can be reduced to water by either GSH or the thiol-specific peroxidases (Prx). The H_2O_2 can also be converted to the hydroxyl radical (•OH), a strong oxidant, in the presence of Fe^{2+b}via the Haber-Weiss Fenton reaction.

In eukaryotes, DNA is packed in chromatins, thus it is important to establish how repair mechanisms work in vivo under such conditions. A recent work shows that a quasistochastic remodeler-driven access to BER initiation glycosylases through transient chromatin "fluidification" by "remosomes" (accessible metastable nucleosomes) is viable [303]. Based on recent findings of the recruitment of the histone chaperone complex FACT (facilitates chromatin transcription) and the remodeler CSB gene at oxidatively generated DNA damage it is suggested that another mechanism to overcome chromatin barrier to BER initiation is the pervasive (intergenic) transcription that could play a role as both DNA damage sensor and/or chromatin "permeabilizer" [304]. In essence, these scientists suggest the idea that the cheapest way for the cell to proceed with the removal of a stochastically generated DNA damage is to use stochastically induced repair initiators such as nucleosome remodelers and RNA polymerases. This stochastic chromatin dynamicity is essential to carry out the surveillance of the DNA by making it accessible to repair.

The phenomenon is dramatic in neurodegenerative diseases [305,306]. Cells generate unpaired electrons, typically via oxygen- or nitrogen-based by-products during normal cellular respiration and under stressed situations. These prooxidant molecules are highly unstable and may oxidize surrounding cellular macromolecules. Under normal conditions, the reactive oxygen or nitrogen species can be beneficial to cell survival and function by destroying and degrading pathogens or antigens. However, excessive generation and accumulation of the reactive prooxidant species over time can damage proteins, lipids, carbohydrates, and nucleic acids. Over time, this oxidative stress can contribute to a range of aging-related degenerative diseases such as cancer, diabetes, macular degeneration, AD, and PD. It is well accepted that natural compounds, including vitamins A, C, and E, β-carotene, and minerals found in fruits and vegetables, are powerful antioxidants that offer health benefits against several different oxidative stress-induced degenerative diseases, including AD. There is increasing interest in developing antioxidative therapeutics to prevent AD. There are contradictory and inconsistent reports on the possible benefits of antioxidative supplements; however, fruits and vegetables enriched with multiple antioxidants (e.g., flavonoids and polyphenols) and minerals may be highly effective in attenuating the harmful effects of oxidative stress. As the physiological activation of either protective or destructive prooxidant behavior remains relatively unclear, it is not straightforward to relate the efficacy of dietary antioxidants in disease prevention [307].

The accumulation of 8-oxoG in DNA is not always an accidental and potentially harmful phenomenon that leads to mutations and carcinogenesis, and alternative scenarios can be delineated. Thus in some cases a localized formation of 8-oxoG in the presence of lysine-specific histone demethylase, which produces stoichiometrically H_2O_2 as a by-product occurs and this activates or enhances the transcription of the affected genes. In another scenario, 8-oxoguanine DNA glycosylase-1 (OGG1) forms a tight complex with the excised free base that acts as a guanine nucleotide exchange factor for small GTPase [307–311].

Mispair due to regular bases appears to be more difficult to correct than those due to modified nucleobases. However, cells have specific DNA repair pathways [312–314].

The lesion caused by UV-B light arises, as mentioned, from $2+2$ cycloaddition, with the Paternò Büchi adduct being formed in about a 10-fold lower yield than the cyclobutane. Light activates enzymes able to repair the lesions exist in humans and models have been studied in detail. Thus a single-stranded all-thymine oligodeoxynucleotide $(dT)_{18}$ and the number of dimers formed with each laser pulse was examined. In this DNA model system, every absorbed photon excited a residue capable of dimerization. Quantum yields in the closely related systems poly(dT) (0.033) and $(dT)_{20}$ (0.028) are among the highest reported for any DNA compound [315,316]. In contrast, the dimerization quantum yield is over 30times lower in double-stranded, genomic DNA. This large reduction is due to the low frequency of TT doublets and absorption by nonthymine bases in mixed-sequence DNA. The glycosylase/AP lyase T4 endonuclease V is an enzyme from bacteriophage T4 which recognizes, binds, and incises DNA containing a UV-induced pyrimidine dimer. Alternatively, it may be that a base is flipped out of the helix in the dimer, as it has been observed in a *cis–syn* dimer for the A opposite the 5′-T of the dimer. In a NMR study of a duplex, an experiment with T4 denV endo V bound to a 49 base-pair oligonucleotide was carried out containing a site-specific cyclobutane thymine dimer.

The effect of an Escherichia coli DNA photolyase, which assists photorepair has been investigated and it has been observed that photolyase facilitates the removal of pyrimidine dimers but not that of other DNA adducts resulting from UV-RABC excision nuclease [317–320].

Photolyases contain a stoichiometric amount of flavin adenine dinucleotide (FAD) that operate as redox active cofactor through one (FADH•) and two ($FADH^-$) electron redox forms. The enzymatic activity of photolyases depends, however, on the photoexcited singlet two electrons reduced state of FAD ($^*FADH^-$). The absorption of $FADH^-$ in the region near 400 nm is too weak to ensure their normal activity and a second chromophore (5,10-methenyltetrahydrofolylpolyglutamate, MTHF or 8-hydroxy-5-deazaflavin, 8-HDF) is present and act as an antenna to absorb light and transfer energy to $FADH^-$. The repair involves transient electron transfer from $FADH_2$ and protonation (Schemes 2.38–2.40).

The determining factor is the rate of the process, since this has to be large enough to avoid back electron transfer to FAD [321].

A related mechanism has been proposed for the other classes of pyrimidine dimers, in particular the 6,4-pyrimidine-pyrimidone and its Dewar isomer. In these cases, an enzyme bound oxetane intermediate is formed in the intermediate complex. Two histidines exchange a proton with the hydroxypyrimidone heterocycle rings (Scheme 2.41).

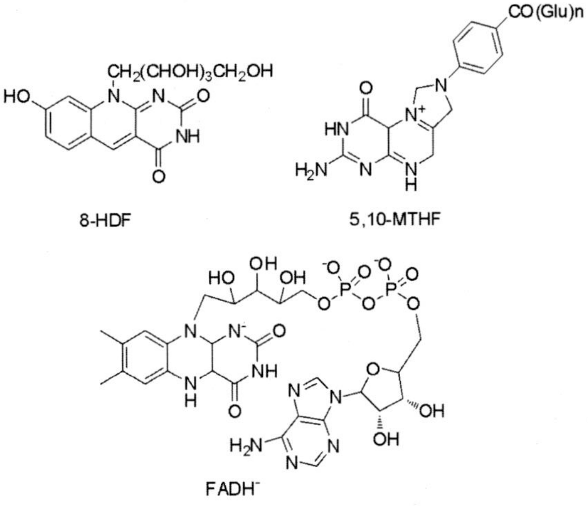

SCHEME 2.38 5,10-Methenyltetrahydrofolylpolyglutamate and 8-hydroxy-5-deazaflavin transfer energy to flavin adenine dinucleotide anion ($FADH^-$).

2.19 Implications for mutagenesis

More than 120 genes are involved in the control of pigmentation, and the most important action is due to melanocortin-1-receptors. These trigger the cellular response to the pigment formation, where eumelanin appears to have a better UV-protective effect than pheomelanin, and the presence of variants of melanocortin correlates with the type of pigmentation, up to 80% for red hair color for fair skin or hair, 20% for brown or black hair, and less than 4% in people with good tanning response. This correlates also with the risk of cutaneous melanoma that increases by 3.9-fold [322].

The deoxynucleoside triphosphates (dNTPs) are determining precursors in the DNA synthesis. Any change in dNTPs concentration has a mutagenic effect. In an in vitro study with a fragment of DNA polymerase, it was

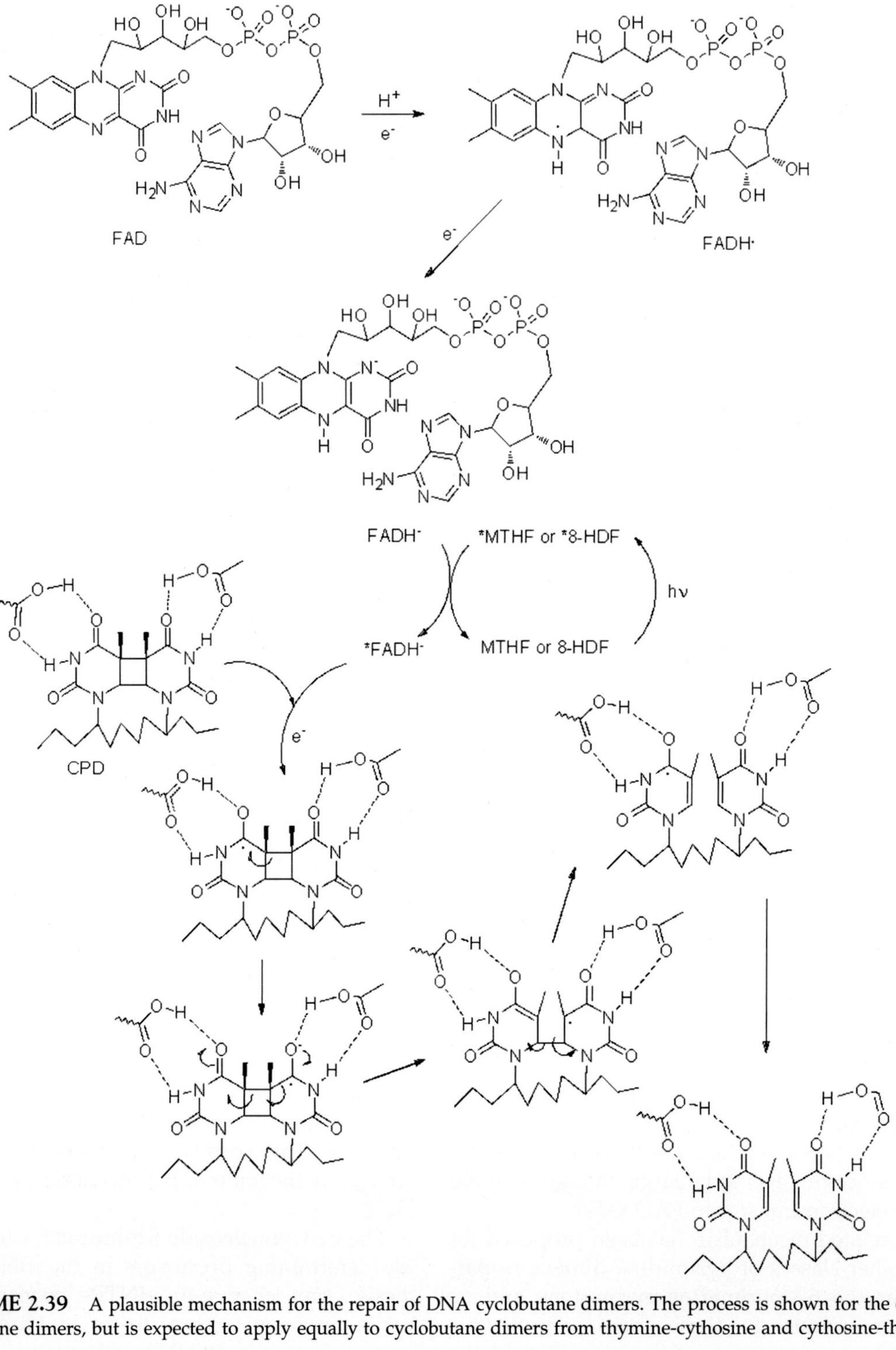

FAD

FADH·

FADH⁻

*MTHF or *8-HDF

hν

*FADH⁻

MTHF or 8-HDF

CPD

SCHEME 2.39 A plausible mechanism for the repair of DNA cyclobutane dimers. The process is shown for the case of thymidine dimers, but is expected to apply equally to cyclobutane dimers from thymine-cythosine and cythosine-thymine dimers.

found that DNA synthesis terminated opposite the A to the 3′ side of the *cis–syn* thymine dimer. At higher dNTP concentrations, synthesis advanced opposite the 3′-T of the dimer, with a large preference for the incorporation of A over G. Since termination mainly occurs opposite the 3′-T, and only to a small extent opposite the 5′-T of the dimer, it appears that the rate-determining step in bypass is the incorporation of a nucleotide opposite this site. The same was true for the 3′→5′ exonuclease-deficient olygomers [323,324].

The RAS protein is a GTPase that regulates signaling cascades involved in a various of cellular processes. The mutation in RAS protein is a frequent genetic alteration found in human cancers. This protein acts like a switch, which is turned on and off by the guanosine phosphate (GTP and GDP), and defects in RAS signaling may cause malignant transformation [325,326].

Receptors tyrosine kinases (RTKs) are one of the main effector pathways of RAS, as are signal through effector proteins, such as phosphatidylinositol 3-kinase (PI3K), RAF kinases, and guanine nucleotide exchange factors of the RAS-like (Ral). Moreover, PI3K offers one of the main effector pathways of RAS, intervenes in the regulation of cell growth, cell cycle entry, cell survival, cytoskeleton reorganization, and metabolism [327,328].

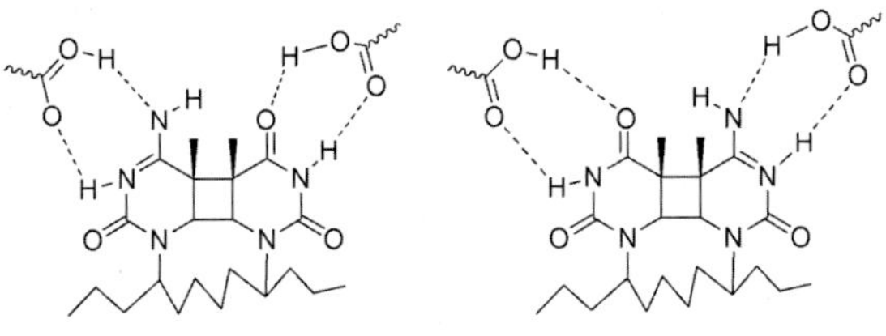

SCHEME 2.40 Cytosine-thymine and thymine-cytosine cyclobutane dimers.

Melanoma arises from malignant transformation of melanocytes. Binding of α-melanocyte stimulating hormone to melanocortin 1 receptor (MC1R) on melanocytes and eumelanin is the photoprotective pigment that attenuates UVR. cAMP production and CREB-mediated transcriptional activation of MITF are stimulated by MC1R. MITF promotes transcription of pigment synthesis genes and melanin production; therefore loss-of-function polymorphisms result in impaired eumelanin production, with large effects on alleles producing red hair and fair skin. Acquired pigmentation can be elicited by stimuli such as UVR (Fig. 2.33) [329].

BRAF and NRAS point mutations were discovered in two-thirds of melanomas. Most oncogenic BRAF mutations cause valine-to-glutamic acid substitutions at codon 600 (V600E) that constitutively activate the kinase domain. Sequencing studies have identified numerous novel melanoma genes involved in regulation of MAPK and other signaling pathways (Fig. 2.34) [330].

However, most oncogenic BRAF and NRAS mutations are not UVR signature mutations.

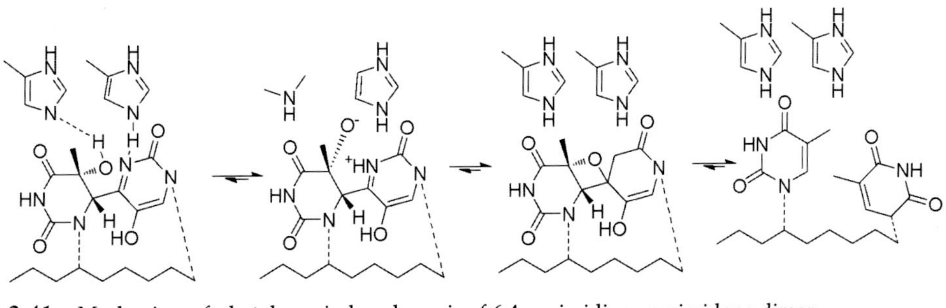

SCHEME 2.41 Mechanism of photolyase-induced repair of 6,4-pyrimidine–pyrimidone dimer.

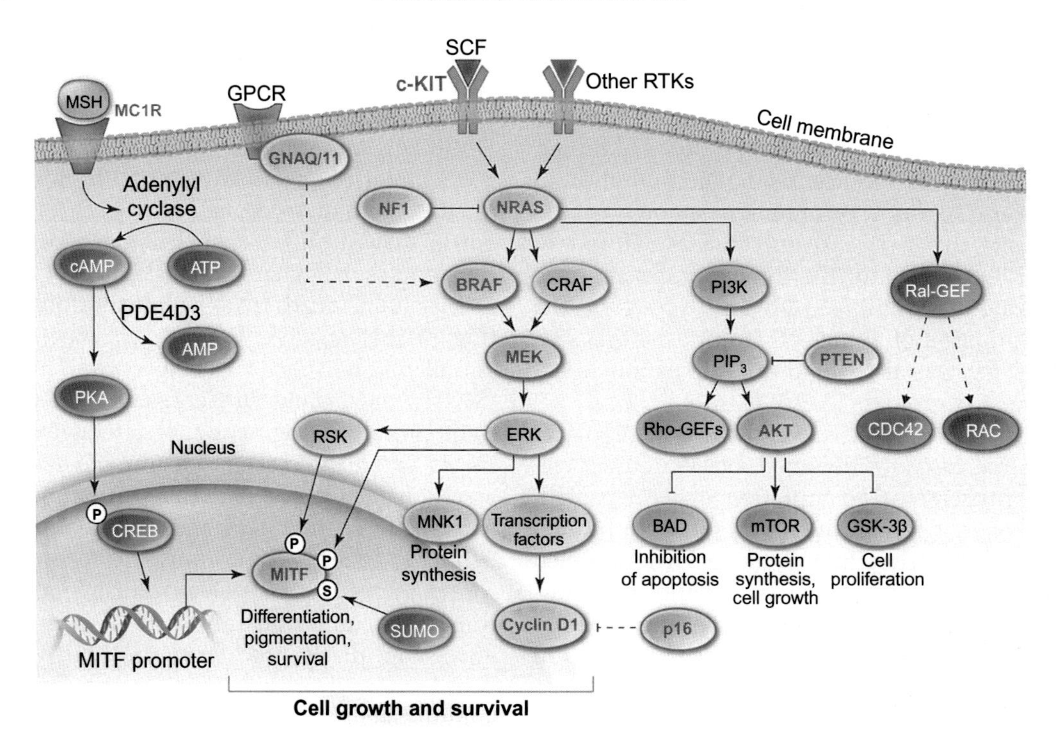

FIGURE 2.33 Signaling pathways in melanoma. Mitogen-activated protein kinase (MAPK) signaling promotes cell growth and survival and is constitutively active in most melanomas. RAS family members are activated by RTKs and signal through effector proteins including PI3K, RAF kinases, and Ral-GEFs. Oncogenic BRAF and NRAS are found in 40%–60% and 10%–30% of melanomas, respectively. c-KIT signaling is essential for melanocyte development and is associated with melanomas arising on acral, mucosal, and chronically sun-damaged skin. Mutations in GNAQ and GNA11, two G protein α-subunits involved in MAPK signaling, are the dominant genetic lesions in uveal melanomas. MITF, the master transcriptional regulator in melanocytes and lineage-specific oncogene, is expressed in response to MC1R signaling. Loss-of-function variants of MC1R are associated with the red hair/fair skin phenotype and increased melanoma susceptibility. Known melanoma oncogenes and tumor suppressors are labeled in red. Dotted lines represent omitted pathway components [329].

Other mutagenic mechanisms participate in melanoma development. For instance, ROS are known to produce a host of various oxidative DNA base lesions. While sunscreens have shown significant protection against melanoma, such protection may be incomplete, particularly for UVR-independent melanomagenesis [329].

Noteworthy, UV-C-induced programed cell death in HaCaT cells and effects of ROS was abrogated when the cells were treated immediately after irradiation with pyridoxamine, supporting that pyridoxamine could be used as anti-UV-C agent [331,332]. Oral retinoids treatment has shown to be effective in preventing reoccurrence of cutaneous SCC, but they cause significant side-effects and resistance. Acute UV-B increases the synthesis of retinoic acid, as well as degradation proteins in the stratum granulosum, but reduces the retinoid storage protein lectin-retinyl acyltranferase in the epidermis. This suggests that repair of the epidermis after acute UV-B exposure required endogeneous retinoid acid synthesis [333].

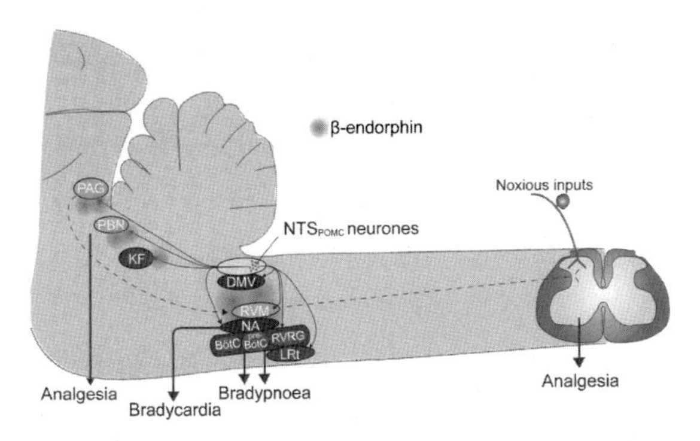

FIGURE 2.34 Signaling pathways in melanoma. MAPK signaling promotes cell growth and survival and is constitutively active in most melanomas. RAS family members are activated by RTKs and signal through effector proteins including PI3K, RAF kinases, and Ral-GEFs. Oncogenic BRAF and NRAS are found in 40%–60% and 10%–30% of melanomas, respectively. c-KIT signaling is essential for melanocyte development and is associated with melanomas arising on acral, mucosal, and chronically sun-damaged skin. Mutations in GNAQ and GNA11, to G protein α-subunits involved in MAPK signaling, are the dominant genetic lesions in uveal melanomas. MITF, the master transcriptional regulator in melanocytes and lineage-specific oncogene, is expressed in response to MC1R signaling. Loss-of-function variants of MC1R are associated with the red hair/fair skin phenotype and increased melanoma susceptibility. Known melanoma oncogenes and tumor suppressors are labeled in red. Dotted lines represent omitted pathway components [330].

2.20 Skin catabolism

Although KCs are relatively resistant to UVR-induced damage, repeated UVR exposure results in accumulated DNA mutations that can lead to epidermal malignancies. KCs play a central role in elaborating innate responses that lead to inflammation and influence the generation of adaptive immune responses in skin. Apart from the minor cellular constituents of the epidermis, specifically LCs and melanocytes, KCs are the major source of cytokines. UVR exposure stimulates KCs to secrete abundant proinflammatory IL-1-family proteins, IL-1α, IL-1β, IL-18, and IL-33. Normal skin contains only low levels of inactive precursor forms of IL-1β and IL-18, which require caspase-1–mediated proteolysis for their maturation and secretion. However, caspase-1 activation is not constitutive, but dependents on the UV-induced formation of an active inflammasome complex. IL-1 family cytokines can induce a secondary cascade of mediators and cytokines from KCs and other cells resulting in wide range of innate processes including infiltration of inflammatory leukocytes, induction of immunosuppression, DNA repair or apoptosis. Thus the ability of KCs to produce a wide repertoire of proinflammatory cytokines can influence the immune response locally as well as systematically, and alter the host response to photodamaged cells.

The skin target of protecting the body from outside agents is reached with the help of a number of structural fibrous proteins, such as keratin and filaggrin, as well as enzymes (proteases) and AMPs (defensins). In particular, KCs produce more and more keratin and undergo terminal differentiation. The fully cornified KCs that form the outermost layer are constantly shed off and replaced by new cells [334].

Collagen is the major insoluble fibrous protein in the extracellular matrix and in connective tissue. In fact, it is the single most abundant protein in the animal kingdom. There are at least 16 types of collagen, but 80%–90% of the collagen in the body consists of types I, II, and III (of fiber structure, in contrast to type IV, two dimensional). Collagens are secreted by fibroblasts in connective tissue, as well as by numerous epithelial cells (Table 2.2) [335–337].

Young skin is perfectly homogeneous with no blemish and perfect elasticity. The appearance to the eye under visible light (400–700 nm) is mainly determined by the concentration and distribution of melanin, hemoglobin, and collagen. In particular, a strong difference in the concentration and homogeneity of chromophores is apparent

TABLE 2.2 Major collagen molecules [319,320].

Type	Molecule composition	Structural features	Representative tissues
Fibrillar collagens			
I	$[\alpha1(I)]_2[\alpha2(I)]$	300-nm-long fibrils	Skin, tendon, bone, ligaments, dentin, interstitial tissues
II	$[\alpha1(II)]_3$	300-nm-long fibrils	Cartilage, vitreous humor
III	$[\alpha1(III)]_3$	300-nm-long fibrils; often with type I	Skin, muscle, blood vessels
V	$[\alpha1(V)]_3$	390-nm-long fibrils with globular N-terminal domain; often with type I	Similar to type I; also cell cultures, fetal tissues
Fibril-associated collagens			
VI	$[\alpha1(VI)][\alpha2(VI)]$	Lateral association with type I; periodic globular domains	Most interstitial tissues
IX	$[\alpha1(IX)][\alpha2(IX)][\alpha3(IX)]$	Lateral association with type II; *N*-terminal globular domain; bound glycosaminoglycan	Cartilage, vitreous humor;
Sheet-forming collagens			
IV	$[\alpha1(IV)]_2[\alpha2(IV)]$	Two-dimensional network	All basal laminaes

between young and aged skin. The young healthy skin has a fresh glowing complexion and the dense network of superficial microcapillaries that bring nutrients and oxygen to the skin are not visible, as is the case also for pigmentation, since melanin has negligible visual irregularities. Stimulation of melanin synthesis by exposure to the sun provides a homogeneous pigmentation of the skin. With age and repeated daily minor stresses, the skin loses its radiance due to the impairment of chromophore balance, while KCs raise their ability to phagocyte melanosomes and an increase in lentigines and hyperpigmentation occur. Locally, the vascularization becomes visible to the eye either in the form of vessels or in the form of diffuse red areas of variable intensity. In addition, it is known that, with age and stress, the proteins of the extracellular matrix are produced less and exhibit greater degradation. This weakens the supporting network of the skin, in an irregular way and collagen degradation makes the skin thinner and makes visible defects [337].

The damage induced to the skin by 254-nm irradiation may be attributed to direct absorption by either tyrosine/phenylalanine or, in general, to the peptide bonds, but the photoreactions arising from the irradiation of collagen via solar UV wavelengths seem to involve several age- and tissue-related photolabile collagen fluorophores that absorb at longer wavelengths. A comparison of fluorescence emission and absorption spectra and fluorescence fading indicates the presence of at least four photolabile compounds, revealed also by their weak fluorescence, including some due to the formation of π complexes between aromatics.

Human skin and its immune cells provide essential protection to the human body from injury and infection. Recent studies reinforce the importance of KCs as sensors of danger through alert systems such as the inflammasome (Fig. 2.35). In addition, newly identified $CD103^+$ dendritic cells (DCs) are positioned for cross presentation of skin-tropic pathogens and accumulating data highlight a key role of tissue-resident rather than circulating T cells in skin homeostasis and pathology [338].

Epidermal stem cells reside in the lower part of the epidermis (stratum basale) and undergo division and amplification, while migrating toward the surface of the epidermis until they become part of the stratum spinosum, stratum granolosum and eventually become corneocytes (that is cells that have lost nucleus and cytoplasmatic organelles) in the stratum corneum and are finally shed off through desquamation as new cells come in. During such process, KCs express specific keratins as well as other markers, such as involucrin, loricin, transglutamiase, and caspase-14 [27,339,340].

Calcium is the major regulator of KC differentiation in vivo and in vitro. A calcium gradient within the epidermis promotes the sequential differentiation of KCs as they traverse the different layers of the epidermis to form the permeability barrier of the stratum corneum. Calcium promotes differentiation by both outside–in and inside–out signaling. A number of signaling pathways involved with differentiation are regulated by calcium, including the formation of desmosomes, adherens junctions, and tight junctions, which maintain cell–cell adhesion and play an important intracellular signaling role through their activation of various kinases and phospholipases that produce second messengers that regulate intracellular free calcium and PKC activity, critical for the differentiation process. The required calcium in part comes from intracellular stores, and in part comes from transmembrane calcium influx, through both calcium-sensitive chloride channels and voltage-independent cation channels permeable to calcium. Vitamin D3 regulates KC proliferation and differentiation mostly by modulating calcium concentrations and regulating the expression of genes involved in KC differentiation. KCs are the only cells in the body with the entire vitamin D metabolic pathway from vitamin D production to catabolism. When microrganisms penetrate into the cells of body, $CD4^+$ T cells, often indicated as "T helper" (Th) cells are activated. Other immune cells modulate the body's response. On the other hand, $CD8^+$ T are known as "T cytotoxic" (Tc) cells and destroy/kill cells infected with foreign invading microorganisms. Both the above cells types are important in autoimmunity, asthma, and allergic responses as well as in tumor immunity. During T-cell receptor activation in a particular cytokine milieu, $CD4^+$ T cells and $CD8^+$ T cells may differentiate into one of several lineages of Th or Tc, including Th1/Tc1, Th2/Tc2, Th17/Tc17, and iTreg (induced regulatory T cells, T regulatory cells induced from CD25--cells in vitro).

The innate immune system comprises DCs, macrophages, neutrophils, natural killer cells, granulocytes, basophils, eosinophils, and MCs, while the adaptive immune system consists predominantly of T and B cells. In addition to these cellular components various chemokines, cytokines, immunoglobulins, and other soluble factors aid in the immune defense mechanism. Roughly, the interplay between antigen-presenting cells (APCs, DCs, macrophage) and the effector cells (T cell, B cell) is an important determinant in eliciting an antitumor immune response. While the innate response is primarily mediated through the production of pro-inflammatory cytokines and inflammation, induction of cell death and production of antimicrobial peptides (AMP), the adaptive immunity is T-cell mediated and is immunosuppressive. UVR absorption activates the acute signaling pathways including platelet activating factor (PAF) by oxidization of membrane lipids of the keratinocytes, and signals

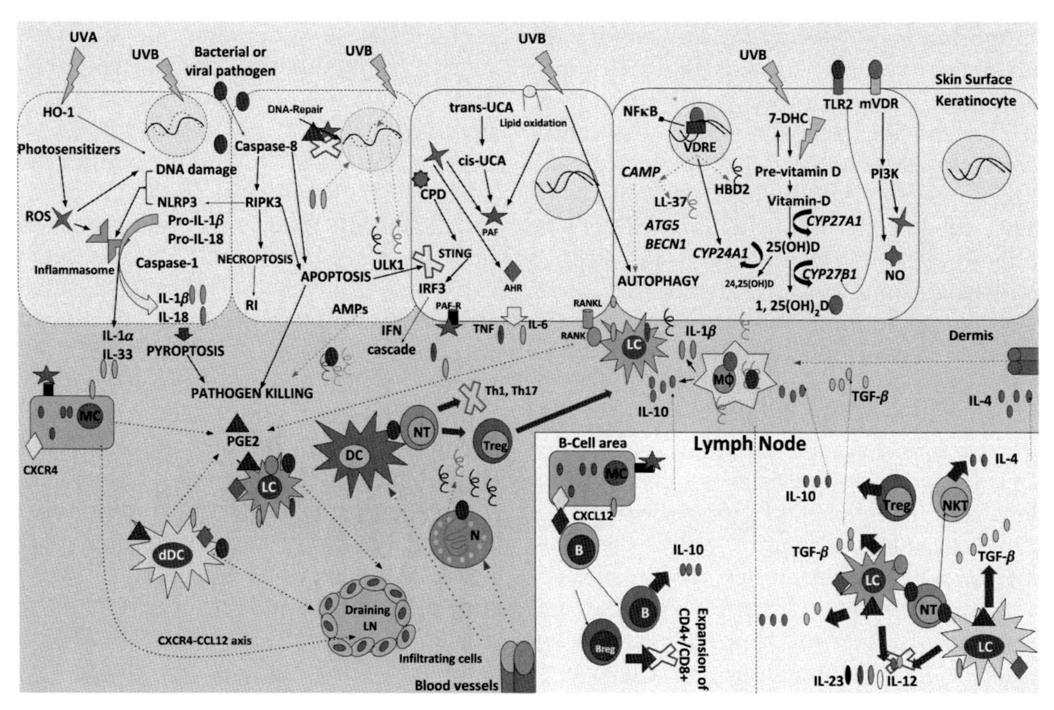

FIGURE 2.35 Immune modulation by ultraviolet radiation (UVR) in the skin. UVR induces both innate and adaptive arms of immunity independent of and dependent on vitamin D production. The innate response is primarily mediated through the production of proinflammatory cytokines leading to inflammation, induction of cell death via various pathways (necroptosis, pyroptosis, or apoptosis) and production of antimicrobial peptides (AMP). While adaptive immunity is T-cell mediated and is immunosuppressive. UVR absorption by chromophores (e.g., *trans*-UCA) leads to the activation of acute signaling pathways including platelet activating factor (PAF). PAF is generated by oxidization of membrane lipids of the keratinocytes, and signals via PAF receptor (R). Many cells express the PAF-R including monocytes, mast cells (MC) and keratinocytes. UVR-generated reactive oxygen species (ROS) activates PAF production and cytoplasmic aryl hydrocarbon receptor (AHR), both of which signal via membrane-expressed epidermal growth factor, which leads to the release of prostaglandin E_2 (PGE2). ROS can also activate inflammasome assembly leading to secretion of IL-1 family of cytokines. UVR-induced cyclobutane pyrimidine dimers (CPD) can stimulate STING (stimulator of interferon genes) leading to the activation of the interferon cascade, via interferon regulatory factor 3 (IRF3). Upregulation of epidermal receptor activator of NF-κB-ligand (RANKL) via UVR-induced PGE2 from keratinocytes leads to the activation of epidermal dendritic cells (DCs). These DCs express RANK, the receptor for RANKL, and receptor activation results in the migration of these cells to the draining lymph nodes (dLN). The effect of PAF, PGE2, and proinflammatory cytokines (TNF, IL-6, and IL-1β) enhances cell-to-cell communication and a change is the local cell population of the skin. UVA can also induce the antioxidant heme oxygenase-1 (HO-1), which can be protective against UVB damage. The activated MC, dermal DC (dDC) and Langerhans cells (LC) migrate to the dLN in response to UV-induced-keratinocyte-derived IL-33. UV-induced keratinocyte-derived PAF activates upregulation of the chemokine receptor CXCR4 on MC and induces the expression of CXCL12 (the ligand for CXCR4) in the B cells present in the dLN. CXCL12 has chemoattractant properties and aids the activated CXCR4 + MC to migrate to the dLN. There, these MCs stimulate the IL-10-producing B cells (B) which mediate immunosuppression and downregulate antibody secretion by interfering with B-cell maturation by blocking germinal center formation, and T follicular helper cell function, hampering the generation of immunological memory to the presented antigen. There regulatory B cells (Breg) which respond to UVB inhibit the activation of immunity by DC. LC and DC, the major APC of the epidermis, present antigens to naïve T cells (NT) in the LN, which then differentiate and proliferate to give a population of activated, antigen-specific T cells. These LC and DC secrete high levels of TGF-β, and when they interact with T cells, lead to the production of Foxp3 + regulatory T cells (Tregs). These Tregs are cytotoxic for APC,

via PAF receptor (R), expressed in many cells including monocytes, mast cells (MC) and keratinocytes. UVR generated reactive oxygen species (ROS) activates PAF production and cytoplasmic aryl hydrocarbon receptor (AHR), both of which signal *via* membrane-expressed epidermal growth factor, in turn leading to the release of Prostaglandin E_2 (PGE2). ROS also activates inflammasome assembly leading to secretion of IL-1 family of cytokines and UVR-induced cyclobutane pyrimidine dimers (CPD) stimulate STING (stimulator of interferon genes) leading to the activation of the interferon cascade, *via* interferon regulatory factor 3 (IRF3). The most important effect is due to UV-B that generates AHR ligands from tryptophan. The absorption of UV-A by endogenous photosensitizers such as flavins, porphyrins, and melanin, leads to the generation of ROS. In contrast, UV-A can induce the antioxidant heme oxygenase-1 (HO-1), in turn able to protect against UV-B damage [338]. A schematic view of the immune modulation induced by UVR in the skin is shown in Fig. 2.36.

As shown in Fig. 2.9 UVR causes DNA damage and generates photoproducts, leading to production and activation of many mediators (including cytokines, AMPs, and alarmins). At the same time homeostatic mechanisms leading to DNA repair, failure of which causes the activation of immunosuppressive mediators are activated. Both UVR and vitamin D lead to immunosuppression via the adaptive immune system and are protective when they activate the innate immune system, especially via the induction of AMP (Fig. 2.37).

The independent and overlapping effects of UV-R and vitamin D on the cellular response to infection. Individual components of solar radiation, UV-A, UV-B, and heat generation, are linked to their downstream immune function, inducing vitamin D production. While some effects are exclusive to UVR (orange lines) and some are exclusive to vitamin D (blue lines) there is a great degree of overlap between various components (black lines), particularly for antimicrobial peptide (AMP) induction and T helper (Th)2 and regulatory T cell (Treg) differentiation. The effect of UVR is primarily localized to the skin, while the vitamin D effects can occur in the skin and systemically in immune cells and in various tissue sites, culminating in pathogen killing and immunosuppression. *Bregs*, regulatory B cells; *DC*, dendritic cells; *MMP*, matrix metalloproteinase; *NO*, nitric oxide; *ROS*, reactive oxygen species.

Most studies suggest that physiological and modest doses of UV-B suppress adaptive T-cell immunity through effects on LCs, leading to the induction of antigen-specific Tregs. An action on innate immune cells promotes the synthesis of AMPs that have a protective effect against bacterial and viral infections. Vitamin D is further known to enhance responsiveness to steroids. The impact of vitamin D on lung

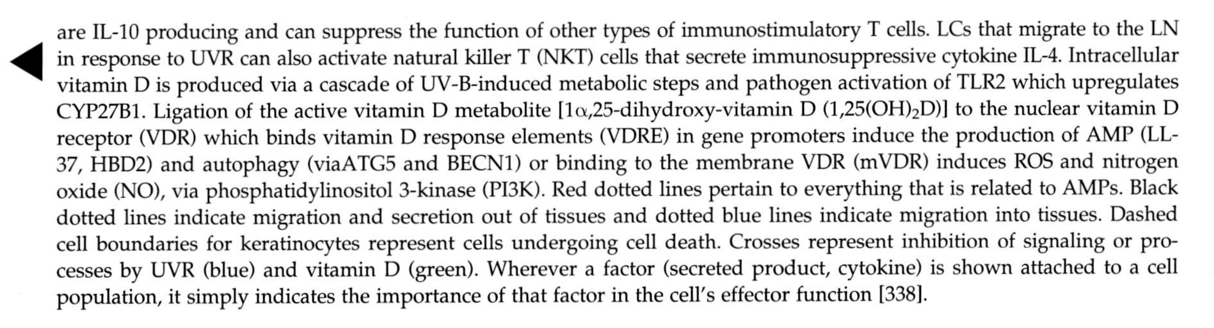

◀ are IL-10 producing and can suppress the function of other types of immunostimulatory T cells. LCs that migrate to the LN in response to UVR can also activate natural killer T (NKT) cells that secrete immunosuppressive cytokine IL-4. Intracellular vitamin D is produced via a cascade of UV-B-induced metabolic steps and pathogen activation of TLR2 which upregulates CYP27B1. Ligation of the active vitamin D metabolite [1α,25-dihydroxy-vitamin D (1,25$(OH)_2$D)] to the nuclear vitamin D receptor (VDR) which binds vitamin D response elements (VDRE) in gene promoters induce the production of AMP (LL-37, HBD2) and autophagy (viaATG5 and BECN1) or binding to the membrane VDR (mVDR) induces ROS and nitrogen oxide (NO), via phosphatidylinositol 3-kinase (PI3K). Red dotted lines pertain to everything that is related to AMPs. Black dotted lines indicate migration and secretion out of tissues and dotted blue lines indicate migration into tissues. Dashed cell boundaries for keratinocytes represent cells undergoing cell death. Crosses represent inhibition of signaling or processes by UVR (blue) and vitamin D (green). Wherever a factor (secreted product, cytokine) is shown attached to a cell population, it simply indicates the importance of that factor in the cell's effector function [338].

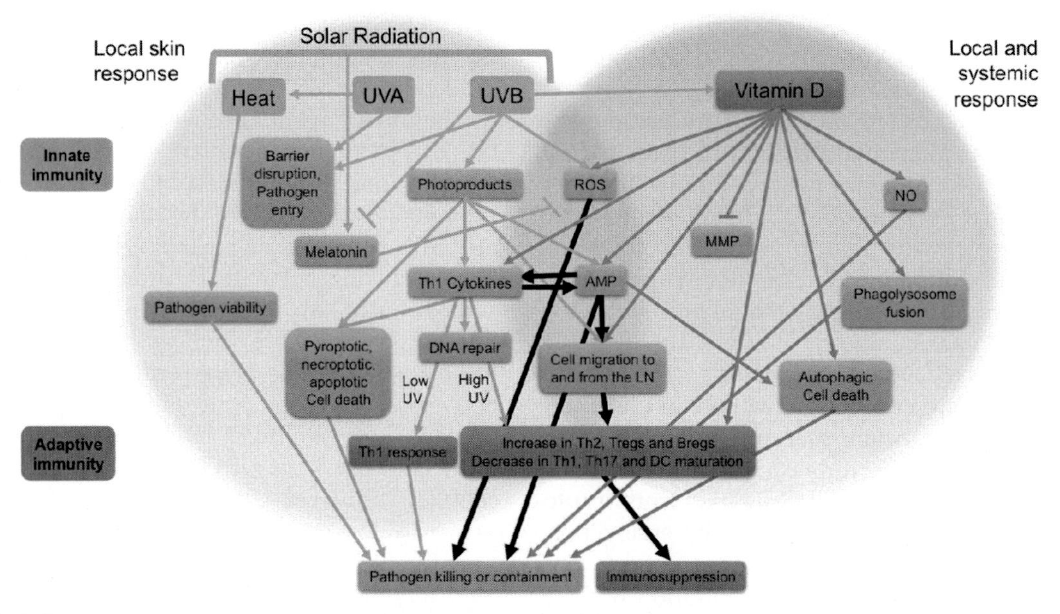

FIGURE 2.36 The independent and overlapping effects of UVR and vitamin D on the cellular response to infection. Individual components of solar radiation, UVA, UVB and heat generation, are linked to their downstream immune function, inducing vitamin D production. While some effects are exclusive to UVR (orange lines) and some are exclusive to vitamin D (blue lines) there is a great degree of overlap between various components (black lines), particularly for antimicrobial peptide (AMP) induction and T helper (Th)2 and regulatory T cell (Treg) differentiation. The effect of UVR is primarily localised to the skin, while the vitamin D effects can occur in the skin and systemically in immune cells and in various tissue sites, culminating in pathogen killing and immnosuppression. Abbreviations: Bregs, regulatory B cells; DC, dendritic cells; MMP, matrix metalloproteinase; NO, nitric oxide; ROS, reactive oxygen species, reproduced with permission [338,341].

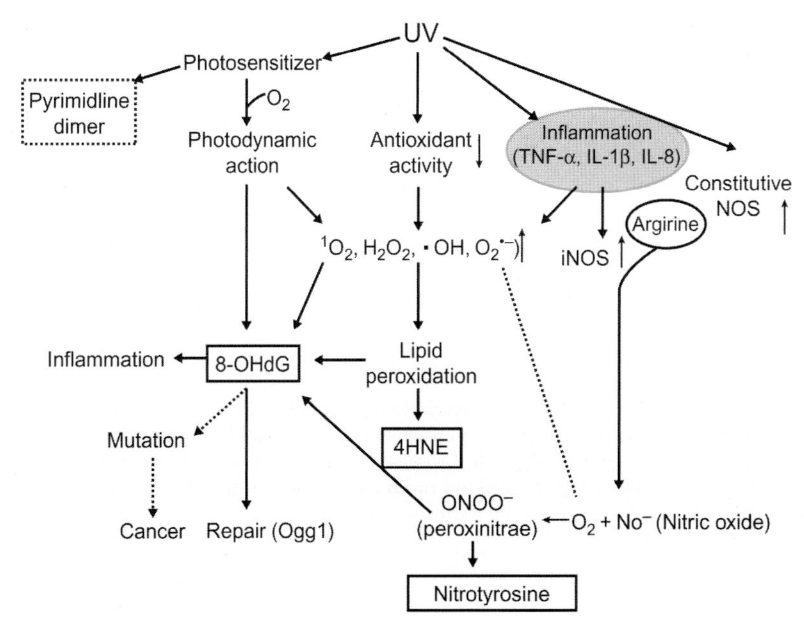

FIGURE 2.37 Schematic pathway of ultra violet (UV)-induced inflammation and the formation of oxidative stress. UV radiation causes the generation of reactive oxygen species (ROS) through various pathways, and ROS induces lipid peroxidation through the interaction with lipid derived from biomembranes, leading to the production of 4-hydroxy-2-nonenal (HNE). In contrast, constitutive nitric oxide synthase (NOS) activated by UV exposure and iNOS is induced by the inflammatory cytokines, and this leads to the production of nitric oxide (NO), followed by peroxynitrite ($ONOO^-$) formation, which modifies the protein and results in the formation of nitrotyrosine [342].

function, respiratory tract infections, and asthma is supported by clinical trials performed in pregnancy, pediatric, and adult cohorts. The studies with vitamin D overall suggest positive effects and ask the question of whether safe, modest, and physiological levels of UV-B irradiation will afford similar protection [343].

In a comprehensive systematic review of epidemiological studies, van der Rhee and colleagues found that possible cancer protecting roles for sunlight could not fully be explained by vitamin D alone [344].

Getting old seems to be an inescapable fate of any living thing, until the death of the organism. Among the theories supplied for understanding how cells and organs become aged, the most comprehensive is probably the one invoking a growing role of radicals, in particular of ROS. These species interact with nucleic acids, both mitochondria and nucleus DNA, causing mutations that predispose to DNA strand breaks, in a continuous process when exposed to light, in particular for the case of mitochondrial derivatives that lack the protection by histones available for nuclear compounds. Hydroxyl radicals react with DNA bases to form oxidative products such as 8-oxo-guanine or with the deoxyribosyl backbone of DNA, leading to strand breaks. Furthermore, phospholipid cellular membranes are attacked, thus causing premature cell death, although the very fact that a large number of identical genomes are present makes that mitochondria tolerate very high levels (up to 90%) of damaged mitochondrial DNA through complementation of the remaining wild-type DNA.

On the other hand, as long as the degradation is limited, ROS accumulate and have important roles in cell signaling and homeostasis. Indeed, ROS-induced damage has the potential to alter the electron transfer chain and decrease the efficiency of ATP production. As it has been expressed, ROS-mediated damage appears to be a mechanism to amplify ROS-stressing cancer cells. Microsomal DNA mutations caused by ROS accumulate within the cell, leading to impaired respiratory chain, thereby generating more ROS, which in turn causes higher microsomal DNA mutation rates. Thus irradiation causes release of the cytochrome c and Smac/DIABLO proteins from the mitochondrial space to the cytosol and activates effector caspases and induces apoptosis.

Many types of human malignancy such as colorectal, liver, breast, pancreatic, lung, prostate, bladder, and skin cancer have been shown to harbor somatic mtDNA mutations [345].

Skin aging mainly results as side path during the energy-generating phosphorylation reactions in mitochondria. In Scheme 2.42 the respiratory chain is shown.

Free radicals in the body cause oxidative damage to cellular components and thus an altered cellular function compromises tissue and organ function, ultimately causing death. Oxygen from the atmosphere is used in the body to produce energy via oxidative phosphorylation in mitochondria, and the radical intermediates cause an oxidative stress on proteins, DNA, and lipids, a phenomenon that becomes more and more important with age and thus may overwhelm the natural repair systems in the elderly. This gives a major contribution to diseases associated with aging [346].

In particular, the development of atherosclerosis depends on the balance between proinflammatory, antiinflammatory, and antioxidative defense mechanisms. Vascular proliferation and inflammation are closely linked, and excessive proliferation of vascular cells plays an important role in the pathology of vascular occlusive disease. Free radicals are considered to play a causal role in this process because ROS lead to the oxidation of LDLs (OxLDLs), and these accumulate within plaques, and contribute to the inflammatory state of atherosclerosis and play a key role in its pathogenesis. OxLDLs lead to endothelial dysfunction, and can result in either cell growth or apoptotic cell death and can cause vasoconstriction. Free radicals have also been implicated in congestive heart failure (CHF), the annual incidence of which is one to five per 1000 persons, and the relative incidence doubles for each decade of life after the age of 45 years. Experimental evidence suggests a direct link between free radicals production and CHF and the presence of ROS in circulating blood is also the key intermediary related to vascular injury and organ dysfunction. A progressive decline in the efficiency of biochemical and physiological processes after the reproduction phase of life affects all of the characteristic diseases. In particular, ischemic stroke is associated with free radicals arising from sources such as xanthine oxidase, cyclooxygenase, inflammatory cells, and mitochondria, and thus to neuronal death. The mitochondrial electron transport chain is altered during ischemia and reperfusion and this is another source of free radicals. The accumulation of blood borne inflammatory cells such as neutrophils and monocytes/macrophages, which can occur during reperfusion, can also promote further oxidative stress. Increased levels of oxidative damage to DNA and evidence for LPO has also been demonstrated in ischemic stroke patients and the increased levels of ROS can make the brain more susceptible to oxidative stress [191]. The AD is the most common neurodegenerative disorder and is characterized by loss of neurons and synapses resulting in cognitive impairment and a gradual loss of memory, language skills, and reasoning leading to dementia and finally death. The onset of this syndrome is gradual, with clinical symptoms appearing between 60 and 70 years of age and is characterized by both synaptic loss and nerve cell loss. Oxidative damage may also play a role in amyloid deposition in AD, and oxidizing conditions can cause protein cross-linking and aggregation of β-amyloid protein and also contribute to aggregation of tau, and other cytoskeletal proteins, as well as cause the oxidation of the nonsaturated carbohydrate side chains of membrane lipids, which leads to the disintegration of the neural membrane thus resulting in cell lysis. LPO has also been assessed in patients with AD and increased brain levels of 4-hydroxy-2-nonenal glutathione conjugates [347]. HD causes uncontrollable movements and restlessness as well as irritability and depression. The role of a defect in oxidative phosphorylation in HD patients is supported by the discovery of a threefold increase in lactate concentrations in the occipital cortex and in the basal ganglia. As for the PD, this is a progressive neurodegenerative movement disorder form of motor system degeneration affecting approximately 1% of the population over the age of 65 years. Clinical symptoms include bradykinesia, rigidity, postural instability, and resting tremor [348]. Experimental evidence supports the involvement of free radicals in the pathogenesis of this disease. It has been observed that the oxidation of dopamine yields potentially toxic semiquinones and that the accelerated metabolism of dopamine by monoamine-oxidase-B may induce an excessive formation of hydrogen peroxide, superoxide anions, and hydroxyl radicals. Further evidence of the involvement of free radicals comes from the fact that oxidative stress is responsible for the initiation of nigra dopamine neuron loss. The substantia nigra has a high metabolic rate combined with both a high content of oxidizable species, including dopamine and dopamine-derived ROS, neuromelanin, polyunsaturated fatty acids, iron, and a low content of antioxidants. Thus oxidative stress can dominate and results in the production of ROS, which serve both to maintain the oxidative stress level, and to initiate/propagate apoptosis of the

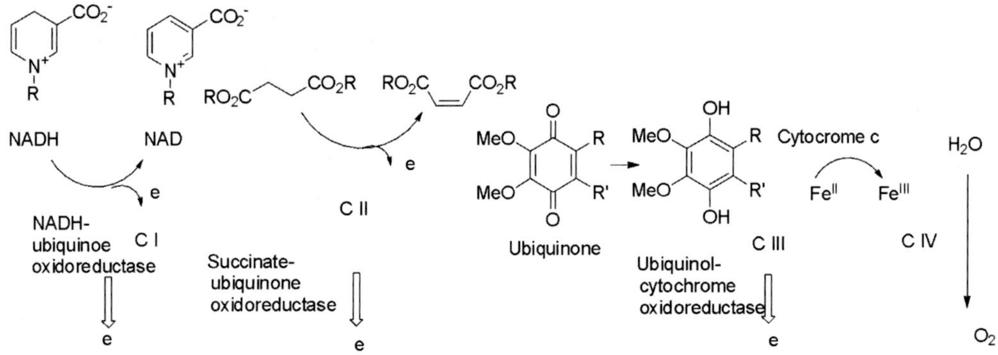

SCHEME 2.42 The respiratory chain and the different sources of electrons across the cell membrane. Leakage from complexes C I, C II, and C III causes an imbalance between the destructive characteristics of reactive oxygen species and the antioxidants protections paths.

dopaminergic neurons. PD has also been found to be associated with increased oxidative damage to DNA proteins and lipids, and further signs of oxidative damage in PD patients are supported by the finding that elevated levels of the prooxidant iron are present [349]. There is increasing evidence that free radical–induced damage also plays a significant part in the development of insulin resistance, β-cell dysfunction, impaired glucose tolerance, and type 2 diabetes mellitus. Hyperglycemia can induce oxidative stress, which increases with age, via several mechanisms including glucose autooxidation, the formation of advanced glycation end-products (AGE), and activation of the polyol pathway. Other circulating factors that are elevated in diabetics such as free fatty acids and leptin also contribute to increase ROS. There is a significant increase in protein glycation (AGE) with age, which is also increased in diabetics. The accumulation of AGE leads to an increase in the microvascular lesions, which are present in diabetic retinopathy, and is also responsible for cardiovascular complications, which are seen in diabetic patients [350]. The damage caused by ROS has also been implicated in primary open angle glaucoma, which is the leading cause of irreversible blindness and the second most common cause of all blindness after cataracts [191].

Epigenetic changes are reversible heritable changes, and can be thus used for the prevention of various types of cancer. DNA methylation is the most characterized epigenetic mechanism that can be inherited without changing the DNA sequence. When exposed to UVR, tumor suppressor genes in the epidermis lead to photocarcinogenesis, which is actually associated with a network of epigenetic modifications, such as alterations in DNA methylation, DNA methyltransferases, and histone acetylations. Various bioactive dietary components have been shown to protect the skin from UVR-induced skin tumors in animal models. Bioactive dietary components such as (-)-epicatechins from green tea and proanthocyanidins from grape seeds have been assessed in chemoprevention of UV-induced skin carcinogenesis and underlying epigenetic mechanism in vitro and in vivo animal models. These bioactive components block UV-induced DNA hypermethylation and histone modifications in the skin required for the silencing of tumor suppressor genes (e.g., Cip1/p21, p16) [228,351].

The globally rising of skin cancer rates has fostered studies carried out under ISO standards (ISO 24444:2010), where a comprehensive determination of the properties of various active ingredients has been carried out. It turned out that barriers about the type of action are not absolute and different mechanisms operate. Visible and IR radiations, as well as heat accumulation, are able to activate different signal transduction pathways and cause an enhanced oxidative stress and premature aging. Different cellular compartments are involved (i.e., mitochondrial cytochrome c complex that absorbs IR-A, lipids in the cell membranes, DNA in the nuclei for UV-B), while the temperature is increased to 40°C and above when exposed to the sun, and a synergic effect may result. Obviously, it is important that the correct properties are clearly indicated. Recent studies report several parameters of sunscreens, viz the universal SPF factor (i.e., sensitive also to UV-A) and the quite sensitive measurement of the EPR signals of radicals are often adopted for a better characterization. The spectroscopic universal sun protection factor is measured by spreading the prescribed amount of the preparation and taking it away by using a tape. The radical formation ratio via EPR spectroscopy is measured by using porcine ear skin ex vivo. All of these quantities are evaluated in order to characterize (and appropriately label for commercial purposes) the preparations [352].

2.21 Autoimmune diseases

Psoriasis is histologically characterized by KC hyperproliferation, inflammation, and increased angiogenesis, but the pathological factor responsible for these symptoms is unknown. Here a neuroendocrine peptide (prokineticin 2, PK2), is highly expressed in human and mouse psoriatic skins but no significant change in other autoimmune diseases, suggesting that PK2 is a psoriasis-specific factor. Bacterial products significantly upregulated PK2, implying that infection induces PK2 overexpression. PK2 promoted KC and macrophage to produce IL-1, the central player of inflammation and psoriasis, which acts on adjacent fibroblast to induce inflammatory cascades and KC hyperproliferation. IL-1 feeds back on macrophages to induce PK2 production to perpetuate PK2-IL-1-positive feedback loop. PK2 also promoted angiogenesis, another psoriatic symptom. In mouse models, PK2 overexpression aggravated psoriasis while its knock-down inhibited pathological development. The results indicate that PK2 overproduction perpetuates psoriatic symptoms by creating PK-2-IL-1 vicious loop. PK2 is a central player in psoriasis and a promising psoriasis-specific target [353].

Solar light is proved to induce cancer, but this involves a variety of mechanisms that occur in parallel. Thus, pyrimidine dimers are generated by irradiation, but furthermore oxygen sensitization leads to ROS and thus to peroxidation reactions. Attack to the *bis*-allyl position in polyunsaturated lipids from membranes cleaves such molecules (exemplified by linoleic and arachidonic acid) and produce oxidized intermediates (in particular, 4-HNE). 8-Hydroxyguanosine is formed in the process and causes inflammation, mutation, contrasted by repair mechanism [354].

On the other hand, constitutive nitric oxide synthase (NOS) activated by UV exposure and iNOS is induced by the inflammatory cytokines, and this leads to the production of nitric oxide (NO), followed by peroxynitrite ($ONOO^-$) formation, which modifies the protein and results in the formation of nitrotyrosine [354]. From the clinical point of view, it has been demonstrated that living at higher latitudes and having less sun exposure increases the risk for many chronic illnesses, infectious diseases, and mortality. Even the birth month affects lifetime disease risk. A recent study reported widespread seasonal gene expression demonstrating a marked annual difference in genes affecting immunity [355]. Risk factors for melanoma include a number of sun burn experiences as a child and young adult, genetic predisposition, and red hair color [356].

LE is a multifactorial autoimmune disease where the immune system becomes aggressive and attacks healthy cells. Different systems can be hit by the disease, including the skin, kidney, blood cells, heart, and the lungs. Skin lesions develop upon UV (A and B) light exposure. This causes the aberrant induction of apoptosis in KCs and thus the accumulation of apoptotic cells in the skin. These cannot be removed by phagocytoses and release proinflammatory compounds and potential autoantigens. In this way an inflammatory microenvironment is generated and the lesions developed. Molecular and cellular mediators that cause a long-lasting photosenstivity include mediators of inflammation, such as cytokines and chemokines. In particular, interferons have an important role, especially during the first phase. Lesions similar to those caused by LE are also caused by interferon of type 1 (α and β) [357].

Intense pulse light from a laser has been used for revaluating the possibility of its using for vascular dermatosis. The treatment was successful, at least for teleangiectasis, arterial spider nevus, and strawberry nevus there was no obvious side-effect [358]. Topical delivery of conveniently impaired RNAi robed with ionic liquids can be used for treating skin diseases in an efficient and safe way, as confirmed by their ability to limit breakdown of elastin, otherwise a major cause of premature aging [359].

ROS increase cytosolic calcium level as well as the expression of some microRNAs and one may wonder on which the relations between the two mechanisms are. In a recent review, three groups of microRNAs have been distinguished: those that are modulated by both Ca^{2+} and ROS, those that are modulated by ROS and have an effect on Ca^{2+}, and those that modulated neither ROS nor Ca^{2+} [360].

In atopic dermatite, the activity of immune-inflammatory cells is increased, and the release of neuromediators by sensory nerve endings, leading to neurogenic inflammation and pruritus. In addition, skin cells also produce neural factors, and thus they can communicate with adjacent nerves to increase inflammatory and pruritic responses.

Peripheral nerve branches reach the dermis and move upward into the epidermis, and the nerve growth factor and the other members of the neurotrophin family are synthesized and released by KCs and then transported "antidromically" to the dorsal root ganglia where they regulate the synthesis and differentiation of neurons, as well as the production and release of neuropeptides in the skin, resulting in neurogenic inflammation and/or pruritus. For example, calcitonin gene–related peptide, substance P, and pituitary adenylate cyclase–activating peptide are expressed differently in several inflammations where they contribute either by a direct action on blood vessels or by stimulation of MCs to release histamine and other inflammatory mediators. The scratching behavior significantly correlates with high nerve growth factor levels may be due to the interaction of this factor, released in high levels from KCs in atopic dermatite patients and further enhanced by calcitonin gene related in an autocrine manner, as well as neuropeptide induced histamine and trypase release from MCs, contribute to pruritus in atopic dermatite. On the other hand, pure immunological models thus far fail to integrate the recognized interplay between the epidermis and the nervous system. In addition, given the reduced expression of filaggrin, it is tempting to speculate that the first trigger in the pathogenesis of atopic dermatite could take place [361,362].

NER defect shows its presence by affecting the phenotype. Three different inborn diseases are characterized by hypersensitivity of the skin to UV light, although showing a remarkable clinical and genetic heterogeneity. The prototype repair syndromes are XP, Cockayne's syndrome, and a photosensitive form of the brittle hair trichothiodystrophy. In all three conditions is evident an extreme sensitivity to sunlight. In XP patients, a great increased frequency of sunlight-induced skin cancer is exhibited [363].

It has been reported above that exposure to solar light causes reddening and itching that may then evolve all the way to different forms of cancer. The actual situation is much more complex and a variety of syndromes of different seriousness have been described to result from the exposure to solar light, although it is outside the scope of the present discussion to consider a detail of these disorders. A typical example is urticaria, of which here are many types. Representative manifestations include wheals, disorders of fleeting nature characterized by a central swelling of variable size, almost invariably surrounded by a reflex erythema, to which an itching, or sometimes, a burning sensation is associated, but where the skin returns to the normal within a few days. Collectively, these disorders are referred to as dermatoses or dermatitis when involving an inflammation (actually the two terms are exchanged in the everyday language).

On the other hand, an angioedema is characterized by a sudden, pronounced swelling of the lower dermis and subcutis associated with pain rather than itching and often involves of the tissue below mucous membranes and a slower resolution than with wheals, up to a few days. The symptoms of hives raised itchy bumps, either red or skin-colored, and -dilatation of the postcapillary venules and lymphatic vessels of the upper dermis. As for the biological frame, skin affected by wheals practically always shows upregulation of endothelial adhesion molecules. In turn, this leads to an inflammatory perivascular infiltrate of variable intensity, consisting of neutrophils and/or eosinophils, macrophages, and T cells and, in some case, a mild-to-moderate increase of MC numbers [364,365]. This syndrome leads to a variety of diagnosis, which vary among eczema, asthma, allergic rhinitis, and food allergy in children [366]. The most recent examination [367] carried out shows that there is no difference in the types of a clinical description related to sex or race, but a large dependence on the age at which such phenomena manifest themselves. As it is apparent from Table 2.3.

2.21.1 Disorders related to erythema

It has been reported above that exposure to solar light causes reddening and itching, that may then evolve to different forms of cancer. The actual situation is much more complex and a variety of syndromes of different seriousness have been described to result from the exposure to solar light, although it is outside the scope of the present discussion to consider a detail of these disorders. A typical example is urticaria, of which here are many types. Representative manifestations include wheals, disorders of fleeting nature characterized by a central swelling of variable size, almost invariably surrounded by a reflex erythema, to which an itching, or sometimes, a burning sensation is associated, but where the skin returns to the normal state within a few days. Collectively, these disorders are referred to as dermatoses or dermatitis when involving an inflammation (actually the two terms are exchanged in the everyday language).

Furthermore, there is a high risk of developing asthma among people who has had a diagnosis or food allergy, and an even stronger one among those that had multiple allergy [368]. The interconnection between food allergy and asthma goes further than a simple comorbidity. Thus there is a peculiar predisposition by food allergy to asthma and the mutual interactions range from respiratory symptoms and bronchial hyperreactivity during food-induced anaphylaxis to severe asthma due to cross-reactive food allergens and to occupational asthma upon exposure to airborne allergens, and possibly even severe and sometimes fatal anaphylactic reactions.

Urticarial lesions are polymorphic, round, or irregularly shaped pruritic wheals that range in size from a few millimeters to several centimeters and can develop anywhere on the body, in the most unfortunate cases they are spread by scratching, combining into large, fiery-red patches.

Angioedema, which can occur alone or with urticaria, is characterized by nonpitting, nonpruritic, well-defined, edematous swelling that involves subcutaneous tissues (e.g., face, hands, buttocks, genitals), abdominal organs, or the upper airway (i.e., larynx). Angioedema tends to occur on the face and may cause significant disfigurement. Laryngeal angioedema is a medical emergency requiring prompt assessment. Acute intestinal and stomach swelling may initially be confused with symptoms of an abdominal surgical emergency. As for the time course, wheals and flare reactions last fewer than six weeks in patients with acute urticaria, but chronic urticaria can persist for months or years. Chronic urticaria of unknown etiology often shows spontaneous remission within one year. Patients with physical urticaria (i.e., lesions produced by physical stimuli) had more persistent disease, with only 16% undergoing spontaneous remission (see Table 2.4) [369].

Time course differentiates subcategories of urticaria. Wheal and flare reactions last fewer than six weeks in patients with acute urticaria, but chronic urticaria can persist for months or years. In a recent study, 47% of patients with chronic urticaria of unknown etiology had spontaneous remission after 1 year [370]. Patients with physical urticaria (i.e., lesions produced by physical stimuli) had more persistent disease, with only 16% undergoing spontaneous remission.

There are different manifestations of urticaria, including wheals, disorders of fleeting nature characterized by a central swelling of variable size almost invariably surrounded by a reflex erythema, to which an itching or sometimes a burning sensation is associated, but ceases where the skin returns to the normal state.

On the other hand, an angioedema is characterized by a sudden, pronounced swelling of the lower dermis and subcutis associated with pain rather than itching and the frequent involvement below mucous membranes and a slower resolution than with wheals up to a few days. Such an edema of the upper and mid-dermis, forms with dilatation of the postcapillary venules and lymphatic vessels of the upper dermis. Skin affected by wheals virtually always exhibits upregulation of endothelial adhesion molecules and a mixed inflammatory perivascular infiltrate of variable intensity, consisting of neutrophils and/or eosinophils, macrophages, and T cells and, in some case, a mild-to-moderate increase of MC numbers.

Solar urticaria, light-induced urticaria [371], is a rare occurrence among those that have a physical origin, and has to be distinguished from heat urticaria, that originates from humidity on the skin. It should be diagnosed differently treated with antihistamines and sun-screens, if sensitivity is to UV wavelengths. Sensitivity to visible light wavelengths is particularly difficult as symptoms can occur indoors as well as outdoors. Delayed pressure urticaria is an exception where symptoms more closely resemble chronic spontaneous urticaria (with which it is commonly associated) and responds poorly to antihistamines [372]. It can be treated with cyclosporine or, perhaps, with the antibody omalizumab.

Histamine and the histamine receptors are important regulators of a plethora of biological processes, including immediate hypersensitivity. As part of an immune response to foreign

TABLE 2.3 Comorbidities associated with psoriasis [348].

Classic	Psoriatic arthritis
	Inflammatory bowel disease
	Psychological and psychiatric disorders
	Uveitis
Emerging	Metabolic syndrome and its components
	Cardiovascular diseases
	Atherosclerosis
	Nonalcoholic fatty liver disease
	Lymphomas
	Sleep apnea
	Chronic obstructive pulmonary disease
	Osteoporosis
	Parkinson's disease
	Celiac disease
	Erectile dysfunction
Related to lifestyle	Smoking habit
	Alcoholism
	Anxiety
Related to treatment	Dyslipidemia (acitretin and cyclosporine)
	Nephrotoxicity (cyclosporine)
	Hypertension (cyclosporine)
	Hepatotoxicity (methotrexate, leflunomide, and acitretin)
	Skin cancer (PUVA)

pathogens, histamine is produced by basophils and by MCs found in nearby connective tissues. Histamine increases the permeability of the capillaries to white blood cells and some proteins, to allow them to engage pathogens in the infected tissues. These may be contrasted by specific drugs, known as antihistamincs that are chose within a large family of amines, in most cases tertiary benzylamines. In the case of photoallergy a combination of light (to avoid absorption of the high energy part of the spectrum and antihistaminics gives the best combination) [373].

A score system used for comparing the symptoms patents present add their evolution [374], based on the number of wheals and heir annoying or troublesome nature.

The symptoms of solar urticaria are a red rash that appears on the skin after exposure to sunlight, along with itching, a stinging pain a rash that disappears within a few hours of ceasing sun exposure. Urticaria can occur in all age groups. Acute spontaneous urticaria is common in infants and young children, in the atopic variant.

The symptoms may sometimes restrict normal daily life. Treatment with cyclosporin A worked as the strongest treatment for patients with antihistamine, PUVA, or chloroquine phosphate without effect. Cyclosporin was given in a dose of 4.5 mg/kg body weight/day. Photo testing before, during, and after treatment showed a decreased light sensitivity to UVA, UVB, and visible light during cyclosporin treatment compared with phototesting before therapy. The patient could be out in the sun for at least 1 h with minimal urticaria during cyclosporin therapy compared with only a few minutes previously. However, one to two weeks after cyclosporin therapy was discontinued, skin symptoms returned. Cyclosporin therapy is a possible treatment in severe cases of solar urticaria [375] where other treatments have failed, especially in countries where treatment is necessary only for a few months during summer.

A comparison between different treatments of solar urticaria showed that a short course of PUVA therapy produced a marked increase in the minimal dose of radiation required to produce urticaria and resulted. Ina greatly increased tolerance to sun-exposure. Chlorpheniramine, an antihistamine, produced a slight increase in the minimal dose of radiation necessary to produce urticaria but its effectiveness was limited by side-effects. Indomethacin, an inhibitor of prostaglandin synthetase, produced no beneficial effect. Chronic urticaria is idiopathic in 80%-90% of cases and in severe cases, people with this condition may experience secondary symptoms, such as light-headedness, headache, nausea, and vomiting.

Treatment with antihistaminic eliminate the symptoms, but not the cause of the disease, which requires in depth examination through appropriate tests.

Carrying out diagnostic phototest, where small areas of skin are exposed to different strengths of UV light and it can be established whether specific rash reaction seen in solar urticaria develops. Alternatively, photopatch test [376], where small patches containing different allergens, is applied to the skin. When they are removed, the skin is exposed to light, and specific reactions are noted for any reaction(s).

Photoprovication test: Patches of skin are exposed to different types and strengths of UV

TABLE 2.4 Causes of anaphylaxis [354].

Common

- Foods: Most commonly peanuts, tree nuts, egg, fish, shellfish, cow's milk, and wheat
- Medications: Most commonly antibiotics and NSAID
- Allergen immunotherapy
- Insect stings (bees and wasps)
- Unidentified (no cause found; idiopathic anaphylaxis)

Less common

- Exercise
- Natural rubber latex
- Semen
- Hormonal changes: Menstrual factors
- Topical medications (e.g., chlorhexidine, polysporin)
- Transfusions

light over several daysand physicians take note of any reactions.

These drugs help combat the histamines that cause the rash and can reduce the redness, stinging, and itching. Antihistamines are the best way to treat solar urticaria in the short term as they provide quick relief from the main symptoms (Scheme 2.43).

Desensitization is a long-term treatment that aims to prevent the allergic skin reaction from happening. Desensitization involves treating the skin with a course of UV light exposure (phototherapy) to try to desensitize it. Over time this can prevent rash developing on exposure to light, or make it less severe. Immunosuppressant (see formulae at the bottom of Scheme 2.43) drugs suppress the immune response that occurs when the skin reacts, preventing histamine being produced. These are potent drugs that may have other side-effects. So this course of treatment is only ever short term and is only recommended for extreme cases. It could be that something in a person's diet is aggravating the reaction to sunlight. They may find removing potential allergens from their diet helps. However, this is a complementary treatment as opposed to the first course of action and more research needs to be conducted to say whether diet is a key factor in this condition [377].

Recommendations for avoiding the negative consequences of exposure to sunlight include wearing loose, dark clothing that covers as much skin as possible; wearing hats with wide brims; carrying a parasol when sitting in the open. A papular urticaria, on the other hand, is the reaction of the skin or better of the immunological system to beating to the

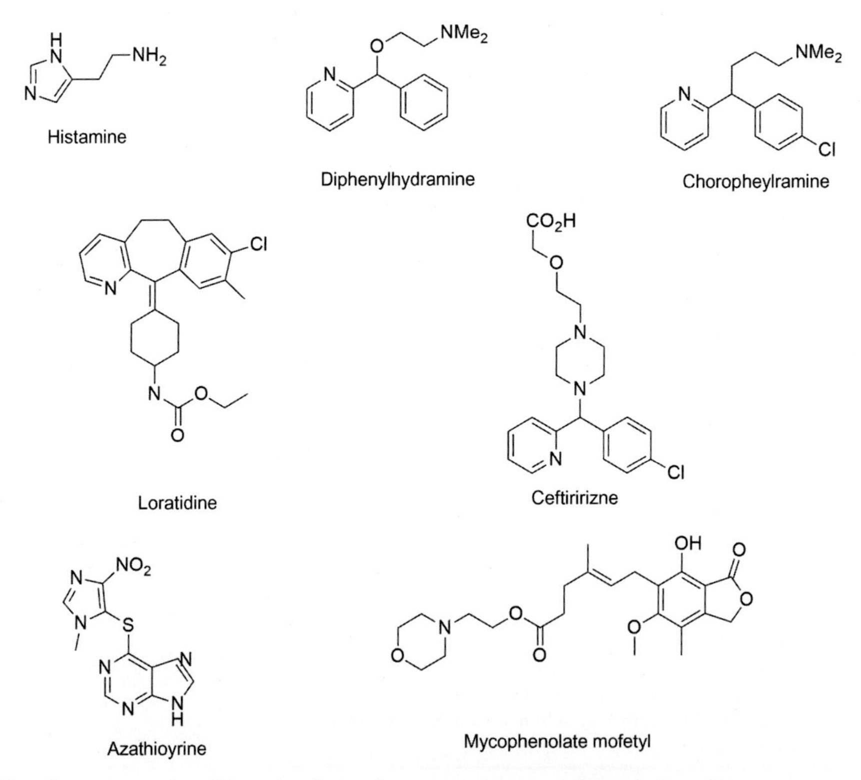

SCHEME 2.43 Histamine and antihistamine drugs. Immunosuppressant drugs, bottom line.

reaction of the hypothetic assault. An insect bite or sting often causes a small, red lump on the skin, which may be painful and itchy[378].

Many bites will clear up within a few hours or days and can be safely treated at home.

It can be difficult to identify which insect bit the patient, but this is of no worry since the treatment for most bites and stings is quite similar. A wasp or hornet sting causes a sudden, sharp pain at first. A swollen red mark may then form on the skin, which can last a few hours and may be painful and itchy.

Sometimes a larger area around the sting can be painful, red, and swollen for up to a week [379]. This is a minor allergic reaction that is not usually anything to worry about.

A few people may experience a serious allergic reaction (anaphylaxis), causing breathing difficulties, dizziness, and a swollen face or mouth.

The sting can cause pain, redness, and swelling for a few hours. As with wasp stings, some people may have a mild allergic reaction that lasts up to a week.

Serious allergic reactions can also occasionally occur, causing breathing difficulties, dizziness, and a swollen face or mouth.

Acute urticaria also is differentiated from chronic urticaria by ongoing or longstanding urticarial symptoms that trigger an IgE-mediated process [380]. Chronic urticaria and angioedema tend to be idiopathic, with no identifiable cause, or to be precipitated by a multitude of endogenous or exogenous factors that can be immunologic or nonimmunologic. The evaluation of patients with urticaria requires a thorough history of travel, recent infection, occupational exposure, medications (prescription drugs and herbal and vitamin supplements), ingestion of foods, timing, and onset of lesions, morphology, and associated symptoms. Family medical history, preexisting allergies, and exposure to physical stimuli should be documented. A comprehensive physical examination can uncover important diagnostic clues that may help diagnose comorbidities. Physicians should ensure that proper health maintenance testing is up to date and consider diagnostic testing directed by history and physical examination findings, especially in patients with chronic urticaria [381].

It should also be taken into account that although physically initiated disorders are usually less severe, the role of solar irradiation cannot been discounted [382]. Acute urticaria is self-limited and requires minimal laboratory evaluation. In asymptomatic patients with chronic urticaria and minimal history or physical examination findings, clinical practice guidelines suggest consideration of a complete blood analysis.

A recent systematic review of more than 6000 patients with urticaria and angioedema found that routine laboratory screening tests independent of the patient's history and physical examination should be discouraged, because these tests are of little value in discovering the cause of the reactions [383]. Clinical scenarios may offer diagnostic clues. For example, an IgE-mediated reaction would be suspected in patients with a history of acute urticaria within an hour after ingesting a food or drug. Skin testing or radioallergosorbent testing may document a causal relationship.

Histamine released from cutaneous MCs and basophils in response to inciting stimuli is the primary mediator of urticaria. In this process, specific IgE antibodies cross-link the IgE receptors bound to MCs and stimulate the production of preformed and newly generated inflammatory mediators. Complement anaphylatoxins also may induce MC histamine release, as can certain medications or physical stimuli through direct nonimmunologic MC activation.

Immunologic urticaria and angioedema are a result of IgE antibody-mediated reactions that usually occur within 1 h of exposure to the allergen. Type I IgE-mediated allergic reactions can be caused by drugs (most notably penicillin and cephalosporin), insect venom, foods (e.g., fish, shellfish, eggs, nuts, legumes, milk, soy, wheat), preservatives, latex, and aeroallergens (e.g., dust mites, molds, pollens, animal dander) [384]. Upregulation of TNF-α and IL-3 expression in lesional and uninvolved skin in different types of urticaria.

A rarer cause of acute urticaria is a type II hypersensitivity reaction mediated by cytotoxic antibodies and complement activation. An example is a transfusion reaction in which IgG and IgM anti-red-cell antibodies activate complement and cause cell lysis [385]. Serum sickness caused by drugs or proteins is an example of a type III hypersensitivity (antigen-antibody complex—mediated) reaction. Clinical presentation may include urticaria of several weeks' duration, arthralgias, fever, and glomerulonephritis [386]. Urticaria also has been associated with herpes virus, cytomegalovirus, Epstein-Barr virus, and chronic hepatitis infections, and with bacterial, fungal, and parasitic infections. Patients with chronic urticaria.

Papular urticaria, an eruption of erythematous wheals, that is a common and often annoying disorder manifested by chronic or recurrent papules caused by a hypersensitivity reaction to the bites of mosquitoes, fleas, bedbugs, and other insects. Individual papules may surround a wheal and display a central punctum.

Urticarial vasculitis should be considered if either a single urticaria lesion lasts longer than 24 h, if lesions are burning or painful, if they are more common in the lower extremities, and if they leave an area of hemosiderin pigment after they have resolved [387]. Infection, drug sensitivity, serum sickness, chronic hepatitis, hypocomplementemia urticarial vasculitis syndrome, and systemic LE may cause urticarial vasculitis. A 4-mm punch biopsy may confirm neutrophilic infiltration consistent with leukocytoclastic vasculitis [388]. Patients with urticarial vasculitis should be referred to a dermatologist for further evaluation and management, which may include immunomodulatory therapy to downregulate the inflammatory response.

Leukocytoclastic vasculitis is due to the vascular damage caused to the cells by the debris of the infiltrating neutrophils, and classically presents a palpable purpura [389]. Nonimmunologic responses to the contact with allergens are usually manifested fast, but are likewise fast to get away.

Herpesviridae are a family of viruses widely spread in nature that can infect a wide variety of species. After the primary infection, the human alpha herpes virinae subfamily remains quiescent in the nerve ganglia from which it can periodically reactivate, causing clinical manifestations. Although spontaneous recurrences are possible, a wide variety of internal and external triggers may lead to transformation of herpes simplex and varicella-zoster viruses from a dormant to a proliferative state. Sunlight is a potent stimulus for the alpha herpesvirinae reactivation. The purpose of this study is to analyze various features of this correlation and several steps you can take to lower your risk of triggering a herpes outbreak after sun exposure. Learning how to reduce the recurrence is extremely important and it is necessary: to perform a gradual and progressive sun exposure; to know what garments to wear; to know the environmental conditions of exposure; to know each skin phototype; to use a protective product against UVB and UVA with sun protection factor suitable for each phototype and environmental conditions.

Angioedema, which can occur alone or with urticaria, is characterized by nonpitting, nonpruritic, well-defined, edematous swelling that involves subcutaneous tissues, e.g., face (with significant disfigurement), hands, buttocks, genitals, abdominal organs, or the upper airway (i.e., larynx). Laryngeal angioedema is a medical emergency requiring prompt assessment. Acute intestinal and stomach swelling

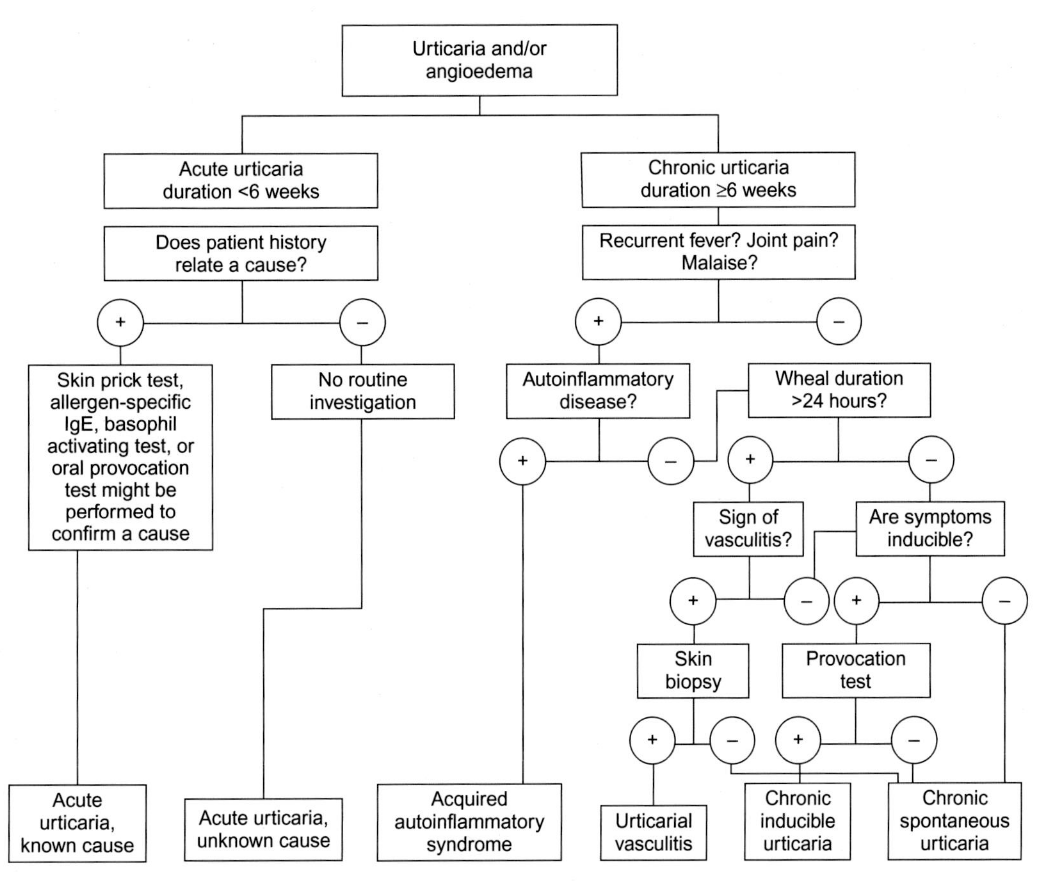

FIGURE 2.38 Algorithm for diagnosis and investigation of subtypes of urticaria. *IgE*, immunoglobulin E [391].

may initially be confused with symptoms of an abdominal surgical emergency.

Hives, also known as urticaria, affects about 20% of people at some time during their lives. When an allergic reaction occurs, the body releases a protein called histamine [390]. It is hypothesized that a photoallergen, produced from a skin "chromophore" on absorption of radiation of an appropriate wavelength, is recognized by specific IgE that binds to MCs, causing degranulation with release of histamine and other mediators. Chromophores, detected in serum samples from some patients who react to intradermal injection of their own irradiated serum (intradermal test), have been isolated but not yet identified. Each has a characteristic absorption spectrum. The action spectrum of Solar Urticartia (SU) is usually broad but can vary widely between patients and within an individual over time. With such heterogeneity, it would appear that a range of chromophores is responsible (Fig. 2.38).

In a cohort of 87 subjects, 25 underwent further phototesting at least 1 year after diagnosis. Six showed resolution, all with solar urticaria alone. The probability of phototest resolution was of 30% at 5 years and 15% at 10 years. Of the 23 patients who were still photosensitive at their second visit, the action spectrum was unchanged (or no further wavelengths were tested) in 8, broadened in 9, and narrowed in 6.

In a recent study, 47% of patients with chronic urticaria of unknown etiology had spontaneous remission after one year [391]. Patients with physical urticaria (i.e., lesions produced by physical stimuli) had more persistent disease, with only 16% undergoing spontaneous remission. In chronic urticaria ongoing or longstanding urticarial symptoms trigger an IgE-mediated process and tend to be idiopathic, with no identifiable cause, or to be precipitated by a multitude of endogenous or exogenous factors that can be immunologic or nonimmunologic. The evaluation of patients with urticaria and angioedema requires a thorough history of travel, recent infection, occupational exposure, medications (prescription drugs and herbal and vitamin supplements), ingestion of foods, timing and onset of lesions, morphology, and associated symptoms. Family medical history, preexisting allergies, and exposure to physical stimuli should be documented. A comprehensive examination can uncover important diagnostic clues that may help diagnose comorbidities. Physicians should ensure that proper health maintenance testing is up to date and consider diagnostic testing directed by history and physical examination findings, especially in patients with chronic urticaria, usually less severe when resulting from a physically initiated disorder.

Acne vulgaris is a very common disorder that has a prevalence of >90% and has a substantial impact on a patient's quality of life, affecting both self-esteem and psychosocial development, the worse side is sustained by advertisement, which leads both patients and physicians to more or less credible targets through the use of many over-the-counter and prescription acne treatments that makes choosing the most effective therapy confusing. The worst outcome is the formation of scars (~10% of the cases), which can be treated in various way, but still may be felt as an obstacle to normal societal contacts.

Acne (what is usually meant is acne vulgaris, the most common disorder of this group) is an inflammatory disorder of the pilosebaceous units that is prevalent in adolescence. The characteristic lesions are indicated in the literature as open (black) and closed (white) comedones (the technical word for pimples), inflammatory papules, pustules, nodules, and cysts, which may lead to scarring and pigmentary changes. A largely used scoring system has Grade 1: Simple noninflammatory acne—comedones and a few papules. Grade 2: Comedones, papules and a few pustules. Grade 3: Larger inflammatory papules, pustules, and a few cysts; a more severe form involving the face, neck, and upper portions of the trunk. Grade 4: More severe, with cysts becoming confluent. The pathogenesis of acne is multifactorial and includes abnormal follicular keratinization, increased production of sebum secondary to hyperandrogenism, proliferation of bacteria. These comorbidities will easy led to serious complications and in that case an important way out is modern technology for healing the scars based on light treatment (see Table 2.5) [392,393].

For mild cases of acne it is suggested to treat them with topical retinoids. For moderate cases, systemic drugs are always needed, including oral antibiotics, hormonal therapy, and oral retinoids. However, for severe or resistant moderate acne, isotretinoin is the treatment of choice. And indeed, isotretinoin is a vitamin A-derivative 13-*cis*-retinoic acid, which is the most effective therapy for acne to date. It targets all four symptoms during acne development, including normalization of follicular desquamation, reduction of sebaceous gland activity, inhibition of the proliferation of propionibacterium acnes and antiinflammatory effects. The metaanalysis suggested that isotretinoin cured around 85% of patients after an average treatment course of 4 months. Furthermore, there has been some discussion on whether the association with the use of

TABLE 2.5 Differential diagnosis of acne.

Diagnosis	Distinguishing features
Bacterial folliculitis	Abrupt eruption; spreads with scratching or shaving; variable distribution
Drug-induced acne	Use of androgens, adrenocorticotropic hormone, bromides, corticosteroids, oral contraceptives, iodides, isoniazid, lithium, phenytoin (Dilantin)
Hidradenitis suppurativa	Double comedo; starts as a painful boil; sinus tracts
Miliaria	"Heat rash" in response to exertion or heat exposure; nonfollicular papules, pustules, and vesicles
Perioral dermatitis	Papules and pustules confined to the chin and nasolabial folds; clear zone around the vermilion border
Pseudofolliculitis barbae	Affects curly haired persons who regularly shave closely
Rosacea	Erythema and telangiectasias; no comedones
Seborrheic dermatitis	Greasy scales and yellow-red coalescing macules or papules

Compiled from the data in A. Daglish, G. Waller, Clinician and patient characteristics and cognitions that influence weighing practice in cognitive-behavioral therapy for eating disorders, Int. J. Eat. Disord. 52 (9) (2019) 977–986; E. Halpert, E., Borrero, M. Ibanez-Pinilla, P. Chaparro, J. Molina, M. Torres, et al., Prevalence of papular urticaria caused by flea bites and associated factors in children 1–6 years of age in Bogotá, D.C, World Allergy Organ. J. 10 (2017) 36 [377,378].

topical topical isoretinal, known as the most active medication against acne may lead to a largely improved depression syndrome and although this does not appear clearly established at the moment, this point should be taken into account [394].

A different case is the use of other chemicals, in particular for doping, where the improper handing may lead to serious consequences [395].

The use (or misuse) (testosterone enantate 250 mg plus metandienone 30 mg twice weekly) in athletes, led to the diagnosis of a severe acne conglobata that were formed by anabolic–androgenic steroid and were accompanied a substantial impairment in sperm concentration and reduced testicular volume. Skin lesions showed rapid improvement after discontinuation of anabolic–androgenic steroid abuse and with antiseptic and antibiotic therapy. However, the extensive scarring is likely to remain with the young man for the rest of his life.

And yet another one is the treatment of chronic wounds, with the aim of removing any sloughy, necrotic, devitalized tissue to prevent wound infection and delayed healing. Biosurgery [396] then becomes a promising adjunct to the whole spectrum of topical treatment methods, in particular for debridement. The term "biosurgery" refers to the use of living maggots on wounds where a devitalized tissue has formed, thus achieving the above targets.

XP [397,398] is a genetic photosensitive disorder in which patients are highly susceptible to skin cancers on the sun-exposed body sites. In Japan, more than half of patients (30% worldwide) with XP show complications of idiopathic progressive, intractable neurological symptoms with poor prognoses. Therefore this disease does not merely present with dermatological symptoms, such as photosensitivity, pigmentary change and skin cancers, but is difficult to tract as neurological and dermatological disease. To the neurocutaneous syndromes that are subject to government research initiatives for overcoming intractable diseases. XP is one of the extremely serious photosensitive disorders in which patients easily develop multiple skin cancers if they are not completely protected from UVR. XP patients thus need to be strictly shielded from sunlight throughout their lives, and they often experience idiopathic neurodegenerative complications that markedly reduce the quality of life for both the patients and their families. Hospitals in Japan often see cases of XP as severely photosensitive in children, and as advanced pigmentary disorders of the sun-exposed area with multiple skin cancers in adults (aged 20–40s), making XP an important disease to differentiate in everyday clinical practice. It was thus decided that there was a strong need for clinical practice guidelines dedicated to XP. This process led to the creation of new clinical practice guidelines for XP.

2.21.2 Skin disorders not involving erythema

Some others dermatites manifest themself with the depigmentation rather than pigmentation (lichen, vitiligo, psoriasis, and Darier's keratitis). Lichen striatus is characterized by a form similar to a lichen in white on the darker skin, but there are further disorders [399,400]. Vitiligo is a chronic disorder, per se benign, but emotionally frustrating for both patients and physicians, since there is no treatment, that is, satisfactory in a short time or a few sittings. Vitiligo is an autoimmune disorder of depigmentation, with an incidence of about 2%–3%. Phototherapy is the firstline treatment in many cases, which needs to be given at frequent sittings for long periods of time. In this case, UV irradiation is the most common choice, obviously after filtering off the more oncogenic component and the method is known as photochemical UVA [401,402]. Psoriasis (from the Greek psora, itch; probably one of the "impure" diseases for the Hebraic tradition that imposed being restrained in the leprosy) is a chronic inflammation of the skin, not infective nor contagious often present independently on sex and age, although infants and adults are certainly not rarely attacked, is a disorder of complex pathogenesis, which include autoimmunity, genetic, and environmental factors. Important comorbidities are some CVDs [403,404], Crohn's disease, and depression. Darier's disease affects both men and women and is not contagious. The disease often starts during or later than the teenage years, typically by the third decade and appears to be more frequent among short stature people. The symptoms of the disease are thought to be caused by an abnormality in the desmosome–keratin filament complex leading to a breakdown in cell adhesion.

The symptoms may seriously restrict normal daily life. Treatment with cyclosporin A worked as the strongest treatment patients for antihistamine, PUVA, and chloroquine phosphate without side-effects [405]. It is known from a long time that some citrus fruits contain a strong medicament that is active under irradiation, in particular the oil of bergamot. The active derivatives are electron-rich hetero aromatics, bearing ketones or esters, able to abstract hydrogen from the triplet state. This principal has been developed with the synthesis of variety of compounds that absorb a large fraction of sun light and forming more persistent radical adduct in reactions where UV-A light is exploited (PUVA) [406,407].

References

[1] R. Kortum, E. Sevick, Quantitative optical spectroscopy for tissue diagnosis, Ann. Rev. Phys. Chem. 47 (1996) 555–606.

[2] C Ash, M Dubec, M Donne, K Bashford, Effect of wavelength and beam width on penetration in light-tissue interaction using computational methods, Proc. Phys. Soc. 32 (2017) 1909–1918.

[3] J.R. Pfaffin, E.N. Ziwegler, Encyclopedia of Envirnromntal Science and Engineering, vol. 1 A-L, CRC Press, Boca Raton, 2006.

[4] https://en.wikipedia.org/wiki/Cross_section_(physics).

[5] L.M. Chaudhari, R. Nathuram, Absorption coefficient of polymers (polyvinyl alcohol) by using gamma energy of 0.39 MeV, Bulg. J. Phys. 37 (2010) 232–240.

[6] C. Campillo, R. Fortes, M. Del Henar Prieto, Solar radiation effect on crop production, in: E.B. Babatunde (Ed.), Solar Radiation, InTech, Rijeka, 2012. Ch 11.

[7] M. Uyuklu, M. Canpolat, H.J. Meiselman, O.K. Baskurt, Wavelength selection in measuring red blood cell aggregation based on light transmittance, J. Biomed. Opt. 16 (2011) 117006.

[8] https://basicmedicalkey.com/the-skin-and-subcutaneous-tissue/#bru_ch16fg1fig2.

[9] Montreal Protocol, United Nations – Treaty Series (1989). 1522, I-26369, 1987.

[10] https://basicmedicalkey.com/the-skin-and-subcutaneous-tissue/#bru_ch16fg1.

[11] A.G. Hearn, The absorption of ozone in the ultraviolet and visible regions of the spectrum, Proc. Phys. Soc. 78 (1961) 932–940.

[12] C.E. Williamson, E.P. Overholt, J.A. Brentrup, R.M. Pilla, T.H. Leach, S.G. Schladow, et al., Sentinel responses to droughts, wildfires, and floods: effects of UV radiation on lakes and their ecosystem services, Front. Ecol. Environ. 2016 (14) (2016) 102–109.

[13] C.E. Williamson, E.P. Overholt, R.M. Pilla, T.H. Leach, J.A. Brentrup, L.B. Knoll, Ecological consequences of long-term browning in lakes, Sci. Rep. 2015 (2015) 18666.

[14] J. Kim, H.-K. Cho, J. Mok, H.D. Yoo, N. Cho, Effects of ozone and aerosol on surface UV radiation variability, J. Photochem. Photobiol. B Biol. 119 (2013) 46–51.

[15] K.V. Gorshelev, A. Serdyuchenko, M. Weber, W. Chehade, J.P. Burrows, High spectral resolution ozone absorption cross-sections – Part 1: measurements, data analysis and comparison with previous measurements around 293K, Atmos. Meas. Tech. 7 (2014) 609–624.

[16] S.O. Andersen, M.L. Halberstadt, N. Borgford-Parnell, Stratospheric ozone, global warming, and the principle of unintended consequences. An ongoing science and policy success story, J. Air Waste Manage. Assoc. 63 (2013) 607–647.

[17] F.S. Rowland, M.J. Molina, Ozone depletion: 20 years after the alarm, Chem. Eng. News 72 (1994) 8–13.

[18] Federal Register Protection of stratospheric ozone: The 2014 and 2015 critical use exemption from the phaseout of methyl bromide. (2014). r 79, 13006–13017.

[19] R. Muller, Introduction to stratospheric ozone depletion, in: R. Muller (Ed.), Stratospheric Ozone Depletion and Climate Change., RSC, UK, 2012, pp. 1–32.

[20] A.H. Soedergren, G.E. Bodeker, S. Kremser, M. Meinshausen, A.J. McDonald, A probabilistic study of the return of stratospheric ozone to 1960 levels, Geophys. Res. Lett. 43 (2016) 9289–9297.

[21] S. Solomon, J. Haskins, D.J. Ivy, F. Min, Fundamental differences between Arctic and Antarctic ozone depletion, PNAS 111 (2014) 6220–6225.

[22] A. Andrady, P.J. Aucamp, A.T. Austin, A.F. Bais, C.L. Ballaré, P.W. Barnes, et al., Environmental effects of ozone depletion and its interactions with climate change: progress report, 2016, Photochem. Photobiol. Sci. 16 (2017) 107–145.

[23] S.R. Yates, D. Wang, J. Gan, F.F. Ernst, W.A. Jury, Minimizing methyl bromide emissions from soil fumigation, Geophys. Res. Lett. 25 (1998) 1633–1636.

[24] A.F. Bais, R.L. McKenzie, G. Bernhard, P.J. Aucamp, M. Ilyas, S. Madronich, et al., Ozone depletion and climate change: impacts on UV radiation, Photochem. Photobiol. Sci. 14 (2015) 19–52.

[25] A. Robock, Volcanic eruption climate, Rev. Geophys. 38 (2000) 191–219.

[26] O. Boucher, Stratospheric ozone, ultraviolet radiation and climate change, Weather 65 (2010) 105–110.

[27] D.D. Bikle, Z. Xie, C.-L. Tu, Calcium regulation of keratinocyte differentiation, Rev. Endocrinol. Metab. 7 (4) (2012) 461–472.

[28] M. Bodenstein, C. Wagner, Ein Vorschlag fur die Bezeichnung der Lightmenge in der Photochemie, Z. Phys. Chem. 3 (B) (1929) 456–458.

[29] G.N. Lewis, The conservation of photons, Nature 118 (1926) 874–875.

[30] A. Einstein, Deduction thermodynamique de la loi de l'équvalence photochimique, J. Phys 3 (1913) 277–281.

[31] A.J. Allmand, Einstein's law of photochemical equivalence. Introductor address to part I, Trans. Faraday Soc. 21 (1926) 438–452.

[32] F. Daniells, A table of quantum yields in experimental photochemistry, J. Phys. Chem. 42 (1938) 713–732.

[33] A. Einstein, Über die Erzeugung und Verwandlung des Lichtes betreffenden heuristischen Gesiktpunkt, Ann. Phys. 17 (1905) 132–148.

[34] J.P. Ortonne, Photoprotective properties of skin melanin, Brit. J. Dermatol. 146 (2002) 7–10.

[35] B.H. Mahmoud, C.L. Hexsel, I.H. Hamzavi, H.W. Lim, Effects of visible light on the skin, Photochem. Photobiol. 84 (2008) 450–462.

[36] R. Schwartz, Squamous skin carcinoma, in: R. Schwartz (Ed.), Skin Cancer: Recognition and Management, Springer, NY, 1988, pp. 36–47.
[37] R.C. Romanhole, J.A. Ataide, P. Moriel, P.G. Mazzola, Update on ultraviolet A and B radiation generated by the sun and artificial lamps and their effects on skin, Int. J. Cosmet. Sci. 37 (2015) 366–370.
[38] J. Moan, M. Grigalavicius, Z. Baturaite, A. Dahlback, A. Juzeniene, The relationship between UV exposure and incidence of skin cancer, Photodermatol. Photoimmunol. Photomed. 31 (2015) 26–35.
[39] P.M. Sapkal, J.R. Pantwalawalkar, T.M. Kashalikar, Skin cancer: an overview, Int. J. Pharm. Pharm. Res. 5 (2016) 64–76.
[40] D.M. Holman, M. Watson, Correlates of intentional tanning among adolescents in the United States: A systematic review of the literature, J. Adolesc. Health. 52 (2013) S52–S59.
[41] F. Solano, Melanins: skin pigments and much more–types, structural models, biological functions, and formation routes, New J. Sci. (2014) 28.
[42] S.-H. Tseng, P. Bargo, A. Durkin, N. Kollias, Chromophore concentrations, absorption and scattering properties of human skin *in-vivo*, Opt. Express. 17 (2009) 14599–14617.
[43] C.R. Simpson, M. Kohl, M. Essenpreis, M. Cope, Near-infrared optical properties of ex vivo human skin and subcutaneous tissues measured using the Monte Carlo inversion technique, Phys. Med. Biol. 43 (1998) 2465–2478.
[44] R.L. Olson, R.M. Sayre, M.A. Everett, Effect of anatomic location and time on ultraviolet erythema, Arch. Dermatol. 93 (1968) 211–215.
[45] X.-M. Shen, F. Zhang, G. Dryhurst, Oxidation of dopamine in the presence of cysteine: characterization of new toxic products, Chem. Res. Toxicol. 10 (1997) 147–155.
[46] S. Sachdeva, R. Fitzpatrick, Skin typing: applications in dermatology, Indian J. Dermatol. Venerolog. Lepros. 75 (2009) 93–96.
[47] C. Battie, M. Verschoore, Cutaneous solar ultraviolet exposure and clinical aspects of photodamage, Indian J. Dermatol. Venereol. Leprol. 78 (Suppl 1) (2012) S9–S14.
[48] C. Battie, S. Jitsukawa, F. Bernerd, S. Del Bino, C. Marionnet, M. Verschoore, New insights in photoaging, UV-A induced damage and skin types, Exp. Dermatol. 23 (Suppl 1) (2014) 43–46.
[49] L.Kong, S.Sprigle, D.Yi, F.Wang, C.Wang, F.Liu, Developing handheld real time multispectral imager to clinically detect erythema in darkly pigmented skin, in: Proc. SPIE 7557, 2010, 75570G-1.
[50] M. Randha, I.S. Seo, F. Liebel, M.D. Southall, N. Kollias, E. Ruvolo, Visible light induces melanogenesis in human skin through a photoadaptive response, PLoS One (2015). Available from: https://doi.org/10.1371/journal.pone.0130949.
[51] L.R. Sklar, F. Almutawa, H.W. Lim, I. Hamzavi, Effects of ultraviolet radiation, visible light, and infrared radiation on erythema and pigmentation: a review, Photochem. Photobiol. Sci. 12 (2013) 54–64.
[52] C.M. Lerche, P.A. Philipsen, H.C. Wulf, UV-R: sun, lamps, pigmentation and vitamin D, Photochem. Photobiol. Sci. 16 (2017) 291–301.
[53] T. Kondo, V.J. Hearing, Update on the regulation of mammalian melanocyte function and skin pigmentation, Expert. Rev. Dermatol. 6 (2011) 97–108.
[54] N.K. Haass, M. Herlyn, Normal human melanocyte homeostasis as a paradigm for understanding melanoma, J. Invest. Dermatol. Symp. Proc. 10 (2005) 153–163.
[55] P.M. Plonka, T. Passeron, M. Brenner, D.J. Tobin, S. Shibahara, A. Thomas, et al., What are melanocytes really doing all day long...? Exp. Dermatol. 18 (2009) 799–819.
[56] T. Tadokoro, Y. Yamaguchi, J. Batzer, S.G. Coelho, B. Z.Z. Zmudzka, S.A. Miller, et al., Mechanisms of skin tanning in different racial/ethnic groups in response to ultraviolet radiation, J. Invest. Dermatol. 124 (2005) 1326–1332.
[57] Yuji Yamaguchi, et al., J. Biol. Chem. 282 (2007) 27557–27561.
[58] C. Delevoye, Melanin transfer: the keratinocytes are more than gluttons, J. Invest. Dermatol. 134 (2014) 877–879.
[59] H. Ando, Y. Niki, M. Ito, K. Akiyama, M.S. Matsui, D. B. Yarosh, et al., Melanosomes are transferred from melanocytes to keratinocytes through the processes of packaging, release, uptake, and dispersion, J. Invest. Dermatol. 132 (2012) 1222–1229.
[60] P. Talalay, J.W. Fahey, Z.R. Healy, S.L. Wehage, A.L. Benedict, C. Min, et al., Sulforaphane mobilizes cellular defenses that protect skin against damage by UV radiation, PNAS 104 (2007) 17500–17505.
[61] Y. Miyamura, S.G. Coelho, K. Schlenz, J. Batzer, C. Smuda, W. Choi, The deceptive nature of UV-A tanning versus the modest protective effects of UV-B tanning on human skin, Pig. Cell Melanoma R. 24 (2011) 136–147.
[62] M. Cichorek, M. Wachulska, A. Stasiewicz, A. Tymińska, Skin melanocytes: biology and development, Postep. Derm. Alergol. 30 (2013) 30–41.
[63] J.P. Ebanks, R.R. Wickett, R.E. Boissy, Mechanisms regulating skin pigmentation: the rise and fall of

complexion coloration, Int. J. Mol. Sci. 10 (2009) 4066–4087.

[64] H. Van der Rhee, H. De Vries, C. Coomans, P. Van de Velde, J.W. Coebergh, A review of positive and negative effects of sun exposure, Cancer Res. Front. 2 (2016) 156–183.

[65] I. Ferreira dos Santos Videira, D.F. Lima Moura, S. Magina, Mechanisms regulating melanogenesis, An. Bras. Dermatol. 88 (2013) 76–83.

[66] V.J. Hearing, Biogenesis of pigment granules: a sensitive way to regulate melanocyte function, J. Dermatol. Sci. 37 (2005) 3–14.

[67] Z. Klimová, J. Hojerová, M. Beránková, Skin absorption and human exposure estimation of three widely discussed UV filters in sunscreens – in vitro study mimicking real-life consumer habits, Food Chem. Toxicol. 83 (2016) 237–250.

[68] M. Ichihashi, H. Ando, The maximal cumulative solar UV-B dose allowed to maintain healthy and young skin and prevent premature photoaging, Exp. Dermatol. 23 (Suppl 1) (2014) 43–46.

[69] B.L. Diffey, P.M. Farr, Sunscreen protection against UV-B, UV-A and blue light: an in vivo and in vitro comparison, Br. J. Dermatol. 124 (1991) 258–263.

[70] P.M. Farr, B.L. Diffey, How reliable are sunscreen protection factors? Br. J. Dermatol. 112 (1985) 113–118.

[71] ISO/TR 26369, Cosmetics—Sun Protection Test Methods — Review and Evaluation of Methods to Assess the Photoprotection of Sun Protection Products, and of Advanced Studies on the Practical Aspects of Applying Sun Creams, 2009.

[72] S.G. Otman, C. Edwards, B. Gambles, A.V. Anstey, Validation of a semiautomated method of minimal erythema dose testing for narrow band ultraviolet B phototherapy, Br. J. Dermatol. 155 (2006) 416–421.

[73] B.L. Diffey, A mathematical optics for ultraviolet optics in the skin, Phys. Med. Biol. 28 (1938) 647–657.

[74] CIE 2014. CIE 209, Rationalizing Nomenclature for UV Doses and Effects on Humans, Vienna, 22p., 2014.

[75] N. Maddodi, A. Jayanthy, V. Setaluri, Shining light on skin pigmentation: the darker and the brighter side of effects of UV radiation, Photochem. Photobiol. 88 (2012) 1075–1082.

[76] Y. Zhang, K.L. Helke, S.G. Coelho, J.C. Valencia, V.J. Hearing, S. Sun, et al., Essential role of the molecular chaperone gp96 in regulating melanogenesis, Pig. Cell Melanoma R. 27 (2014) 82–89.

[77] B.L. Diffey, When should sunscreen be reapplied? J. Am. Acad. Dermatol. 45 (2001) 882–885. 200.

[78] F.P. Gasparro, M. Mitchnick, J.F. Nash, A review of sunscreen safety and efficiency, Photochem. Photobiol. 68 (1998) 243–256.

[79] S. Seite, A. Fourtanier, D. Moyal, A.R. Young, Photodamage to human skin by suberythemal exposure to solar ultraviolet radiation can be attenuated by sunscreens: a review, Br. J. Dermatol 163 (2010) 903–914.

[80] C. Marionnet, C. Tricaud, F. Bernerd, Exposure to non-extreme solar UV daylight: spectral characterization, effects on skin and photoprotection, Int. J. Mol. Sci. 2015 (16) (2015) 68–90.

[81] D. Garoli, M.G. Pelizzo, P. Nicolosi, A. Peserico, E. Tonin, M. Alaibac, Effectiveness of different substrate materials for in vitro sunscreen tests, J. Dermatol. Sci. 56 (2009) 89–98.

[82] J.P. Townley, K.A. Greive, Protection from skin damage due to *visible light*, Aust. J. Dermatol. 54 (suppl. 2) (2013) 40.

[83] C.A. Langley, D. Belcher, Pharmaceutical Compounding and Dispensing, second ed., Pharmaceutical Press, London UK, 2012.

[84] M.E. Burnett, S.Q. Wang, Current sunscreen controversies: a critical review, Photodermatol., Photoimmunol. Photomed. 27 (2011) 58–67.

[85] M.T. Ignasiak, C. Houe-Levin, G. Kciuk, T. Pedzinski, A reevaluation of the photolytic properties of 2-hydroxybenzophenone-based UV sunscreens: are chemical sunscreens inoffensive? Chem. Phys. Chem. 16 (2015) 628–633.

[86] S.N. Nisakorn, A. Jimtaisong, Photoprotection of natural flavanoids, J.Appl. Pharm. Sci. 3 (2013) 129–141.

[87] J.A. Nichols, K. Santosh, Skin photoprotection by natural polyphenols: anti-inflammatory, anti-oxidant and DNA repair mechanisms, Arch. Dermatol. Res. 2302 (2010) 71.

[88] D.J. Tobin, Aging of the hair follicle pigmentation system, Int. J. Tricol. 1 (2009) 83–93.

[89] R. Colomer, A. Sarrats, R. Lupu, T. Puig, Natural polyphenols and their synthetic analogs as emerging anticancer agents (2017), Curr. Drug Targets (2017) 18147–18159.

[90] M.S. Latha, J. Martis, V. Shobha, H. Sudhakar Bangera, B. Shantala Bellary, B. Varughese, et al., Sunscreening agents. A review, J. Clin. Aestheth. Dermatol. 6 (2013) 16–26.

[91] D. Dondi, A. Albini, N. Serpone, Interactions between different solar UV-B/UV-A filters contained in commercial suncreams and consequent loss of UV protection, Photochem. Photobiol. Sci. (2006) 835–843.

[92] S.M.C.M. Fernandes, A. Alonso-Varona, T. Palomares, V. Zubillaga, J. Labidi, V. Bulone, Exploring mycosporines as natural molecular sunscreens for the fabrication of UV absorbing green materials, ACS Appl. Mater. Interfaces 7 (2015) 16558–16564.

[93] Q. Gao, F. Garcia-Pichel, Microbial ultraviolet sunscreens, Nat. Rev. Microbiol. 2011 (2011) 791–802.
[94] W.M. Bandaranayake, Mycosporines: are they nature's sunscreens? Nat. Prod. Synth. 15 (1998) 159–172.
[95] C.A. Peyman Derikvand, S.P. Llewellyn, Cyanobacterial metabolites as a source of sunscreens and moisturizers: a comparison with current synthetic compounds, Eur. J. Phycol. 50 (2017) 43–56.
[96] L.R. Comini, M. Vieyra, P.L. Mignone, M.L. Páez, B. S. Mugas, J.J. Konigheim, et al., Reaction of dihydroxyacetone (DHA) with human skin callus and amino compounds, J. Invest. Dermatol. 36 (2017) 283–286.
[97] F. Liebel, S. Kaur, E. Ruvolo, N. Kollias, M.D. Southall, Irradiation of skin with visible light induces reactive oxygen species and matrix-degrading enzymes, J. Invest. Dermatol. 132 (2012) 1901–1907.
[98] H.Y. Wang, H.Q. Wei-RongYa, Melanoidins produced by the Maillard reaction: structure and biological activity, Food Chem. 128 (2011) 573–584.
[99] M. Sohn, H.U.G. Herzog, Calculation (of the sun protection factor of sunscreens with different vehicles using measured film thickness distribution—comparison with the SPF in vitro, J. Photochem. Photobiol. 159 (2016) 74–81.
[100] A.G. Chittiboyina, C. Avonto, D. Rua, I.A. Khan, Alternative testing methods for skin sensitization: NMR spectroscopy for probing the reactivity and classification of potential skin sensitizers, Chem. Res. Toxicol. 28 (2015) 1704–1714.
[101] G. Pirotta, An overview of sunscreen regulations in the world, H&PC Today 10 (2015) 17–22.
[102] L. Egli, J. Gröbner, G. Hülsen, L. Bachmann, M. Blumthaler, L. Blumthaler, et al., Quality assessment of solar UV irradiance measured with array spectroradiometers, Atmos. Meas. Tech. 9 (2016) 1553–1567.
[103] M. Og Manley, J. McCavana, Durham MED tester acceptance testing and calibration using EBT3 Gafchromic film, Phys. Med. 42 (2017) 353–354.
[104] D. Larouche, D.H. Kim, G. Ratté, C.L. Beaumont, Effect of intense pulsed light treatment on human skin in vitro: analysis of immediate effects on erde mal papillae and hair follicle stem cells, Br. J. Dermatol. 169 (2013) 859–868.
[105] T. Desmond, B. Vladimir, P. Ralf, C.L. Beaumont, The Fate of Hair Follicle Melanocytes During the Hair Growth Cycle. The journal of investigative dermatology. Symposium proceedings/the Society for Investigative Dermatology, Inc. [and], European Society for Dermatological Research 4 (2000) 323-332.
[106] P. Mistriotis, P. Stelios, T. Andreadis, Hair follicle: a novel source of multipotent stem cells for tissue engineering and regenerative medicine, Tissue Eng. B Rev. 19 (2013) 265–278.
[107] H. Rhee, L. Polak, E. Fuchs, Lhx2 maintains stem cell character in hair follicles, Science 312 (5782) (2006) 1946–1949.
[108] P. Serrano-Grau, A. Campo-Voegeli, D. Romero, Photodepilation, Actas Dermosifiliogr 100 (2009) 351–361.
[109] S. Vano-Galvan, P. Jaen, Complications of nonphysician-supervised laser hair removal. Case report and literature review., Can. Fam. Physician 55 (2009) 50–52.
[110] C.M. Littler, Hair removal using an Nd-YAG laser system, Dermatol. Clin. 17 (1999) 401–430.
[111] C.A. Nanni, T.S. Alster, Laser-assisted hair removal: side effects of Q-switched Nd:YAG, long-pulsed ruby, and alexandrite lasers, Dermatol. Surg. 26 (2) (2000) 109–113.
[112] D.J.Goldberg, J.A.Samady, Evaluation of a long-pulse Q-switched Nd:YAG laser for hair removal, Dermatol Surg, 26, 2000, 119-113.
[113] M.C. Polderman, S. Pavel, S. le Cessie, J.M. Grevelink, R.L. van Leeuwen, Efficacy, tolerability, and safety of a long-pulsed ruby laser system in the removal of unwanted hair, Dermatol. Surg. 26 (2000) 240–243.
[114] R.R. Blanche, B.S. Richard, P. Chiacchierini, E.R. Kazmirek, J.A. Sklar, The growth of human scalp hair in females using visible red light laser and LED sources, MD Lasers Surg. Med. 46 (8) (2014) 601–607.
[115] P. Avci, G.K. Gupta, J. Clark, N. Wikonkal, M.R. Hamblin, Low-level laser (Light) therapy (LLLT) for treatment of hair loss, Lasers Surg. Med. 46 (2014) 144–151.
[116] M.R. Avram, N.E. Rogers, The use of low level light for hair growth: part I, J. Cosmet. Laser Ther 11 (2009) 110–117.
[117] E.K. Nishimura, S.R. Granter, D.E. Fisher, Mechanisms of hair graying: incomplete melanocyte stem cell maintenance in the niche, Science 307 (2005) 720–724.
[118] S. Murao, C. Kirdmanee, K. Sera, S. Goto, C. Takahashi, Detection of lead in human hair: a contribution of PIXE to the lead-elimination issue, Int. J. PIXE 23 (1 & 2) (2013) 31–37.
[119] D.J. Tobin, R. Paus, Graying: gerontobiology of the hair follicle pigmentary unit, Exp. Gerontol. 36 (2001) 29–54.
[120] M. Sulieman, E. MacDonald, J.S. Rees, R.G. Newcombe, M. Addy, Tooth bleaching by different

concentrations of carbamide peroxide and hydrogen peroxide whitening strips: an *in vitro* study., J. Esthet. Restor. Dent. 18 (2006) 93–100.

[121] C.M. Carey, Tooth whitening, what we now know, J. Evid. Based Dental. Pract. 14 (2014) Suppl, 70–76.

[122] W.G. Chang, C.C. Wen, T.K. Huang, J.L. Wang, P.S. Fu, J.H. Chen, et al., Evaluating the accuracy of tooth color measurement by combining the Munsell color system and dental colorimeter, Kaohsiung J. Med. Sci. 28 (2012) 490–494.

[123] S.J. Chu, R.D. Trushkowsky, R.D. Paravina, Dental color matching instruments and systems. Review of clinical and research aspects, J. Dent. 38 (2010). suppl- e 2 – e 16.

[124] M. Sulieman, M. Addy, J.S. Rees, Development and evaluation of a method in vitro to study the effectiveness of tooth bleaching, J. Dentistry 31 (2003) 415–422.

[125] L. Wang, R.T. Tang, T. Bonstein, P. Bush, G.H. Nancollas, Enamel demineralization in primary and permanent teeth, J. Dent. Res. 85 (2006) 359–363.

[126] A.L. Boskey, Mineralization of bones and teeth, Elements 3 (2007) 387–393.

[127] A. Kadam, M. Ganachari, M. Kumar, M. Gurunath, Drug induced tooth discoloration, Internet J. Dent. Sci. 7 (2008) 2.

[128] L. Adorini, G. Penna, Dendritic cell tolerogenicity: a key mechanism in immunomodulation by vitamin D receptor agonists, Hum. Immunol. 70 (2009) 345–352.

[129] B. Farrerons, R. Rodriguez, L.-N. Yoldi, Moragas clinically prescribed sunscreen (sun protection factor 15) does not decrease serum vitamin D concentration sufficiently either to induce changes in parathyroid function or in metabolic markers, Brit. J. Dermatol. 139 (1998) 422–427.

[130] M.F. Holick, Sunlight and vitamin D for bone health and prevention of autoimmune diseases, cancers, and cardiovascular disease, Am. J. Clin. Nutr. 80 (2004) 1678S–1688SS.

[131] S. Patience, Vitamin D deficiency in at-risk groups, Commun. Pract. 86 (3) (2013) 38–40.

[132] G.D. Zhu, W.H. Hokamura, Synthesis of vitamin D, Chem.Rev. 95 (1995) 1877–1952.

[133] H.F. DeLuca, Overview of general physiologic features and functions of vitamin D, Am. J. Clin. Nutr. 80 (6 Suppl) (2004) 1689S–1696SS.

[134] C.-W. Pan, D.-J. Qian, S.-M. Saw, Time outdoors, blood vitamin D status and myopia: a review., Photochem. Photobiol. Sci. 16 (2017) 426–452.

[135] J. Reichrath, R. Saternus, T. Vogt, Challenge and perspective: the relevance of untraviolet radiation on the vitamin D endocrine system (VDS) for psoriasis and other inflammatory skin diseases, Photochem. Photobiol. Sci. (2017).

[136] D. Feldman, J.W. Pike, J.S. Adams (Eds.), Vitamin D, third Ed, Academic Press, Whashington, 2011.

[137] J. Reichrath, K. Berg, S. Emmert, J. Lademann, G. Seknmeyer, L. Zastrow, et al., Biologic effects of light: an enlighting prospective., Anticancer Res. 36 (2016) 1339–1343.

[138] P.G. Lindqvist, A.T. Silva, S.A. Gustafsson, S. Gidlö, A.T. Silva, S.A. Gustafsson, et al., Maternal vitamin D deficiency and fetal distress/birth asphyxia: a population nested control-case study, BMJ Open 6 (2016) e009733.

[139] D.D. Bikle, Vitamin D metabolism, mechanism of action, and clinical applications, Chem. Biol. 21 (3) (2014) 319–329.

[140] B. Lehmann, P. Knuschke, M. Meurer, The UV-B-induced synthesis of vitamin D3 and 1α,25-dihydroxyvitamin D_3 (calcitriol) in organotypic cultures of keratinocytes: effectiveness of the narrowband Philips TL-01 lamp (311nm), J. Steroid Biochem. Mol. Biol 103 (2007) 682–685.

[141] J.S. Adams, M. Hewison, Update in vitamin D, J. Clin. Endocrinol. Metab. 95 (2010) 471–478.

[142] M. Bär, D. Domaschke, A. Meye, B. Lehmann, M. Meurer, Wavelength-dependent induction of CYP24A1-mRNA after UV-B-triggered calcitriol synthesis in cultured human keratinocytes, J. Invest. Dermatol. 127 (2007) 206–213.

[143] B. Lehman, S. Abraahm, M. Meurer, Role for tumor necrosis factor-alpha in UV-B-induced conversion of 7-dehydrocholesterol to 1α 25-dihydroxyvitamin D_3 in cultured keratinocytes, J. Steroid Biochem. Mol. Biol. 89-90 (2004) 561–565.

[144] M. Mao-Qiang, A.J. Fowler, M. Schmuth, P. Lau, S. Chang, B.E. Brown, et al., Peroxisome-proliferator-activated receptor (PPAR)-c activation stimulates keratinocyte differentiation, J. Invest. Dermatol. 123 (2004) 305–312.

[145] M. Grabacka, W. Placha, K. Urbanska, P. Laidler, P. M. Płonka, K. Reiss, PPARγ regulates MITF and β-catenin expression and promotes a differentiated phenotype in mouse melanoma S91, Pigment Cell Melanoma Res. 21 (2008) 388–396.

[146] C. Couteau, L. Coiffard, Overview of skin whitening agents: drugs and cosmetic products, Cosmetics 3 (2016) 27.

[147] J.P. Ebanks, R.R. Wickett, R.E. Boissy, Mechanism regulating skin pigmentation: the rise and fall of complexation coloration, Int. J. Mol. Sci. 10 (2009) 4066–4087.

[148] R. Sarkar, P. Arora, K.V. Garg, Cosmeceuticals for hyperpigmentation: what is available? J. Cutan. Aesthet. Surg. 6 (2013) 4–11.
[149] F. Miao, Y. Shi, Z.-F. Fan, S. Jiang, S.-Z. Xu, T.-C. Lei, Deoxyarbutin possesses a potent skin-lightening capacity with no discernible cytotoxicity against melanosomes, PLoS One 11 (2016) e0165338.
[150] H.Y. Park, J. Lee, S. González, M.A. Middelkamp-Hup, S. Kapasi, S. Peterson, et al., Topical application of a protein kinase C inhibitor reduces skin and hair pigmentation, J. Invest. Dermatol. 122 (2004) 159–166.
[151] K.A.U. Gonzales, E. Fuchs, Skin and its regenerative powers: an alliance between stem cells and their niche, Dev Cell. 43 (4) (2017) 387–401.
[152] S.V. Ramagopalan, A. Heger, A.J. Berlanga, N.J. Maugeri, M.R. Lincoln, A. Burrell, et al., A ChIP-seq-defined genome-wide map of vitamin D receptor binding: associations with disease and evolution, Genome Res. 20 (2010) 1352–1360.
[153] M. Abboud, M.S. Rybchyn, R. Rizk, D.R. Fraser, R.S. Mason, Sunlight exposure is just one of the factors which influence vitamin D status, Photochem. Photobiol. Sci. 16 (2017) 302–303.
[154] Australian Government, Department of Health, Therapeutic goods administration. 2016 literature review on the safety of titanium dioxide and zinc oxide nanoparticles in sunscreens, Sci. Rev. Rep. (2016) 1–24.
[155] K.M. Dixon, S.S. Deo, A.W. Norman, J.E. Bishop, G. M. Halliday, V.E. Reeve, et al., In vivo relevance for the photoprotection by the vitamin D rapid response pathway, J. Steroid Biochem. Molecular Biol. 103 (2007) 451–456.
[156] P.H. Itin, M.R. Pittelkow, R. Kumar, Effects of vitamin D metabolites on proliferation and differentiation of cultured human epidermal keratinocytes grown in serum-free or defined culture medium, Endocrinology 135 (5) (1994) 1793–1798.
[157] K. Rueter, A. Siafarikas, S.L. Prescott, D.J. Palmer, In utero and postnatal vitamin D exposure and allergy risk, Expert. Opin. Drug Saf. 13 (2014) 1–11.
[158] I.S. Saidi, S.L. Jacques, F.K. Tittel, Mie and Rayleigh modeling of visible-light scattering in neonatal skin, Appl. Opt. 34 (1995) 7409–7418.
[159] M. Di Rosa, M. Malaguarnera, F. Nicoletti, L. Malaguarnera, Vitamin D3: a helpful immunomodulator, Immunology 134 (2) (2011) 123–139.
[160] E. Acheson, A.C. Bacharach, The distribution of multiple sclerosis in U. S. veterans by birthplace, Am. J. Hyg. (1960) 88–99.
[161] H.F. DeLuca, L. Plum, UV-B radiation, vitamin D and multiple sclerosis, Photochem. Photobiol. Sci. 16 (2017) 411–415.
[162] A. Gupta, T.C. Arora, A. Jindal, A.S. Bhadoria, Efficacy of narrowband ultraviolet B phototherapy and levels of serum vitamin D3 in psoriasis: a prospective study, Indian Dermatol. Online J. 7 (2) (2016) 87–92.
[163] K.P. Liao, M. Gunnarsson, H. Källberg, B. Ding, R.M. Plenge MD, L. Padyukov, et al., A specific association exists between type 1 diabetes and anti-CCP positive rheumatoid arthritis, Arthritis Rheum. 60 (2009) 653–660.
[164] P.G. Lindqvist, A.T. Silva, S.A. Gustafsson, S. Gidlö, Maternal vitamin D deficiency and fetal distress/ birth asphyxia: a poulation nested control-case study, BMJ Open 6 (2019) e009733.
[165] S. Gorman, R.M. Lucas, A. Allen-Hall, N. Fleury, M. Feelisch, Ultraviolet radiation, vitamin D and the development of obesity, metabolic syndrome and type-2 diabetes, Photochem. Photobiol. Sci. 16 (3) (2017) 362–373.
[166] P.P. Toh, M. Bigliardi-Qi, A.M. Yap, G. Sriram, P. Bigliardi, Expression of peropsin in human skin is related to phototransduction of violet light in keratinocytes, Exp. Dermatol. 25 (12) (2016) 1002–1005.
[167] H.J. Kim, E.D. Son, J.Y. Jung, H. Choi, T.R. Lee, D.W. Shin, Violet light down-regulates the expression of specific differentiation markers through Rhodopsin in normal human epidermal keratinocytes, PLoS One 8 (9) (2013) e73678.
[168] R.B. Weller, The health benefits of UV irradiation exposure through vitamin D production or non-vitamin D pathways. Blood pressure and cardiovascular disease, Photochem. Photobiol. Sci. 16 (3) (2017) 374–380.
[169] C.S. He, X.H. Aw Yong, N. Walsh, M. Gleeson, Is there an optimal Vitamin D status for immunity in athletes and military personnel? Exerc. Immunol. Rev. 22 (2016) 41–62.
[170] P.G. Lindqvist, L. Landin-Olssonb, The relationship between sun exposure and all-cause mortality, Photochem. Photobiol. Sci. 16 (2017) 354–361.
[171] L. Korkina, Metabolic and redox barriers in the skin exposed to drugs and xenobiotics,, Expert Opin. Drug Metab. Toxicol. 12 (4) (2016) 377–388.
[172] F. Kiss, A.V. Anstey, A review of UV-B-mediated photosensitivity disorders, Photochem. Photobiol. Sci. 12 (2013) 37–46.
[173] P. Lehmann, T. Schwarz, Photodermatoses: diagnosis and treatment, Dtsch Arztebl Int. 108 (9) (2011) 135–141.
[174] O. Kučera, M. Cifra, Radiofrequency and microwave irradiation between biomolecular systems, J. Biol. Phys. 42 (1) (2016) 1–8.

[175] P. Boscol, M.B. Di Sciascio, S. D'Ostilio, A. Del Signore, M. Reale, P. Conti, et al., Effects of electromagnetic fields produced by radiotelevision broadcasting stations on the immune system of women, Sci. Total Environ. 273 (2001) 1–10.
[176] N.R. Kumar, S. Sangwan, P. Badotr, Exposure to cell phone radiations produces biochemical changes in worker honey bees, Toxicol. Int. 18 (2011) 70–72.
[177] R. Pandel, B. Poljšak, A. Godic, R. Dahmane, Skin photoaging and the role of antioxidants in its prevention, ISRN Dermatol. 2013 (2013). Article ID 930164, 11 pages.
[178] S. Trombino, Strategies for vitamin E transdermal delivery, human health handbooks, in: V.R. Preedy (Ed.), Handbook of Diet, Nutrition and the Skin, 2, Wageningen Academic Publishers, 2012, pp. 129–143.
[179] S. Kang, J.H. Chung, H. Joo, G.J. Fisher, S. Yiang, E. A. Duel, et al., Topical N-acetyl cysteine and ganistein prevent ultraviolet-light-induced singaling that leads to photoaging in human skin in vivo, J. Invest. Dermatol. 120 (2003) 835–841.
[180] J. Xu, R.C. Spitale, L. Guan, R.A. Flynn, E.A. Torre, R. Li, et al., Novel gene expression profile of women with intrinsic skin youthfulness by whole transcriptome sequencing, PLoS One. 11 (11) (2016) e0165913.
[181] S. Lippä, R. Saffrich, W. Ahnsorge, D. Bohmann, Differential regulation of c-Jun by ERK and JNK during PC12 cell differentiation, EMBO J. 17 (1998) 4404–4413.
[182] W. Chen, G.T. Bowden, Activation p38 MAP kinase and ERK are required for ultraviolet-B induced c-fos gene expression in human keratinocytes, Oncogene 18 (1999) 7469–7476.
[183] D. Barolet, F. Christiaens, M.R. Hamblin, Infrared and skin: friend or foe, J. Photochem. Photobiol. B 155 (2016) 78–85.
[184] G.J. Fisher, S. Kang, J. Varani, et al., Mechanisms of photoaging and chronological skinaging, , Arch. Dermatol. 138 (2002) 1462–1470.
[185] N. Puizina-Ivić, Skin aging, Acta Dermatovenerol. Alp. Pannonica Adriat 17 (2) (2008) 47–54.
[186] L. Zastrow, J. Lademann, Light - instead of UV protection: new requirements for cancer prevention, Anticancer Res. 36 (3) (2016) 1389–1394.
[187] L. Zastrow, N. Groth, F. Klein, D. Kockott, J. Lademann, L. Ferrero, UV, sichtbares Licht, Infrarot. Welche Wellenlängen produzieren oxidativen Stress in menschlicher Haut? Hautarzt 60 (2009) 310–317.
[188] L. Zastrow, J. Lademan, Instead of UV protection: new requirements for skin cancer prevention, Anticancer Res. 36 (2016) 1389–1394.
[189] C.-Y. Chen, P.E. Milbury, K. Lapsley, J.B. Blumberg, Flavonoids from almond skins are bioavailable and Ac, and G. Rajeev, *Role of Reactive Oxygen Species and Antioxidants in Atopic Dermatitis*, J. Clin. Diagn. Res. 7 (12) (2013) 2683–2685.
[190] B.H. Mahmoud, C.L. Hexsel, I.H. Hamzavi, H.W. Lim, Effects of visible light on the skin, Photochem. Photobiol. 84 (2008) 450–462.
[191] K. Rahman, Studies on free radicals, antioxidants, and co-factors, Clin. Interv. Aging. 2 (2007) 219–236.
[192] L.A. Pham-Huy, H. He, C. Pham-Huy, Free radicals, antioxidants in disease and health, Int. J. Biomed. Sci. 4 (2) (2008) 89–96.
[193] V. Manova, D. Gruszka, DNA damage and repair in plants – from models to crops, Front. Plant Sci. 6 (2015) 885.
[194] J. Bouayed, T. Bohn, Exogenous, antioxidants—double-edged swords in cellular redox state health beneficial effects at physiologic doses versus deleterious effects at high doses, Oxid. Med. Cell Longev. 3 (4) (2010) 228–237.
[195] M. Conrad1, S.G. Moreno, F. Sinowatz, F. Ursini, S. Kölle, A. Roveri, et al., The nuclear form of phospholipid hydroperoxide glutathione peroxidase is a protein thiol peroxidase contributing to sperm chromatin stability, Mol. Cell Biol. 17 (2005) 7627–7644.
[196] E. Lubos, J. Loscalzo, D.E. Handy, Glutathione peroxidase-1 in health and disease: from molecular mechanisms to therapeutic opportunities, Antioxid. Redox. Signal. 15 (7) (2011) 1957–1997.
[197] F. Nogales, M.L. Ojeda, M. Fenutria, M.L. Murillo, O. Carreras, Role of selenium and glutathione peroxidase on development, growth, and oxidative balance in rat offspring, Reproduction (Bristol, United Kingdom) 6 (2013) 659–667.
[198] S.-J. Kim, P. Cheresh, R.P. Jablonski, L. Morales-Nebreda, Y. Cheng, E. Hogan, et al., Mitochondrial catalase overexpressed transgenic mice are protected against lung fibrosis in part via preventing alveolar epithelial cell mitochondrial DNA damage, Free Radic Biol. Med. 101 (2016) 482–490.
[199] C.E. Schaar, D.J. Dues, K.K. Spielbauer, E. Machiela, J.F. Cooper, M. Senchuk, et al., Mitochondrial and cytoplasmic ROS have opposing effects on lifespan, PLoS Genetics 11 (2) (2015) e1004972/1–e1004972/24.
[200] T. Rui, G. Cepinskas, Q. Feng, P.R. Kvietys, Delayed preconditioning in cardiac myocytes with respect to development of a proinflammatory phenotype: role of SOD and NOS, Cardiovasc. Res. 59 (4) (2003) 901–911.
[201] S.L. Marklund, Extracellular superoxide dismutase in human tissues and human cell lines, J. Clin. Invest. 74 (4) (1984) 1398–1403.

[202] K.E. Szabo-Taylor, E.A. Toth, A.M. Balogh, B.W. Sodar, L. Kadar, K. Paloczi, et al., Monocyte activation drives preservation of membrane thiols by promoting release of oxidised membrane moieties via extracellular vesicles,, Free Radic. Biol. Med. 108 (2017) 56–65.
[203] E.A. Sisein, Biochemistry of free radicals and antioxidants, Sch. Acad. J. Biosci. 2 (2) (2014) 110–118.
[204] O. Zitka, S. Skalikova, J. Gumulec, M. Masarik, V. Adam, J. Hubalek, et al., Redox status expressed as GSH:GSSG ratio as a marker for oxidative stress in paediatric tumour patients, Oncol. Lett. 4 (2012) 1247–1253.
[205] R. Masella, R. Di Benedetto, R. Varì, C. Filesi, C. Giovannini, Novel mechanisms of natural antioxidant compounds in biological systems: involvement of glutathione and glutathione-related enzymes, J. Nutr. Biochem. 16 (10) (2005) 577–586.
[206] E.S. Arnér, A. Holmgren, Physiological functions of thioredoxin and thioredoxin reductase, Eur. J. Biochem. 267 (20) (2000) 6102–6109.
[207] M.T. Islam, Antioxidant activities of dithiol alpha-lipoic acid, Bangladesh J. Med. Sci. 8 (3) (2009).
[208] M.V. Trivedi, J.S. Laurence, T.J. Siahaan, The role of thiols and disulfides in protein chemical and physical stability, Curr. Protein Pept. Sci. 10 (6) (2009) 614–625.
[209] J.-C. Preiser, A. Van Gossum, J. Berré, J.-L. Vincent, FCCM, Y. Carpentier, Enteral feeding with a solution enriched with antioxidant vitamins A, C, and E enhances the resistance to oxidative stress,, Crit. Care Med. 28 (12) (2000) 3828–3832.
[210] S.C. Morton, L.G. Hilton, W. Tu, D. Valentine, P.G. Shekelle, Antioxidants vitamin C and vitamin E for the prevention and treatment of cancer, J. Gen. Intern. Med. 21 (7) (2006) 735–744.
[211] E.L. Abner, F.A. Schmitt, M.S. Mendiondo, J.L. Marcum, R.J. Kryscio, Vitamin E and all-cause mortality: a meta-analysis, Curr. Aging Sci 4 (2) (2011) 158–170.
[212] M.O. Grimm, J. Mett, T. Hartmann, The impact of vitamin e and other fat-soluble vitamins on Alzheimer's disease, Int. J. Mol. Sci 17 (11) (2016) pii: E1785.
[213] S.Z.A.S. Shariat, S.A. Mostafavi, F. Khakpour, Antioxidant effects of vitamins C and E on the low-density lipoprotein oxidation mediated by myeloperoxidase, Iran Biomed. J. 17 (1) (2013) 22–28.
[214] S. Parthasarathy, A. Raghavamenon, M.O. Garelnabi, N. Santanam, Oxidized low-density lipoprotein, Methods Mol. Biol. 610 (2010) 403–417.
[215] T. Heitzer MD, S.Y. Herttuala MD, E. Wild, J. Luoma MD, H. Drexler MD, Effect of vitamin E on endothelial vasodilator function in patients with hypercholesterolemia, chronic smoking or both, J. Am. Coll. Cardiol. 33 (2) (1999) 499–505.
[216] S. Kinlay, D. Behrendt, C.F. James, D. Delagrange, J. Morrow, J.L. Witztum, et al., Long-term effect of combined vitamins E and C on coronary and peripheral endothelial function, J. Am. Coll. Cardiol. 43 (4) (2004) 629–634.
[217] H. Boorsook, G. Keighley, Oxidation reduction potential of ascorbic acid (vitamin C), PNAS 19 (1933) 875–883.
[218] C.-Y. Chen, P.E. Milbury, K. Lapsley, J.B. Blumberg, Flavonoids from almond skins are bioavailable and act synergistically with vitamins C and E to enhance hamster and human LDL resistance to oxidation, J. Nutri. 6 (2005) 1366–1373.
[219] M. Dietrich, G. Block, N.L. Benowitz, J.D. Morrow, M. Hudes, P. Jacob 3rd, et al., Vitamin C supplementation decreases oxidative stress biomarkerf2-isoprostanes in plasma of nonsmokers exposed to environmental tobacco smoke, Nutr. Cancer. 45 (2) (2003) 176–184.
[220] G. Agati, M. Tattini, Multiple functional roles of flavonoids in photoprotection, New Phytol. 186 (2010) 786–793.
[221] N. Panche, A.D. Diwan, S.R. Chandra, Flavonoids: an overview, J. Nutr. Sci. 5 (2016) e47.
[222] WHO Study Group on Epidemiology and Prevention of Cardiovascular Diseases in the Elderly, Epidemiology and Prevention of Cardiovascular Diseases in Elderly People: Report of a WHO Study Group, Geneva, Switzerland World Health Organization1995, WHO Technical Report Series No. 853.
[223] J. Chen, S. Mangelinckx, A. Adams, Z.T. Wang, W.L. Li, N. De Kimpe, Natural flavonoids as potential herbal medication for the treatment of diabetes mellitus and its complications, Nat. Prod. Commun. 10 (2015) 187–200.
[224] P. Yang, Y. Yang, Z. Gao, Y. Yu, Q. Shi, G. Bai, Combined effect of total alkaloids from Feculae Bombycis and natural flavonoids on diabetes, J. Pharm. Pharmacol. 59 (2007) 1145–1150.
[225] R.M. van Dam, N. Naidoo, R. Landberg, Dietary flavonoids and the development of type 2 diabetes and cardiovascular diseases: review of recent findings, Curr. Opin. Lipidol. 24 (2013) 25–33.
[226] S.B. Doshi, A. Agarwal, The role of oxidative stress in menopause, J. Midlife Health. 4 (3) (2013) 140–146.
[227] S.K. Katiyar, T. Singh, R. Prasad, Q. Sun, M. Vaid, Epigenetic alterations in ultraviolet radiation-induced skin carcinogenesis: interaction of bioactive dietary components on epigenetic targets, Photochem. Photobiol. 88 (2012) 1066–1074.
[228] E. Groenniger, B. Weber, O. Heil, N. Peters, F. Staeb, H. Wenck, et al., Aging and chronic sun exposure

cause distinct epigenetic changes in human skin, PLoS Genet. 6 (5) (2010) e1000971.

[229] B. Demmig-Adams, W.W. Adams 3rd, Antioxidants in photosynthesis and human nutrition, Science. 298 (5601) (2002) 2149–2153.

[230] A.C. Kaliora, G.V. Dedoussis, H. Schmidt, Dietary antioxidants in preventing atherogenesis, Atherosclerosis. 187 (1) (2006) 1–17. Epub 2005 Nov 28.

[231] K. Rahman, Effects of garlic on platelet biochemistry and physiology, Mol. Nutr. Food Res. 51 (11) (2007) 1335–1344.

[232] I. Peluso, M. Palmery, M. Serafini, Effect of cocoa products and flavanols on platelet aggregation in humans: a systematic review, Food Funct. 6 (2015) 2128–2134.

[233] B. Bordeaux, L.R. Yanek, T.F. Moy, L.W. White, L.C. Becker, N. Faraday, et al., Casual chocolate consumption and inhibition of platelet function, Prev. Cardiol. 10 (4) (2007) 175–180.

[234] D. Fusco, G. Colloca, M.R.L. Monaco, M. Cesari, Effects of antioxidant supplementation on the aging process, Clin. Interv. Aging. 2 (3) (2007) 377–387.

[235] G.P. Hubbard, S. Wolffram, J.A. Lovegrove, J.M. Gibbins, The role of polyphenolic compounds in the diet as inhibitors of platelet function, Proc. Nutr. Soc. 62 (2) (2003) 469–478.

[236] W.S. Hambright, R.S. Fonseca, L. Chen, R. Na, Q. Ran, Ablation of ferroptosis regulator glutathione peroxidase 4 in forebrain neurons promotes cognitive impairment and neurodegeneration, Redox Biol. 12 (2017) 8–17.

[237] R. Remington, C. Bechtel, D. Larsen, A. Samar, R. Page, C. Morrell, et al., Maintenance of cognitive performance and mood for individuals with Alzheimer's disease following consumption of a nutraceutical formulation: a one-year, open-label study, J. Alzheimers Dis. 51 (4) (2016) 991–995.

[238] J.M. van der Burg, M. Björkqvist, P. Brundin, Beyond the brain: widespread pathology in Huntington's disease, Lancet Neurol. 8 (8) (2009) 765–774.

[239] P. Dayalu, R.L. Albin, Huntington disease: pathogenesis and treatment, Neurologic. Clin. 33 (1) (2015) 101–114.

[240] R.R. Korać, K.M. Khambholja, Potential of herbs in skin protection from ultraviolet radiation, Pharmacogn. Rev. 5 (10) (2011) 164–173.

[241] R. Bridi, F.P. Crossetti, V.M. Steffen, A.T. Henriques, The antioxidant activity of standardized extract of Ginkgo biloba (EGb 761) in rats, Phytother. Res. 15 (5) (2001) 449–451.

[242] R. Hardeland, S.R. Pandi-Perumal, Melatonin, a potent agent in antioxidative defense: actions as a natural food constituent, gastrointestinal factor, drug and prodrug, Nutri. Metab. 2 (2005) 22.

[243] M. Smeyne, R.J. Smeyne, Glutathione metabolism and Parkinson's disease, Free Radic. Biol. Med. 62 (2013) 13–25.

[244] P. Seyeon, The effects of high concentrations of vitamin C on cancer cells, Nutrients. 5 (9) (2013) 3496–3505.

[245] M.D. AhmetUnlu, M.D. OnderKirca, M.D. Mustafa Ozdogan, D. Erdinç NayırM, High-dose vitamin C and cancer, J. Oncol. Sci. 1 (2016) 10–12.

[246] W.C. You, L. Zhang, M.H. Gail, et al., Gastric dysplasia and gastric cancer: *Helicobacter pylori*, serum vitamin C, and other risk factors, J. Natl. Cancer Inst. 92 (2000) 1607–1612.

[247] M. Hosokawa, M. Kudo, H. Maeda, H. Kohno, T. Tanaka, K. Miyashita, Fucoxanthin induces apoptosis and enhances the antiproliferative effect of the PPARg ligand, troglitazone, on colon cancer cells, Biochim. Biophys. Acta 1675 (2004) 113–119.

[248] S. Takeda, H. Okazaki, E. Ikeda, S. Abe, Y. Yoshioka, K. Watanabe, et al., Down-regulation of cyclooxygenase-2 (COX-2) by cannabidiolic acid in human breast cancer cells, J. Toxicol. Sci. 39 (5) (2014) 711–716.

[249] K. Sudheer, A. Mantena, S.K. Katiyar, Grape seed proanthocyanidins inhibit UV-radiation-induced oxidative stress and activation of MAPK and NF-κB signaling in human epidermal keratinocytes, Free Radic. Biol. Med. 40 (2006) 1603–1614.

[250] T.M. Vance, J. Su, E.T.H. Fontham, S.I. Koo, O.K. Chun, Dietary antioxidants and prostate cancer: a review, Nutr. Cancer. 65 (6) (2013). Available from: https://doi.org/10.1080/0163558.

[251] P.S. Hedge, N.S. Rajasekaran, T.S. Chandra, Effects of the antioxidant properties of millet species on oxidative stress and glycemic status in alloxan-induced rats, Nutr. Res. 25 (2005) 1109–1120.

[252] H. Yin, E. Niki, K. Uchida, Recent progress in lipid peroxidation based on novel approaches: lipid oxidation in the skin, Free Radic. Res. 49 (2015) 813–815.

[253] L.M. Gaetke, C.K. Chow, Copper toxicity, oxidative stress, and antioxidant nutrients toxicity, Toxicology 189 (2003) 147–163.

[254] J.A. Enriquez, G. Lenaz, Coenzyme Q and the respiratory chain: coenzyme Q pool and mitochondrial supercomplexes, Mol. Syndromol. 5 (3-4) (2014) 119–140.

[255] J. Garrido-Maraver, M.D. Cordero, M. Oropesa-Ávila, A.F. Vega, M. de la Mata, A.D. Pavón, et al., Coenzyme Q_{10}therapy, Mol. Syndromol. 5 (3-4) (2014) 187–197.

[256] M. Disse, H. Reich, P.K. Lee, S. Schram, A review of the association between Parkinson disease and malignant melanoma, Dermatol. Surg. 42 (2016) 141–146.

[257] H. Rich, M. Odlyha, U. Cheema, V. Mudera, L. Bozec, Effects of photochemical riboflavin-mediated

crosslinks on the physical properties of collagen constructs and fibrils, J. Mater. Sci. Mater. Med. 25 (1) (2014) 11–21.
[258] K. Zhang, M.J. Huentelman, F. Rao, E.I. Sun, J.J. Corneveaux, A.J. Schork, et al., Genetic implication of a novel thiamine transporter in human hypertension, J. Am. Coll. Cardiol. 63 (15) (2014) 1542–1555.
[259] M.M. Berger, A. Shenkin, J.-P. Revelly, E. Roberts, M. C. Cayeux, M. Baines, et al., Copper, selenium, zinc, and thiamine balances during continuous venovenous hemodiafiltration in critically ill patients, Am. J. Clin. Nutr. 80 (2) (2004) 410–416.
[260] R. Silvério, A. Laviano, F. Rossi Fanelli, M. Seelaender, L-Carnitine induces recovery of liver lipid metabolism in cancer cachexia, 2011, Amino Acids 42 (2012) 1783–1792.
[261] H. Zhong, H. Yin, Role of lipidperoxidation derived 4-hydroxynonenal(4-HNE) in cancer: focusing on mitochondria, Redox Biol. 4 (2015) 193–199.
[262] M.E. Zujko, A.M. Witkowska, A. Waśkiewicz, I. Mirończuk-Chodakowska, Dietary antioxidant and flavonoid intakes are reduced in the elderly, Oxid. Med. Cellular Longev. (2015). Article ID 843173, 8 pages, 2015.
[263] A. Jalikov, C. Zhang, E.L.G. Samuel, W.K.A. Sikkema, G. Wu, V. Berka, et al., Mechanistic study of the conversion of superoxide to oxygen and hydrogen peroxide in carbon nanoparticles, ACS Appl. Mater. Interfaces 8 (24) (2016) 15086–15092.
[264] E.L. Samuel, D.C. Marcano, V. Berka, B.R. Bitner, G. Wu, A. Potter, et al., Highly efficient conversion of superoxide to oxygen using hydrophilic carbon clusters, Proc. Natl. Acad. Sci. U S A. 112 (8) (2015) 2343–2348.
[265] A. Carr, B. Frei, Does vitamin C act as a pro-oxidant under physiological conditions? FASEB J. 13 (9) (1999) 1007–1024.
[266] E.H. Sarsour, M.G. Kumar, L. Chaudhuri, A.L. Kalen, P.C. Goswami, Redox control of the cell cycle in health and disease, Antioxid. Redox Signal. 11 (2009) 2985–3011.
[267] R. Stocker, Y. Yamamoto, A.F. McDonagh, A.N. Glazer, B.N. Ames, Bilirubin is an antioxidant of possible physiological importance, Science 235 (1987) 1043–1046.
[268] H. Sakurai, M. Nitao, K. Matsuura, Y. Iuchi, Uric acid, an important antioxidant contributing to survival in termites, PLoS One 12 (2017) e0179426.
[269] R. Mailloux, W. Willmore, S-glutathionylation reactions in mitochondrial function and disease, Front. Cell Dev. Biol. 2 (2014) 68.
[270] K.T. Bieging, S.S. Mello, L.D. Attardi, Unravelling mechanisms of p53-mediated tumour suppression, Nat. Rev. Cancer 14 (2014) 359.
[271] C. Rauer, J.J. Nogueira, P. Marquet, L. González, Cyclobutane thymine photodimerization mechanism revealed by nonadiabatic molecular dynamics, J. Am. Chem. Soc. 138 (2016) 15911–15916.
[272] A. Banyasz, T. Douki, R. Improta, T. Gustavsson, D. Onidas, I. Vaya, et al., Electronic excited states responsible for dimer formation upon UV absorption directly by thymine strands: joint experimental and theoretical study, J. Am. Chem. Soc. 134 (2012) 14834–14845.
[273] H. Yokoyama, R. Mizutani, Structural biology of DNA (6-4) photoproducts formed by ultraviolet radiation and interactions with their bindingproteins, Int. J.Mol. Sci. 15 (11) (2014) 20321–20338.
[274] J. Cadet, T. Douki, Formation of UV-induced DNA damage contributing to skin cancer development, Photochem. Photobiol. Sci. 17 (12) (2018) 1816–1841.
[275] P.J. Rochette, J.-P. Therrien, R. Drouin, D. Perdiz, N. Bastien, E.A. Drobetsky, E. Sage, UV-A-induced cyclobutane pyrimidine dimers form predominantly at thymine–thymine dipyrimidines and correlate with the mutation spectrum in rodent cells, Nucleic Acids Res. 31 (11) (2003) 2786–2794.
[276] A. Gulia, A.M.G. Brunasso, C. Massone, Dermoscopy: distinguishing malignant tumors from benign, Expert Rev. Dermatol. 7 (2012) 439–445.
[277] J. Cadet, K.J.A. Davies, Oxidative DNA damage & repair: an introduction, Free Radic. Biol. Med. 107 (2017) 2–12.
[278] M.R. Hussein, Ultraviolet radiation and skin cancer: molecular mechanisms, J. Cutan. Pathol. 32 (2005) 191–205.
[279] F. Trautinger, Mechanisms of photodamage of the skin and its functional consequences for skin ageing, Clin. Exp. Dermatol. 26 (2001) 573–577.
[280] A. Collins, N. El Yamani, M. Dusinska, Sensitive detection of oxidative damage to DNA induced by nanomaterials, 5-hydroxymethylcytosine;5-Hydroxy metylcytosine, 5-formylcytosine; and 5-carboxy cytosine, Free Radic. Biol. Med. 107 (2017) 69–76.
[281] A. Klugland, A.B. Robertson, Oxidized C5-methyl cytosine bases in DNA: 5-hydroxymethylcytosine, 5-formylcytosine and 5-carboxycytosine, Free Radic. Biol.Med. 107 (2017) 62–68.
[282] R. Brem, M. Guven, P. Karran, Oxidatively-generated damage to DNA and proteins mediated by UV-A photo-sensitization, Free Radic.Biol. Med. 107 (2017) 101–109.
[283] A.P. Schuch, N.C. Moreno, N.J. Schuch, C.F.M. Menck, C. Carrião, M. Garcia, et al., Consequences of sunlight in cellular DNA: focus on the effects of oxidatively generated DNA damage, Free Radic. Biol. Med. 107 (2017) 101–109.

[284] J. Richard Wagner, J. Cadet, Oxidation reactions of cytosine DNA components by hydroxyl radical and one-electron oxidants in aerated aqueous solutions., Acc. Chem. Res. 43 (4) (2010) 564–571.

[285] M. Morikawa, K. Kino, T. Oyoshi, M. Suzuki, T. Kobayashi, H. Miyazawa, Analysis of guanine oxidation products in double-stranded dna and proposed guanine oxidation pathways in single-stranded, double-stranded or quadruplex DNA, Biomolecules 4 (2014) 140–159.

[286] Z. Zhang, B. Cheng, Y.C. Tse-Dinh, Crystal structure of a covalent intermediate in DNA cleavage and rejoining by *Escherichia coli* DNA topoisomerase I, PNAS 108 (2011) 6939–6944.

[287] K. Kaarniranta, E. Pawlowska, J. Szczepanska, A. Jablkowska, J. Blasiak, Role of mitochondrial DNA damage in ROS-mediated pathogenesis of age-related macular degeneration (AMD), Int. J. Mol. Sci. 20 (2019) 2374.

[288] M. Nita, A. Grzybowski, The role of the reactive oxygen species and oxidative stress in the pathomechanism of the age-related ocular diseases and other pathologies of the anterior and posterior eye segments in adults, Oxid. Med. Cell. Longev. 6 (2016) 3164734.

[289] S.W. Ryter, R.M. Tyrrel, The heme synthesis and degradation pathways: role in oxidant sensitivity: Heme oxygenase has both pro- and antioxidant properties, Free Radic. Biol. Med. 28 (2000) 289–309 (Fig 3).

[290] L. Daya Grosjean, A. Sarasin, The role of UV induced lesions in skin carcinogenesis: an overview of oncogene and tumor suppressor gene modifications in xeroderma pigmentosum skin tumors, Mutat. Res. 571 (2005) 43–56.

[291] Y.J. Kim, D.M. Wilson, Overview of base excision repair biochemistry, Curr. Mol. Pharmacol. 5 (2012) 3–13 (Fig 1.

[292] E. Sage, N. Shikazono, Radiation-induced clustered DNA lesions: repair and mutagenesis, Free Radic. Biol Med. 107 (2016) 125–135 (Fig 1).

[293] A.M. Fleming, C.J. Burrows, Formation and processing of DNA damage substrates for the hNEIL enzymes, Free Radic. Biol.Med. 107 (2017) 35–52.

[294] V. Shafirovich, N.E. Geacintov, Removal of oxidatively generated DNA damage by overlapping repair pathways, oxidized C5-methyl cytosine bases in DNA, Free Radic. Biol. Med. 107 (2017) 62–68.

[295] B. Tudek, D. Zdżalik-Bielecka, A. Tudek, K. Kosicki, A. Fabisiewicz, E. Speina, Lipid peroxidation in the face of DNA damage, DNA repair and other cellular processes, Free Radic. Biol. Med. 107 (2017) 77–89.

[296] A.J. Lee, S.S. Wallace, Hide and seek: how do DNA glycosylases locate oxidatively damaged DNA bases amidst a sea of undamaged bases? Free Radic. Biol. Med. 107 (2017) 70–78.

[297] S.T. Sarvestani, J.L. McAuley, The role of the NLRP3 inflammasome in regulation of antiviral responses to influenza A virus infection, Antiviral Res. 148 (2017) 32–42 (Fig !).

[298] P.J. Brooks, The cyclopurine deoxynucleosides: DNA repair, biological effects, mechanistic insights, and unanswered questions, Free Radic. Biol. Med. 107 (2017) 90–100.

[299] I. Kuraoka, C. Bender, A. Romieu, J. Cadet, R.D. Wood, T. Lindahl, Removal of oxygen free-radical-induced 5′, 8-purine cyclodeoxynucleosides from DNA by the nucleotide excision-repair pathway in human cells, Proc. Natl. Acad. Sci. USA. 97 (2000) 3832–3837.

[300] J. Wang, C.L. Clauson, P.D. Robbins, L.J. Niedernhofer, Y. Wang, The oxidative DNA lesions 8,5′-cyclopurines accumulate with aging in a tissue-specific manner, Aging Cell. 11 (4) (2012) 714–716.

[301] R.A. Caston, B. Demple, Risky repair: DNA-protein crosslinks formed by mitochondrial base excision DNA repair enzymes acting on free radical lesions, Free Radic. Biol. Med. 107 (2017) 146–150.

[302] D.M. Banda, N.N. Nuñez, M.A. Burnside, K.M. Bradshaw, S.S. David, Repair of 8-oxoG:A mismatches by the MUTYH glycosylase: mechanism, metals and medicine, Free Radic. Biol. Med. 107 (2017) 202–215.

[303] M. Saki, A. Prakash, DNA damage related crosstalk between the nucleus and mitochondria., Free Radic. Biol. Med. 107 (2017) 216–227.

[304] H. Menoni, P.D. Mascio, J. Cadet, S. Dimitrov, D. Angelov, Chromatin associated mechanisms in base excision repair – nucleosome remodeling and DNA transcription, two key players,, Free Radic. Biol. Med. 107 (2017) 159–169.

[305] Y. Nakabeppu, E. Ohta, N. Abolhassani, MTH1 as a nucleotide pool sanitizing enzyme: friend or foe? Free Radic. Biol. Med. 107 (2017) 151–158.

[306] Y. Nakabeppu, D. Tsuchimoto, H. Yamaguchi, K. Sakumi, Oxidative damage in nucleic acids and Parkinson's disease, J. Neurosci. Res. 85 (2007) 919–934.

[307] S. Boiteux, F. Coste, B. Castaing, Repair of 8-oxo-7,8-dihydroguanine in prokaryotic and eukaryotic cells: properties and biological roles of the PFG and Ogg1 8-oxoguanine DNA N-glycosylases, Free Radic.Biol. Med. 107 (2017) 179–201.

[308] A. Thapa, N.J. Carrol, Dietary modulation of oxidative stress in Alzheimer's disease, Int.J. Mol. Sci. 18 (7) (2017).

[309] R. Abbotts, D.M. Wilson III, Coordination of DNA single strand break repair, Free Radic.Biol. Med. 107 (2017) 228–244.

[310] D. Crouch, R.M. Brosh Jr., Mechanistic and biological considerations of oxidatively damaged DNA for helicase-dependent pathways of nucleic acid metabolism, Free Radic. Biol. Med. 107 (2017) 245–257.

[311] M. Seifermann, B. Epe, Oxidatively generated base modifications in DNA: not only carcinogenic risk factor but also regulatory mark? Free Radic. Biol. Med. 107 (2017) 258–265.

[312] I. Talhaoui, B.T. Matkarimov, T. Tchenio, D.O. Zharkov, M.K. Saparbaev, Aberrant base excision repair pathway of oxidatively damaged DNA:implications for degenerative diseases, Free Radic. Biol. Med. 107 (2017) 266–277.

[313] J.K. Horton, H.J. Seddon, M.-L. Zhao, N.R. Gassman, A.K. Janoshazi, D.F. Stefanick, et al., Role of the oxidized form of XRCC1 in protection against extreme oxidative stress, Free Radic. Biol. Med. 107 (2017) 292–300.

[314] M. D'Errico, E. Parlanti, B. Pascucci, P. Fortini, S. Baccarini, V. Simonelli, et al., Single nucleotide polymorphisms in DNA glycosylases: from function to disease, Free Radic. Biol. Med. 107 (2017) 278–291.

[315] W.J. Schreier, T.E. Schrader, F.O. Koller, P. Gilch, C. E. Crespo-Hernández, V.N. Swaminathan, et al., Thymine dimerization in DNA is an ultrafast photoreaction, Science 315 (5812) (2007) 625–629.

[316] Y.K. Law, J. Azadi, C.E. Crespo-Hernández, E. Olmon, B. Kohler, Predicting thymine dimerization yields from molecular dynamics simulations, Biophys. J. 94 (9) (2008) 3590–3600.

[317] V. Labet, N. Jorge, C. Morell, T. Douki, A. Grand, J. Cadet, et al., UV-induced formation of the thymine–thymine pyrimidine (6-4) pyrimidone photoproduct – a DFT study of the oxetane intermediate ring opening, Photochem. Photobiol. Sci. 12 (2013) 1509.

[318] K. McAteer, Y. Ling, J. Kao, J.S. Taylor, M.A. Kennedy, Solution-state structure of a DNA dodecamer duplex containing a Cis-Syn thymine cyclobutane dimer, the major UV photoproduct of DNA, J. Mol. Biol. 282 (5) (1998) 1013–1032.

[319] A. Sancar, Structure and function of DNA photolyase, Biochemistry 33 (1999) 2–9.

[320] B.B. Wenke, L.N. Huiting, E.B. Frankel, B.F. Lane, M. E. Núñez, Base pair opening in a deoxynucleotide duplex containing a *cis-syn*thymine cyclobutane dimer lesion, Biochemistry 52 (51) (2013) 9275–9285.

[321] S. Weber, Light-driven enzymatic catalysis of DNA repair: a review of recent biophysical studies on photolyase, Biochim. Biophys. Acta 1707 (2005) 1–23.

[322] D. Udayakumar, H. Tsao, Moderate to low-risk variants alleles of cutaneous malignancies and nevi: lessons from genome-wide association studies, Genome Med. 1 (2009) 95–99.

[323] J. Cadet, S. Mouret, J.-L. Ravanat, T. Douki, Photoinduced damage to cellular DNA: direct and photosensitized reactions, Photochem. Photobiol. 88 (2012) 1048–1065.

[324] D. Kumar, A.L. Abdulovic, J. Viberg, A.K. Nissoln, T.A. Kunkel, A. Chabes, Mechanism of mutagenesis *in vivo* due to imbalanced dNTP pools, Nucleic Acid Res. 39 (2011) 1360–1371.

[325] C.W. Han, M.S. Jeong, S.B. Jang, Structure, signaling and the drug discovery of the Ras oncogene protein, BMB Rep. 50 (7) (2017) 355–360.

[326] K. Rajalingam, R. Schreck, U.R. Rapp, Š. Albert, Ras oncogenes and their downstream targets, Biochim. Biophys. Acta 1773 (2007) 1177–1195.

[327] N.F. Neel, T.D. Martin, J.K. Stratford, T.P. Zand, D.J. Reiner, C.J. Der, The RalGEF-Ral effector signaling network. The road less traveled for anti-Ras drug discovery, Genes Cancer. 2 (3) (2011) 275–287.

[328] E. Castellano, J. Downward, RAS interaction with PI3K more than just another effector pathway, Genes Cancer. 2 (3) (2011) 261–274.

[329] J.A. Lo, D.E. Fisher, The melanoma revolution: from UV Carcinogenesis to a new era in therapeutics, Science 346 (6212) (2014) 945–949.

[330] S. Cerritelli, S. Hirschberg, R. Hill, N. Balthasar, A.E. Pickering, Activation of brainstem pro-opiomelanocortin neurons produces opioidergic analgesia, bradycardia and bradypnoea, PLoS One 11 (2016) e0153187 (Fig 11).

[331] S.C. Wang, H.X. Ji, C.L. Hsiao, T.C. Wang, Y.R. Syu, C.E. Miao, et al., Protective effects of pyridoxamine against UV-C-induced programmed cell death in HaCaT cells, In vivo (Athens, Greece) 29 (2015) 379–383.

[332] K.L. Gressel, F.J. Duncan, T.M. Oberyszyn, K.M. La Perle, H.B. Everts, Endogenous retinoic acid required to maintain the epidermis following ultraviolet cell death in SKH-1 hairless mice, Photochem. Photobiol. 91 (2015) 901–908.

[333] F.O. Nestle, P.D. Meglio, J.-Z. Qin, B.J. Nickoloff, Skin immune sentinels in health and disease, Nat. Rev. Immunol. 9 (10) (2009) 679–691.

[334] H. Lodish, A. Berk, S.L. Zipursky, et al., Molecular Cell Biology., fourth edition, W. H. Freeman, New York, 2000.

[335] K. Kuhn, in: R. Mayne, R. Burgeson (Eds.), Structure and Function of Collagen Types, Academic Press, 1987, p. 2.

[336] M. van der Rest, R. Garrone, Collagen family of proteins, FASEB J. 5 (1991) 2814–2823.

[337] K. Lintner, N. André, E. Doridot, O. Gracioso, P. Criton, P. Mondon, Chromophore mapping reveals skin ageing delay, Cosmet. Sci. Technol. (2011) 27.

[338] Abhimanyu, A.K. Coussens, The role of UV radiation and vitamin D in the seasonality and outcomes of infectious disease,, Photochem. Photobiol. Sci. 16 (2017) 314–338.

[339] J.M. Menter, J.D. Williamson, K. Carlyle, C.L. Moore, I. Willis, Photochemistry of type I acid-soluble collagen: dependence on excitation wavelength, Photochem. Photobiol. 62 (1995) 402–408.

[340] T.H. Nasti, L. Timares, Inflammasome activation of IL-1 family mediators in response to cutaneous photodamage, Photochem. Photobiol 88 (2012) 1111–1125.

[341] Y.Y. Zheng, B. Viswanathan, P. Kesarwani, S. Mehrotra, Dietaryagents in cancer prevention: an immunological perspective, Photochem. Photobiol. 88 (2012) 1083–1098.

[342] C. Nishisgori, Current concept of photocarcinogenesis, Photochem. Photobiol. Sci. 14 (2015) 1713–1721 (Fig 1).

[343] K.A. Morgan, E.H. Mann, A.R. Young, C.M. Hawrylowicz, ASTHMA – comparing the impact of vitamin D versus UV-R on clinical and immune parameters, Photochem. Photobiol. Sci. 16 (2017) 399–410.

[344] H.J. van der Rhee, E. de Vries, J.W. Coebergh, Does sunlight prevent cancer? A systematic review, Eur. J. Cancer. 42 (2006) 2222–2232.

[345] H. Pelicano, D. Carney, P. Huang, ROS stress in cancer cells and therapeutic implications, Drug Resist. Updat. 7 (2004) 97–110.

[346] K. Kandasamy, S.M. Srinivasula, E.S. Alnemri, C.B. Thompson, S.J. Korsmeyer, J.L. Bryant, R.K. Srivastava, Involvement of proapoptotic molecules Bax and Bak in tumor necrosis factor-related apoptosis-inducing ligand (TRAIL)-ligand induced mitochondrial disruption and apoptosis: differential regulation of cytochrome c and Smac/DIABLO release, Cancer Res. 63 (2003) 1712–1721.

[347] D.M. Holtzman, C.M. John, A. Goate, Alzheimer's disease: the challenge of the second century, Sci. Transl. Med. 3 (77) (2011) 77sr1.

[348] L.V. Kalia, A.E. Lang, Parkinson's disease, Lancet 386 (2015) 896–912.

[349] V. Dias, E. Junn, M.M. Mouradian, The role of oxidative stress in Parkinson's disease, .J. Parkinsons Dis 3 (4) (2013) 461–491.

[350] G. Ferdinando, M. Brownlee, Oxidative stress and diabetic complications, Circ. Res. 107 (9) (2010) 1058–1070.

[351] H. Qian, X. Xu, Reduction in DNA methyltransferases and alteration of DNA methylation pattern associate with mouse skin ageing, Exp. Dermatol. 23 (5) (2014) 357–359.

[352] D.H. McDaniel, I.H. Hanizavi, J.A. Zeichner, S.G. Fabi, V.W. Bucay, J.C. Harper, et al., Total defense + repair: a novel concept in solar protection and skin rejuvenation, J. Drugs Dermatol. 14 (7, Suppl.) (2015) S3–S11.

[353] X. He, C. Shen, Q. Lu, J. Li, Y. Wei, L. He, et al., Prokineticin 2 plays a pivotal role in psoriasis, EBioMedicine 13 (2016) 248–261.

[354] M.F. Holick, Biological effects of sunlight ultraviolet radiaton, visible light infrared radiation, and vitamin D for health, Anticancer Res. 36 (2016) 1345–1357.

[355] M.R. Boland, Z. Shahn, D. Madigan, G. Hripcsak, N. P. Tatonetti, Birth month affects lifetime disease risk: a phenome-wide method, J. Am. Med. Inform. Assoc. 22 (2015) 1045–1053.

[356] A. Kuhn, J. Wenzel, H. Weyd, Photosensitivity, apoptosis and cytokines in the pathogenesis of lupus erythematosus: a critical review, Clin. Rev. Allergy Immunol. 47 (2014) 148–162.

[357] Q. Li-Jun, X. Li-Wei, Y.Y. Li, J.Q. Yang, J.X. Gi, Clinical therapeutic effect of intense pulse light PhotoDerm for wacular dermatosis, Poc SPIE (2009) 720811/1–728011/6.

[358] M. Zakrewsky, S. Miragotri, Therapeutic RNAi robed with ionic moieties as a simple, scalable prodrug platform for treating skin disease, J. Control Release 242 (2016) 80–88.

[359] A. Magenta, E. Dellambra, R. Ciarrapicas, M.C. Capogrossi, Oxidative stress, microRNAs and cytosolic calcium homeostasis, Cell Calcium 60 (2016) 207–217.

[360] C. Pincelli, M. Steinhoff, Recapitulating atopic dermatitis in three dimensions: cross talk between keratinocytes and nerve fibers, J.Invest. Dermatol. 133 (2013) 1465–1467.

[361] D. Roggenkamp, S. Kopnick, F. Stab, et al., Epidermal nerve fibers modulate keratinocyte growth via neuropeptide signaling in an innervated skin model, J. Invest. Dermatol. 133 (2013) 1620–1628.

[362] D. Bootsma, G. Weeda, W. Vermeulen, H. VanVuuren, C. Troelstra, P. Van Der Spek, et al., Nucleotide excision repair syndromes: molecular basis and clinical symptoms, Philos. Trans. R. Soc. B Biol. Sci. 347 (1319) (1995) 75–81.

[363] E.H. Grattan Clive, S.S. Saini, Urticaria and angioedema without wheals Middleton's allergy essentials, (2017) 249–263.

[364] M. de Fátima Santos Paim de Oliveira, B. de Oliveira Rocha, G. Vieira Duarte, Psoriasis: classical and emerging comorbidities, An. Bras.Dermatol. 90 (2015) 9–20.
[365] A. Zutavern, I. Brockow, B. Schaaf, A. von Berg, U. Diez, et al., Pediatrics timing of solid food introduction in relation to eczema, asthma, allergic rhinitis, and food and inhalant sensitization at the age of 6 years: results from the prospective birth cohort study, LISA 27 (2008) 235–238.
[366] D. De Silva, M. Geromi, S. Halken, A. Host, S.S. Panesar, A. Muraro, et al., Primary prevention of food allergy in children and adults: systematic review., Allergy 69 (2014) 581–589.
[367] G. Longo, I. Berti, W. Burks, B. Krauss, E. Barbi, IgE-mediated food allergy in children, Lancet 382 (9905) (2013) 1656–1664.
[368] R.S. Gupta, C.M. Warren, B.M. Smith, J. Jiang, J.A. Blumenstock, M.M. Davis, et al., Prevalence and severity of food allergies among US adults, JAMA 2 (2019) e185630.
[369] M. Ferrer, J. Bartra, A. Giménez-Arnau, I. Jauregui, M. Labrador-Horrillo, J. Ortiz de Frutos, et al., Management of urticaria: not too complicated, not too simple, Clin. Exp. 45 (2015) 731–743.
[370] P.R. Criado, R. Facchini Jardim Criado, C. Wakisaka Maruta, V.M. Silva dor reis, Chronic urticaria in adults: state-of-the-art in the new millennium, An. Bras. Dermatol. 90 (2015) 74–79.
[371] T. Zuberbier, W. Aberer, R. Asero, A.H. Abdul Latiff, D. Baker, B. Ballmer-Weber, et al., EAACI/GA^{2}LEN/EDF/WAO guideline for the definition, classification, diagnosis and management of urticaria, Allergy 73 (2018) 1393–1414.
[372] P. Schaefer, Acute and chronic urticaria: evaluation and treatment, Am. Fam. Phys 95 (2017) 717–724.
[373] M.A. Pathack, Sunscreens: topical and systemic approaches for protection of human skin against harmful effects of solar radiation, J. Am. Acad. Dermatol. 7 (1982) 285–312.
[374] K. Hollis, C. Proctor, D. McBride, M.M. Balp, L. McLeod, S. Hunter, et al., Comparison of urticaria activity score over 7 days (UAS7) values obtained from once-daily and twice-daily versions: Results from the ASSURE-CSU study, Am. J. Clin. Dermatol. 19 (2018) 267–274.
[375] A.M. Ros, D.V. Edström, Cyclosporin A therapy for severe solar urticaria, Photodermatol. Photoimmunol. Photomed 13 (1997) 61–63.
[376] R. RaI, M. Thomas, Photopatch and UV-irradiated patch testing in photosensitive dermatitis, Indian Dermatol. Online J. 7 (2016) 12–16.
[377] A. Daglish, G. Waller, Clinician and patient characteristics and cognitions that influence weighing practice in cognitive-behavioral therapy for eating disorders, Int. J.Eat.Disord. 52 (9) (2019) 977–986.
[378] E. Halpert, E. Borrero, M. Ibanez-Pinilla, P. Chaparro, J. Molina, M. Torres, et al., Prevalence of papular urticaria caused by flea bites and associated factors in children 1–6 years of age in Bogotá, D.C, World Allergy Organ. J. 10 (2017) 36.
[379] R.D. Pesek, R.F. Lockey, Management of insect sting hypersensitivity: an update, Allergy Asthma Immunol.Res. 5 (2013) 129–137.
[380] N. Akdogan, P. Incel-Uysal, A. Oktem, Y. Hayran, B. Yalcin, Educational level and job status are the most important factors affecting compliance with oral antihistamine therapy for patients with chronic urticaria, J.Dermatol. Treart. 30 (2019) 183–188.
[381] J.M. Negro-Alvarez, J.C. Miralles-Lopez, Chronic idiopathic urticaria treatment, Allergolo. Immunopathol. 29 (2001) 129–132.
[382] M. Abajian, A. Mlynel, M. Maurer, Physical urticaria, Curr Allergy Asthma Rep. 12 (2012) 281–287.
[383] B.A. Muller, Urticaria and angioedema: a practical approach, Am.Fam. 46 (2004) 645–657.
[384] F. Braido, P.J. Bousquet, Z. Brzoza, G.W. Canonica, E. Compalati, A. Fiocchi, Specific recommendations for PROs and HRQoL assessment in allergic rhinitis and/or asthma: a GA(2)LEN taskforce position paper, Allergy 65 (2010) 959–968.
[385] M. Noris, G. Remuzzi, Overview of complement activation and regulation, Semin. Nephrol. 33 (2013) 479–492.
[386] C. Seikrit, P. Ronco, H. Debiec, Factor H autoantibodies and membranous nephropathy, N. Engl. J.Med. 379 (2018) 2479–2481.
[387] M. de Brito, G. Huebner, D.F. Murrel, P. Bullpitt, K. Hartmann, Normocomplementaemic urticarial vasculitis: effective treatment with omalizumab, Clin. Transl. Allergy 8 (2018) 37.
[388] K. Bouiller, S. Audia, H. Devillieers, E. Collet, M.H. Aubriot, V. Leguy-Seguin, et al., Etiologies and prognostic factors of leukocytoclastic vasculitis with skin involvement, Medicine (Baltimore) 95 (2016) e4238.
[389] L. Da Dalt, C. Zerbinati, M.S. Strafella, S. Renna, L. Riceputi, P. Di Pietro, et al., Henoc-Schönlein purpura and drug and vaccine use in childhood : a case-control study, Ital. J.Pediatr. 42 (2016) 60.
[390] P.E. Beattie, R.S. Dawe, S.H. Ibbotson, J. Ferguson, Characteristics and prognosis of idiopathic solar urticaria. A cohort of 87 cases, Arch. Dermatol. 139 (2003) 1149–1154.

[391] T. Boonpiyathad, W. Mitthamsiri, P. Pradubpongsa, A. Sangasapaviliya, Urticaria diagnosis, Eur.Med. J. 3 (2018) 98–105.

[392] K. Wolff, T.B. Fitzpatrick, R.A. Johnson, Acne vulgaris (common acne) and cystic acne, Fitzpatrick's Color Atlas and Synopsis of Clinical Dermatology, sixth ed, McGraw-Hill, New York, NY, 2009, pp. 2–6.

[393] S. Titus, J. Hodge, Diagnosis and treatment of acne, Am. Fam. Phys. 86 (2012) 734–740.

[394] A. Sparavigna, B. Tenconi, I. De Ponti, L. La penna, An innovative approach to the topical treatment of acne,, Clin. Cosmet.Investig. Dermatol. 8 (2015) 179–185.

[395] P.A. Gerber, G. Kukova, S. Meller, N. Neumann, B. Homey, The dire consequences of doping, Lancet 372 (2008) 656.

[396] J.J. Mao, M. Stosich, E.K. Moioli, C.H. Lee, S. Fu, B. Bastian, et al., Facial reconstruction by biosurgery: cell transplantation versus cell homing, Tissue Eng. B Rev. 16 (2010) 257–262.

[397] S. Moriwaki, F. Kanda, M. Hayashi, D. Yamashita, Y. Sakai, C. Nischigori, Xeroderma pigmentosum clinical practice guidelines, J.Dermatol. 44 (2017) 10878–11096.

[398] K.H. Kraemer, J.J. DiGiovanna, Forty years of research on xeroderma pigmentosum at the US National Institutes of Health, Photochem.Photobiol. 91 (2015) 452–459.

[399] M.J. Harries, K. Meyer, I. Chaudhry, J.E. Kloepper, E. Poblet, C.E.M. Griffiths, et al., Lichen planopilaris is characterized by immune privilege collapse of the hair follicle's epithelial stem cell niche, J. Pathol. 231 (2013) 236–247.

[400] E.H. Van Fer Mij, K.P. Schepman, D.R. Plonait, T. Axèll, I. Van Der Vaal, Interobserver and intraobserver variability in the clinical assessment of oral lichen planus, J. Oral Pathol. Med. 31 (2002) 95–98.

[401] S.A. Birlea, P.R. Fain, T.M. Ferrara, S. Ben, S.L. Riccardi, J.B. Cole, et al., Genome-wide association analyses identify 13 new susceptibility loci for generalized vitiligo, Nat. Genet. 44 (2012) 676–680.

[402] C. Vrijman, M.W. Kroon, J. Limpens, M.M.G. Leeflang, R.M. Luiten, J.P.W. van der Veen, et al., The prevalence of thyroid disease in patients with vitiligo: a systematic review, Br. J.Dermatol. 167 (2012) 1224–1235.

[403] T. Bhattacharya, B. Nardone, A. Rademaker, M. Martini, A. Amin, H.M. Al-Mudaimeagh, et al., Co-existence of psoriasis and melanoma in a large urban academic centre population: a cross-sectional retrospective study, J. Eur. Acad. Dermatol. Venereol. 30 (2015) 83–85.

[404] C.E.M. Griffiths, J.M. van der Walt, D.M. Ascroft, C. Flohr, L. Naldi, T. Nijsten, et al., The global state of psoriasis disease epidemiology: a workshop report, Br. J.Dermatol. 177 (2017) e4–e7.

[405] C. Feliciani, A. Zampetti, P. Forleo, L. Cerritelli, P. Amerio, G. Proietto, et al., Nail psoriasis: combined therapy with systemic cyclosporin and topical calcipotriol, J. Cutan. Med. Surg 8 (2004) 122–125.

[406] F.M. Garritse, E. Roekevisch, J. van der Schaft, J. Deinum, P.I. Spils, M.S. de Bruin Weller, Ten years experience with oral immunosuppressive treatment in adult patients with atopic dermatitis in two academic centres, J. Eur. Acad. Dermatol. Venereol. 29 (2015) 1905–1912.

[407] E. Piva, L. Sciacovelli, M. Zaninotti, M. Laposata, M. Plebani, Evaluation of effectiveness of a computerized notification system for reporting critical values, Am. J.Clin. Pathol. 131 (2009) 432–441.

CHAPTER

3

Contact with the environment: sight

3.1 Theory of light and colors: historic aspects

Of the senses that put man in contact with what is around him, vision is the most informative, but also in a way the most ambiguous and a short excursus through the theoretical problems that this sense confronts may be useful. From its earliest time, philosophy was indeed concerned with the interaction between man and what is outside of him through the senses. The situation is illustrated by the well-known example presented by Plato in "Republic," where he narrates that people can be likened to prisoners chained in the bottom of a cave in such a way that they can look only forward, and see the interior of the cave, while the shadow of people passing outside, as projected by a fire behind them is the only message from outside,. Uneducated people take such images as actual living beings and give names to them, while philosophers well understand that nothing is real in this impression. Classical philosophers were concerned in educating good citizens much more than on the way science depicted the word. However, ontology, the study of what things actually are, has always been a key part of philosophy, or more precisely, of metaphysics. A brief summary of the ancient Greeks point of view about how sight works shows that Empedocles thought that as a lamp *emanates* rays of light, so does the eye, which contains an eternal fire shut up within it and is thought to be able to emerge through channels in the surrounding water. Earth and air can easily be inserted into this description and allow putting this view in the general framework of the four elements (air, fire, water, earth). The colors, meant as a mixture of white and black, could be paired with each of the elements. The theory of *emanation* applies to all of the senses, not only to sight, and emanations from the eye are not corporeal (even though light is); however, the particles that make rays of light are not materials in the sense of something dead that can be accumulated; rather they are compounded of the active principles of the four elements—ultimately, of divine forces. In contrast, Democritus attempted to explain the characteristics of each particular color by referring to the characteristics of the atoms constituting it. Thus the color called white has perfectly smooth atoms, which cast no shadows, somewhat similar to the mother of pearl (which suggests not only light but also luster, both categories being associated in the image this scientist had). Black, on the contrary, is a combination of rough-surfaced atoms, and finally red is allowed to consist directly of fire (light) atoms. Warmth is associated with red, not merely because fire was the only source of

Light, Molecules, Reaction and Health
DOI: https://doi.org/10.1016/B978-0-12-811659-3.00003-7

energy known to the ancient world, but also because it denoted for Democritus the interchangeability of the atoms associated with warmth, soul, mind, and movement. Thus the element fire was not defined as a specific force in nature but as an unstable constellation of atoms. Democritus conceived colors as many quantities of energy, ranging from a pure form of it (white) to a total lack of it (black). Like Democritus, Plato also reckoned with self-radiating objects, but he rather thought that their rays met and mingled with the pure fire (rays) placed in all human eyes by the gods. Thus seeing (or not seeing) depends on the size, strength, and speed of the rays emanating from the objects, and perception of the various objects depended also on that process, thus on the speed of light. In retrospect, it may sound unlikely that such position was taken, but Greek natural philosophers were always connected more with the physiological aspect of the theory of sight (how does the eye receives impressions?), rather than on a speculative matter (what is the ultimate nature of light and color, etc.?). For whatever reason, Plato simply accepted the tradition, dealt with the problem in relatively short order and did not go beyond. Aristotle rejected the notion that a fiery ray emanated from the eye and was reflected back from the objects to create sight on the grounds that, if this were so, night vision would be undisturbed. By the same token he objected to the theory of emanations from objects, since the eye does not perceive them when the objects are pressed against the closed eye. Still, he did not attempt to eliminate the idea of physical context altogether, for he postulated the necessity of a medium between the eye and its percept—the first idea of ether, the invisible matter through which light propagates that dominated science for many centuries and reached back to the Pre-Socratic translucence (*diaphanes*), which exists in water, air, and translucent objects. Light is the agent (*energeia*) that reveals translucence as an incorporeal state ranging from bright to dark. Insofar as this flows into objects, it ceases being mere light and reveals the objects as well as their substantiality. The color of the object, in turn, puts the medium itself in motion and this is transmitted to the eye. Thus like light itself, the color is immaterial and is a state or form of energy. Light is what makes this process possible, but Aristotle attributes no movement to it, whereas the resulting color is an activator (*kinetikon*) of the medium. In the final form, Aristotle's view suggested that visual sensation passed from the eye to the heart, which was at that time considered the center of sensation and psychic function, as well as a cooling device. This cardiocentric nature of sensation continued into the Middle Ages depicted by an ancient illustration in Fig. 3.1, despite the direct experimental evidence reported by Galen that pressing on the heart in human subjects did not lead to loss of consciousness or loss of sensation, while severing the spinal cord in animals abolished sensory responses after brain stimulation).

At the height of classic Greek age, a four-color theory emerged, in parallel to the much better known four elements theory. The descriptive determination of such momentary states lies within two pairs of opposing conditions: hot–cold and wet–dry. These qualities in effect give the parameters of two of the elements, fire and water, whereby it can be concluded that fire and water have a particular axial quality, a central governing position in the total concept of four. The most obvious and striking aspect of this relationship is, as already suggested, the uncontested polarity of fire and water. The archenemy of fire is water; equally, fire opposes water but with much less immediate impact and finality. Fire is quenched by water; water is evaporated (goes into air) by fire. This stronger quality of water may allow determining how to picture their relationship. Since the inalienable tendency of water is to seek the horizontal, a horizontal

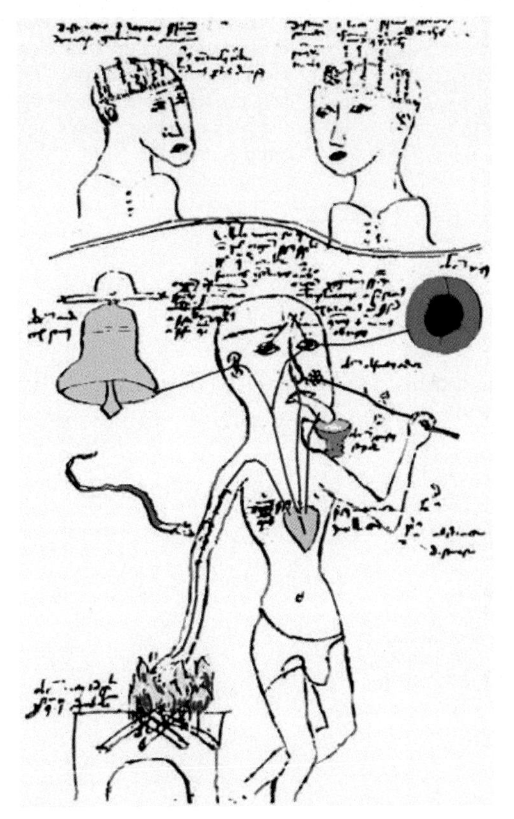

FIGURE 3.1 Aristotelian concept of five senses projecting to the heart either directly or via the sensus communis in the anterior part of the head (lower panel). The upper panel shows the four (Galen's and Avicennas's) or five (Albertus Magnus's) brain compartments [1,2].

line may be used, whereby the placement of fire and water to left or right is still to be discussed. After this, a second step can be placed as less dramatic, but equally inescapable polarity, and involves liquefaction that opposes combustion, viz earth and air. The normal relationship of these elements is to be contiguous, with earth below and air above. Their difference in density results in the phenomenon of gravity, which would not be observable without a contrasting medium through which things can fall and, for that matter, rise. If gravity is a force—as science proposes—beyond earth itself, then dialectically an opposing force, levity, must also be postulated. This relationship is logically illustrated by a vertical line: condensation opposes rarefaction. According to Empedocles, each of the colors has a different function, each has its own characteristic and in the rounds of time they take their turn of being dominant, each one after the other. Earth is always stable to the extent that it remains the darker part in any condition. In principle, yellow is the color of dispersal, black of concentration, red of intensity or arrested movement, and white of nonphysicality or minimal physicality. In a brilliant work on the theory of the Greek philosophers, J.L. Benson found that a series of paradigms can be established, as represented in the following figures where the relation between the four elements and the four basic colors, the four seasons and the four man's temperaments is highlighted (Fig. 3.2) [3,5].

In accord with Aristotle was the great Ibn al-Haytham, best known in the West as Alhazen, who along with many other Arab scientists, made transmission of the texts by the Greek philosophers to the modern eve possible until after the Renaissance, when the optics began to be elaborated in mathematical terms. According to him, vision is a passive experience, a representation by which he distances himself decisively from Euclid and Galen. At the same time, however, he maintains the geometrical basis of the extramission theory and thinks that every point on the surface of an object radiates diverging bundles of light rays in many directions, some of which fall on the eye. Each individual point reaches the eye by means of light rays, enters the pupil and comes together in the middle of the eye. Alhazen uses what he had experimentally studied on the *camera obscura*, the modern pinhole camera, and thinks that rays from various directions can meet at one point without colliding with each other and conserving their momentum. This scientist formulates the key objection to the view of objects radiating, how

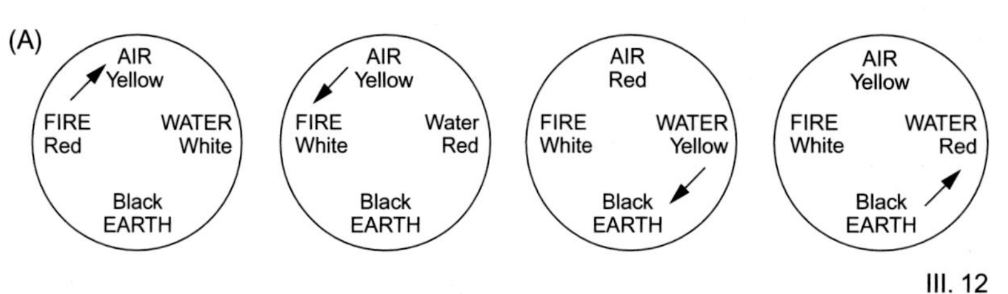

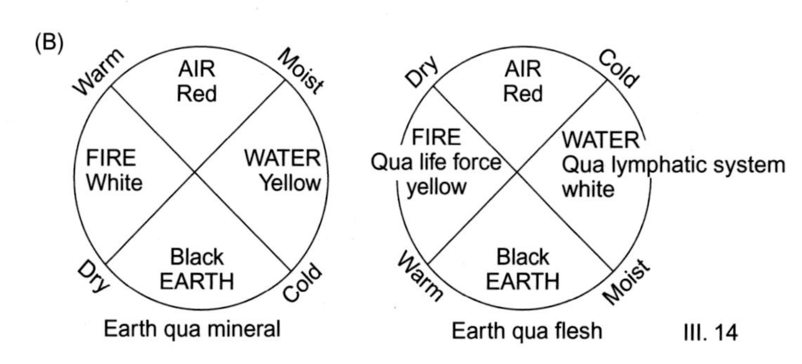

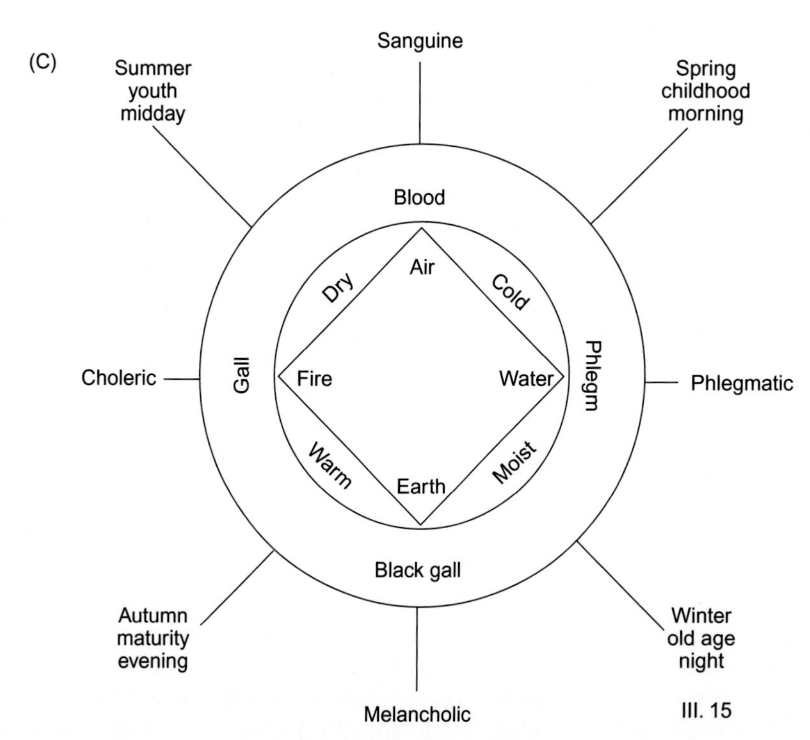

FIGURE 3.2 (A)–(C) Different presentations of the four elements: combustion, rarefaction, liquefaction, condensation; relations between the four elements, man's age, and temperament [3].

can an object come to a thousand people at the same time? Do forms and colors detach themselves—thus and fold—from the perceived object, so extremely reduced in size that they can enter our eyes? This sounds unlikely. Actually, the idea of an optic image in the eye and its use for getting information about what is around us forms only slowly, not before the year 1000 AC [6,7]. A passionate reader of Alhazen was Kepler, who in 1604 offered the first theory of the retinal image and stated that vision occurs through a picture of the visible things on the white, concave surface of the retina and the eye was indeed best described as a pinhole camera [8], an idea that well fitted a time where perspective was everything. In 1583 the Swiss physician Felix Platter argued that the optic nerve had to be viewed as the "primary organ of vision." He relegated the crystalline humor to a secondary role, elevating the status of the retina (Figs. 3.3 and 3.4) [9].

The well-known experiences by Newton of the prism decomposition of light into colors was an *experimentum crucis*, a key experiment that allowed to accept or refuse a statement, particularly in that it showed that a second prism regenerated white light [10]. Thus colors were not, as many thought, in the prism, but in the white light ray. Reasoning about colors has to confront several questions, however, such as: are colors an objective property? Are they the same color that our senses perceive? Or what we see is only an impression, as are Plato's shadows in the cavern?

3.2 Theory of colors and light: linguistic aspects

Actually, the problem of what is light is made worse when colors are considered. Everyday experience tells us that objects have different colors, and that they are distinguished by this property. However, colors differ from properties such as weight and form that can be depicted in an indisputable way (a rectangle is a rectangle of that measure and that weight), and are at most an impression, which is perceived only when there is someone able to do that (Galileo arrived to state that the objects were colored only when someone was looking at them, otherwise the objects fell back to black and white). At best, colors were an impression, relative to the light wavelength(s) reflected or refracted by that object, or a word, with the additional trouble that consensus about a color

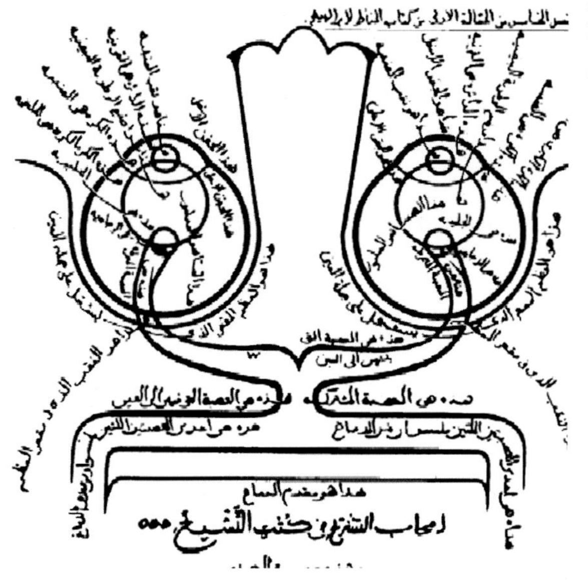

FIGURE 3.3 The visual system according to Alhazen. The hollow optic nerves meet in the chiasma and then diverge [7].

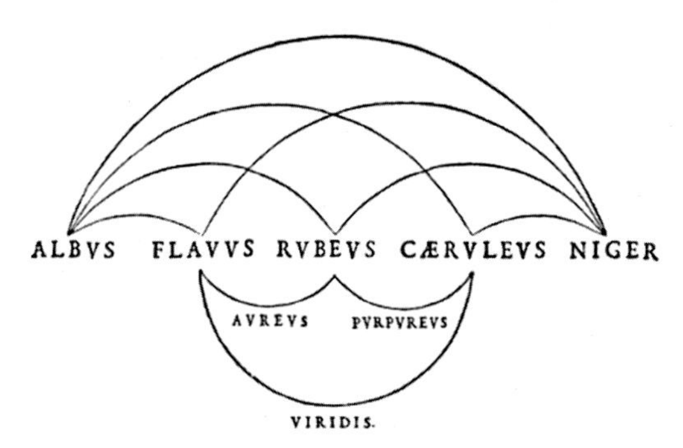

FIGURE 3.4 Aguilonius' linear scheme of colors (1613). From left to right: white, yellow, red, blue, black. Mixed colors: gold, green, purple [7,8].

was not easily reached (different blue objects were not blue in the same way, or under different conditions). Nevertheless, man and animals do use light to move in the environment and distinguish between forms and colors, although it is mainly the difference in colors that makes the borders distinguishable. The physics of the eye, that closely follows modern pinhole camera functions, by focusing (and reversing) an image on the back of the eye was known at about the same time. For what concerns the linguistic aspects of colors, an important debate followed the publication of a book by B. Berlin and P. Kay in 1969, which was a part of a long-established debate, in particular by the school of W. Gladstone [11,12].

This debate discussed the possible relation between color names in different languages. An example of what means building such relations is given by the few lines in Table 3.1, where terms for colors in some languages (the first known eight languages in alphabetical order, spoken really in any part of the globe) are listed. Berlin and Kay observed that in a given language the number of terms indicating a color varied largely, and examining 20 languages, found that this number changed from 2 to 20, and the introduction of new terms recognizing more and more precisely the hue design occurred during the evolution and followed a specific trend, the same in every case, or at least in the majority of the languages. They identified 11 possible basic color categories: white, black, red, green, yellow, blue, brown, purple, pink, orange, and gray. To be included in a basic color category, the term for the color in each language had to meet certain criteria, viz a term must be both monomorphemic and monolexemic (e.g., blue, but not bluish); its signification must not be included in that of any other color term (e.g., crimson is a type of red). Furthermore, its application must not be restricted to a narrow class of objects (e.g., blonde is restricted to hair, wood, and beer); it must not be psychologically salient for all of the informants (e.g., "the color of grandma's freezer" is not psychologically salient for all of the speakers). In case of doubt, a number of "subsidiary criteria" were implemented. The doubtful form should have the same distributional potential as the previously established basic terms (e.g., you can say reddish but not salmonish). Terms that are also the name of an object characteristically having that are suspect, for example: gold, silver, and ash; recent foreign loan words may be a suspect. In cases where the lexemic status is difficult to assess, morphological complexity is given some weight as a secondary criterion (e.g., red-orange might be questionable). Berlin and Kay also found that, in languages with fewer than the maximum eleven color categories, the colors followed a specific *evolutionary* pattern, as indicated below.

All languages contain terms for black and white.
If a language contains three terms, then it contains a term for red.
If a language contains four terms, then it contains a term for either green or yellow (but not both).
If a language contains five terms, then it contains terms for both green and yellow.
If a language contains six terms, then it contains a term for blue.

TABLE 3.1 Name of the colors in a few languages.

	Black	White	Red	Yellow	Blue	Green	Brown	Pink	Orange	Gray	Violet
Abaza		шкIвакIва	къапщы	гIвашь	йчIыхву чIыхь хIвыхIвчва	йачIва удзышва	гIвабджя	Чвапщ	къапщыра гIачвыцIххуа гIаважь	гIвабджя	чIыхв йачIва лилова
Abenaki	mkazawi	wôbi	Mkui	wizôwi	wlôwi	askaskui	wdamôôbame			wibgui	minôbowi
Acehnese	braʔa	Iĵu	Pirã	Iĵu	prãna	prãna					
Achuar-Shiwiar	šuwín	puhu	kapántin	Yaŋkú	wiŋkiá	samék					
Achumawi	hookíci	tiwici	taxtaxi	Makmaki	samtal	misuqadi					
Adyghe	шIуцIэ	фыжьы	плъыжьы	гъожьы	шхъуантIэ	уцышъо	хьаплъы	Шэплъы	гъожьы- шэплъ	ежьашъо, шъо	шхъуантIэ
Afrikaans	swart	blank wit	Rooi	geel	blou	groen	bruin donker	pienk ligrooi	Oranje	grys vergrys	Purper
Aguacatec	q̣eq	Saq	Kyaq	Q̣an	č̣aʔs	č̣aʔs					

The name of the colors in some languages (the first eight languages in alphabetical order). Abaza, spoken in the Caucasus mountain in Russia; Abenaki, a language of the Eastern Algonquian languages spoken by a group of Native Americans in New England and Quebec; Acenhese, a Malayo-Polynesian language spoken in the coastal region of Sumatra Island as well as in some parts of Malaysia; Achuar-Sciviar a Jivaroan language spoken along the Pastaza and Bobonaza rivers in Ecuador; Achumawi, a language spoken by some tribes of Native Americans living along the Pit river in what is now Notheastern California; Adyghe, a language spoken in the Northwest Caucasus mountains in Russia; Afrikaans, of the German group, one of the official languages in South Africa, spoken there and in Nanibia; Aguacatec, a Mayan language spoken in Guatemala, primarily in Huehuetenango and in the province of Aguacatán.

If a language contains seven terms, then it contains a term for brown.
If a language contains eight or more terms, then it contains terms for purple, pink, orange, or gray.

In addition to following this evolutionary pattern absolutely, each one of the languages studied also selected virtually identical focal hues for each category present. For example, the term for "red" in each of the languages corresponded to roughly the same shade in the Munsell system. Consequently, they posited that the cognition, or perception, of each color category is also universal [13–15]. Later study supported this universal, physiological theory, in particular, in a further paper it was demonstrated that 4-month-old children reacted in the same way to different frequencies corresponding to different shades [16]. The habituation time was found to be longer when the infant was presented with successive hues surrounding a certain focal, rather than with successive focal. This pattern of response is what is expected when the infants are distinguishing between the focals, but not distinguishing between successive hues (i.e., different shades of red are all "red" but "blue" and "red" focals are different) [17]. This is to say that infants respond to different hues of color in much the same way as adults do, demonstrating the presence of vision at an age younger than previously expected. Thus some scientists claimed that the ability to perceive the same distinct focals is present even in infants and thus is innate [18,19]. As for the theory of colors, there are two formal sides to consider, that may be successful in understanding the following debate, the universalist and the relativist [20,21]. The universalist side claimed that the biology of all human beings is everywhere the same, and thus the development of terminology obeys to an absolute universal constraint. The relativist side claims that the variability of terms cross linguistically (from language to language) points to more culture-specific phenomena. Because it exhibits both biological and linguistic aspects, it has become a deeply studied domain that addresses the relationship between language and thought [22].[1] Actually, English has, as mentioned above, 11 terms for indicating colors, black, white, red, blue, green, yellow, orange, brown, pink, gray, and purple, but other languages can have both more and less (as an example, both Greek and Russian have 12, since they distinguish light and deep blue, in the same way as English distinguish red and pink and Irish distinguish between different types of red, while Himba has only four terms: *zuzu:* dark shades of blue, red, green, and purple; *vapa:* white and some shades of yellow; *buru:* some shades of green and blue; and *dambu:* some other shades of green, red, and brown and some only two). Languages evolve with time and new terms may be added (whether or not from a different language, but in any case, through an additive evolution). Thus a fifth term was added to Himba, *serandu* for some shades of red, orange, and pink, but notice that pink entered in the English language only in the middle of the 18th century (it was originally the name of the flower, and seemingly the same phenomenon is going to occur with peach at present time). In modern science, this position has evolved to a theory of names. This theory is based on the fact that, in general, the development of social phenomena can be expressed according to the Zipf's law. The assigning of a sound (a word) to a color, as in general the emergence of a complex language is one of the fundamental events of human evolution, and the way in which it appears depends on several remarkable features as the presence of fundamental principles of organization and are common to all languages. The best example

[1] The advantage of an historically oriented course has been commented [23].

probably is the so-called Zipf's law, an empirical statistical law that suggest an inverse relationship between the frequency of the use of a symbol and its role in the language, for example, in the case of a word use (in a text of a given language, the most used word appears "n" times, the second one $n/2$, the third one $n/3$, etc.) [24].

Forms and colors, the two main aspects humans use for characterizing objects are actually strongly connected, since a difference is essential for distinguishing the borders of an object. Despite the fact that philosophers of the classic age considered the color as an intrinsic property of objects, so that a red object was endowed with a property of redness, there is no such thing as a red object. Rather, the color exists only in connection with the brain that sees it and reacts through internal signals to the wavelength that hits it (only wavelengths that are absorbed cause some effect, a statement sometimes referred to as the Maxwell law, which would then be a version of the first law of photochemistry) (Fig. 3.5) [25–27].

The evolution of the above system, the Munsell, based on three-dimensional axes, was created by Prof. Albert H. Munsell in the first decades of the 20th century and adopted by the United States, Department of Agriculture, as the official system for soil research [25] and still in use, a space that specifies based on three dimensions: the hue, the value or lightness, and the chroma (the purity) each in the 1930s. Previous scientists already had recourse to placing them in three dimensions of different forms, but Munsell was the first to separate hue, value, and chroma into perceptually uniform and independent dimensions, and he was the first to systematically illustrate these in three-dimensional space. Munsell's system, particularly the later renotations, is based on rigorous measurements of human subjects' visual responses to putting it on a firm experimental scientific basis. The Munsel system, or other systems based on the same principle are presently still used.

3.3 Newton's theory of vision

As for physics, in Newton's time the predominant idea was that colors were a mixture of light and darkness, from brilliant red (white with a small amount of darkness, to dull blue, with much darkness, just a step before black) to that of the prisms colored light. Newton demonstrated that white light could be both decomposed and recomposed by the prism, thus white light was a mixture of colored lights, red, orange, yellow, green, blue, and violet, to which he later added indigo. The theory found enemies, first of all Robert Hooke, who presented in 1672 to the Royal Society a strong critique to Newton's theory of light and mocked his choice of the magic number of 7 for the colors, as the borders between them were blurred, and any number of different colors could be chosen [28,29]. Newton actually published his complete experiments only after the death of Hooke and after having succeeded him as president of the Royal Society, and his view of the subject is as follows:

> (Qu. 13) ... Do not several sorts of Rays make Vibrations of several bignesses, which according to their bignesses excite Sensations of several Colors, much after the manner that the Vibrations of the Air, according to their several bignesses excite Sensations of several Sounds? And particularly do not the most refrangible Rays excite the shortest Vibrations for making a Sensation of deep violet, the least refrangible the largest for making a Sensation of deep red, and the several intermediate sorts of Rays, Vibrations of intermediate bigness? (Qu. 14). May not the harmony and discord of Colours arise from the proportions of the Vibrations propagated through the Fibres of the optick Nerves into the Brain, as the harmony and discord of Sounds arise from the proportions of the Vibrations of the Air? For some Colours, if they be view'd together, are agreeable to one another, as those of Gold and Indigo, and others disagree. (Qu 16). When a Man in the dark presses either corner of his Eye with his Finger, and turns his Eye away from his Finger, he will see a Circle of Colours like those in the Feather of a Peacock's Tail. If the Eye and the Finger remain quiet these Colours vanish in a

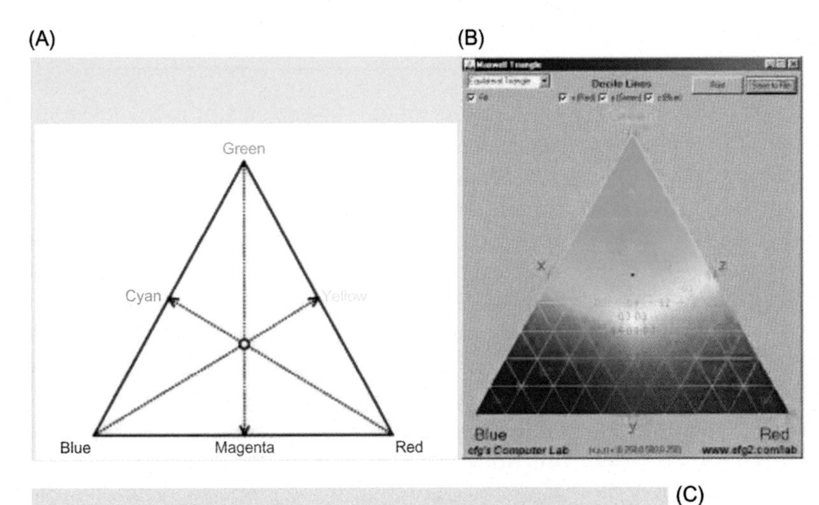

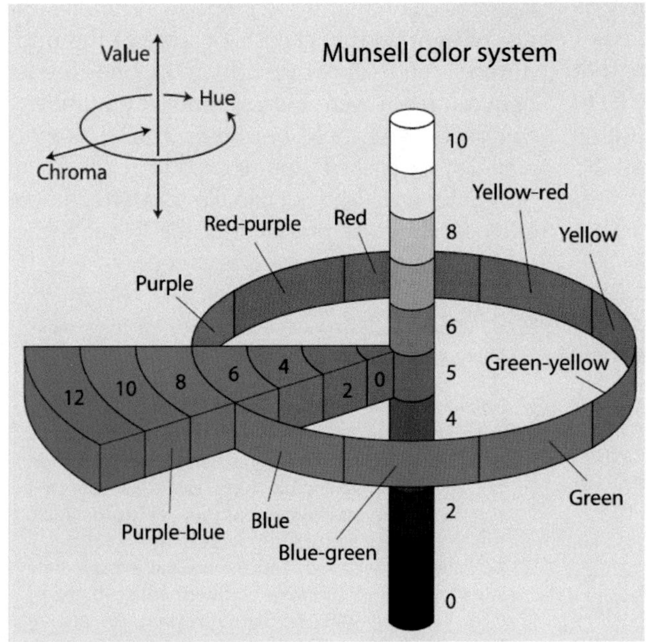

FIGURE 3.5 (A) The Maxwell triangle, empty; (B) pure, primary additive colors (blue, red, and green) sit at the corners, while the subtractive ones sit at the bisection by the height and the side (cyan, yellow, and magenta). All of the combinations within the triangle give different colors, with white, or black for additive or subtractive colors, respectively, black at the center; (C) Munsell color system, classifies colors in a three-dimensional system [25]. The axes express value, chroma, and hue (see text).

second Minute of Time, but if the Finger be moved with a quavering Motion they appear again. Do not these Colours arise from such Motions excited in the bottom of the Eye by the Pressure and Motion of the Finger, as, at other times are excited there by Light for causing Vision? And do not the Motions once excited continue about a Second of Time before they cease? And when a Man by a stroke upon his Eye sees a flash of Light, are not the like Motions excited in the Retina by the stroke? And when a Coal of Fire moved nimbly in the circumference of a Circle, makes the whole circumference appear like a Circle of Fire; is it not because the Motions excited in the bottom of the Eye by the Rays of Light are of a lasting nature, and continue till the Coal of Fire in going round returns to its former place? And considering the lastingness of the Motions excited in the bottom of the Eye by Light, are they not of a vibrating nature? (Qu.17). If a stone be thrown into stagnating Water, the Waves excited thereby continue some time to arise in the place where the Stone fell into the Water, and are propagated from

> thence in concentrick Circles upon the Surface of the Water to great distances. And the Vibrations or Tremors excited in by percussion, continue a little time to move from the place of percussion in concentrick Spheres to great distances. And in like manner, when a Ray of Light falls upon the Surface of any pellucid Body, and is there refracted or reflected, may not Waves of Vibrations, or Tremors, be thereby excited in the refracting or reflecting Medium at the point of Incidence, and continue to arise there, and to be propagated from thence as long as they continue to arise and be propagated, when they are excited in the Eye by the Pressure or Motion of the Finger, or by the Light which comes from the Coal of Fire in the Experiments above mention'd? and are not these Vibrations propagated from the point of Incidence to great distances? And do they not overtake the Rays of Light, and by overtaking them successively, do they not put them into the Fits of easy Reflexion and easy Transmission described above? For if the Rays endeavour to recede from the densest part of the Vibration, they may be alternately accelerated and retarded by the Vibrations overtaking them [29].

Newton more and more used the analogy with musical notes, although this does not hold quantitatively when a unit of measure, for example, the frequency, is chosen, because it seemed untenable that each sensitive point of the retina contained an infinite number of particles, each capable of vibrating in perfect unison with every possible undulation, it was supposed that the number was limited to the three principal colors, red, yellow, and blue, "of which the undulations are related in magnitude as the numbers 8, 7, and 6... for instance, the undulations of green light being nearly in the ratio 6 1/2, will affect equally the particles in unison with yellow and blue."

Most useful for art practitioners was the arrangement of colors on a circumference, in such a way that complementary colors were located directly opposed one to another. The physical aspects of light refraction, the analogy of light and sound propagation, as well as between colors and harmonies, and the rate of diffusion of light were largely studied in the 18th and initial 19th centuries, in the frame of the debate on the vibrational versus particle theory of light, and several proposals of the function of the organs and the role of colors were formulated. The mechanism of vision and the number and relationship of the receivers of information as well as the theory of diffraction of light were first formulated clearly by Young, in a way that is substantially correct [30], based on the physical analysis of the nature of colors and light as formulated by Newton. Among the advancements in describing the physical function of the eye, an important step forward was the studies by Dalton [31], who demonstrated that some people, like himself, could be said to be dichromat, as all of the hues, except some that they perceived as various shades of gray, and not trichromat, as most people were (an often mentioned anecdote refers to young Dalton offering to the beloved, but strongly religious, Mother, for her birthday, a pair of underpants of what he thought a delicate tinge of grey, but most people classified as a lively red), and that by Wheatstone [4,32], demonstrating that binocular vision allows a steroscopical view (Fig. 3.6 and Tables 3.2 and 3.3).

The understanding of light as a vibrating electromagnetic wave by Maxwell sets the movement toward a satisfying resolution of such a story, one that began with the earliest philosophers and remained a hot topic for scientists for centuries and engaged the interest of many of the greatest names of science, some of whom (including Newton, Dalton, Helmholtz, Maxwell, Einstein, and Schrödinger) contributed materially to the theories, was found at the end of the 20th century and then with modern quantum physics [29–31,35]. It was mentioned above that every hue of color resulted from different combinations of three colors. This was initially proposed by Young [30], that is, each hue resulted from a contribution of three different particles sensitive to different parts of the spectrum, mainly at the green, red, and violet. These studies were further

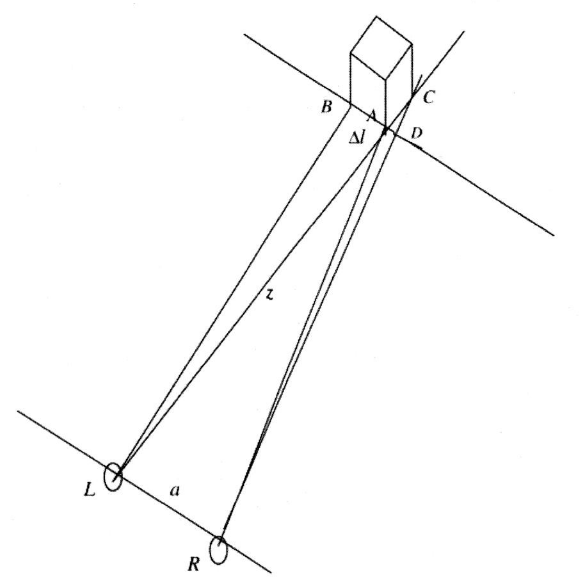

FIGURE 3.6 Pairs of outline figures, calculated to give rise to the perception of objects of three dimensions when placed in the stereoscope in the manner described, are represented. They are one half the linear size of the figures actually employed. As the drawings are reversed by reflection in the mirrors, Wheatstone supposed that these figures were the reflected images to which the eyes are directed in the apparatus; those marked *b* being seen by the right eye, and those marked *a* by the left eye. The drawings are two different projections of the same object seen from two points of sight, the distance between which is equal to the interval between the eyes of the observer; this interval is generally about 2½ inches [4].

TABLE 3.2 The intensity of light perceived by humans depends on its wavelength and is usually communicated by lux [33,34].

Lux	
Full sunlight	103,000
Partly sunny	50,000
Cloudy day	1000–10,000
Full moon under clear conditions	0.1–0.3
Quarter moon	0.01–0.03
Clear starry night	0.001
Overcast night sky	0.00003–0.0001
Operating table	18,000
Bright office	400–600
Most homes	100–300
Main road street lighting (average street level illuminance)	15
Lighted parking lot	10
Residential side street (average street level illuminance)	5
Urban skyglow	0.15

developed by Helmholtz and resulted in the Young–Helmholtz theory of trichromacy, which accounted perfectly for the observed colors. Helmholtz thought that the trichromacy was obtained through appropriate filters, which is indeed a system used by nature, but not in humans, and on this basis proposed the way to make colored pictures (Fig. 3.7) [36,37]

3.3.1 Further advancements

When the Newton wheel is made to rotate rapidly, the impression of the observers is that the colors merge in a uniform white. The proposal by Newton explained this by assuming that complementary colors were placed in the same order as they appeared in the prism decomposed light. Red, green, and blue were considered primary colors and their combination in different amounts indeed gave the various hues (Fig. 3.8).

During the whole of the 18th century, debate on the theory of colors continued, as a part of the theory of light, where the field was strongly divided between corpuscular and ondulatory theories. Newton himself never took an uncompromising position on the question, although he was clearly a corpuscular, while Descartes was among the supporters of the ondulatory theory involving the vibration of a medium and the debate assumed a

TABLE 3.3 SI (J) and conventional (candela) units of measure that take into account the sensitivity of human eyes.

SI photometry quantities					
Quantity		Unit		Dimension	
Name	**Symbol**[a]	**Name**	**Symbol**	**Symbol**[b]	**Notes**
Luminous energy	Q_v[c]	Lumen second	lm · s	**T · J**	The lumen second is sometimes called the *talbot*
Luminous flux, luminous power	Φ_v[c]	Lumen (=candela steradians)	lm (=cd · sr)	**J**	Luminous energy per unit time
Luminous intensity	I_v	Candela (=lumen per steradian)	cd (=lm/sr)	**J**	Luminous flux per unit solid angle
Luminance	*Lv*	Candela per square meter	cd/m^2	**L^{-2} · J**	Luminous flux per unit solid angle per unit *projected* source area. The candela per square meter is sometimes called the *nit*.
Illuminance	E_v	Lux (=lumen per square meter)	lx (=lm/m^2)	**L^{-2} · J**	Luminous flux *incident* on a surface
Luminous existence, luminous emittance	M_v	Lux	lx	**L^{-2} · J**	Luminous flux *emitted* from a surface
Luminous exposure	H_v	Lux second	lx · s	**L^{-2} · T · J**	Time-integrated illuminance
Luminous energy density	ω_v	Lumen second per cubic meter	lm · s/m^3	**L^{-3} · T · J**	
Luminous efficacy (of radiation)	*K*	Lumen per watt	lm/W	M^{-1} · L^{-2} · T^3 · J	Ratio of luminous flux to radiant flux
Luminous efficacy (of a source)	η[c]	Lumen per watt	lm/W	**M^{-1} · L^{-2} · T^3 · J**	Ratio of luminous flux to power consumption
Luminous efficiency, luminous coefficient	*V*			**1**	Luminous efficacy normalized by the maximum possible efficacy

See also: SI · Photometry · Radiometry

[a]*Standards organizations recommend that photometric quantities be denoted with a suffix "v" (for "visual") to avoid confusion with radiometric or photon quantities. For example,* USA Standard Letter Symbols for Illuminating Engineering *USAS Z7.1-1967, Y10.18-1967.*
[b]*The symbols in this column denote dimensions:* **"L", "T,"** *and* **"J"** *are for length, time, and luminous intensity respectively, not the symbols for the units liter, tesla, and joule.*
[c]*Alternative symbols sometimes seen:* W *for luminous energy.* P *or* F *for luminous flux, and* ρ *for luminous efficacy of a source.*
From Wikipedia. Candela (units of measure)

somewhat nationalistic and political flavor. Further "physical" improvements included the understanding of binocular vision and the possibility of obtaining a three-dimensional view in this way [32].

3.3.2 Goethe's contribution

In 1810 Johan Wolfang von Goethe published a book on the theory of colors (*Zur Farbenlehre* [38]) that he always cherished among his most important works. The book

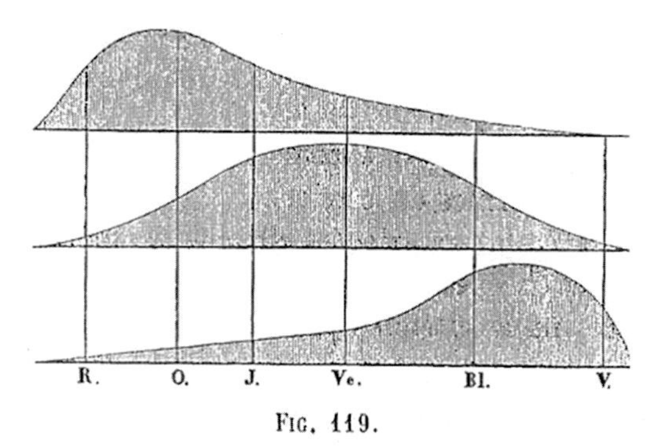

FIGURE 3.7 Helmholtz represented trichromacy as resulting from the filtering of light through three different dye layers [36].

actually originated from the interest of the poet in painting and had a lasting effect on this art, as were other treatises printed earlier [39], although the author rather meant it as scientific polemic with Newton on the color theory, not using the mathematical approach, but a merely experimental one [40]. The key idea was to account for the conditions under which a color is perceived under a large variety of circumstances, not limited to the white light decomposition obtained by Newton (it is important to recall that the original Newton's account found considerable trouble in being universally accepted). Newton's experiment had been poorly described, and in fact, the decomposition of white light into the colors of the spectrum depended on the distance between the prism and the observer (Fig. 3.9). This led Goethe to suggest that Newton's theory referred only to a special case, which in a sense was true, as indeed previously reported by other "professional" scientists (in particular by L. Castel, [41][2]), and only with the much later published Newton's complete reports the theory acquired the final, fully developed (and fully convincing) aspect.

Returning to the likeliness of sound in the notes and light in the spectrum used by Newton, this may serve to illustrate another point. If the particles of light activated the nerves, it was almost impossible to think that every shade of the thousands or millions of conceivable hues of the spectrum finds in the retina exactly the corresponding vibration, first of all because it would have been difficult to find the required space. The solution was that only a limited number of colors were active, actually the three "primary" colors, and their combination originated every possible hue, as indeed proved by experiments (Fig. 3.10) [40,42].

As Ludwig Wittgenstein noticed, Goethe's theory of the constitution of colors of the spectrum had not proved to be an unsatisfactory theory; rather, it really was no theory at all [43,44]. Nothing can be predicted with it. It is a vaguely schematic outline of the sort we find in James' psychology. Nor there was any *experimentum crucis* which could decide for or against the theory. Based on his experiments with turbid media, which he characterized as arising from the dynamic interplay of darkness and light, with light streaming into a dark space finding no resistance from the darkness to overcome, Goethe pictured light and darkness related to each other like the north and south poles of a magnet. The darkness can weaken light in its working power. Conversely, light can limit the energy of the

[2] Apropos L. Castel, it may be useful at this point to mention Castel's harpsichord. The ocular harpsichord was a kind of "thought experiment," a realization of an idea in an imagined instrument, proposed by French mathematician and experimentalist Louis Bertrand Castel in 1725. Based on an analogy between the seven tones of the diatonic scale and the seven colors of the rainbow (which Isaac Newton had advanced in his book on optics), Castel argued that it should be possible to create music for the eyes by replacing the pitches of a harpsichord with colors. Castel did not build an instrument to demonstrate the music he derived theoretically [41]; the point is not making a son et lumière show, but rather making real music with colors in the place of notes.

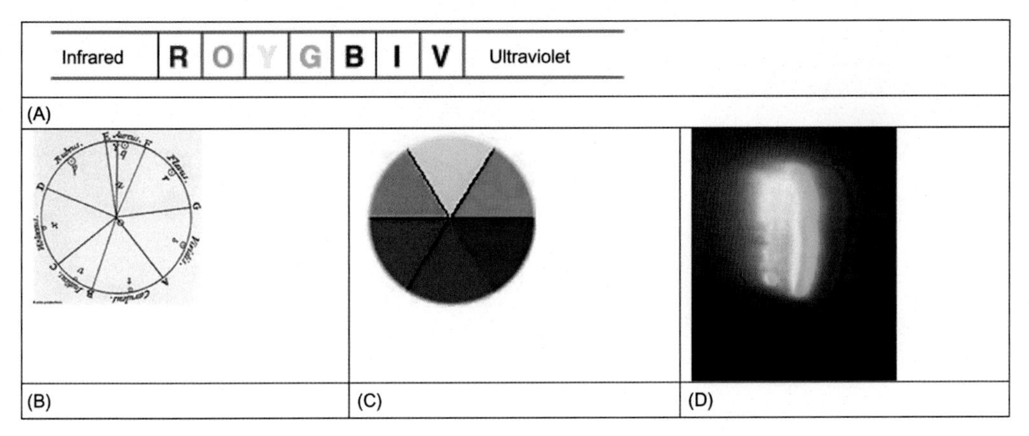

FIGURE 3.8 (A) Linear arrangement of the colors resulting from the decomposition by a prism of white light; (B) the Newton wheel, original form (notice that different areas are allotted to each color); (C) idealized form with all the colors equally spaced, (D) light defracted through a prism (notice that the boundary region between the colors is blurred) [29,36].

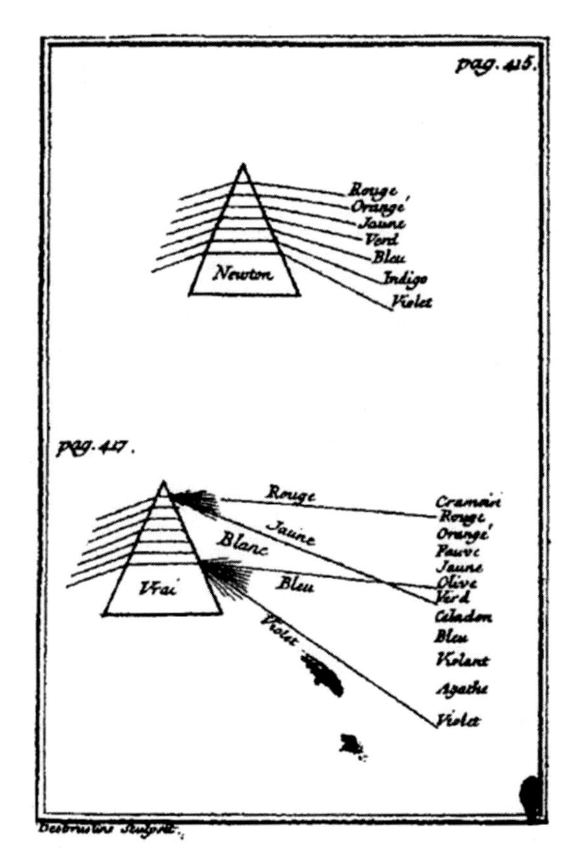

FIGURE 3.9 Dependence of the diffraction of light by prisms on the distance from the observer [41].

darkness. As he expresses it, yellow is a light which has been dampened by darkness; while blue is a darkness weakened by light.

In an earlier study in collaboration with Schiller, Goethe (1798–99) compared 12 colors and matched them with the 12 human occupations or their characters as typified in the 12 temperaments [*tyrants, heroes, adventurers, hedonists, lovers, poets, public speakers, historians, teachers, philosophers, pedants, rulers* (Fig. 3.11) [45]].

Goethe also includes experiments on the esthetic qualities in his color wheel, under the title of "allegorical, symbolic, mystic use of colors" [46] and offered a kind of psychology of colors, where he associated red with the "beautiful," orange with the "noble," yellow to the "good," green to the "useful," blue to the "common," and violet to the "unnecessary." In turn, these six qualities were assigned to four categories of human cognition, the rational (*Vernunft*) to the beautiful and the noble (red and orange), the intellectual (*Verstand*) to the good and the useful (yellow and green), the sensual (*Sinnlichkeit*) to the useful and the common (green and blue), and, closing the circle, imagination (*Phantasie*) to both the unnecessary and the beautiful (purple and

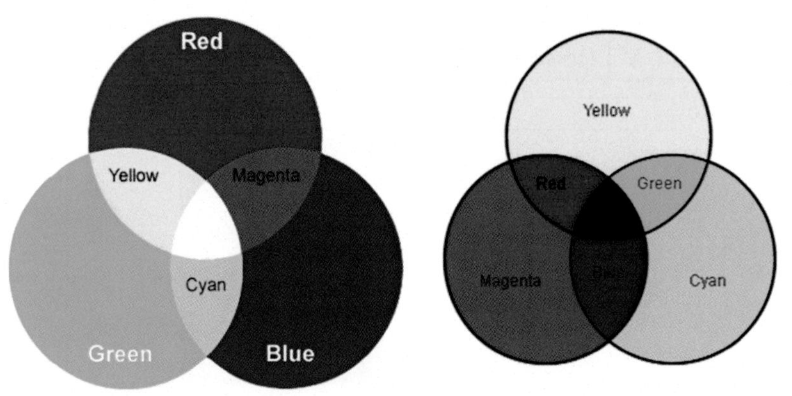

FIGURE 3.10 The additive (right) and subctractive (left) primary colors and their combinations.

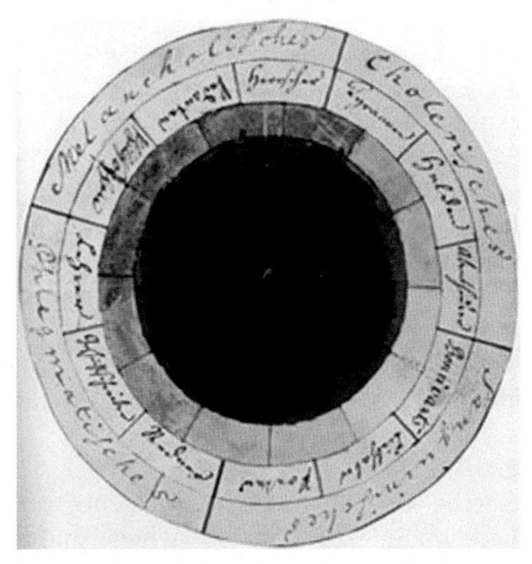

FIGURE 3.11 The "rose of man's temperaments," by Goethe and Schiller [45].

red), overall resulting in the correlation shown in Table 3.4.[3]

Of his own theory, Goethe was supremely confident: "From the philosopher, we believe we merit thanks for having traced the phenomena of colors to their first sources, to the circumstances under which they appear and are, and beyond which no further explanation respecting them is possible" [46]. Goethe's scientific conclusions have, of course, long since been thoroughly demolished because there is no science without mathematics, but his conjectures regarding the connection between philosophical ideas is still of interest, as is the insight it gives into early 19th-century beliefs and modes of thought, and for the flavor of life in Europe just after the American and French revolutions. The work may also be read as an accurate guide to the study of phenomena. Goethe's conclusions have been repudiated, but no one quarrels with his reporting of the facts to be observed. With simple objects, such as vessels, prisms, lenses, and the like, the reader will be led through a demonstration course not only in subjectively produced colors, but also in the observable physical phenomena of color. By closely following Goethe's explanations of the phenomena as they are experimenting them, in a way that is felt as satisfactory even divorced from the wavelength theory (Goethe never even mentions it)

[3] Magenta appeared as a color term only in the mid-19th century, after Goethe's death. Hence, references to Goethe's recognition of magenta are fraught with interpretation. If one observes the colors coming out of a prism—an English person may be more inclined to describe as magenta what in Germany is called Purpur—so one may not lose the intention of the author. However, literal translation is more difficult. Goethe's work uses two composite words for mixed (intermediate) hues along with corresponding usual color terms such as "orange" and "violet."

TABLE 3.4 Psychological values of colors according to Goethe.

German	English	Symbolism
Purpur	Magenta (or purple) see below	*Schön* (beautiful)
Rot	Red	
33Gelbrot	Orange	*Edel* (noble)
Orange		
Gelb	Yellow	*Gut* (good)
Grün	Green	*Nützlich* (useful)
Blau	Blue	*Gemein* (mean, common)
Violett	Violet	*Unnöthig* (unnecessary)
Blaurot		

that he may begin to think about theory relatively unhampered by prejudice, ancient or modern [46,47].

3.4 The eye

The organ of vision (Fig. 3.12), the eye, is a ball-shaped object (2.4 cm diameter) located in and protected by a cavity in the bone of the skull [48,49].

It is filled by the vitreous humor (a transparent gel made of proteins and water) in the anterior part, and the aqueous humor (a dilute solution of proteins) in the posterior part and covered by several membranes. Six muscles are fixed to it and can move the eye bulb in every direction. From the external toward the inner part, the membranes are the conjunctiva, the sclera, and the retina. The conjunctiva is a thin, clear mucous membrane that lines the inner surface of the eyelids and the front, outer surface of the eye. The conjunctiva makes a

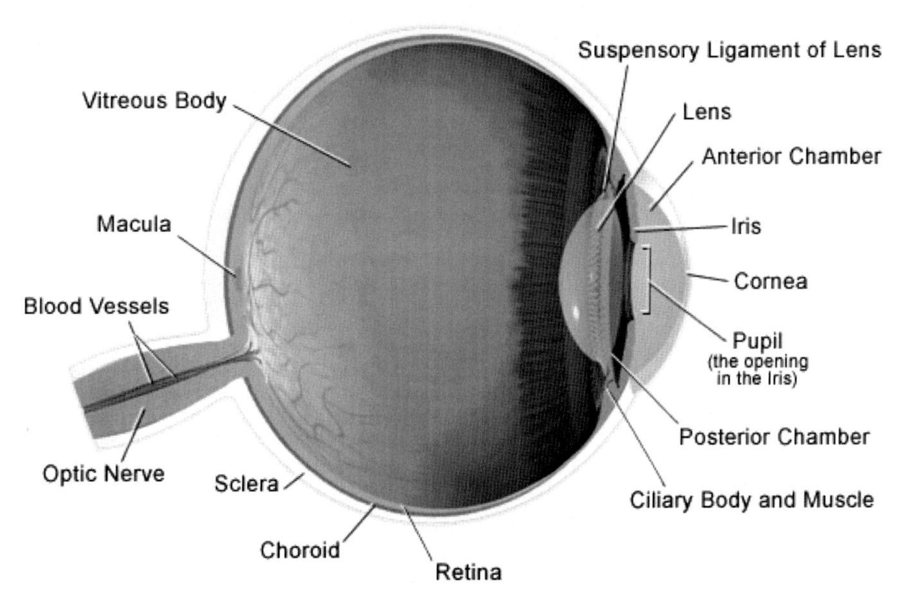

FIGURE 3.12 Human eye anatomy diagram from https://www.stanfordchildrens.org/en/topic/default?id=anatomyoftheeye-85-P00506

mucous that lubricates the eyeball and keeps it moist. The conjunctiva of the fornix is the lining of the sac that is present at the junction between the posterior surface of the eyelid and the anterior part of the globe of the eye. The conjunctiva is relatively thicker in this region and this allows for free movement of the eyeball. The conjunctival sac formed at the transition of palpebral and fornix conjunctiva holds around 7 μL of tear fluids. The sac has a capacity to hold 30 μL of fluid. The bulbar conjunctiva is the thinnest part of the conjunctiva. It covers the cornea and the anterior portion of the eyeball. It is perfectly transparent, so that one can see with the naked eye the underlying white sclera and the blood vessels. The sclera, of white color, is the tough outer coat of the eyeball. It is a thick, opaque membrane that is made up of white fibrous collagenous tissue. It is underlined by the choroid layer. The sclera is continuous with the cornea in the front, and the optic sheath at the back of the eye. The optic sheath encircles the optic nerve as it emerges out from the retina. In humans the color of the sclera contrasts with the smaller size and dark color of the iris because of which it is distinctly visible. In other mammals the sclera gets camouflaged with the larger size of the iris and is therefore not visible. The sclera is responsible for maintaining the shape of the globe and of light but offers resistance to internal and external forces. It also provides a base for attachment for the extra ocular muscles that are responsible for eye movement. The thickness of the sclera varies from 1 mm in most points in the rear face to 0.3 mm just behind the rectus muscle insertions. The sclera is also divided into four membranes: episclera, stroma, lamina fusca, and endothelium.

3.4.1 Micro anatomy of the retina

The interior wall of the eye is lined with a further membrane, the retina, where the absorption of light takes place. This is no uniform layer and several peculiarities are observed, viz the macula lutea, a nonvascolarized yellow spot on the retina, further containing smaller areas that are called the fovea and the foveola, a small cavity at the center of the back of the eye at 7 degrees, or 5 mm below the meridian line. This part of the eye is filled by the aqueous humor (the largest part of the eye, a dilute solution containing some proteins) [50,51]. The retina is approximately 0.5-mm thick and lines the back of the eye. The optic nerve contains the ganglion cell axons running to the brain as well as incoming blood vessels that open into the eye. A radial section of a portion of the retina shows that the ganglion cells (the output neurons of the retina, rods, and cones) lie on the internal part of the eye, closest to the lens and on the front side of the eye. In contrast, the photosensitive cells lie on the external side of the eye that lies outermost in the retina against the pigment epithelium and choroid. Thus the light ray has to cross through the thickness of the retina before striking and activating the rods and cones (Fig. 3.13) [51].

As for the photoactive cells, rods are much more in number than cones, and are relatively homogeneously distributed. On the contrary, cones are accumulated in several superimposed layers in the fovea, where there are no rods. Subsequently, the absorption of photons by the visual pigment of the photoreceptors is translated first into a biochemical message and then into an electrical message that can stimulate all the succeeding neurons of the retina. The retinal message concerning the photic input and some preliminary organization of the visual image into several forms of sensation are transmitted to the brain from the spiking discharge pattern of the ganglion cells. Moving from the frontal part toward the back of the head, one finds (1) the inner limiting membrane; (2) the nerve fiber layer, formed by the axons of the ganglion cell nuclei; (3) the ganglion cell layer, containing the nuclei of

(A)

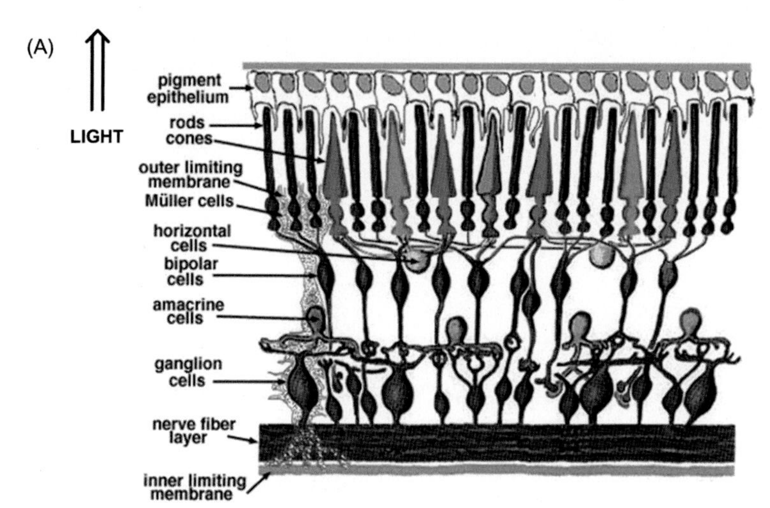

(B)

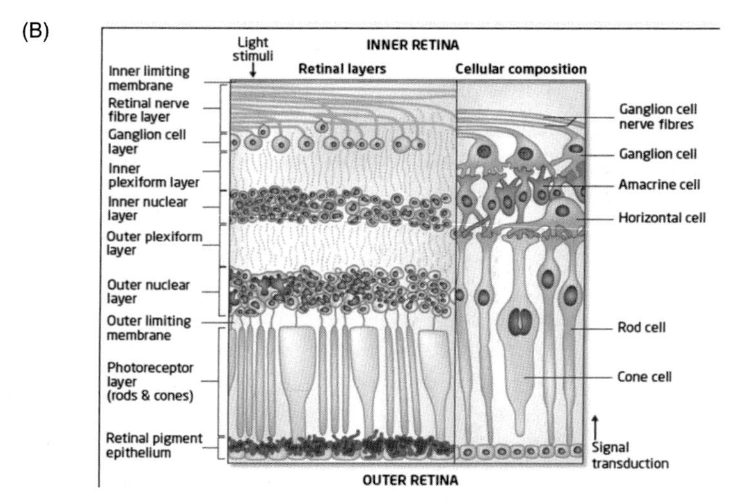

FIGURE 3.13 (A) Simple organization of the retina [51]. (B) An illustration of the layers of the retina. Note that the direction of signal transduction is directed from the photoreceptors (rods and cones) at the outer aspect of the retina toward the retinal nerve fiber layer (RNFL) at the inner aspect of the retina. The fibers of the RNFL exit the retina at the optic disc, where they form the optic nerve. Also note the origin of the nerve fibers of the RNFL from the ganglion cells in the ganglion cell layer. Optic nerve demyelination results in RNFL degeneration, which in turn leads to ganglion cell body death.

such cells; (4) the axons of which become the optical nerve; (5) the inner plexiform layer (IPL), where the synapse between the axons of bipolar cells, the dendrites of the ganglion and the dendrites of bipolar and amacrine cells is located; (6) the inner nuclear layer, which contains nuclei and the close-lying structures (perikarya) of amacrine, bipolar and horizontal cells; (7) the outer plexiform layer (OPL), where the projections of rods (ending in the rod spherule) and cones (ending in the cone pedicle) are located and make synapse with the dendrites of bipolar cells. After the photoresponsive cells, the external limiting membrane that separates the inner segment portion from their cell nucleus follows and then the rod and cone layer, containing such cells is located. The last layer is the retinal pigment epithelium (RPE). The outer nuclear layer contains the bodies of rods and cones (the latter species are present in three types that are characterized by a shift toward the blue of the light absorption, and are called L, M, and S, or long, medium, and short wavelength, or red, green, and blue, with λ_{max} at 564–580, 534–545, and 420–440 nm, respectively (Fig. 3.14 [51]), while rods show a single

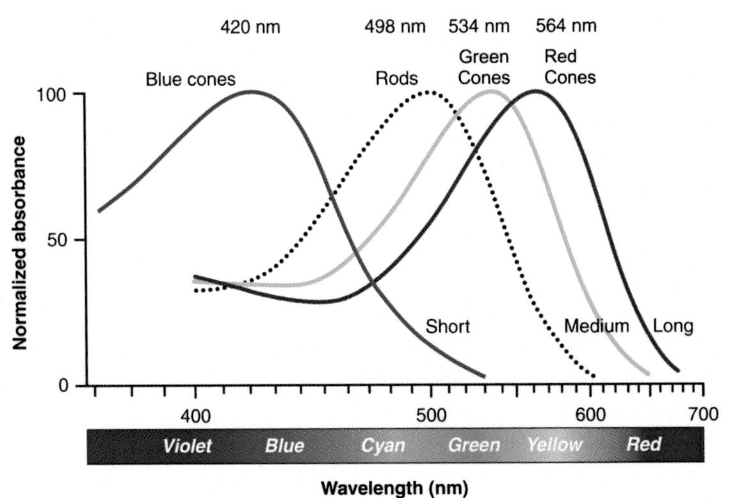

FIGURE 3.14 The vision. The absorption spectra of the four photopigments in the normal human retina. The solid curves indicate the normalized spectra of the three kinds of cone opsins; the dashed curve shows rod rhodopsin for comparison. Absorbance is defined as the log value of the intensity of incident light divided by intensity of transmitted light [51,52]. Source: *Wikipedia*.

maximum of absorption (420 nm) and are much more numerous (9×10^7) than cones (4×10^6).

These characteristics give cones the capacity to mediate high visual acuity, as their density strongly declines while portions far from the fovea are considered (e.g., only 25% of the maximal value at 6 degrees from the median line). This is important, because it allows to obtain some lateral view, that can be followed by movements of the eyes and the head in order to put in the focal spot various parts of what we see, in a typical case, when reading a book or a newspaper. In peripheral retina, S-cone pedicles are bilobed in shape, with synaptic invaginations and ribbons separated to the two lobes. They do not exhibit the long telodendria (the presence of specialized thin endings) typical of other cone pedicles, so they remain rather isolated from gap junction contacts, with L- or M-cones at the level of the OPL.

As mentioned above, most of the active light signaling structures in the human retina are rods. These are thin cells of 2-μm diameter and amount to 93% of the visual cells that are specialized for low-intensity, monochromatic, as is typical of diurnal animals, with only a 3% of the other type of visual cells, cones, larger cells that direct the upper part of their conical shape toward the inner part of the membrane and are more suitable for detecting bright light and distinguish colors; these are located close to the fovea (Fig. 3.15). Rods are characterized by a large and slow response, high sensitivity and a slow dark adaptation. In contrast, cones are characterized by a small and fast response, low sensitivity, and a fast-dark adaptation. These response profiles originate from the functional proteins in them.

A third type of light activated cells was also discovered; these (photosensitive ganglion cells), however, do not contribute to the sight in a direct way, but rather affect the nonvisual functions. The light-sensitive pigment is in any case rhodopsin, although small differences in the opsin result in small differences in λ_{max} mentioned above. Rhodopsin is a purple pigment, discovered by F.C. Boll in 1876 in the frog's retina, along with the information that it was rapidly bleached by ambient light and drawings or letters could be reproduced by projecting the forms on the back of the retina, thus demonstrating the role as visual pigment of these compound [54]. More than 120 years

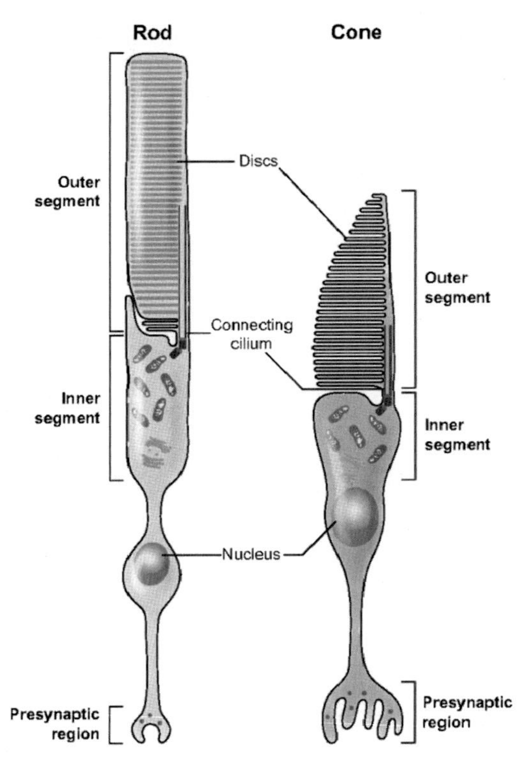

FIGURE 3.15 Schematic structure of human rods and cones [53]. http://thebrain.mcgill.ca/flash/d/d_02/d_02_m/d_02_m_vis/d_02_m_vis.html.

elapsed before the structure of rhodopsin was demonstrated (Fig. 3.16) [56].

The prostetic group is 11-cis retinal, bound through a covalent $C=N^+$ bond to an opsin, pertaining to the family of G-protein-coupled receptors (GPCRs). This comprises the largest family (c.800 proteins in humans, of which 670 are grouped in a single class, that of rhodopsin-like pigments) of such receptors that function as switches, able to respond to a large variety of the environment characteristics, such as physiological and environmental signals including both chemical (hormones, neurotransmitters, olfactory, and gustatory stimuli) and physical (light) changes. Membrane-spanning receptors that connect the extracellular environment and the cell interior, then operate signal transduction. As shown in Fig. 3.16, seven transmembrane helical proteins (TM or H) are placed roughly parallel through the membrane, and an eight one horizontally (HVIII), perpendicularly to the other ones. This architecture is highly conserved among different animal species. A conserved set of residues on the cytoplasmic surface, where G-protein activation occurs, likely undergo a conformational change upon photoactivation of the chromophore that leads to rhodopsin activation and signal transduction.

As it appears from the Fig. 3.17 [57], rhodopsin's polypeptide chain traverses the lipid bilayer seven times as the helix segments HI through HVII. Hydrophilic loop sequences are exposed to the cytoplasmic surface (CI–CIV) and to the intradiscal or extracellular surface (EI–EIII). The amino-terminal sequence is acetylated at M^1 and contains oligosaccharide chains at the asparagine N^2 and N^{15}, cysteines^{110} and C^{187} are linked by a disulfide bridge, and C^{322} and C^{323} are palmitoylated. Retinal is linked via a Schiff base to Lys^{296} in helix VII.

Rhodopsin molecules are packed in the form of a series of helical tubes arranged to form seven trans membrane helices (H or TM). 194 residues are involved overall, precisely 30 residues (35 to 64) for H-I, 30 (71 to 100) for H-II, 33 (107 to 139) for H-III, 23 (151 to 173) for H-IV, 26 (200 to 225) for H-V, 31 (247 to 277) for H-VI, 21 (286 to 306) for H-VII. Furthermore, a number of residues pertain to the extracellular region, in detail 35 residues (1 to 34) in the NH_2-terminal chain, 6 (101 to 106) for E-I, 22 (174 to 199) for E-II, and 9 (278 to 285) for E-III, as well as to the cytoplasmatic region, in detail, 6 (65 to 70) for C-I, 11 (140 to 150) for C-II, 10 (226 to 235) and 7 (240 to 246) for C-III, and finally 21 (307 to 327) and 15 (334 to 348) for the COOH terminal region. The current model for bovine rhodopsin in the membrane lipid bilayer additionally contains two oligosaccarides and two Zn^{2+} and three Hg^{2+} ions and some water molecule per molecule, as well as, obviously, the light-sensitive prostetic group.

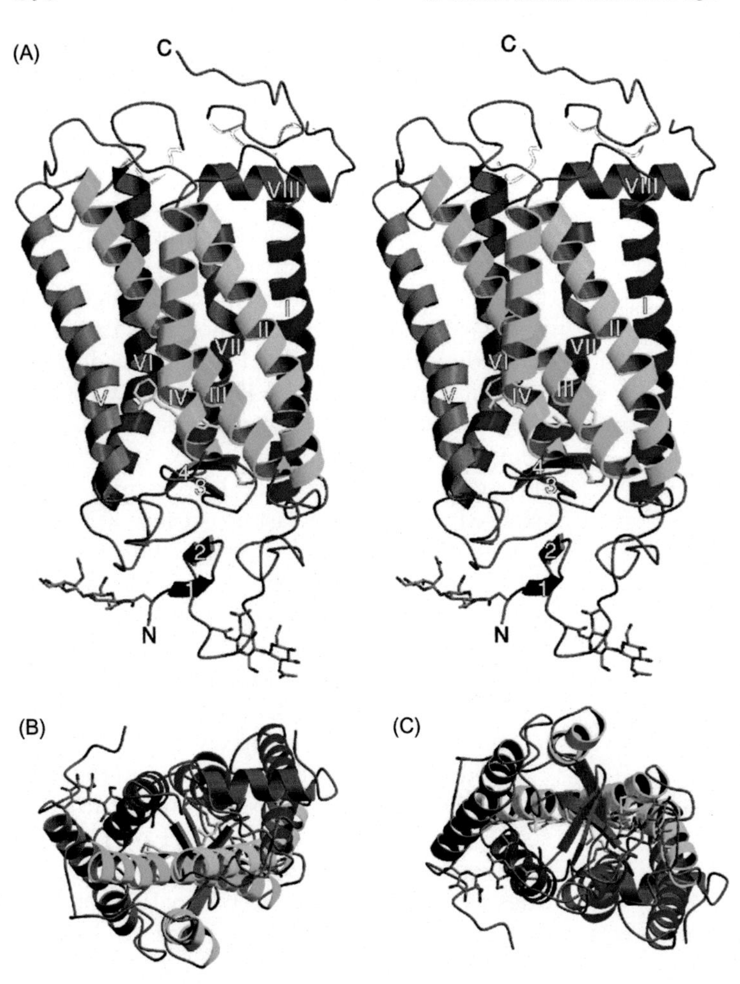

FIGURE 3.16 Ribbon drawings of rhodopsin. (A) Parallel to the plane of the membrane (stereoview). A view into the membrane plane is seen from the cytoplasmic (B) and intradiscal side (C) of the membrane [55].

The first of two antiparallel strands, G^3 to P^{12} form a typical β-sheet-fold running almost parallel to strands three to five (S3–S5) forms along axis of the molecule. S4 connects S^{14} and N^{15} in the NH_2 terminal region of the molecule rophilic loop sequences are exposed to the triangle from F^{13} to P^{14}, with the third strand running just below E-III, almost parallel to the long axis of the molecule.

The NH_2 terminal sequence is acetylated at M^1 and contains oligosaccharide chains at N^2 and N^{15}. Cysteines C^{110} and ^{187} are linked by a disulfide bridge, and C^{322} and C^{323} are palmitoylated. The NH_2-terminal tail of rhodopsin contains five distorted strands. The NH_2-terminus is located just below loop EIII, with the side chain of aspartic acid D^{282} close to that of N^2. The first two antiparallel strands, G^3 to P^{12}, form a typical β-sheet fold (β1 and β2) running almost parallel to the expected plane of the membrane. Strands three to five (S3-S5) form a right triangle from F^{13} to P^{34}, with the third strand running just below E-III, almost parallel to the long axis of the molecule. S4 connects S^{14} and N^{15} in the NH_2-terminal region of the molecule with P^{23}, located close to E-I. S5 from P^{27} to P^{34} runs along the surface of the membrane covering the extracellular (intradiscal) space between H-I and H-II. Oligosaccharides at N^2 and N^{15} extend from the domain and are not included in any significant interactions.

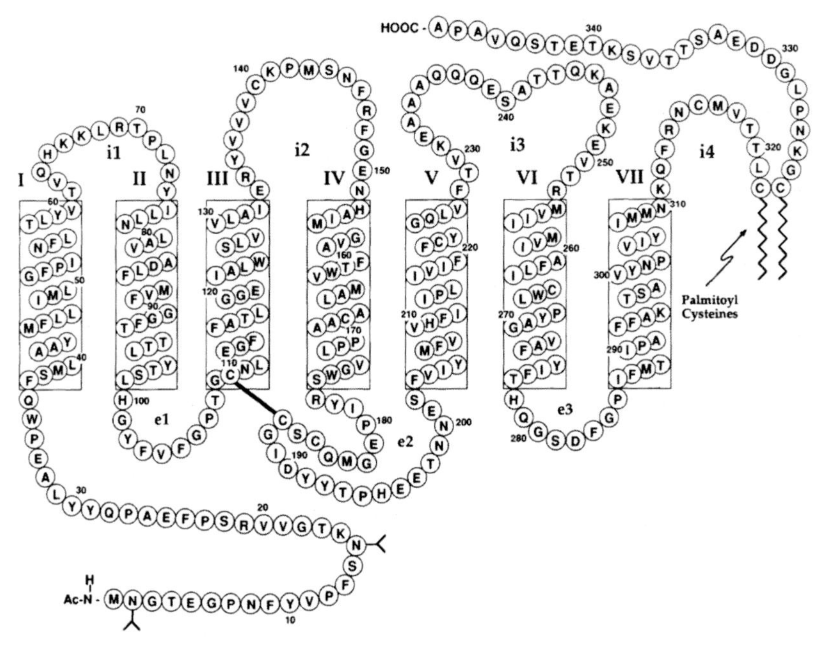

FIGURE 3.17 Two-dimensional model of bovine rhodopsin. Crystal structure of the rhodopsin molecule in the membrane lipid bilayer. Darkened circles indicate main interactions. Rhodopsin's polypeptide chain traverses the lipid bilayer seven times as the helix segments HI through HVII. Hydrophilic loop sequences are exposed to the cytoplasmic surface (CI–CIV) and to the intradiscal or extracellular surface (E1–E3). The amino-terminal sequence is acetylated at M^1 and contains oligosaccharide chains atasparagine N^2 and N^{15}. Cysteines C^{110} and C^{187} are linked by a disulfide bridge, and C^{322} and C^{323} are palmitoylated. Retinal is linked via a Schiff base to K^{296} in helix HVII. Source: *Adapted from P.A. Hargrave, J.H. Dowell, Rhodopsin and phototransduction, Int. Rev. Cytol. 37B (1992) 9–97.*

These side chains are located close together in a region between the S4 and S5 strands and are also close to the side chain of Y^{102} from the E-I loop. Thus these residues may maintain the proper orientation between E-I and the NH_2-terminal domain. The NH_2-terminal domain may also contact the E-III loop in the region of P^{12}.

Although both the E-I and E-III loops run along the periphery of the molecule, a part of E-II folds deeply into the center of rhodopsin. From the extracellular end of H-IV, a long strand from G^{174} to M^{183} crosses the molecule along the membrane surface. The terminal two residues, M^{183} and glutamine, Q^{184}, have two extended side chains. The former points to a hydrophobic pocket around H-I while the latter is surrounded by hydrophilic groups, including a water molecule located close to peptide carbonyl of arginine R^{177}, proline P^{180} and the OH group of Y^{192}. Residues in the middle of this strand, R^{177} to E^{181} (β3), form an antiparallel β-sheet with residues, S^{186} to D^{190} (β4), which is deeper inside the molecule than molecule than β3. The strand β4 is just below the 11-*cis*-retinal and is apart of the chromophore-binding pocket C^{187} forms a disulfide bond with C^{110} at the extracellular end of H-III. This disulfide is conserved in most GPCRs. Residues Y^{191} to N^{200} from E-II form a loop region at the periphery of the molecule, like E-I and E-III. The peptide carbonyl of Y^{191} in E-II and the side-chain amide of Q^{279}, which is at the beginning of E-III, are close to each other, while N^{199} is near to W^{175}, which is one of the initial residues of E-II, thus in proximity to the extracellular end of H-IV. This arrangement places E-II in extensive contact with the extracellular regions and also with retinal.

Transmembrane helices: As expected, helices H-I, H-IV, H-VI, and H-VII are bent at Pro residues, although it is not significant in H-I, while in H-IV, it causes distortion only around the extracellular end. H-V, with Pro^{215} in the middle, is almost straight. There is a significant bend at P^{267} in H-VI. H-VII exhibits irregular helicity, mainly around K^{296} to which retinal is covalently attached. H-II is also kinked around G^{89} and G^{90}, so that in this region it is closer to H-III than to H-I, placing G^{90} close to the residue that interacts with the Schiff base, E^{113}.

The cytoplasmic ends of H-II and H-IV are near each other, but they diverge in the region of W^{161}, one of the residues that are highly conserved among GPCRs. This residue is near the point where H-III penetrates toward H-V between H-II and H-IV. G^{120} and G^{121} do not distort H-III, but the region, E^{134}-R^{135}-Y^{136}, does exhibit a slight deviation from regular helical structure. This cytoplasmic terminal region is surrounded mostly by hydrophobic residues from H-II (P^{71}, L^{72}), C-II (F^{148}), H-V (L^{226}, V^{230}), and H-VI (V^{250}, M^{253}), forming the binding site for a G protein. H-IV and H-V exhibit irregular helicity in the cytoplasmic region and at H^{211}, respectively. The phenolic ring of Y^{223}, which is also highly conserved among GPCRs, partially covers the interhelical region between H-V and H-VI near the lipid interface. The cytoplasmic end of H-VI extends past the putative membrane surface to T^{243}. Three basic residues, K^{245}, K^{248}, and R^{252}, located near the cytoplasmic end of H-VI, extend from the helical bundle, making this region of C-III highly basic. In H-VII, two phenyl rings of F^{293} and F^{294} interact with L^{40} of H-I and C^{264} of H-VI, respectively. This interaction with H-VI is likely to be particularly important because it is facilitated by distortion of H-VI in the region of I^{263}. H-VII is considerably elongated in the region from A^{295} to Y^{301}. This region includes A^{299}, whose peptide carbonyl can hydrogen bond with the side chains of N^{55} in H-I and D^{83} in H-II. A highly conserved NPXXY motif in GPCRs follows this region in a regular helical structure.

From the experimental electron density, the conformation of the retinal chromophore in the Schiff base linkage with K^{296} is 6s-*cis*, 11-*cis*, 12s-*trans*, *anti* C5N. The density for the β-ionone ring exhibits a larger bulge indicating the positions of the two methyl groups connected to C1 and a smaller bulge for the single methyl at C5. Two small bulges along the polyene chain indicate the positions of the C9- and C13-methyl groups. The refined structure of the retinylidene group is consistent with resonance Raman spectroscopy, nuclear magnetic resonance, and chemical analysis. The density of the polyene chain merges with that of the side chain of K^{296}, indicating the presence of a protonated Schiff base bond Schiff base linkage. The retinylidene group is located closer to the extracellular side in the putative lipid bilayer, as suggested previously. The position of the β-ionone ring is mainly covered from the cytoplasmic side by the residues in H-III and H-VI, E^{122}, F^{261}, and W^{265}. The indole ring of W^{265} points down to the retinylidene group near the β-ionone ring, and also comes close to its C13-methyl group with a distance of 3.8 Å. Because deletion of this methyl group is known to cause partial constitutive activity of rhodopsin in the dark, loss of its interaction with W^{265} may be a possible mechanism of this activity. From the β-ionone ring to C11, the retinylidene group runs almost parallel to H-III, which provides many of the side chains for the binding pocket, E^{113}, G^{114}, A^{117}, T^{118}, G^{120}, and G^{121}, mainly around the polyene chain. The side chain of T^{118}, in addition to Y^{268} and I^{189} from the extracellular side, appears to determine the position of the C9-methyl of the retinylidene group. Side chains mostly from H-V and H-VI, M^{207}, H^{211}, F^{212}, Y^{268}, and A^{269} also surround the β-ionone ring. The proximity of F^{261} and A^{269} to the retinylidene group is consistent with information showing that these are responsible for the absorption difference between red and green pigments in humans. Arrangement of the four residues from H-VI appears to be determined by a significant bend around P^{267}. From H-IV, only C^{167} participates covering a part of this pocket. Residues from H-I, H-II, and H-VII and Y^{43}, M^{44}, L^{47}, T^{94}, and F^{293} are part of the region surrounding the Schiff base. Finally, the extracellular side of the polyene chain is covered by a part of the E-II loop, β-sheet β4 from S^{186} to I^{189}. The side chain of E^{181}in β3 of the E-II loop points toward the retinylidene group, supporting the previous results demonstrating that the corresponding amino acid in red/green pigments may be the binding site for chloride ion, which is responsible for the red shift in their absorption compared with rhodopsin. Another amino acid from the E-II loop participating in the retinylidene group binding site is Y^{191}, OH group is also close to that of Y^{268} in H-VI. Since mutation of this residue does not affect the absorption but reduces the ability to activate transducin, it may participate in the transition to the active form of rhodopsin through interaction with Y^{268}. The arrangement around the Schiff base is of particular interest in terms of understanding the mechanism of the primary process in photoactivation of rhodopsin. The direction of the side chain of K^{296}, almost along the long axis of rhodopsin, is supported by two hydrophobic side chains in H-I, M^{44} and L^{47}, and by a nearby peptide bond between F^{293} and F^{294}. This region is stabilized through the two phenyl rings interacting with other helices. Since it is difficult to determine exactly from the current structure how the protonated Schiff base linkage is stabilized in the protein environment, our model cannot discriminate whether any water molecules participate in making a complex counter ion or not. The distances between the carboxylate oxygen atoms of E^{113} and the Schiff base nitrogen are 3.3 and 3.5 Å. In addition, the OH group of T^{94} comes close to one of the oxygen atoms of E^{113} with a distance of 3.4 Å. Any other residues, including the nearby T^{92} and T^{93}, are too far from the Schiff base region to contribute to stabilization of its protonated state. Further improvements in resolution will provide more detailed views of this region.

The structure around the C-I loop exhibits a rigid organization. Of the three basic side chains in this region, K^{67} projects toward the solvent, whereas K^{66} and R^{69} point toward lipid-facing region. Another basic side chain of H^{65} sits closely to C-IV loop (H-VIII). The side chain of K^{67} appears to interact with a part of COOH-terminal tail region, which runs nearly parallel to C-I. The extreme COOH-terminal residues are the most exposed part of rhodopsin molecule and could be involved in vectorial transport of *rhodopsin* to rod outer segment. The region from C^{140} to E^{150} pertains the C-II loop. This loop exhibits an L-shaped structure, when viewed parallel to membrane plane, with a barrel (M^{143} to F^{146}) almost along the main axis of rhodopsin. Four polar side chains in this loop (K^{141}, S^{144}, N^{145}, and R^{147}) form a distinct cytoplasmic border from the transmembrane region. The height of these side chains is roughly comparable to that of the cytoplasmic border of C-III loop. Thus the current model can assign a border corresponding to the major cytoplasmic part of rhodopsin. The extra membranous extension from H-VI, tentatively assigned from $T^{243}$to A^{246}, still exhibits helical structure with no obvious break. In contrast, the cytoplasmic extension of H-V breaks around L^{226}, followed by an S-shaped flat loop structure almost along the surface of membrane. This connection from H-V to H-VI, the C-III loop, reaches close to the lipid-facing side of H-VI at A^{235}, without covering the cytoplasmic surface of the helical bundle of rhodopsin. Thus although the model demonstrates a highly flexible nature of this region and still lacks the tetrapeptide from Q^{236} to E^{239}, it is obvious that C-III does not fold over the helical region at all. On the other hand, two polar side chains of S^{240} and T^{242} in C-III comes close to a part of COOH-terminal tail around S^{334}, making a cluster of OH groups in this region. It should be also noted that the C-III loop is known to vary considerably among related GPCRs, so the flexibility and variability of this region may be critical for functionality and specificity in G-protein activation. The helical structure of the C-IV loop is of particular interest in the cytoplasmic region, considering previous studies of a variety of synthetic peptides and their effects on the activation of G proteins. Direct evidence for interaction of this region with the G-protein transducin has been provided using a synthetic peptide from N^{310} to L^{321} of bovine rhodopsin. The short helix is clearly distinct from H-VII and, via M^{309} to K^{311} linker, lies nearly perpendicular to H-VII. It is also the region that follows the NPXXY motif as a part of a conserved block of residues up to C^{322}. It has also been supposed that a group of peptides called mastoparans, which assume an amphiphilic helical structure and have activity on G proteins, mimic the structure of receptors in this region. From the rhodopsin structure it appears that this short stretch of amino acids is located in a hydrophobic environment, which distribution of side chains along this helix also exhibits an amphiphilic pattern; the charged/polar groups cluster on one side while hydrophobic ones are on the other, suggesting that the latter, F^{313}, M^{317}, and L^{321}, are buried in hydrophobic core of the receptor. F^{313} and R^{314} are the most conserved residues in this region, suggesting that the arrangement of this short helix in rhodopsin may be functionally important. Although we do not include any lipid-like structure in the current model, the side chains of C^{322} and C^{323} project to the outside of rhodopsin, consistent with the probable attachment of palmitic acid residues. The helical structure appears to be terminated by G^{324} and the following COOH-terminal tail changes the direction. Although current model lacks residues from 328 to 333, the positions of 327 and 334 suggest that this missing part runs covering the short H-VIII helix from the solvent region. As a whole, COOH-terminal tail of rhodopsin occupies the space over only a part of the helical bundle, H-I and H-VII. Surface potential of cytoplasmic and extracellular surfaces are shown, in detail, in the Web (Fig. 3.2).

The transmembrane region of rhodopsin is stabilized by a number of interhelical hydrogen bonds and hydrophobic interactions, and most of them are mediated by highly conserved residues in GPCRs. One of the residues that exhibit the highest conservation is N^{55} in H-I. Its side chain is responsible for two interhelical hydrogen bonds to D^{83} in H-II and to the peptide carbonyl of $A^{299}D^{83}$ is in turn connected via a water molecule to the peptide carbonyl of G^{120} in H-III. Another region that mediates constraints for three helices includes N^{78} of H-II, which is hydrogen-bonded to OH groups of S^{127} of H-III and T^{160}, W^{161} of H-IV. Helices H-III, H-IV, and H-V can be also linked through interaction among E^{122}, M^{163}, and H^{211}. The tripeptide E^{134}-R^{135}-Y^{136} is part of a highly conserved (D/E)R(Y/W) motif found in GPCRs. These residues participate in several hydrogen bonds with surrounding residues. The carboxylete group of E^{134} forms salt-bridge with guanidium of the next R^{135}. R^{135} is also connected to E^{247} and T^{251} in H-VI. V^{137}, V^{138}, and V^{139} are also closely located to partly cover the cytoplasmic side of E^{134} and R^{135}. These could be one of the critical constraints keeping rhodopsin in the inactive conformation. This region has high β-values; however, the side chains may assume different orientations. H-VII of most of the GPCRs in the rhodopsin family contains an NPXXY sequence near the cytoplasmic end, but the functional importance of this motif remains unclear. The side chains of the two polar residues in this region, N^{302} and Y^{306} in bovine rhodopsin, project inside the molecule. The OH group of

Y^{306} is close to N^{73}, which is also highly conserved among GPCRs, suggesting the presence of additional interhelical hydrogen-bonding constraints between H-VII and H-II. Although the distance between N^{302} and D^{83} is too long to make a hydrogen bond, it appears possible that the water near D^{83}interacts with the side chain of N^{302}. In this case, this water mediates a contact among H-II, H-III, and H-VII. The energy of light is utilized for photoisomerization of the 11-*cis*-retinal chromophore to an all-*trans*-configuration. This change in conformation would cause multiple effects, including movement of β-ionone toward H-III and/or displacement of Schiff base/C9/C13-methyl regions, ultimately switching the receptor to active conformation, metarhodopsin II. The model of bovine rhodopsin confirms that these effects can change the environment of the salt-bridge between the Schiff base and E^{113}, resulting in its neutralization. Displacement of H-III will result in changing the environment of the ERY motif and its reorientation. Our rhodopsin model also suggests that interaction between β-ionone ring and H-III occurs at E^{122}, which is one of the residues that determine the rate of metarhodopsin II decay. Because E^{122} interacts with H^{211} in rhodopsin, the proposed movement of H-III caused by the b-ionone ring can affect the interaction between these residues in the transition to metarhodopsin II. In addition, the change around the Schiff base region can affect the interaction between the C13-methyl group of retinal and W^{265}. The photoactivation may also cause breakage of some of the three interhelical constraints mediated by A^{299}, N^{302}, and Y^{306}, and hydrophobic constraints via F^{294} to the highly kinked region in H-VI. As a result, rearrangement of the helical bundle may be triggered, and finally lead to the movements of H-III and/or H-VI. The proposed mechanism stresses importance of the chromophore in the activation process, in agreement with the physiology of photoreceptor cells. Ideally, for a complete picture of the activation process, high-resolution structures of intermediates of photolyzed rhodopsin will be necessary. The GPCR family is one of the largest and most diverse groups of proteins encoded by 1%–3% of the genes present in our genome. They are involved in many physiological processes and are attractive targets for pharmacological intervention to modify these processes in normal and pathological states. The crystal structure of rhodopsin reveals a highly organized heptahelical transmembrane bundle with 11-*cis*-retinal as a key cofactor involved in maintaining rhodopsin in the ground state. A set of residues that interacts with the 11-*cis*-retinal chromophore t results in an absorption shift of the chromophore to a longer wavelength. The structure provides insight into the spectral tuning of related receptors, cone pigments. The structure also gives information on the molecular mechanism of GPCR activation. A conserved set of residues on the cytoplasmic surface, where G-protein activation occurs, likely undergo a conformational change upon photoactivation of the chromophore that leads to rhodopsin activation and signal transduction.

Rhodopsin is synthesized in the endoplasmic reticulum and passes to the Golgi membranes where it becomes glycosylated. Rhodopsin-containing vesicles move from the inner to the outer segments, where they fuse with the outer segment plasma membrane. After that morphologically separate disc membranes are formed, the regions of rhodopsin that were facing the outside of the cell become facing the inside surface of the disc membrane. The regions of rhodopsin that were facing the outer segment cytoplasm remain facing the cytoplasm when the plasma membrane forms disc membranes. The protein consists in a bundle of transmembrane helices that encompass a binding pocket for the light-sensitive 11-*cis*-retinal. Progressing toward the inner part of the retina the OPL is found, where connections between rods and cones and vertically running bipolar cells and horizontally oriented horizontal cells occur, further layers are found. The first membrane is a layer composed of the cell bodies of the rods and cones, which is about the same thickness in central and peripheral retina. Further membranes are then found, viz the horizontal, bipolar, and amacrine (=having no axon) cells. These are connected on the outer part with the rods and cones, and to the inner part to the optic nerves. This membrane is again formed by several layers. The synapses by which horizontal cells provide their feedback signals appear to use both conventional and unconventional mechanisms. The horizontal cells have relatively simple structure. Retinal neurons are minuscule cells, but all of the expected 12 types are known, one type of rod bipolar cells and 11 types of cones bipolar cells, "blue cone bipolar," which selectively contacts the short wavelength sensitive cones, as is necessary if the chromatic information is not to be degraded. Symmetrically, some bipolar cells avoid the terminals—they are numerically infrequent—of blue cones. And there is some crosstalk with the rods. β4 is just below the 11-*cis*-retinal and is apart of the chromophore-binding pocket. C^{187} forms a disulfide bond with C^{110} at the extracellular end of H-III. This disulfide is conserved in most GPCRs. Residues of tyrosine Y^{191} to asparagine N^{200} and from glutamic acid E-II form a loop region at the periphery of the molecule, like E-I and E-III. The peptide carbonyl of Y^{191} in E-II and the side-chain amide of glutamine^{279}, which is at the beginning of E-III, are close to each other, while asparagine N^{199} is close to W^{175}, which is one of the initial residues of E-II, thus in proximity to the extracellular end of H-IV. This arrangement places E-II in extensive contact with the extracellular regions and also with retinal. Transmembrane helices H-I, H-IV, H-VI, and H-VII are bent, as one may expect, at proline residues, although it is not significant in H-I, while in H-IV, it causes distortion only around the extracellular end. H-V, where P^{215} in the middle, is almost straight. There is a significant bend at P^{267} in H-VI. H-VII exhibits irregular helicity, mainly around lysine K^{296} to which retinal is covalently attached. H-II is also kinked around G^{89} and G^{90}, so that in this region it is closer to H-III than to H-I, placing close to the residue that interacts with the Schiff base, E^{113}. The of H-II and H-IV are near each other, but they diverge in the region of W^{161}, one of the residues that are highly conserved among GPCRs. This residue is near the point where H-III penetrates toward H-V between H-II and H-IV. G^{120} and G^{121} do not distort H-III, but the region, E^{113}-R^{135}-Y^{136}, does exhibit a slight deviation from regular helical structure. This cytoplasmic terminal region is surrounded mostly by hydrophobic residues from H-II (P^{71}, L^{72}), C-II (F^{148}), H-V (L^{226}, V^{230}), and H-VI (V^{250}, M^{253}), forming the binding site for a G protein. H-IV and H-V exhibit irregular helicity in the cytoplasmic region and at H^{211}, respectively. The phenolic ring of Y^{223}, which is also highly conserved among GPCRs, partially covers the interhelical region between H-V and H-VI near the lipid interface. The cytoplasmic end of H-VI extends past the putative membrane surface to T^{243}. Three basic residues, K^{245}, K^{248}, and R^{252}, located near the cytoplasmic end of H-VI, extend from the helical bundle, making this region of H-III highly basic. In H-VII, two phenyl rings of F^{293} and F^{294} interact with L^{40} of H-I and C^{264} of H-VI, respectively. This interaction with H-VI is likely to be particularly important because it is facilitated by distortion of H-VI in the region of I^{263}. H-VII is considerably elongated in the region from A^{295} to Y^{301}. This region includes A^{299}, whose peptide carbonyl can hydrogen bond with the side chains of N^{55} in H-I and D^{83} in H-II. A highly conserved NPXXY motif in GPCRs follows this region in a regular helical structure. The 11-*cis*-retinal chromophore. From the experimental electron density, the conformation of the retinal chromophore in the Schiff base linkage with K^{296} is 6s-*cis*, 11-*cis*, 12s-*trans*, and *anti* C_5N. The density for the β-ionone ring exhibits a larger bulge indicating the positions of the two methyl groups connected to C_1 and a smaller bulge for the single methyl at C_5. Two bulges along the polyene chain indicate the positions of the C_9- and C_{13}-methyl groups. The retinylidene group is located closer to the extracellular side in the putative lipid bilayer, as suggested previously. The position of the β-ionone ring is mainly covered from the cytoplasmic side by the residues in H-III and H-VI, E^{122}, F^{261}, and W^{265}. Because deletion of this methyl group is known to cause partial constitutive activity of rhodopsin in the dark, loss of its interaction with W^{265} may be a possible mechanism of this activity. From the β-ionone ring to C_{11}, the retinylidene group runs almost parallel to H-III, which provides many of the side chains for the binding pocket, E^{113}, G^{114}, A^{117}, T^{118}, G^{120}, and G^{121}, mainly around the polyene chain. The side chain of T^{118}, in addition to Y^{268} and I^{189} from the extracellular side, appears to determine the position of C_9-methyl of the retinylidene group. Side chains mostly from H-V and H-VI, M^{207}, H^{211}, F^{212}, Y^{268}, and A^{269}, also surround the β-ionone ring. As indicated above, rhodopsin's polypeptide chain traverses the lipid bilayer seven times as the helix segments HI through HVII. Hydrophilic loop sequences are exposed to the cytoplasmic surface (C1–C4) and to the intradiscal or extracellular surface (E1–E3). The amino-terminal sequence is acetylated at M^1 and contains oligosaccharide chains at the Asn^2 and N^{15}. C^{110} and C^{187} are linked by a disulfide bridge, and C^{322} and C^{323} are palmitoylated. Retinal is linked via a Schiff base to K^{296} in helix VII. In vertebrate rod cells, the outer rod segment contains a stack of rhodopsin-containing disc in membranes connected to an inner segment via a ciliary process. Rhodopsin is synthesized in the endoplasmic reticulum and passes to the Golgi membranes where it becomes glycosylated. Rhodopsin-containing vesicles move from the inner to the outer segments, where they fuse with the outer segment plasma membrane. After that morphologically separate disc membranes are formed, the regions of rhodopsin that were facing the outside of the cell become facing the inside surface of the disc membrane. The regions of rhodopsin that were facing the outer segment cytoplasm remain facing the cytoplasm when the plasma membrane forms disc membranes. The protein consists in a bundle of transmembrane helices that encompass a binding pocket for the light-sensitive 11-*cis*-retinal.

Rhodopsin's polypeptide chain traverses the lipid bilayer seven times as the helix segments I through VII. Hydrophilic loop sequences are exposed to the cytoplasmic surface (I–IV) and to the intradiscal or extracellular surface (EI–EIII). The amino-terminal sequence is acetylated at M^1 and contains oligosaccharide chains at N^2 and N^{15}, while C^{110} and C^{187} are linked by a disulfide bridge, and C^{322} and C^{323} are palmitoylated. Retinal is linked via a Schiff base to Lysine K^{296} in helix VII. Activation of rhodopsin by light absorption causes hydrolysis of the prostetic group and is under control by an enzyme. While both the EI and EIII loops run along the periphery of the molecule, a part of EII folds deeply into the center of rhodopsin. From the extracellular end of HIV, a long strand from G^{174} to M^{183} crosses the molecule along the membrane surface. The terminal two residues, M^{183} and Q^{184}, have two extended side chains. The former points toward a hydrophobic pocket around HI, while the latter is surrounded by hydrophilic groups, including a water molecule located close to peptide carbonyl of P^{180} and OH group of Y^{192}. Residues in the middle of this strand, R^{177} to E^{181} (β3), form an antiparallel β-sheet with residues, S^{186} to E^{190} (β4), which is deeper inside the molecule than β3, β4 is just below the 11-*cis*-retinal and is a part of the chromophore-binding pocket. C^{187} forms a disulfide bond with C^{110} at the extracellular end of HIII. This disulfide is conserved in most GPCRs. Residues Y^{91} to N^{200} from EII form a loop region at the periphery of the molecule, like EI and EIII. The peptide carbonyl of Y^{191} in EII and the side-chain amide of Q^{279}, which is at the beginning of EIII, are close to each other, while N^{199} is close to W^{175}, which is one of the initial residues of EII, thus in proximity to the extracellular end of HIV. This arrangement places EII in extensive contact with the extracellular regions and also with retinal. As one may expect, helices HI, HIV, HVI, and HVII are bent at P residues, although this is not significant in HI, while in HIV, it causes distortion only around the extracellular end and HV, where P^{215} is in the middle, is almost straight. There is a significant bend at P^{267} in HVI and HVII exhibits irregular helicity, mainly around K^{296} to which retinal is covalently attached. HII is also kinked around G^{89} and G^{90}, so that in this region it is closer to HIII than to HI, placing G^{90} close to the residue that interacts with the Schiff base, E^{113}. The cytoplasmic ends of HII and HIV are near each other, but they diverge in the region of T^{161}, one of the residues that are highly conserved among GPCRs. This residue is near the point where HIII penetrates toward HV between HII and HIV. $G^{120}G^{121}$do not distort HIII, but the regionE^{13}–R^{135}–Y^{136} does exhibit a slight deviation from regular helical structure. This cytoplasmic terminal region is surrounded mostly by hydrophobic residues from HII (P^{71}, L^{72}), CII (F^{148}), HV (L^{226}, V^{230}), and HVI (V^{250}, M^{253}), forming the binding site for a G protein. HIV and HV exhibit irregular helicity in the cytoplasmic region and at H^{211}, respectively. The phenolic ring of Y^{223}, which is also highly conserved among GPCRs, partially covers the interhelical region between HV and HVI near the lipid interface. The cytoplasmic end of HVI extends past the putative membrane surface to T^{243}. Three basic residues, K^{245}, K^{248}, and R^{252}, located near the cytoplasmic end of HVI, extend from the helical bundle, making this region of HIII highly basic. In HVII, two phenyl rings of F^{293} and F^{294} interact with L^{40} of HI and C^{264} of HVI, respectively. This interaction with HVI is likely to be particularly important because it is facilitated by distortion of HVI in the region of I^{263}. HVII is considerably elongated in the region from A^{295} to Y^{301}. This region includes A^{299}, whose peptide carbonyl can hydrogen bond with the side chains of N^{55} in HI and N^{83} in HII. A highly conserved NPXXY motif in GPCRs follows this region in a regular helical structure. From the experimental electron density, the conformation of the retinal chromophore in the Schiff base linkage with K^{296} is 6s-*cis*, 11-*cis*, 12s-*trans*, and *anti* C_5N. The density for the β-ionone ring exhibits a larger bulge indicating the positions of the two methyl groups connected to C_1 and a smaller bulge for the single methyl at C_5. Two bulges along the polyene chain indicate the positions of the C_9- and C_{13}-methyl groups. The retinylidene group is located closer to the extracellular side in the putative lipid bilayer, as suggested previously. The position of the β-ionone ring is mainly covered from the cytoplasmic side by the residues in HIII and HVI, E^{122}, F^{261}, and W^{265}. The indole ring of W^{265} points down to the retinylidene group near the β-ionone ring, and also comes close to its C_{13}-methyl group with a distance of 3.8 Å. The loss of the interaction of this group with W^{265} is a possible mechanism of this activity. From the β-ionone ring to C_{11}, the retinylidene group runs almost parallel to HIII, which provides many of the side chains for the binding pocket, E^{113}, G^{114}, A^{117}, T^{118}, G^{120}, and G^{121}, mainly around the polyene chain. The side chain of T^{118}, in addition to Y^{268} and I^{189} from the extracellular side, appears to determine the position of the C_9-methyl of the retinylidene group. Side chains mostly from HV and HVI, M^{207}, His^{211}, F^{212}, Y^{268}, and A^{269}, also surround the β-ionone ring [53].

Regions in the extracellular domain of rhodopsin (NH_2-terminal and interhelical loops EI, EII, and EIII) associate to form a compact structure. The NH_2-terminal tail of rhodopsin contains five distorted strands. The NH_2-terminus is located just below loop EIII, with the side chain of D^{282} close to that of N^2. The first two antiparallel strands, G^3 and P^{12}, form a typical β-sheet fold (β1 and β2) running almost parallel to the expected plane of the membrane. Strands three to five (S3–S5) form a right triangle from F^{13} to P^{34}, with the third strand running just below EIII, almost parallel to the long axis of the molecule. S4 connects S^{14}–N^{15} in the NH_2-terminal region of the molecule with P^{23}, located close to EI. S5 from P^{27} to P^{34} runs along the surface of the membrane covering the extracellular (intradiscal) space between HI and HII. Oligosaccharides at N^2 and N^{15} extend from the domain and are not included in any interactions. These side chains are located close together in a region between the S4 and S5 strands and are also close to the side chain of Y^{102} from the EI loop. Thus these residues may maintain the proper orientation between EI and the NH_2-terminal domain. The NH_2-terminal domain may also contact the EIII loop in the region of P^{12} [57].

The complexity of the retina structure is exploited in the complex functions that it has to operate. In depth studies demonstrated that the signal from any individual cone is decomposed into 12 different components, each of which is separately transmitted to the inner retina by a structurally and molecularly distinct path. The thus activated bipolar cell channels are sampled by different sets of retinal ganglion cells; the partially selective responses mediated by bipolar cells are refined by amacrine cells—a few per ganglion cell type—to create arrays of precisely specific ganglion cell subtypes.

Light transduces the visual pigment via an enzyme cascade, namely photons, rhodopsin-activated rhodopsin (metarhodopsin II)—a GTP binding protein (transducin)—an enzyme hydrolyzing cGMP [cGMP-phosphodiesterase (PDE)]—closes a membrane bound cGMP-gated cation channel. The most important synaptic interactions that occur at the OPL are the splitting of the visual signal into two separate channels of information flow, one for detecting objects lighter than background and one for detecting objects darker than background and the instillation of pathways to create simultaneous contrast of visual objects.

In the first synaptic interactions, the channels of information flow are known as the basis of successive contrast, or ON and OFF pathways, respectively, whereas the second interaction puts light and dark boundaries in simultaneous contrast and forms a receptive field structure, with a center contrasted to an inhibitory surround. These two necessary visual information processes are created by synaptic interactions at the OPL.

In human retina, 11 different bipolar cell types have been detected. Ten types are for cones, and one type is for rods. The rod bipolar dendritic terminals end into a rod spherule as the central invaginating dendrite.

Among diffuse-cone bipolar cells, some have a wide dendritic spread, 70–100 μm, and are indicated as giant. These connect with as many as 15–20 cones, while the smaller diffuse bipolar cells collect information from 5 to 7 cones in the central retina and from 12 to 14 cones in the peripheral retina. The midget bipolar cells contact single cones, but two different varieties of them per cone are found, depending on their contact with the cone pedicle.

All mammalian retinas have two types of horizontal cells (HCs) as the laterally interconnecting neurons in the OPL. In primates it was thought to be only three types of HCs, HI, HII, and HIII. In peripheral retina, the HI cells have much bigger dendritic trees, and their radiating dendrites contact as many as 18 cones. The HI cell has a single, thick axon that passes laterally in the OPL and terminates more than

1 mm away in a thickened axon terminal stalk that bears many terminals arranged in the form of a fan. HI axon terminals end in rod spherules as lateral elements of the ribbon synapses. HIII cells are somewhat larger, but otherwise similar in appearance to HI cells, although everywhere in the retina, HIII cells are one-third bigger in dendritic tree size and typically, particularly in the peripheral retina, asymmetrical in shape (one or two dendrites are much longer than others; Fig. 3.18).

The clusters of terminals contact cones in the same manner as the HI cell terminals, and because of their bigger field size, they contact more cone pedicles (9–12 in the foveal retina, 20–25 in the peripheral retina). HII cells are spidery and intricate in dendritic field characteristics than either of the other types. Their terminals are not clearly seen as clusters approaching cone pedicles, but they are known to end in cone pedicles. HII cells also bear an axon, but this is quite different from that of the other two HC types. It is short (100–200 μm), curled instead of straight, and has contacts to cone pedicles by means of small, wispy terminals. Recent findings from electron microscopic studies of Golgi-stained HCs of the human retina show that there is some color-specific wiring going on for the three cell types (Fig. 3.19) [51].

Thus HI cells contact medium- and long-wavelength cones primarily, but with a small number of contacts to any short-wavelength

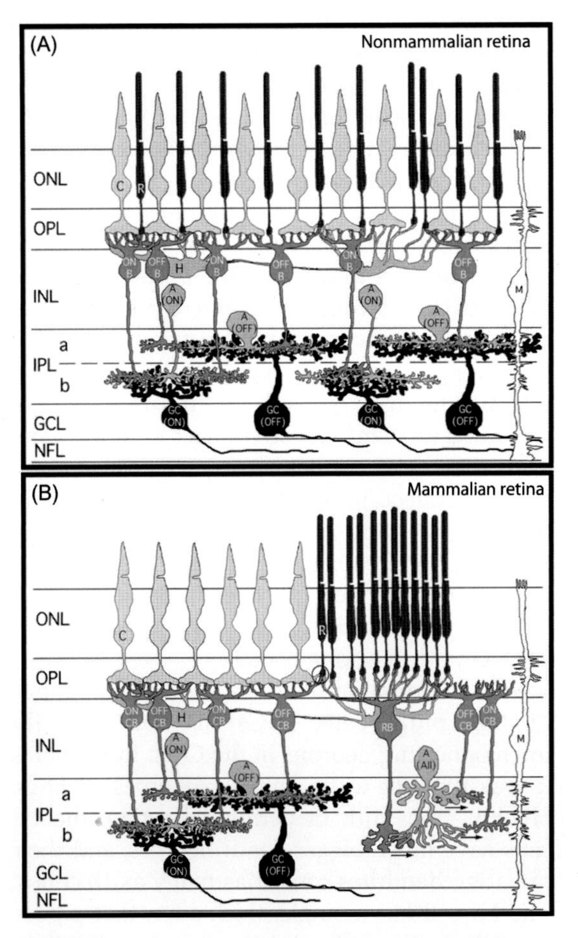

FIGURE 3.18 Rodes and cones organizations [58].

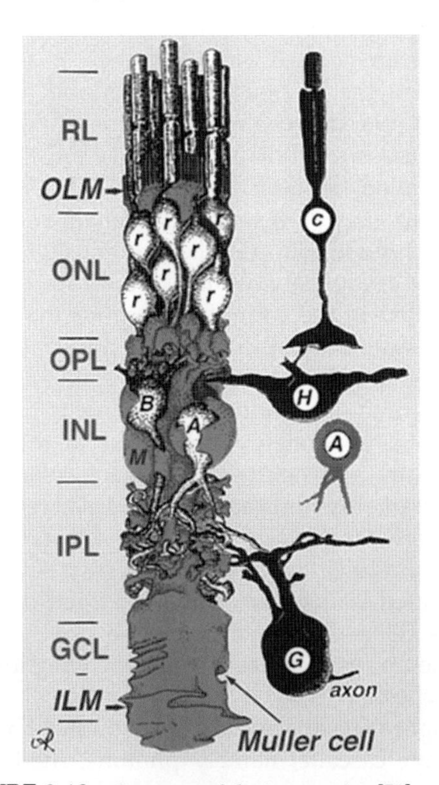

FIGURE 3.19 Cross cut of the inner retina [51].

cones in the dendritic field. HII cells contact short-wavelength cones, directing major dendrites to these cones in their dendritic fields where they occur and contacting with lesser numbers of terminals of other types of nonshort-wavelength cones. The HII cells axon contacts short-wavelength cones only. HIII cells have large dendritic terminals in medium- and long-wavelength cones, seemingly avoiding short-wavelength cones in their dendritic tree. Thus a wiring diagram can be made (Fig. 3.20) that summarizes the present understanding of the spectral connections of the three HC types of the primate retina [51].

Rhodopsin is synthesized in the endoplasmic reticulum and passes to the Golgi membranes where it becomes glycosylated. Rhodopsin-containing vesicles move from the inner to the outer segments, where they fuse with the outer segment plasma membrane. After that morphologically separate disc membranes are formed, the regions of rhodopsin that was facing the outside of the cell becomes facing the inside surface of the disc membrane. The region of rhodopsin that was facing the outer segment cytoplasm becomes facing the cytoplasm when the plasma membrane forms disc membranes.

Progressing toward the inner part of the retina, further layers are found that constitute the OPL, where connections between rods and

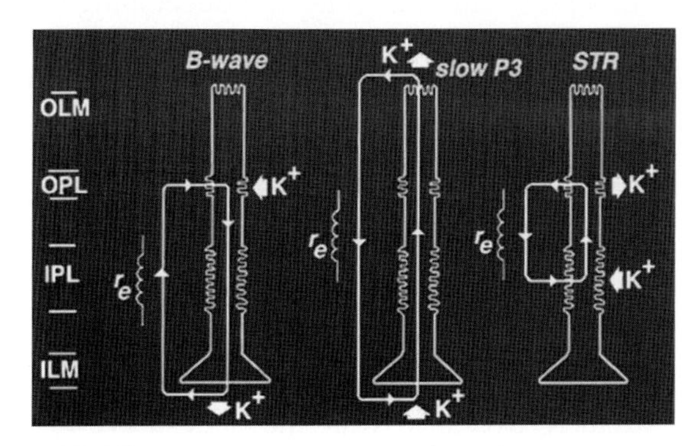

FIGURE 3.20 Regulation of K^+ by Muller cells in Muller cells [51].

cones and cells are able to convert the photochemical reaction into a range of different information. The first membrane is a layer composed of the cell bodies of the rods and cones, which is about the same thickness in central and peripheral retina. Further layers are then found, viz the horizontal, bipolar, and amacrine (=having no axon) cells (see Fig. 3.19).

These are connected on the outer part with the rods and cones, and to the inner part to the optic nerves. Several layers can then distinguished; the horizontal cells have relatively simple structure. The synapses by which horizontal cells provide their feedback signals appear to use both conventional and unconventional (pH sensitive) mechanisms [59]. Ganglion cells are the final output neurons of the vertebrate retina. The ganglion cells collect the electrical messages concerning the visual signal from the two layers of nerve cells preceding it in the retinal wiring scheme. A great deal of preprocessing has been accomplished by the neurons of the vertical pathways (photoreceptor to bipolar and to ganglion cell chain), and by the lateral pathways (photoreceptor to horizontal cell, to bipolar, to amacrine, and to ganglion cell chain), before presentation to the ganglion cell, and so it represents the ultimate signaler to the brain of retinal information. Ganglion cells are larger on average than most preceding retinal inter neurons and have large diameter axons capable of passing the electrical signal, in the form of transient spike trains, to the retinal recipient areas of the brain many millimeters or centimeters distant from the retina. The optic nerve collects all of the axons of the ganglion cells, and this bundle of more than a million fibers (in humans, at least) then passes information to the next relay station in the brain for sorting and integrating into additional information-processing channels. Important in this phase is the direction selective ganglion cells (DSGCs) that detect motion along the space [60]. DSGCs can be classified into separated types based on their functional properties. They either respond to both light onset and offset (ON–OFF), or just to the former (ON). Furthermore, they prefer different directions of motion. Distributions of directional preferences reveal four types of ON–OFF DSGCs and three types of ON DSGCs. The ON–OFF DSGCs (ooDSGCs) detect motion in one of four cardinal axes while the ON DSGCs detect movement in the dorsal, ventral, and nasal directions [61,62].

Müller cell bodies sit in the inner nuclear layer and project irregularly thick and thin processes in either direction to the outer limiting membrane and to the inner limiting membrane. Müller cell processes insinuate themselves between cell bodies of the neurons in the nuclear layers and envelope groups of neural processes in the plexiform layers. In fact, retinal neural processes have direct contact, without enveloping Müller cell processes, only at their synapses. A single progenitor cell gives rise to both Müller cells and retinal neurons, although apparently in two phases. The earliest phase neurons born at the apical margin of the neuroepithelium, adjacent to it and produces primary neurons consisting of cone cells, horizontal cells, and ganglion cells. The second phase of cells, also born at the apical margins, produces Müller cells and rod photoreceptors, bipolar cells and amacrine cells. All of the developing neurons and the Müller cells have to migrate inward to their final position, and it is thought that the Müller cell processes and trunks guide much of the neuron migrations and direct the neurite differentiations [63].

The junctions forming the outer limiting membrane are between Müller cells and other Müller cells with photoreceptor cells as sturdy desmosomes' coronial adherents. In some species, gap junctions (specialized membrane associations and channels that allow passage of small molecules and ions) or tight junctions are part of these Müller cell junctions but not so in mammalian species, where no dye coupling has ever been observed.

Müller cells have a range of functions, all of which are vital to the health of the retinal neurons. Müller cells function in a symbiotic relationship with the neurons [64]. Thus Müller cell functions include supplying end products of anaerobic metabolism (breakdown of glycogen) to fuel aerobic metabolism in the nerve cells, mops up neural waste products, such as carbon dioxide and ammonia, and recycle spent amino acid transmitters. They further protect neurons from exposure to excess neurotransmitters, such as glutamate, using well-developed uptake mechanisms to recycle this transmitter. They are particularly characterized by the presence of high concentrations of glutamine synthase and may be involved in both phagocytosis of neuronal debris and release of neuroactive substances such as γ-aminobutyric acid (GABA), taurine, and dopamine, while synthesizing retinoic acid from retinol.

These cells also control homeostasis and protect neurons from deleterious changes in their ionic environment by taking up extracellular K^+ and redistributing it and further contribute to the generation of the electroretinogram (ERG) β-wave, the slow P3 component of the ERG, and the scotopic threshold response, a result obtained from the regulation of K^+ distribution across the retinal vitreous border, across the whole retina, and locally in the IPL of the retina.

Astrocytes are not glial cells of the retinal neuroepithelium, but enter the developing retina from the brain along the developing optic nerve. They have a characteristic morphology of a flattened cell body and a fibrous series of radiating processes. Intermediate filaments fill their processes, and thus they stain with antibodies against glial fibrillary acidic protein GFAP. Astrocyte cell bodies and processes are almost entirely restricted to the nerve fiber layer of the retina. Their morphology changes from the optic nerve head to the periphery: from extremely elongated near the optic nerve to a symmetrical stellate form in the far peripheral retina [63].

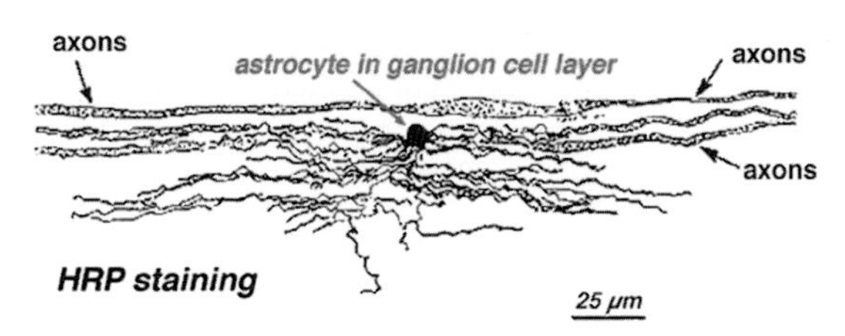

FIGURE 3.21 Horseradish peroxidase (HRP) staining of astrocytes in the ganglion cell layer [51,65].

In Golgi and immunocytochemical staining, they look like cell bodies with fibrous tangles aligned along the ganglion cell axons coursing through the nerve fiber layer. In distribution, astrocytes reach their peak on the optic nerve head and have a fairly uniform decline in density in radiating rings from the nerve head. They are not present in the avascular fovea or ora serrata. Horseradish peroxidase staining has been used to measure intracellular pH and passive electrical properties of cortical astrocytes (Fig. 3.21) [65].

Thick and thin astrocytes have been distinguished on the basis of staining with antibodies to GFAP. Thus astrocytes are arranged over the surface of the ganglion cell axon bundles as they course into the optic nerve head, forming a tube through which the axons run. Gap junctions and zonula adherent junctions have been described between astrocytic processes in cat retina.

Similar to Muller cells, astrocytes contain much glycogen and supply glucose to the cells. And further serve a role in ionic homeostasis in regulating extracellular potassium levels and metabolism of neurotransmitters like GABA (Fig. 3.22) [66].

Finally, the microglial cells, arise from the mesoderma, enter the retina coincident with the mesenchymal precursors of retinal blood vessels in development. Microglial cells are ubiquitous in the human retina, and have an important role in the case of a lesion to the retina, where they sweep away the debris.

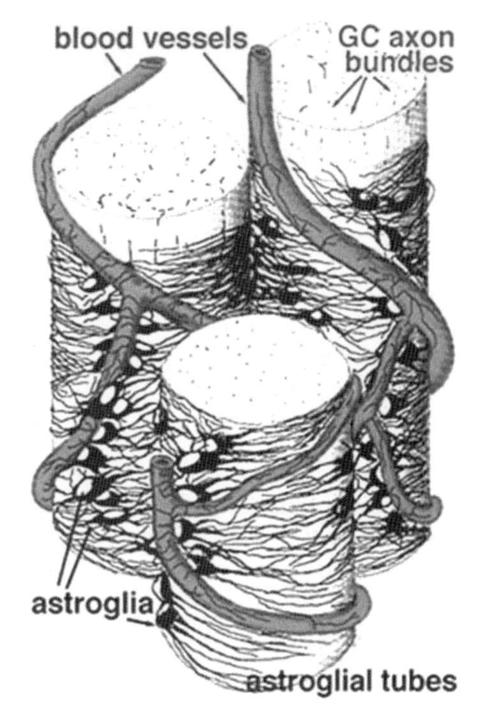

FIGURE 3.22 Three-dimensional block of astrocytes arranged over the surface of ganglion cell axon bundles. Source: *From H. Kolb, Simple anatomy of the retina, in: H. Kolb, E. Fernandez, R. Nelson (Eds.), Webvision (2005): The Organization of the Retina and Visual System [Internet], University of Utah Health Sciences Center, Salt Lake City (UT), 1995. https://www.ncbi.nlm.nih.gov/books/NBK11518/ and A. Trivino, J.M. Ramirez, J.J. Salazar, A.I. Ramirez, J. Garcia-Sanchez, Immunohistochemical study of human optic nerve head astroglia. Vis. Res. 36 (1996) 2015–2028, reproduced with permission.*

In the retina of mammals, a single type of bipolar cells collecting information from rods and 11 types of bipolar cells connected to cones are present. Although cells with mixed rod–cone connections have been identified, a rule is that rod bipolar cells do not contact cone photoreceptors, while bipolar cells exist that receive their input predominantly from cones [67,68] (Fig. 3.23).

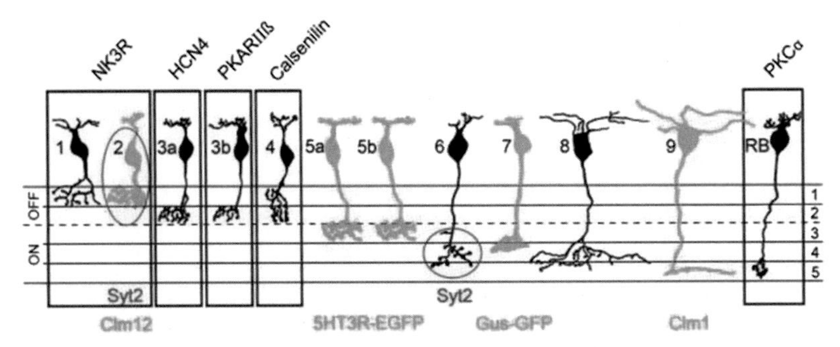

FIGURE 3.23 Summary diagram of the bipolar cell types of the mouse retina. A total of 11 cone bipolar cells and one rod bipolar cell types can be distinguished by immunocytochemical staining or by GFP/chamaleon expression in transgenic mouse lines [67,68].

3.4.2 Physiology of vision

The eye is the organ that works in cooperation with the brain to provide us with the three-dimensional sense of sight or vision. It works much like a modern camera. The main function of the eye is to collect light and turn it into electric signals, which are sent to the brain. The brain then turns those signals into a visual image or picture for us to see. We have two eyes, so two pictures are usually created, which guarantees three-dimensional sight. A light ray passes through the cornea and then through the pupil, where the iris adjusts the amount of light entering the eye. The light then passes through the lens of the eye and is focused onto the retina, where it is changed into a signal that is transmitted to the brain by the optic nerve. The signal is received and interpreted by the brain as a visual image. Blood supply is ensured by two systems, the first one is the hyaloid circulation, composed of, from posterior to anterior, the hyaloid artery, the vasa hyaloidea propria, the tunica vasculosa lentis, and the papillary membrane that develops under the action of the vascular endothelium growth factor (VEGF). The hyaloid artery enters the optic cup from the optic stalk via the optic fissure and grows toward the lens expanding into the dense capillary system of the tunica vasculosa lentis [66] and the pupillary membrane. In addition, the tunica vasculosa lentis form radial anastomoses with the annular vessel, which are fed by the choroidal circulation of the optic cup. The second one is the retinal vascularization that commences from the optic disc in the fetus and proceeds centrifugally toward the retinal periphery, a process that is completed only during the last month of gestation. An important molecular mediator of this process is again VEGF, which acts as an endothelial mitogen and survival factor. The retina circulation is expressed by retinal astrocytes and Müller cells and is tightly controlled by calcium ion level [69]. The retinal ganglion axon fibers travel within the optical nerve fiber layer to converge on the optic disc while obeying strict retinotopic organization. Because all retinal ganglion cell axons and all retinal blood vessels have to pass through the optic disc, relatively small lesions at the optic disc and in the optic nerve can have devastating clinical effects. Pathology affecting these ganglionic axons along their anatomic course between the retina and the brain gives rise to characteristic visual field defects. Light irradiation in a high-oxygen environment results in the production of free radicals and damage to disc lipid membranes in the photoreceptor outer segments. Being nondividing permanent cells, the photoreceptors are particularly vulnerable to the accumulation of peroxidation damage, a problem which nature has solved by regularly shedding their outer segments [70]. Outer segments' shedding exhibits diurnal variation and is faster for rods than for cones. In the monkey, the time required to replace the entire rod outer segment is 9 days in the peripheral retina and 13 days in the parafoveal retina. Each RPE cell is in contact with an average of 45 photoreceptors in the human retina. RPE microvilli phagocytose the distal tips of the outer segments. Undigested material is egested and cleared from Bruch's membrane via the choriocapillaris. Large amounts of outer segment lipid membranes are thus disposed of by the RPE cell—up to 4000 discs per day. The transmembranous protein, opsin, anchors the photopigment into the outer segment membrane. In the visual process, a small G protein, rhodopsin, that has 11-*cis* retinaldehyde as the prostetic group and chromophore (bound as a Schiff base) is isomerized to the corresponding all-*trans* derivative. During phototransduction, 11-*cis*-retinaldehyde is converted to all-*trans*-retinol. All-*trans*-retinol is transported back to the RPE cell where it is isomerized to 11-*cis*-retinol and then reoxidized to 11-*cis*-retinaldehyde. The strongly electrophilic aldehyde group allows recombination with the opsin and is converted to the Shiff base, thus completing its regeneration in the photoreceptor cell. The retinol and retinal molecules are chaperoned by special binding proteins. In the subretinal space, interphotoreceptor retinoid-binding protein shuttles retinol from the photoreceptor to the RPE and retinal back to the photoreceptor, whilst cellular retinol-binding protein and cellular retinaldehyde-binding protein protect retinol and retinal inside the cell [71]. The transduction process, during which the formation of a photoproduct is transformed into an electric signal, may be sketched as indicated below. The perikarya of the light-sensitive cells are located in the inner nuclear layer with cell processes [72] that span the entire neuroretina. The proximal extensions of Müller cells originate the inner limiting membrane, while the outer limiting membrane results from a series of junctional complexes from the distal face. Müller cells play a crucial role in maintaining the local environment that allows the visual process to function optimally, but there are also astrocytes arising from stem cells in the optic nerve and microglia from the circulation [73]. These cells are phagocytic and part of the reticuloendothelial system. They are usually found in small numbers in the nerve fiber layer, but are mobile and can reach any part of the retina. The choriocapillaris is separated from the epithelium by the Bruch's membrane, an elastic membrane composed of five layers, the basement membrane of the choriocapillaris, an outer collagenous layer, a central elastic layer, an inner collagenous layer, and the basement membrane of the RPE. It stretches from the optic disc at the posterior pole to the ora serrata anteriorly and varies in thickness between 2 and 4 μm at the posterior pole and 1–2 μm at the ora serrata. With age, Bruch's membrane grows thicker and its ultrastructure becomes less distinct. The chemical composition of this membrane is complex and consists of elastin, different types of collagens, as well as several adhesive glycoproteins, including fibronectin and laminin, which help to anchor cells to Bruch's membrane. RPE cells are numerous, about 3.5 million in each eye. These are held together by junctional complexes and thus appear as a continuous epithelial monolayer. The tight complexes separate the choriocapillaris from the photoreceptors of the outer retina. In this way an outer blood–retina barrier is formed. Part of the neuroretina is, however, constituted by the thin layer of rods and cones that are tightly stacked together into a single layer. In such OPL, the photoreceptor cells of the outer nuclear layer form connections with the bipolar and horizontal cells of the inner nuclear layers [74]. The bipolar cells (c.390,000 cells/mm^2 [75]) operate transduction of the molecular signal into an electric signal. The rule followed would seem to be that every bipolar cell contacts all of the cone terminals within the spread of its dendritic arbor, although the functions are different. This is a geographically simple rule, but understanding their function is not as easy. Functionally, however, this arrangement allows something more sophisticated. The characteristics of the synapses cone-to-bipolar cells can in fact be tuned, and different analyses of the cone's output can be transmitted to the bipolar cells, where a characteristic set of receptors, ion channels, and intracellular signaling systems further elaborate the signal. The depolarizing response to light of both rod and ON cone bipolar cells is due to the interaction of photoreceptor-released glutamate with a particular type of metabotropic receptor, named mGluR6 that is present only in the retina. Retinal ON bipolar cells make up over 70% of all bipolar cells [76]. These comprise all rod bipolar as well as some cone bipolar cell. Glutamate, released from photoreceptors in the dark, binds to mGluR6 receptors localized to the dendritic tips of rod bipolars and ON cone bipolar cells and activates a specific splice variant of a G protein, named G0 alpha, which ultimately closes the nonspecific cationic channel. The postsynaptic channel that mediates synaptic transmission from photoreceptor to ON bipolar cells appear to be a member of the family of transient receptor potential channels and suitable proteins (GRS) direct the reaction course. The transduction channel is composed of transient receptor positive cation channel, subfamily M1 (TRPM1), either as a homomer or in associated with other TRP channels. Genetic disruption of either RGS7 or RGS11 causes delays in the β-wave of the gene ERG, but with different outcomes, since in the cone pathway, RGS11 plays an accelerating role in the deactivation of G0 alpha, which comes before activation of the depolarizing current in ON bipolar cells, thus acting as variable regulators of kinetics of ON bipolar cell responses [77].

As for the intensity, usually measured in photometric units that take into account both the intensity of the light ray and the eye sensitivity to that λ and are used when psychic quantities are concerned, it must be recalled that in a well-lit environment, the human eye is maximally sensitive to light of about 555 nm (yellow-green light) and relatively insensitive to far-red and far-blue light. The function describing the response to different wavelengths is known as the relative luminous efficiency function. The parameters required are the intensity of a ray (expressed in $cd.m^2$, or in lumen, see Table 3.3) hits an object and the ray emitted by the lamp, the intensity of the ray emitted and reflected in direction of the observer that receives the signal. The acuity of sight is measured under photopic conditions (rod dominated), not scotopic (cone dominated) conditions [78].

The actual ability of the eyes of each individual is determined through a range of psychophysical procedures that allow to determine thresholds, including the visual field analysis. For a perfect observer, a threshold is the point where the stimulus can just be detected or where it just cannot be detected. However, since humans are not perfect observers, often thresholds are defined in probabilistic terms: for example, half the points presented would be detected and half would not. Under certain psychophysical techniques, threshold is the point where 50% of the stimuli are detected [79].

Color and form characterize objects and our brain elaborates the signals into color perception. The progress of science is going to be able to allow to model the neural circuits that underlie the perception of color and form.

When we look at an object with two eyes, we perceive it as a singular object, like we do with other parts of the visual scene stimulating points on our retina that share a common visual direction. These points are termed "retinal corresponding points" and fall on an area called the "horopter." Points outside the horopter fall on slightly different retinal areas and so do not have the identical visual direction and lead to "retinal disparity," the basis of our depth discrimination. This retinal image disparity occurs due to the lateral displacement of the eyes. The region in visual space over which we perceive single vision is known as "Panum's fusional area," with objects in front and behind this area being in physiological diplopia (i.e., double vision) [80,81]. Our visual system suppresses this diplopia, and hence we do not perceive double vision under normal viewing conditions. To understand the discussion on the horopter and Panum's fusional space, the sense of direction will be introduced. Two terms describing direction sense are oculocentric and egocentric visual direction. Retinal image size allows us to judge distance based on our past and present experience and familiarity with similar objects. As the car drives away, the retinal image becomes smaller and smaller. We interpret this as the car getting further and further away. This is referred to as size constancy. A retinal image of a small car is also interpreted as a distant car and the receptor is thus sign inverting [81]. When light causes less glutamate to be received from the photoreceptor terminal, cation channels open and the cell depolarizes [82]. Similarly, the distinction between sustained and transient bipolar cells is caused by the expression of rapidly or slowly inactivating glutamate receptors [83]. This creates four classes of bipolar cells: ON-sustained, ON-transient, OFF-sustained, and OFF-transient. In detail, the different structural/molecular types of bipolar cells show a wide diversity of response waveforms in response to light; aside from the simple tonic versus phasic dimension, these responses display complex mixtures of the two [84]. The functional meanings of these are only beginning to be understood. A case in point is the expression of differing sets of regulation of G protein signaling (RGS) proteins, which control the kinetics of the response to synaptic input in ON bipolar cells [85]. Another is a type of bipolar cell that generates Na^+ action potential. Na^+ currents have been known to occur from studies of many retinas, but their functions are unclear [86].

The dendrites of the retinal ganglion cells intertwine with the inner nuclear layer and then through the ganglion cells arrive to the optical nerve and the vitreous cavity. They form a dense reticulum of fibrils, indicated as the IPL. When light strikes a molecule of rhodopsin in the rod cell, a photon is absorbed, causing isomerization of rhodopsin's 11-*cis*-retinal to the all-*trans* isomer. This causes a change in the conformation of rhodopsin. The photochemical reaction is an ultrafast isomerization (<200 ps) that leads through various equilibrations to metharodopsin (Meta II state), a blue shifted rhodopsin which binds and activates the G protein transducing, and involve proton transfer from the Schiff base function to its counter ion as well as a torsion outward of TM6 and a second proton transfer, this time from water to G^{134}, an intermolecular step that makes the protonation pH-dependent [87,88].

Hydrolysis of the Schiff's base linkage generates opsin and free all-*trans* retinal (Fig. 3.24) [88].

The structure of rhodopsin and of intermediates from its irradiation, bathorhodopsin, isorhodopsin, and lumirhodopsin has been determined by taking advantage of their different stability, as assessed by X-ray irradiation in frozen crystals (Fig. 3.25).

The photoisomerization gives a highly strained all-trans retinal chromophore. The time scale is too fast for a normal rearrangement and calorimetric studies show that bathorodopsin has stored c.35 kcal/mol of the absorbed light energy. The protein–retinal complex decades thermally through a series of distinct, spectrally defined intermediates. In contrast, a large structural transition takes place in the conversion of metarhodopsin I (Meta I) to metarhodopsin II (Meta II). Meta II corresponds to the active state of rhodopsin, formed through an increase in enthalpy in the Meta I to Meta II transition, that is, compensated by a large increase of entropy [89].

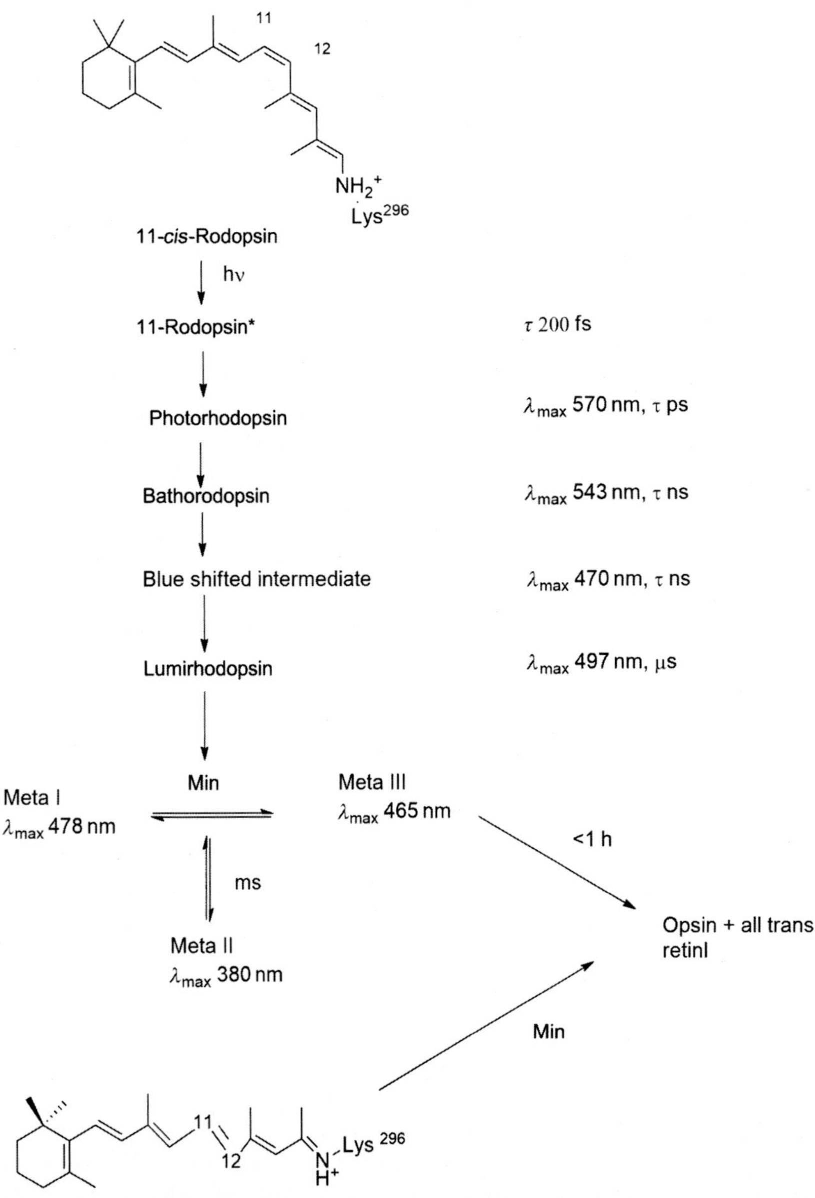

FIGURE 3.24 Rhodopsin photoreaction, Electronic excitation of 11-*cis*-retinal chromophore causes isomerization to the all-*trans* configuration within 200 fs and then relaxes thermally through a series of spectrally well-defined intermediates to the active Meta II state (in min). The last isomer binds and activates the protein transducin. Hydrolysis of the Schiff base linkage generates opsin while setting free all-trans retinal. Source: *Modified from S.O. Smith, Structure and activation of the visual pigment rhodopsin. Annu. Rev. Biophys. 39 (2010) 309–328.*

As visual cells consume a lot of energy for their function, the mitochondria, the producers of energy in the cell, have a quite important role in sight. Within the helix bundle, opsin shows only small changes relative to rhodopsin for TM1–TM4, reminiscent of the proposed stable TM1–TM4 core structure that does not change much on receptor activation. The most prominent change in this core is the formation of a short helical turn in loop CI of opsin. Larger changes are observed for TM5–TM7, especially dominant at the cytoplasmic ends of

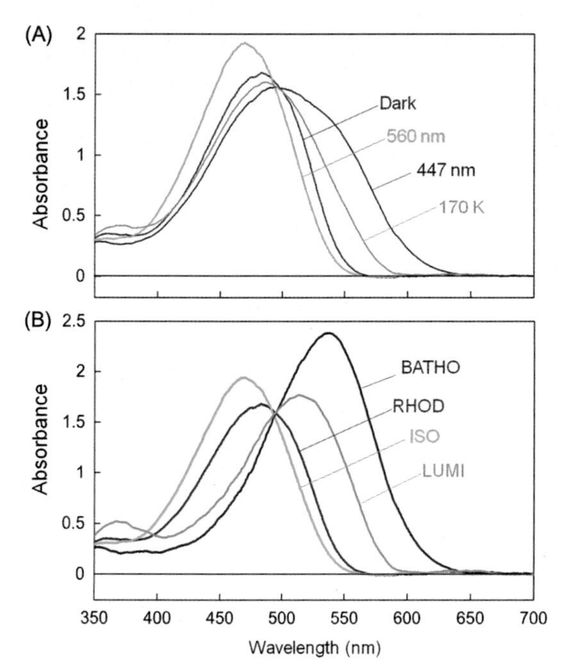

FIGURE 3.25 Absorption spectra of rhodopsin (RHOD), bathorhodopsin (BATHO), isorhodopsin (ISO), and lumirhodopsin (LUMI) in a frozen crystal at 100 K. (A) Red line: the absorption spectrum of a frozen crystal that was kept in the dark (Dark). Dark blue line: the spectrum recorded after one minute of illumination at 100 K with blue light at 447 nm. Yellowish-green line: the spectrum recorded after one minute of illumination at 100 K with yellow light at 560 nm. Cyan line: the spectrum recorded after the crystal was illuminated at 100 K with blue light, warmed in the dark to 170 K and recooled to 100 K. (B) Calculated absorption spectra of RHOD (red line), BATHO (blue line), ISO (green line), and LUMI (cyan line) [89].

these helices, causing rearrangement of loops CII and CIII and the kink region between TM7 and HVIII. A number of charges particularly involving TM3 and TM6 occur in the so-called ionic lock (Fig. 3.26) [90,91].

The retinal rod cell functions as a very sensitive single-photon detector that primarily functions in dim light (e.g., moonlight). However, rod cells must routinely survive light intensities more than a billion times greater (e.g., bright daylight). One serious challenge to rod cell survival in daylight is the massive amount of all-*trans*-retinal that is released by Meta II. All-*trans*-retinal is toxic, and its condensation products have been implicated in eye diseases. It further appears that rod arrestin (arrestin-1), which terminates Meta II signaling, has an additional role in protecting rod cells from the consequences of bright light by limiting free all-*trans*-retinal. Arrestin-1 serves as both a single-photon response quencher as well as an instrument of rod cell survival in bright light. Arrestin-1 is the second most abundant protein in rod photoreceptors and is nearly equimolar to rhodopsin. Its well-recognized role is to "arrest" signaling from light-activated, phosphorylated rhodopsin, a prototypical GPCR. In doing so, arrestin-1 plays a key role in the rapid recovery of the light response. This protein has the ability to arrest signaling from light-activated, phosphorylated rhodopsin, a prototypical GPCR. In doing so, arrestin-1 plays a key role in the rapid recovery of the light response. Arrestin-1 exists in a basal conformation that is stabilized by two independent sets of intramolecular interactions [92]. However, such intramolecular constraints are disrupted by encountering active conformation of the rhodopsin and phosphates attached to the rhodopsin. The fact that these two events are required ensures its highly specific high-affinity binding to phosphorylated, light-activated rhodopsin. In the dark-adapted state, further organization occurs and dimers and tetramers are formed. Arrestin-1 appears to have further functions, beyond that described above for rod cells, such as the translocation from the inner segment to the outer segment of light-activated rhodopsin, probably for the protection against damaging, and several further ones [92].

Three distinct functional modules of vision are important: signal transduction, the retinoid cycle, and protein translocation. The isomerization activates the phototransduction enzymatic cascade and leads to the generation of a light

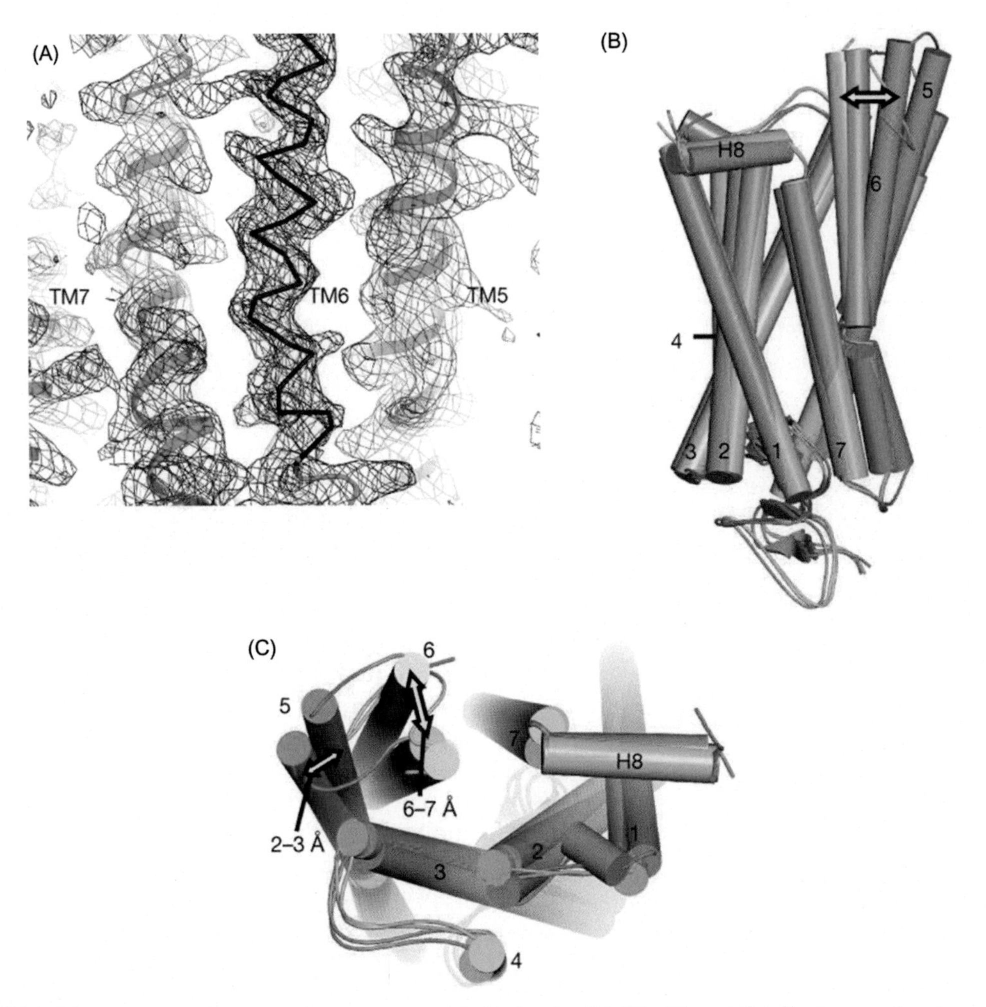

FIGURE 3.26 Structure of opsin and comparison with rhodopsin. (A) 2Fo−Fc and Fo−Fc electron density maps contoured at 1.0 s (blue mesh) and 2.0 s (red mesh), respectively. Electron density maps were calculated using data to 4.0 A° resolution and initial molecular replacement phases, which were obtained using a truncated rhodopsin model (PDB accession 1U19) lacking both extracellular and cytoplasmic regions and TM6. The truncated initial model is shown as a green cartoon and the first Ca trace of TM6 is shown as a black ribbon. (B, C) Comparison of helix orientations: superposition of Ca traces of opsin (orange) and different rhodopsin structures [PDB accessions 1F88 (lime-green); 1U19 (green); and 1GZM (forest-green)]. Helices are shown as rods; loop regions were smoothened for clarity. The side view (B) and the view from the cytoplasmic side (C) are shown. The yellow double arrows indicate different positions of cytoplasmic segment of TM6, with Trp 265 as the pivot point of movement [90].

response. Resetting the activated visual pigment to its inactive ground state occurs by the release of all-*trans*-retinal and its recycling back to 11-*cis*-retinal in a process known as the visual or retinoid cycle. For rod photoreceptors, the supply of recycled chromophore appears to be the rate-limiting step of visual pigment regeneration that controls the kinetics of dark adaptation [93]. This canonical visual cycle involves light-independent processing of the all-*trans*-chromophore by the RPE followed by transfer of the recycled 11-*cis*-retinal to both rod and cone photoreceptors in the adjacent retina [94]. Pigment regeneration is critical for the function of cone photoreceptors in bright and rapidly changing light conditions. This process is facilitated by the recently characterized retina visual cycle, in which Müller cells recycle spent all-*trans*-retinol visual chromophore back to 11-*cis*-retinol. This 11-*cis*-retinol is oxidized selectively in cones to the 11-*cis*-retinal used for pigment regeneration. However, the enzyme responsible for the oxidation of 11-*cis*-retinol remains unknown.

The phototransduction cascade involves the series of biochemical reactions following the absorption of a photon and resulting in vision. The molecular machinery of phototransduction is located in cylindrical subcellular compartments containing a stack of intracellular membranes called discs. The most abundant protein of the disc membranes is rhodopsin. Rhodopsin's covalently attached chromophore, 11-*cis*-retinal, photoisomerizes to its all-*trans* form, and the protein undergoes a conformational change to an active state, metarhodopsin II (Meta II, R*).Meta II catalyzes GDP–GTP exchange on the α subunit of the heterotrimeric G protein, transducin (G_t) at a rate of several hundred per second. Each of the activated transducin α subunits ($G_t\alpha^*$) binds the γ-subunit of the PDE6, producing an active effector complex ($G_t\alpha$-PDE*) and a large increase of the rate of hydrolysis of cyclic GMP. The decrease in cGMP concentration due to $G_t\alpha$-PDE* activity rapidly causes the closure of cGMP-gated cation channels in the plasma membrane. The consequent decrease in inward cation current hyperpolarizes the cell, and the rate of glutamate released from the photoreceptor terminal is reduced. Calculation of the total number of absorptions by the average molecule, adding up absorptions while in the rhodopsin state and those while in the MII state upon 30 min of blue light exposure (403 nm at 300 μW/cm^2) gives an amazing total of 55. The quantum efficiency for rhodopsin bleaching is 0.7, and thus 1.4 photons must be absorbed, on average, to bleach one molecule. In contrast, the photoreversal of MII has a quantum efficiency of 0.22 for producing one rhodopsin; therefore 4.5 ($1/0.22 = 4.5$) photons, on average, must be absorbed by MII to produce one rhodopsin [89,95,96].

The phototransduction cascade has long been recognized to produce a photoresponse with remarkably short latency, while having a slower offset that is approximately exponential in nature. The time course of the mammalian rod photoresponse is roughly 10-fold faster than that of amphibians. In mouse retina, the time to peak of the dim flash response of a healthy rod is ~100 ms and the recovery time constant, ~200 ms. In vivo ERG recordings in mice have revealed nearly identical time to peak and recovery time constant. The kinetics of the photoresponse are remarkably consistent within a given rod from trial to trial and across a wide range of flash strengths. In a mouse rod, the response to a single photon typically reaches a peak amplitude of approximately 0.5 pA. Brighter flashes produce responses that are larger in amplitude, until all the cGMP channels are closed, and the response reaches a maximum, or saturating, amplitude. Further increases in flash strength produce more cascade activity, but no additional increase in amplitude. Rather, the responses remain in saturation for longer times. Plotting the time that a bright-flash response remains in saturation as a function of the natural log of the flash strength (the so-called Pepperberg plot) yields a linear relation for up to ~3000 photoisomerized rhodopsin molecules in mouse rods. The slope of this linear relation is the dominant recovery time constant, τ_D, which is remarkably similar (~200 ms) to the time constant fitted empirically to the final falling phase of the response to dim flashes (so-called τ_{rec}).

The correspondence of τ_{rec} and τ_D suggests that the same first-order deactivation step rate-limits recovery from both dim and bright flashes. The molecular identity of this slowest deactivation step was the subject of much study and debate for more than 15 years. Identification of the biochemical steps that are essential for recovery was necessary before it could be determined which step was the slowest and rate-limiting one. In order for the electrical response to recover, the cGMP-dependent channels must reopen, and for this to occur, the cGMP concentration must be restored. This requires that the rate of cGMP hydrolysis by PDE must decrease. In addition, cGMP must be resynthesized by guanylate cyclase (GC-1 and GC-2 or GC-E and GC-F in mouse). The rate of cGMP synthesis increases during the photoresponse. In normal rods, calcium feedback to guanylate cyclase sufficiently speeds the rate of cGMP synthesis so that the flash response at late times well-approximates the time course of decline of the overall PDE activity, rather than being limited by the rate of cGMP synthesis [97].

After photoisomerization, rhodopsin becomes phosphorylated and the protein arresting (ARR1) binds with high affinity. Experiments on mice have shown that these deactivation steps occur on the time scale of the flash response and either the absence of rhodopsin's C-terminal phosphorylation sites or the absence of rhodopsin kinase (GRK1) led to single-photon responses that were larger than normal. This finding indicated that R* activity was normally reduced by GRK1 within 70 ms. Single-photon responses generated by unphosphorylated R*s typically maintained this larger amplitude for several seconds before abruptly turning off. On average, dim flash responses of rods lacking R* phosphorylation showed a τ_{rec} of 2–5 s. In response to bright flashes, final recovery was slower still, with a time constant of approximately 40 s. Together, these results indicate that phosphorylation of R*'s C-terminal residues are absolutely essential for normal response recovery, and that this phosphorylation must be mediated by GRK1 within 100 ms of the flash. Whether other kinases contribute to R* phosphorylation on other time scales or illumination conditions remains unknown. Both phosphorylation by GRK1 and the binding of ARR1 are essential for normal recovery of the rod flash response.

Transducin remains active until the α subunit hydrolyzes its bound GTP to GDP. In isolation, this GTP hydrolysis occurs far too slowly to account for the time constant of recovery of the flash response. In the 1990s, it was discovered that GTP hydrolysis by $G_t\alpha^*$ is catalyzed by a photoreceptor-specific protein called RGS9–1 (regulator of G protein signaling, ninth family member, first splice variant) [98,99]. RGS9 also binds to the G protein β subunit Gβ5-L and R9AP (RGS9 anchoring protein), which holds RGS9/Gβ5-L with high affinity on the disc membrane. Deleting any one of these three genes (RGS9/Gβ5/R9AP) abolishes expression of the entire complex and the GTPase stimulating activity for $G_t\alpha^*$ in vitro. The single-photon responses of each of these knockout rods are all very similar, recovering roughly 10 times slower than normal. Thus the RGS9 complex is absolutely essential for the normal deactivation of $G_t\alpha$-PDE* and recovery of the light response in rods.

Despite the requirement for RGS9 in stimulating GTP hydrolysis, the fastest RGS9-hydrolysis occurs specifically when $G_t\alpha^*$ is bound to PDEγ. The requirement of PDEγ for rapid $G_t\alpha^*$ deactivation was proposed to increase the gain of transduction by assuring that every $G_t\alpha^*$ produced would bind and activate the effector before turning off. Indeed, mutations in PDEγ that interfered with the ability of PDEγ to stimulate GTP hydrolysis also interfered with the ability of $G_t\alpha$ to bind PDE and resulted in lower transduction gain and slow photoresponse recovery [97].

When the standard theoretical framework for rod phototransduction incorporates a "Michaelis module" to describe the RGS9-dependent decay of $G_t\alpha/E^*$ activity, solutions of the differential equations were shown to be able to precisely account for the dominant recovery rate over the 20-fold range of RGS9 expression levels in the rods of the a recent study [98]. As a matter of fact, the dominant time constant of recovery followed the predicted tail-phase kinetics for the rate of the decline of substrate $G_t\alpha$-PDE* of a standard Michaelis scheme: the rate of recovery, $\nu(=1/\tau_D)$, was equal to V_{max}/K_m for the RGS9 reaction. Thus the Michaelis module for RGS9-mediated deactivation of $G_t\alpha$-PDE* was able to precisely account for the RGS9-concentration dependence of τ_D that was experimentally observed (Fig. 3.27) [97–100]. A schematic representation on the activation of vertebrate rod transduction is depicted in Fig. 3.28 [101].

3.5 Chemistry of vision

Nature has evolved a rather large variety of systems for the sense of sight, although the basic chemistry remains the same. Thus the photoreceptor cells of invertebrate animals differ from those of vertebrates in morphology and physiology. Rhodopsin is converted by light into a meta-rhodopsin which is thermally stable and is usually reisomerized by light in invertebrates, while in vertebrates photoisomerization causes the dissociation of the chromophore from opsin, and a metabolic process is required to regenerate the pigment [102].

The absorption of light leads to an isomeric change in the retinal molecule to form the all-*trans* isomer (Scheme 3.1).

Rhodopsin is a protein containing a polyenal, 11-*cis*-retinal, covalently bound as a protonated Schiff base to opsin, a G protein. As for the

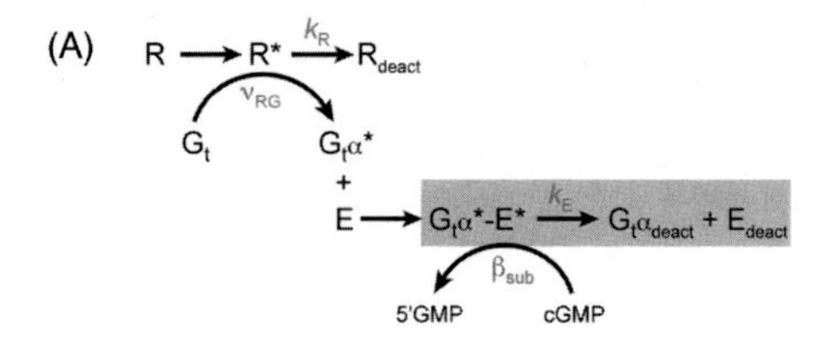

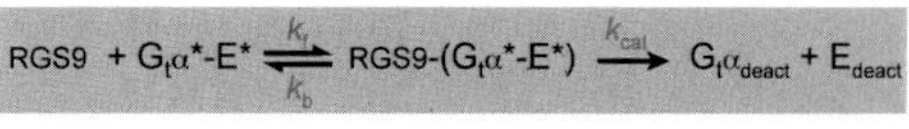

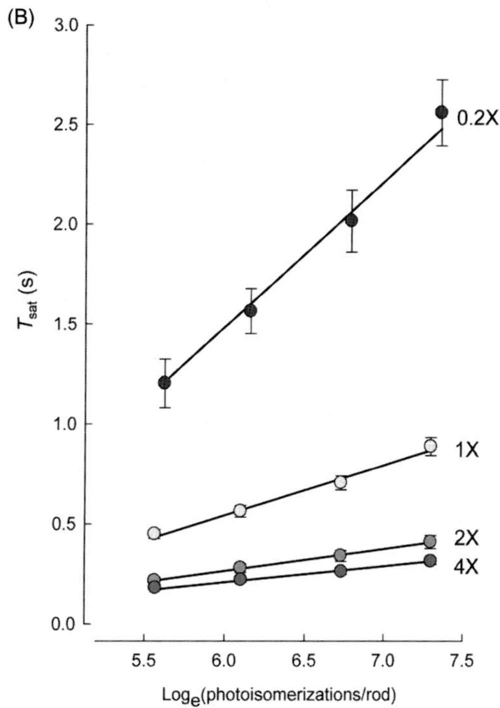

FIGURE 3.27 Michaelis module for the RGS9-dependence of $G_t\alpha$-PDE* deactivation reveals the rate constants of RGS9 binding and catalysis, and constrains R* lifetime. (A) Standard scheme for phototransduction, in which the Michaelis module for RGS9-mediated GTP hydrolysis was substituted for the first order decay of $G_t\alpha$-PDE*. (B) The time that flash responses remained in saturation (T_{sat}) as a function of the natural log of the number of R* (photoisomerizations) produced by each flash for mouse rods expressing a 20-fold range of RGS9 complex [98]. Error bars represent SEMs. *Straight lines* are the best-fitting curves produced using simplex searches of the solutions to the differential equations representing the expanded scheme in (A). Parameter values were $k_R = 33\ s^{-1}$, $k_f = 0.051\ \mu m^2/s$, $k_b = 13.8\ s^{-1}$ and $k_{cat} = 52.8\ s^{-1}$ [99].

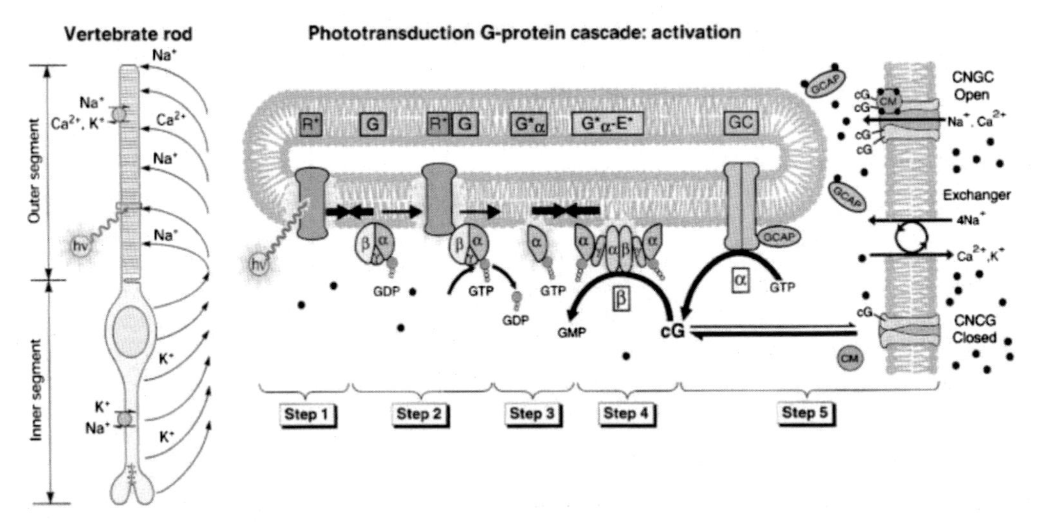

FIGURE 3.28 Schematic of a vertebrate rod and the activation steps of the rod phototransduction cascade. At left, a rod with its normal circulating current, capturing a photon. At right, a magnified view of the five principal steps of activation. Step 1: capture of a photon (hνn) causes rhodopsin to transform into its enzymatically active form, R*. Step 2: R* repeatedly contacts molecules of the G protein, catalyzing the exchange of GTP for GDP, producing the active form G*a (5Ga-GTP). Step 3: G*a subunits bind to the inhibitory g subunits of the phosphodiesterase (E), thereby activating the corresponding "a" and "b" catalytic subunits, forming E*'s. Step 4: E*'s catalyze the hydrolysis of cyclic GMP (cG). Step 5: The consequent reduction in the cytoplasmic concentration of cGMP leads to closure of the cyclic nucleotide-gated channels (CNGCs) and blockage of the inward flux of Na^+ and Ca^{2+}. A generally accepted model of these reactions provides a quantitative account of the activation phase of rod responses. The closure of the CNGCs initiates feedback signals, as the exchanger continues to pump Ca^{2+} out, and the cytoplasmic Ca^{2+} concentration declines, activating at least three distinct mechanisms. Two of these mechanisms are illustrated in this figure: loss of Ca^{2+} from GCAPs allow them to bind to a cytoplasmic domain of the guanylyl cyclase (GC), increasing its activity; loss of Ca^{2+} from calmodulin (CM) causes it to dissociate from the CNGCs, increasing their affinity for cGMP [101].

SCHEME 3.1 Photochemical isomerization of the Schiff base chromophore in rhodopsin.

underlying chemistry, that cones and rods functioned as terminal of the optic nerve was early proposed by anatomists, but that a single pigment was established only in 1877 when Boll began to publish his papers on frog eyes [103]. Boll found that when dark adapted animals were examined and the retina pulled away from the eye basement with fine tweezers it looked intensely red, but it bleached to yellow within 20 min, then to a satin sheen and then became transparent within a few minutes. On the contrary, the red color was not observed when the animals had been exposed to bright light. A year later, Kűhne [104] isolated several pigments from the eyes, which, however, were only slowly reacting photochemically and indeed do not participate in the vision process. The only pigment in mammals is rhodopsin, with a single prostetic group, 11-*cis*-retinal [105]. More precisely, in human pigments of different spectrum are active, as otherwise vision would not be a three chromatic process, but this was due to a minor difference in the apoprotein structure that determine the difference of absorption spectrum. The double $\rangle{=}\mathrm{N}^{+}\mathrm{H}$ bond is formed with the highly conserved nucleophylic lysine^{296} fragment. The isomerization causes hydrolysis of the Schiff base to give retinal and the opsin. Thus retinal was coupled again with opsin, after that it was reduced to the corresponding alcohol, back E/Z isomerized and reoxidized [106]. Key structure features of the rhodopsin molecule are the parallel arrangement of the seven transmembrane helices and the compact structure of the "ionic lock," the interhelical cytoplasmic hydrogen bond network between HIII and HVI [91]. The photoisomerization, however, brings helix 6 (HVI) outward of the helix bundle relative to HIII (Fig. 3.29) [107].[4]

Accurate measurements on the photochemical reaction of rhodopsin showed that the quantum yield at 500 nm was Φ_{500} 0.65 ± 0.01 and the quite small fluorescence quantum yield demands that there is a quite fast reacting path (Fig. 3.30) [108]. The isomerization quantum yield was dependent on the irradiation wavelength, and the quantum yield dropped by about one-third while the excess energy decreased above 550 nm.

Transduction is initiated by a transformation of GPCRs from a quiescent to an active state. All GPCRs share a common architecture comprising seven transmembrane-spanning α-helices, which is well suited for signal propagation from a diverse repertoire of external stimuli across biological membranes to a heterotrimeric G protein. The following generalizations have been found to apply.

The orthosteric ligand-binding pocket of GPCRs exists in many different types of structural forms to accommodate a broad range of external stimuli. Despite such variation, all GPCRs share a common 7-TM architecture and likely a common mechanism through this region to propagate external signals across the membrane.

The activity of GPCRs is often conceptualized within the context of two-state thermodynamic equilibrium models, such as the ternary complex model or extended ternary complex model where the receptor exists in equilibrium between an inactive state (R) and an active state (RG or R*), which can be modulated by ligands, G protein, or mutations. Receptor activity is presumed to be governed by an on-off molecular switch involving rigid-body conformational movements.

Minimal changes are noted between dark state and photoactivated state crystal structures of rhodopsin. This observation suggests that significant conformational changes may not be the sole determinant of receptor activation. Activity, likely partial, may still be present in the absence of large tertiary structural changes. GPCRs can exist in oligomeric clusters and the quaternary changes between each monomeric unit are likely to be important determinants in the mechanism of activation. GPCRs are dynamic molecules. Their activity is better represented within the

[4] The amino acid numbering used in this manuscript incorporates the residue number from the amino acid sequence of the specific receptor being discussed (e.g., K^{296}) and a residue number from a generic numbering system developed by Ballesteros and Weinstein (J.A. Ballesteros, H. Weinstein, Methods, Neurosci. 25 (1995) 366–428) (e.g., Lysine$^{296\ 7.43}$) that gives the position of an amino acid relative to the most conserved amino acid (designated 50) on a specific helix. In this example, Lys296$^{7.43}$ on H7 is seven residues toward the N terminus from the most conserved residue on H7, Proline $^{7.50}$. Because sequence alignments are poor for the extracellular and intracellular loops, as well as for the N and C termini, a generic superscript (e.g., EL2) is used to designate the position of a nontransmembrane residue [107].

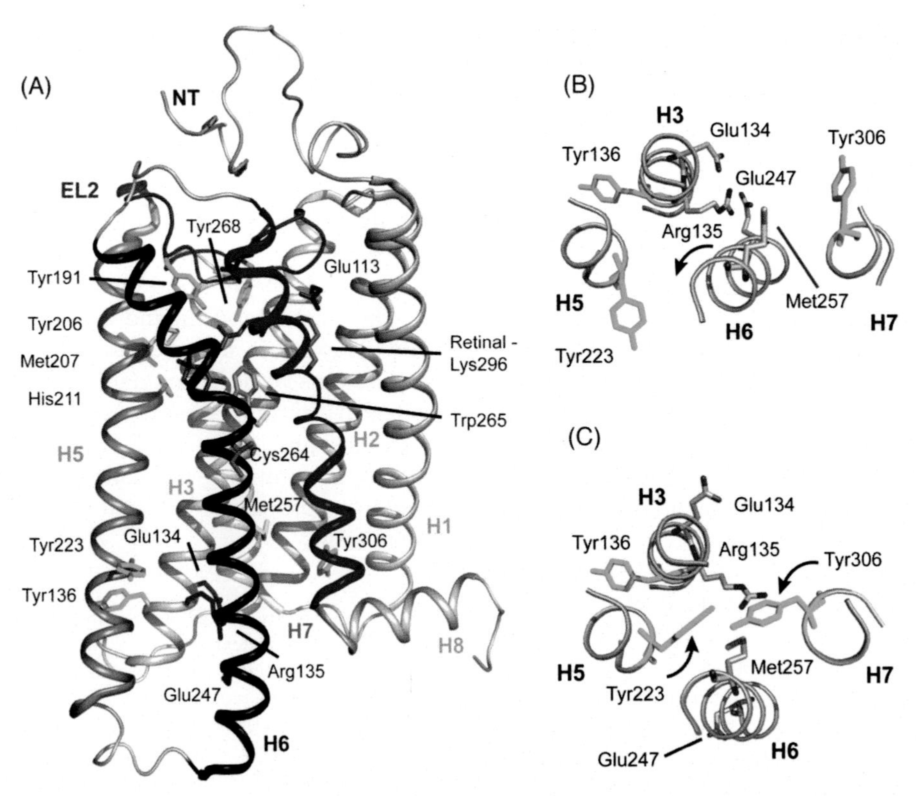

FIGURE 3.29 Crystal structures of rhodopsin and meta II highlighting key residues associated with photoactivation. (A) The full receptor structure is shown. The retinal is buried between the transmembrane helices on the extracellular side of the protein with the second extracellular loop (EL2) forming a cap on the retinal binding site. Motion of helices H5, H6, and H7 during activation opens up the G-protein–binding site on the intracellular side of the receptor, while helices H1–H4 form a tightly packed core. (B) Cross section of the intracellular side of rhodopsin in the region of the R^{135}-E^{247} salt bridge. Only selected helices are shown for clarity. (C) Structure of the intracellular side of meta II in the region of the ionic lock. There are several differences between inactive rhodopsin (B) and meta II (C). Rotation of H6 breaks the R^{135}-E^{247}contact and moves M^{257}into the H3-H6 interface. The intracellular end of H6 tilts outward, allowing the side chain of R^{135} to extend and contact M^{257}. Rotation of H5 and H7 place $Y^{223}$and Y^{306} into contact with R^{135}. Hydrogen bonds between these tyrosines and R^{135}stabilize the active state of the receptor [107].

context of a funnel-shaped energy landscape in which the receptor exists in an ensemble of states and activation proceeds via multiple pathways" [109].

The availability of new instruments leads to a better characterization of the variuos intermediates. Initially, studies were focused on the determination of the cascade of specroscopically detectable intermediates, all of them containing a all-*trans* retinal structure. In the gas phase trans-cis interconversion is the dominant photoisomerization, although cyclic isomers, not absorbing visible light, also appear during collisional excitation of retinal protonated Schiff base (RPSB). Photodissociation is a very minor channel (*r*, 1%). In the drift tube environment photodissociation products have been observed, although in minimal amount, presumably because the excited RPSB molecules undergo rapid vibrational deactivation through buffer gas collisions (Schemes 3.2 and 3.3) [110].

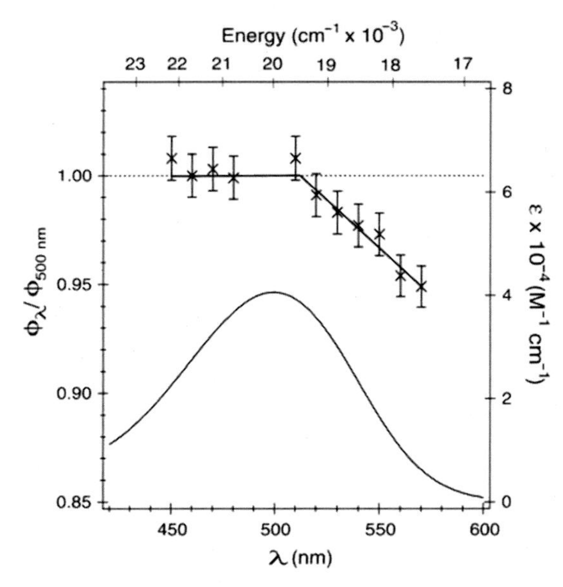

FIGURE 3.30 Summary of results for the wavelength dependence of the quantum yield of rhodopsin. The linear fit to the quantum yield data is added as a guide. Errors are reported as $\pm 1\sigma$. The experimental absorption spectrum for rhodopsin is also displayed (solid curve, right axis) [108].

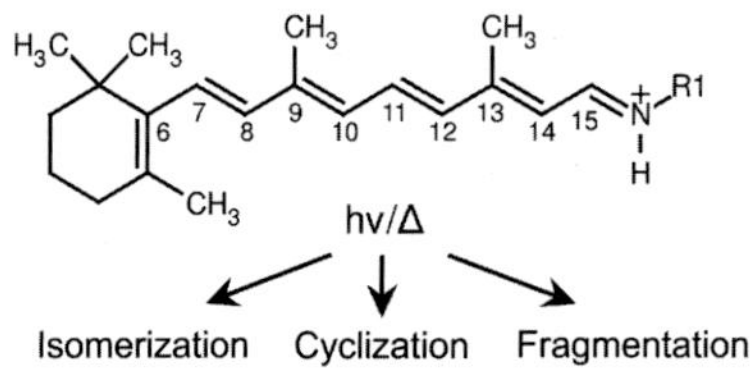

SCHEME 3.2 All-trans retinal protonated Schiff base undergoes structural rearrangement following absorption of a photon or thermal excitation (in this study R1 = n-Bu, in vivo R1 = Lys) [110].

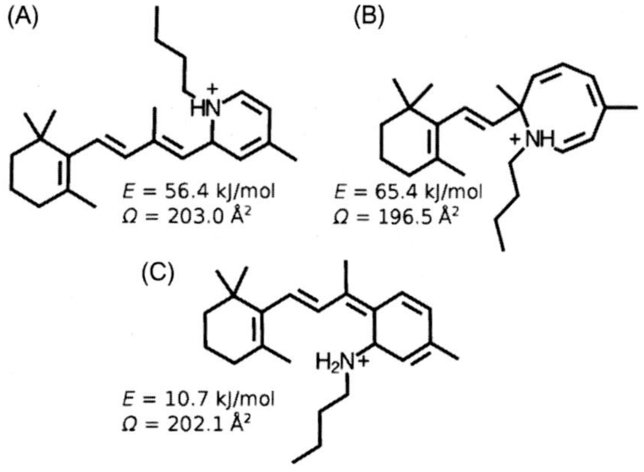

SCHEME 3.3 Possible cyclic isomers generated from RPSB through collisions or photo excitation. Energies of the different isomers are given with respect to that of 6s-*cis* (−) conformer of *trans* RPSB [110].

This picture of protein reaction dynamics is unique in many ways. First there is a breakdown of the Born-Oppenheimer approximation; the electronic wave function is a rapidly changing function of nuclear geometry in the crossing region and as a result the isomerization proceeds via a nonadiabatic pathway. Second a Fermi Golden Rule analysis of the internal conversion in rhodopsin is inapplicable because there is insufficient time for excited-state equilibration or relaxation before product formation. Finally the cis-to-trans isomerization occurs so rapidly that the equipartition limit is not reached and the excited state vibrational phase space is only partially occupied. This results in a *dynamic* coupling between ground and excited states where the strength of the coupling depends intimately on the kinetic energy of the wave packet along reactive torsional degrees of freedom as it enters the surface crossing region. This description of a coherent nonadiabatic path toward photoproduct is entirely consistent with the presence of a conical intersection between the S_1 and S_0 potential energy surfaces. In the vicinity of such a surface funnel, extremely strong and localized nonadiabatic coupling can result in rapid and efficient internal conversion prior to excited-state equilibration. Such an excited-state surface funnel has been reported in studies of retinal, too, and further contributes to the rationalization and prediction of the dynamical behavior of biomolecules [110] Thus in a femtosecond Raman spectrum study it was found that the reaction was satisfactorily described as a two concerted hydrogen-out-of-plane (HOOP) wagging motion of the 11 and 12

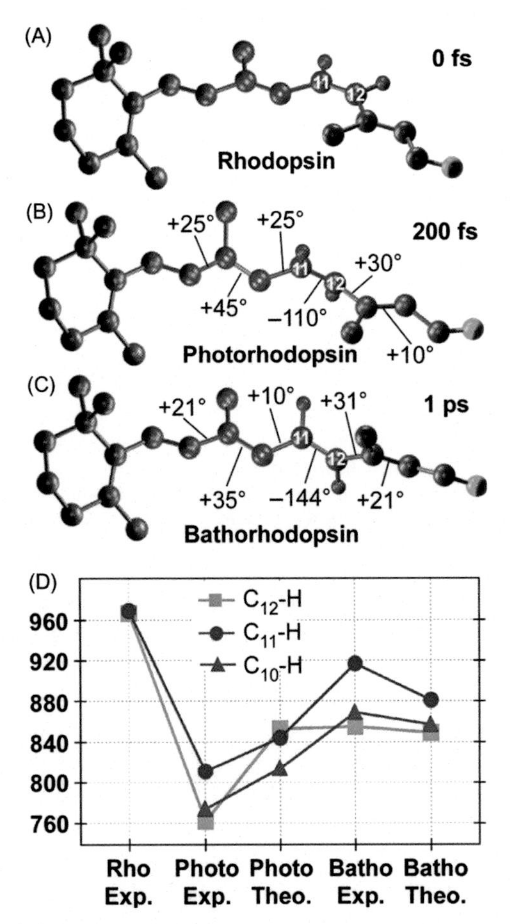

FIGURE 3.31 (A–C) Retinal chromophore structures for reactant rhodopsin and for photorhodopsin and bathorhodopsin that reproduce the observed hydrogen wagging frequencies. Backbone dihedral twist angles from the rhodopsin reactant are indicated. (D) Comparison of density functional theory calculated (Theo.) and experimental (Exp.) hydrogen wagging frequencies for the photo and bathorhodopsin structures [111].

hydrogen atoms [111]. This allows to follow the primary events in vision, from initial isomerization of 11-*cis*-retinal in rhodopsin the all-trans bathorhodopsin and the ensuing changes from 200 fs to 1 ps (Fig. 3.31).

There is an ongoing outward potassium current through nongated K^+-selective channels. This outward current tends to hyperpolarize the photoreceptor at around −70 mV (the equilibrium potential for K^+). There is also an inward sodium current carried by cGMP-gated sodium channels. This so-called "dark current" depolarizes the cell to around −40 mV. Note that this is significantly more depolarized than most other neurons. A high density of Na^+-K^+ pumps enables the photoreceptor to maintain a steady intracellular concentration of Na^+ and K^+.

Under high illuminations, rods are in saturation and photopic vision is initiated by the outputs of three cone photoreceptor classes (L-, M-, and S-cones) that have spectral sensitivities suitable to provide trichromatic color perception. Under intermediate light flux, mesopic illuminations intervenes when rods gradually become sensitive and cones are still active. In this case subtle changes and a reduction in both the perceptual quality and gamut of perceivable colors occur. Under dim, scotopic illuminations, only rods are active and color perception is still possible by different physiological computations rather than the trichromatic system. Visual neuroscience is still active for determination of the mechanisms of rod–cone interaction as well as how these give rise to the altered perceptual experience under mesopic illumination. Methodologies to measure rod and cone signal contributions separately and during rod–cone interaction are still elaborated. These methodologies are typically based on known functional differences between the rod and cone systems (Fig. 3.32), and apply these methods to study rod–cone interactions arising between the stimulus area and surround (lateral interactions) or within the stimulus area (local interactions).

In the dark, the transduction of a molecular reaction into an electric signal is the job of the cells that populate the following layers. Photoreceptor cells are unusual cells in that they depolarize in response to absence of stimuli or scotopic conditions (darkness). Under photopic conditions (light), photoreceptors hyperpolarize to a potential of c. −60 mV,

	Scotopic			Mesopic		Photopic	
Photoreceptors	Rods only			Rods and Cones		Cones only	
[1] Luminance (log cd.m^{-2})	-8.0	-6.0	-4.0 -4.7 (Starlight)	-2.0 -1.9 (Full moon)	0.0	2.0	4.0
[2] CIE Transition (log cd.m^{-2})				-3.00	~0.47		
[3] Pupil diameter (mm)	7.99	7.95	7.76	6.99	5.00	3.00	2.23
Troland (photopic)	**$5.0*10^{-7}$**	**$4.9*10^{-5}$**	**$4.7*10^{-3}$**	**0.38**	**19.6**	**710**	**$3.9*10^{4}$**
[4] Troland (scotopic)	$1.2*10^{-6}$	$1.2*10^{-4}$	$1.1*10^{-2}$	0.96	48.9	1769	$1.37*10^{5}$
[5] Cone threshold				203±38 photons			
[6] Rod saturation (log sc Td)							3.3-3.6;>3;>3.9
[7] Colour vision		Relational		3 cones + rods			Trichromatic
[8] Critical flicker frequency (Hz) – Rod			8	11 13	21	32	
[8] Critical flicker frequency (Hz) – Cone					52	64	75
[9] Impulse response function tp(ms) – Rod			72 62	53 44	34		
[9] Impulse response function tp(ms) – Cone				48	39	30	
[10] Spatial frequency resolution (c/°) – Rod						6	
[10] Spatial frequency resolution (c/°) – Cone						60	
[11] Spatial integration (area, deg^2)			0.4				.025

FIGURE 3.32 Visual functions under scotopic, mesopic, and photopic illumination. The luminance level (log cd.m^{-2}) is the reference for estimating pupil diameter using the empirical relation between pupil diameter (mm) and field luminance [112].

while in the dark, cGMP levels are high and keep cGMP-gated sodium channels open, allowing a steady inward current, called the dark current. This dark current keeps the cell depolarized at about −40 mV, leading to glutamate release which inhibits excitation of neurons. The depolarization of the cell membrane under scotopic conditions opens voltage-gated calcium channels. An increased intracellular concentration of Ca^{2+} causes vesicles containing glutamate, a neurotransmitter, to merge with the cell membrane, therefore releasing glutamate into the synaptic cleft, an area between the end of one cell and the beginning of another neuron. Glutamate, though usually excitatory, functions here as an inhibitory neurotransmitter. Glutamate that is released from the photoreceptors in the dark binds to metabotropic glutamate receptors (mGluR6) which, through a G-protein coupling mechanism, causes nonspecific cation channels in the cells to close, thus hyperpolarizing the bipolar cell [113].

When a ray of light shines on the cell, the cGMP-gated sodium channels close and the influx of both Na^{+} and Ca^{2+} ions is reduced. Stopping the influx of Na^{+} ions effectively switches off the dark current. Reducing this dark current causes the photoreceptor to hyperpolarize, which reduces glutamate release and the inhibition of retinal nerves, leading to *excitation* of these nerves. This reduced Ca^{2+} influx enables deactivation and recovery from phototransduction, along the visual phototransduction-deactivation of the phototransduction cascade [114,115] (Fig. 3.33).

Deactivation of the phototransduction cascade in light, low cGMP levels close Na^{+} and Ca^{2+} channels, reducing intracellular Na^{+} and Ca^{2+}. During recovery (dark adaptation), the low Ca^{2+} levels induce recovery (termination of the phototransduction cascade), because low intracellular Ca^{2+} makes intracellular Ca-GCAP (guanylate cyclase activating protein, a calcium binding protein) dissociate into Ca^{2+} and GCAP.

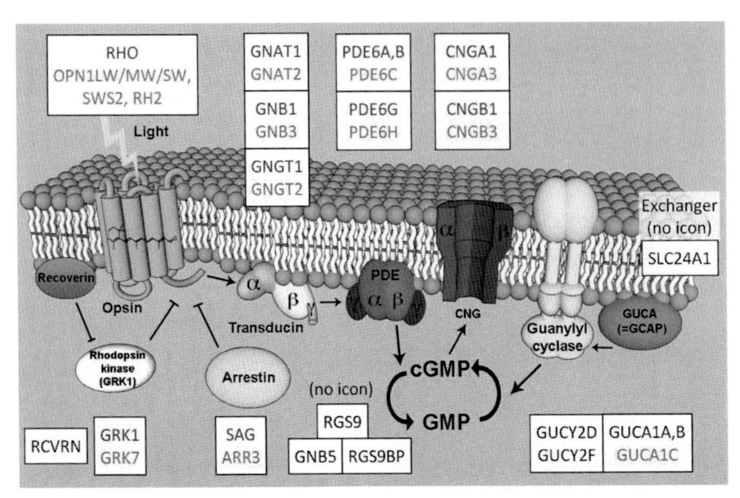

FIGURE 3.33 Schematic role of arresting, transducing, cGMP [116].

Finally the GCAP liberated from this dissociation regenerates depleted cGMP levels. In turn this reopens the cGMP-gated cation channels, which had been closed during the phototransduction process. The protein liberated GAP (GTPase accelerating protein) interacts with transducin and causes its hydrolyze to bound GTP to GDP, interrupting the transformation of cGMP to GMP. The fall of calcium level during phototransduction causes calcium dissociates from recovering [another calcium-binding protein, that is, normally bound to rhodopsin kinase (RK)]. RK is thus released. RK liberated then phosphorylates the metarhodopsin II. This reduces its binding affinity for transducin. Arrestin then completely deactivates the phosphorylated-metarhodopsin II, terminating the phototransduction cascade [97].

Vision is a very efficient sense and a visually guided behavior at its sensitivity limit is achieved by means of single-photon responses originating in a small number of rod photoreceptors. This striking sensitivity has been explored by understanding how the constraints arising from the retinal output signals provided by distinct act on retinal ganglion cell types. In the primate retina ON and OFF parasol ganglion cells. The cell types likely to underlie light detection at the absolute visual threshold, differ both in response to the polarity and in the way, they handle single-photon responses originating in rods. The brain receives a thresholded, low-noise readout and the OFF pathway with a noisy, linear readout. The ON pathway is expensive for the organism, more single-photon responses are lost, and they are propagated with a delay compared with the OFF pathway. However, the response of ON gaglion cells allows a better intensity discrimination in comparison with the OFF-gaglion cell responses when close to the visual threshold [117] (Fig. 3.34).

3.6 Sight in nonhumans

Vision in humans is "middling" at best, both figuratively and literally in the animal visible spectrum of 300–750 nm. This comes as a surprise to many of us as we cannot imagine the need to see more than the millions we can manage.

Many animals have vision that exceeds our red–green–blue (RGB)-based trichromacy. Horses, dogs, some primates, and barracuda, on the other hand, have only two spectral

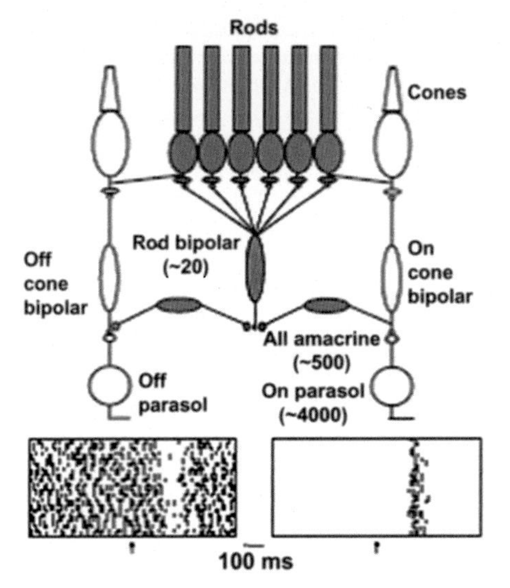

FIGURE 3.34 Schematics of the rod bipolar pathway in the primate retina. Near the absolute threshold, the primary OFF and ON pathways (rod bipolar pathway) share the circuitry up to the AII amacrine cell (highlighted in blue). The synapse between ON cone bipolar cells and ganglion cells not only operates as a thresholding nonlinearity to reduce noise, but also limits information about single photons. The numbers shown in the diagram indicate the number of rods converging on a particular cell type. Spike responses to dim flashes are shown for an OFF (left) and ON parasol cell (right) at the bottom [117].

classes of photoreceptors, and may be likened to red–green. Humans and other primates with cone cells sampling in three different wavelength ranges earn the classification trichromat by combining this information such as the ratio of excitations from different cone types encode (Figs. 3.35–3.37) [118,120,123,124].

As dichromatic space only has two dimensions, hue, saturation and brightness are not independent. A tetrachromatic space (probably possessed by birds and fish) may theoretically add a fourth dimension to (akin to hue or saturation), but the physical interpretation and ecological significance of this are not obvious. While organisms such as butterflies and stomatopods have many more than four types of spectral receptors, it is not clear how they are used. To our knowledge, there is no evidence for "pentachromatic" or higher-dimensional vision. QS, QM, and QL are quantum catches of receptors sensitive to short, medium, and long wavelengths, and S, M, and L in chromatic spaces depict hypothetical eliciting signals in the respective receptor only. NP indicates the neutral point where a spectral light appears the same to the dichromat as broad-spectrum white light (Fig. 3.38).

Note further that the generic name given to an autofluorescent, membrane-bound intracellular material that is distributed widely among the postmitotic cells of different organs of the body. With advancing age, there is a marked increase in the lipofuscin granule content of human RPE cells and there now exists substantial evidence that the bulk of this lipofuscin represents the chemically modified residues of incompletely digested photoreceptor outer segments. The origin of the lipofuscin in the RPE therefore differs significantly from that of the lipofuscin in other tissues [127].

In the late 1900s, the genes for human cone opsins were isolated and sequenced. Subsequent analysis of cone opsin genes from a large number of contemporary animals fostered the development of opsin gene phylogenies. The idea that all of the vertebrate visual photopigments are specified by opsin genes belonging to five gene families, one linked to rod photopigments while the other four underlie cone pigments has received general consensus [128]. All four of the cone opsin gene families emerged together, at a point early in vertebrate evolution, perhaps as long as 540 million years ago [129]. The earliest true mammals, however, evolved from therapsid ancestors during the Early Jurassic, somewhere around 200 million years and are presumed to be nocturnal, and thus to have a capable system for vision. The genes that specify the protein moieties ("opsins") of the three different sensitive pigments in the human eye have been isolated and it has been possible to establish that there is

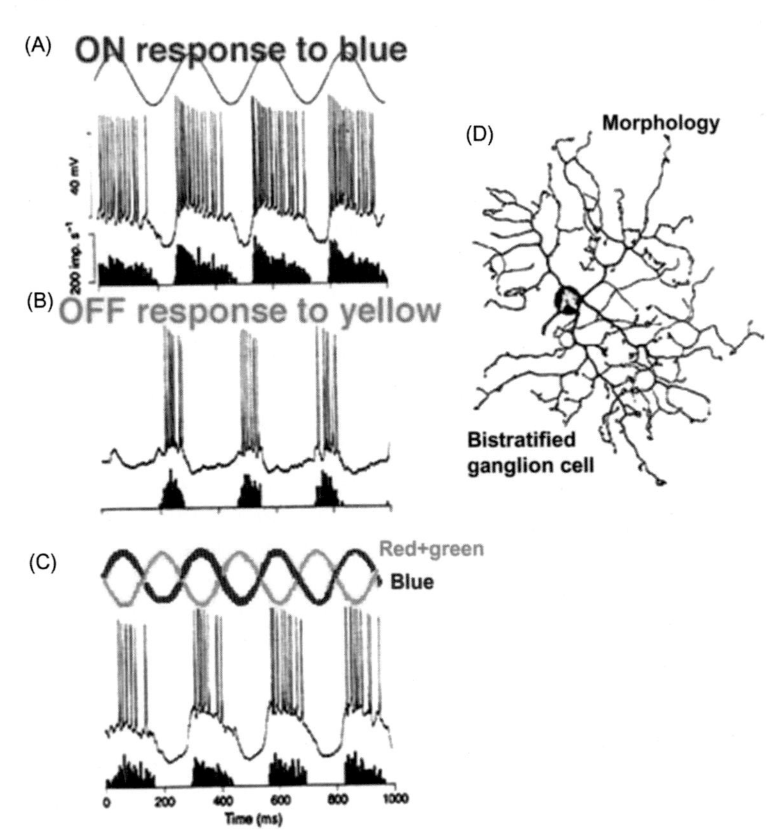

FIGURE 3.35 Blue-ON, yellow-OFF chromatic ganglion cell in monkey retina. (A) Alternating blue and dark stimuli evoke ON responses. (B) Alternating yellow and dark stimuli evoke OFF responses. (C) Alternating blue and yellow colors matched in brightness evoke blue-ON, yellow-OFF responses. (D) Morphological appearance of the small bistratified ganglion cell from which blue-sensitive responses were recorded [118].

indeed a gene that specifies opsins for pigments that recognize primarily red, others that recognize primarily green, and yet others that see primarily blue on the basis of the combination of molecular and genetic analysis that allowed the unambiguous identification of gene (as DNA) with the function. Two of these (the red- and green-sensitive opsins) were located in the X-chromophore and comprise a family of repeated genes on the human X chromosome. The green and red genes are in a tandem array, all oriented with the tail of one gene near the head of the next. Indeed, when the DNA of protanopes (hypothesized by the Young-Helmholtz theory to be missing the ability to see red) was analyzed, loss or alteration of one of the genes (the first in the array) was observed, and it was assumed that this gene specifies the red opsin. Consistently, when deuteranopes (hypothesized to be missing the ability to see green) were examined, the other X-linked genes (the number is variable) were missing or altered [130].

The most complete way of understanding how a sense in general and vision in particular is implemented by the animal under consideration is studying how it behaves in vivo. Such studies are less easy than, for example, comparing the anatomy and cellular component of the retina and further they are exposed to the risk of misinterpretation of the behavior of other animals by superimposing what is known of humans. In fact, we tend to assume, for example, that any species with three spectral sensitivities must be trichromatic, and to often use terms such as dichromacy, trichromacy, and tetrachromacy rather loosely, based

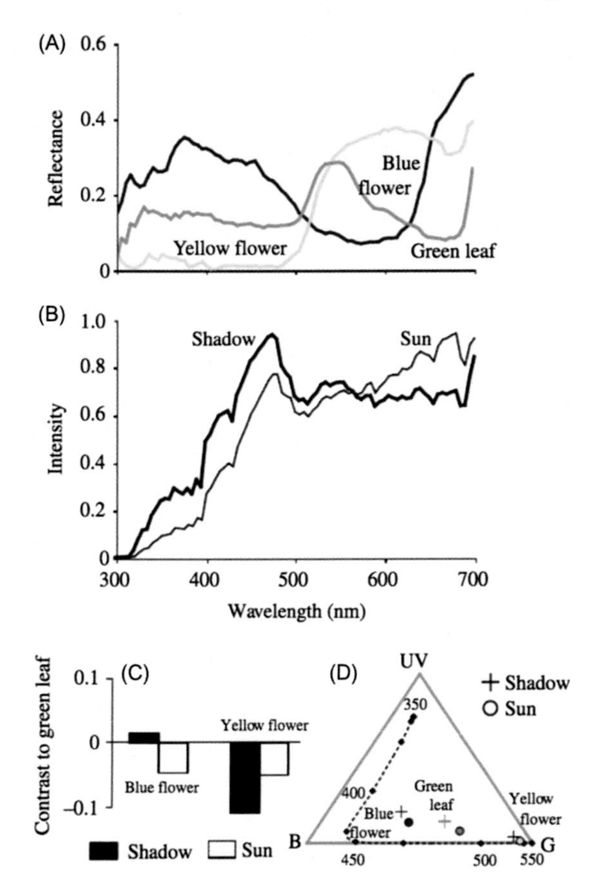

FIGURE 3.36 The use of chromatic vision. (A) The reflectances of a typical yellow and a typical blue flower and a green leaf. (B) The spectral composition of direct sunlight and skylight in a shadow, measured shortly before sunset, on a summer day in Utah, United States. (C) The achromatic contrast between both flowers and the green leaf in both illuminations differ dramatically for the green receptor of *Deilephila elpenor*. (D) The color loci of all three stimuli in the color triangle of *D. elpenor* are rather constant in both illuminations, even without the assumption of color constancy. UV, UV light; B, blue light; G, green light [119–121].

only on the retinal complement of spectral channels. However, a knowledge of how channels combine information is required for the correct labeling of a system. As an example, humans are true trichromats, as we use all three S, M, and L receptors, in combinations such as S−(L + M) or (M−L) to encode relative ratios of excitation, as the cones view in the outside world. It is the different cone response triplets that encode the many colors we identify, while (M + L) is used for luminance or for the blue−green. It thus appears that there is no sufficient contrast information to need more than two spectral channels.

In contrast many animals have color vision different from our RGB-based trichromacy. Birds and reptiles may be termed tetrachromats because they have four color receptors, extending the sensitivity to light at both ends of the human visible spectrum (400−700 nm). Horses, dogs, some primates, and barracuda are dichromats and have only two spectral classes of photoreceptors, corresponding to red−green color blind humans. Insects, smaller fish, most birds, and even mice use the ultraviolet (UV), a part of the spectrum that is toxic for human. Furthermore, there are some animals with the potential for "penta"-chromacy and beyond. As an example, the eyes of stomatopods (mantis shrimp) and butterflies are endowed by up to 12 spectral sensitivities and our mind get confused at the potential for "dodeca"-chromatic color space. Behavioral considerations also allow us to suggest why the waterflea *Daphnia* is unlikely to be tetrachromatic, like a bird. In fact, this diminutive crustacean displays four relatively well-spread spectral channels at 348, 434, 525, and 608 nm, but these act through a low-resolution compound eye (containing only 22 ommatidia, the visual units of the compound eye) and this certainly does not suggest a complex lifestyle. In 2010 it was reported that "four grades of vision on a behavioral scale: first, taxes or light environment seeking; second, wavelength-specific behaviors directed towards objects; third, learning through neural representative of colors; and fourth, colors appearance including colors categorization." This categorization allows to class the use of light by different species, at least as far as we admit that some animals may cross category

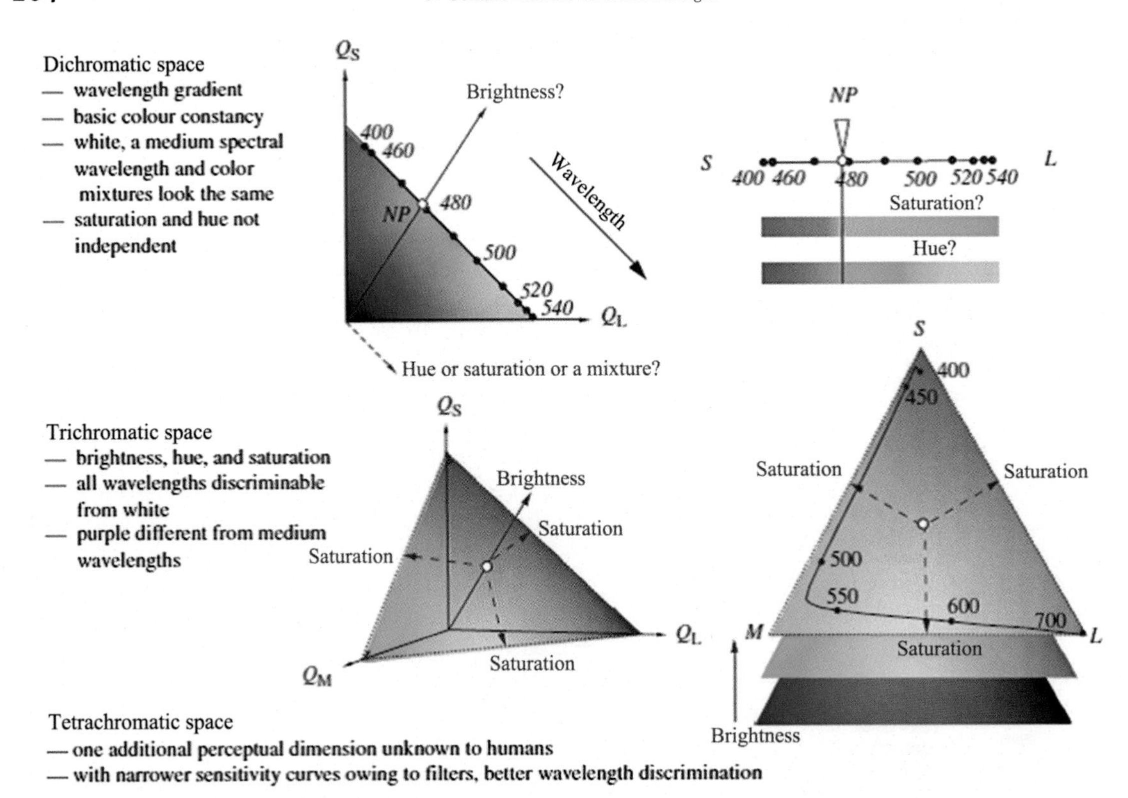
Dichromatic space
— wavelength gradient
— basic colour constancy
— white, a medium spectral wavelength and color mixtures look the same
— saturation and hue not independent
Brightness?
Wavelength
Hue or saturation or a mixture?
NP
Saturation?
Hue?
Trichromatic space
— brightness, hue, and saturation
— all wavelengths discriminable from white
— purple different from medium wavelengths
Brightness
Saturation
Tetrachromatic space
— one additional perceptual dimension unknown to humans
— with narrower sensitivity curves owing to filters, better wavelength discrimination
Specialized systems with multiple receptors
— possibly different sets of receptors for different colour-guided behaviours such as mate, food or host plant detection

FIGURE 3.37 Spaces (receptor spaces, left side) and chromatic spaces (right side) of dichromats and trichromats illustrating the relationship between the human percepts of hue, saturation and brightness and the physiological receptor axes [122].

boundaries or possess multiple grades. As an example, the color world of *Daphnia* seems essentially restricted in the first of these categories [122,126].

The rod system has very low spatial resolution but is extremely sensitive to light; it is therefore specialized for sensitivity at the expense of resolution. Conversely, the cone system has very high spatial resolution but is relatively insensitive to light; it is therefore specialized for acuity at the expense of sensitivity. The properties of the cone system also allow us to see colors. Color discrimination is based on opponent photoreceptor interactions but is limited by receptor noise. In dim light, photon shot noise impairs color vision and in vertebrates the absolute threshold of vision is set by dark noise in cones. In other species the situation might be different, for example, in nocturnal insects, such as moths and nocturnal bees, and vertebrates lacking rods (geckos) have adaptations to reduce receptor noise and use chromatic vision even in very dim light. However, vertebrates with duplex retinae use color-blind rod vision when noisy cone signals become unreliable, and their transition from cone- to rod-based vision is marked by the Purkinje shift, that is, the blue shift of the peak

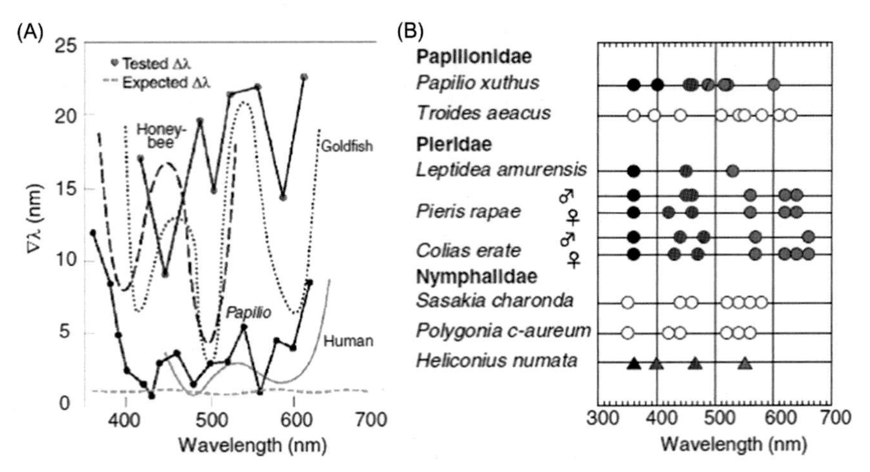

FIGURE 3.38 Comparative wavelength discrimination in different animals and spectral sensitivity variability in butterflies. (A) Wavelength discrimination functions ($\Delta\lambda$) determined through behavioral testing with the animal choosing between narrow-band stimuli. Colored data point (red and green) are for the stomatopod *Haptosquilla trispinosa;* other discrimination functions are labeled according to species [124,125]. The very different observed and expected result and apparently 'poor' discrimination in stomatopod indicate a form of color signal processing different to other species, including the butterfly. (B) Butterflies whose spectral receptors are identified. Positions of circles indicate spectral sensitivity peak wavelengths determined by electrophysiology coupled with cell marking. Positions of triangles are absorption peaks of visual pigments predicted from spectral sensitivity of the entire retina. Symbols that are colored indicate absorption spectra of visual pigments whose gene expression has been shown in respective photoreceptors. Black and gray correspond to ultraviolet. Note that some photoreceptors coexpress two visual pigments with different absorption spectra [126].

luminance of the human eye occurring upon low illumination. Rod–cone interactions have not been shown to improve color vision in dim light but may contribute to color vision in mesopic light low intensity (10^{-2} to 10^{-3} cd/m^2). Frogs and toads that have two types of rods use opponent signals from these rods to control phototaxis even at their visual threshold. However, for tasks such as prey or mate choice, their color discrimination abilities fail at brighter light intensities, similar to other vertebrates, probably limited by the dark noise in cones [131–133].

It is possible to test our visual sensitivity by using a very low level light source in a dark room. The experiment was first done successfully in 1942 [134], and it was possible to conclude that the rods can respond to a single photon during scotopic vision. In their experiment they allowed human subjects to have 30 min to get used to the dark. They positioned controlled light source 20 degrees to the left of the point on which the subject's eyes were fixed, so that the light would fall on the region of the retina with the highest concentration of rods. The light source was a disk that subtended an angle of 10 min of arc and emitted a faint flash of 1 ms to avoid too much spatial or temporal spreading of the light. The wavelength used was about 510 nm (green light). The subjects were asked to respond "yes" or "no" to say whether or not they thought they had seen a flash. The light was gradually reduced in intensity until the subjects could only guess the answer. They found that about 90 photons had to enter the eye for a 60% success rate in responding. As only about 10% of photons arriving at the eye actually reach the retina, this means that about 9 photons were actually required at the receptors. As the photons would have been spread over about 350 rods, the experimenters were able to conclude statistically that the rods must be responding to single photons, even if the subjects were not able to see such photons when they arrived too infrequently [135]. As for the perception of the intensity of light, animals have two types of photoreceptors, based on ciliary (activating a PDE that changes the concentration of cyclic GMP in the cell), rhabdometric [based on the activation of phospholipase C and the inositol phosphate (IP3) pathway] approach, respectively. Vertebrates have a peculiar situation where two distinct ciliar receptors are present and operate together over the whole range of vision from single photon to bright light. It has been thought that the main advantage of ciliary receptors is due to the fact that they consume less ATP than rhabdometric photosensors, a difference that provided a sufficient selection pressure for the development of a completely ciliary eye. Rods, however, evolved quite early, much earlier than jaws. In addition, the outer segment did not contribute that largely to rod sensitivity, but rather may be evolved to increase the efficiency of protein renewal, and throughout the expression of novel proteins evolution of the rod was incremental and multifaceted, produced by the formation of several novel protein isoforms and by changes in protein expression, with no alteration having more than a few-fold effect on transduction activation or inactivation [136].

The sophisticate system of vision provides animals with a refined ability to form a representation of the environment around them, and this almost instantaneously. Humans tend to have an anthropocentric view of vision and assume that they have evolved the "best" visual system, as far as the color discrimination is concerned. The human retina, with its rods and three spectral classes of single cone, looks relatively simple, when compared to that of a diurnal bird or a turtle, where the retina contains not only rods, but four spectral classes of single cone with an additional class of double cone. These cones further possess in their inner segments an array of color filters in the form of oil drop complex containing high concentrations of carotenoids [133].

The appearance of color vision early in the evolution of vertebrates can be deduced from the identification of five families of opsin genes in the most ancient vertebrate lineage, the jawless lamprey. Assuming a parallel evolution of the neural mechanism, these primitive fish had the potential for tetrachromatic color vision at least 540 million years and four spectral classes of cones appears to be the maximum number employed.

Gene duplications and mutations have provided a wide spectral range for all of the four cone opsins and the rod opsin, with considerable spectral overlap between cone classes. In lamprey five opsin genes have been identified, but there is some debate over whether two of the classes are more rod like or cone like. This points to the question of the still uncertain definition of rods and cones, that is clear in many species, but not in other ones. Indeed, in some lamprey and elasmobranchs some of "rods" appear to function at both scotopic and photopic levels. Furthermore, the "green rods" of amphibians and the "rods" of nocturnal geckos contain cone pigments that presumably function primarily at scotopic and/or mesopic levels and are rather likely involved in the color vision. Wavelength discrimination based on rod/cone interactions is all that is available to marine mammals and also possibly to some nocturnal mammals.

The understanding of the color vision in vertebrates has greatly increased due to the ability to identify and isolate opsin genes from a wide range of animals including rare and protected species. This has produced some unexpected and exciting observations such as the presence of multiple opsins in many teleosts and a functional SWS2 cone opsin in monotremes [137].

The basic architecture of eyes was present in the jawless fish (agnaths) of the Cambrian and

Ordovician periods (about 450–550 million years ago), although these animals lived in shallow lagoons probably feeding by shifting food from the muddy substrate where a complex vision was of little importance and their visual sense was primarily directed at identifying the approach of predators simply by detecting the movement of a sudden shadow or an abrupt change in illumination. This may cause surface ripples and waves, as well as reflections from the substrate, and cause continuously flickering and variable luminance, but it was not sufficient for the straightforward identification of a menace. Flickering, a fast variation of luminance, required to be supplemented but a variation of chromaticity in such a way not enough, if the brightness of either the object or the background were highly variable. In contrast, flickering will change the luminance, but not chromaticity, so that an opponent process between two spectrally different receptors can filter out the flicker, but leaves a "color" signal, thus making much easier the detection of objects against the background, even when either of them showed a variable brightness [138]. Clearly, the minimum requirement for color vision is two spectrally distinct classes of photoreceptors combined with a nervous system that can compare the quantum catch of each class of receptors. To summarize, an effective (color) vision system has to compromise between the high sensitivity of opsin-based photosensitive organelles and the high energy demands of the receptors and the complexity of the neural mechanisms required for color perception. Taking into account these limitations, it turns out that the maximum of efficiency is reached with four spectral classes of spectrally different classes of photoreceptors, while adding a fifth spectral class provides little or no advantage [138].

From its unique derivation, the lipofuscin of the RPE differs from that of the brain, heart, testes, and other organs in both electron density and uniformity of granule size. The composition, origin, and possible deleterious effects of RPE cell lipofuscin have been topics of considerable scientific interest and controversy in recent years and this article reviews current knowledge.

Trichromatic vision in humans results from the combination of red, green, and blue photopigment opsins. Despite intensive studies of molecular and psychophysical research on vision abnormalities, it is not yet clear which are the pattern of normal genetic variation of these genes. Recent studies could determinate the nucleotide sequence analysis and test of neutrality for 5.5-kb region of the X-linked long-wave "red" opsin gene (OPN1LW) in 236 individuals from ethnically diverse human populations. Analysis of the recombination landscape across OPN1LW reveals an unusual haplotype structure that is associated with amino acid replacement variation in exon 3. This is consistent with gene conversion, in contrast with the absence of OPN1LW amino acid replacement fixation that happened with the divergence from chimpanzee. From this time the human population exhibits a significant excess of high-frequency OPN1LW replacements. The results suggest that subtle changes in L-cone opsin wavelength absorption may have been adaptive during human evolution [139].

Vision in humans and other Old World primates depends on differences in the absorption properties of three spectral types of cone photoreceptors. Primate cones are linked by gap junctions, but it is not known to what extent the various cone types are electrically coupled through these junctions. Here is shown, by using a combination of dye labeling and electrical recordings in the retina of macaque monkeys, that neighboring red and green cones are homologously and heterologously coupled by nonrectifying gap junctions. This indiscriminate coupling blurs the differences between red- and green-cone signals. The average junctional conductance is about 650 ps. Coupling between red and green cones may cause a modest decrease in human discrimination with a comparable increase in luminance discrimination [139].

The coding region of the human rhodopsin gene is interrupted by four introns, which are located at positions analogous to those found in the previously characterized bovine rhodopsin gene. The amino acid sequence of human rhodopsin, deduced from the nucleotide sequence of its gene, is 348 residues long and is 93.4% homologous to that of bovine rhodopsin. Interestingly, those portions of the polypeptide chain predicted to form loops on the cytoplasmic face of rhodopsin are perfectly conserved between the human and bovine proteins. Comparative wavelength discrimination was found to vary very much in terms of spectral sensitivity. Wavelength discrimination functions ($\Delta\lambda$) are determined through behavioral testing with the animal choosing between narrowband stimuli.

End data points (red and green) are for the stomatopod *H. trispinosa*; other discrimination functions are labeled according to species [120,125]. The large difference observed and expected result and apparently "poor" discrimination in stomatopod indicate a form of signal processing different to other species, including the butterfly. Butterflies whose spectral receptors are identified. The reported data are based on spectral sensitivity peak wavelengths determined by electrophysiology coupled with cell marking. While absorption peaks of visual pigments predicted from spectral sensitivity of the entire retina (Fig. 3.39). The "sensation" corresponding to a color has little to do with an "actual" color, provide that it has a sense. Sensations in previous small spot experiments [140].

Absorption spectra of visual pigments whose gene expression are also reported. Notice that some photoreceptors coexpress two visual pigments with different absorption spectra.

The Na^+ K^+-gated pump is closed when light as bleached rhodopsin and opsin decreases cGMP thus increasing Na^+ polarization, while in the dark rhodopsin is inactive, cGMP is high and ion channels are open (Fig. 3.40) [141]. In the reaction phase retinal recombines with opsin.

The behavior of the photosentive cells under illumination can be reproduced by a wiring diagram as shown in Figs. 3.41 and 3.42 [142].

An overall balance of phototransduction is shown in Fig. 3.43, with participation of rods and cones and the exchange of energy. This process is quite expensive in term of energy

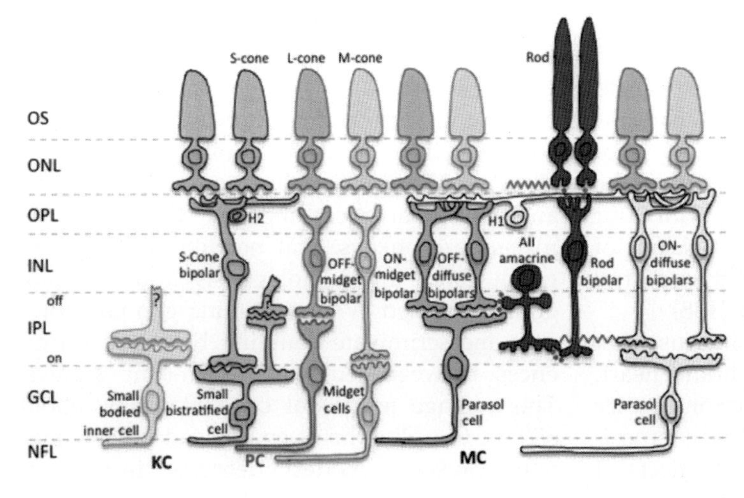

FIGURE 3.39 Neural pathways related to mesopic vision. Rod signals input to all three primary retinogeniculate pathways, namely magnocellular (MC), parvocellular (PC), and koniocellular (KC) pathways. Only rod inputs to the MC pathway are shown in this schematic. Green circles indicate chemical synapses. The red zig–zags indicate electrical synapses. Retinal layers are indicated on the left: *OS*, outer segment; *ONL*, outer nuclear layer; *OPL*, outer plexiform layer; *INL*, inner nuclear layer; *IPL*, inner plexiform layer; *GCL*, ganglion cell layer; *NFL*, nerve fiber layer [112].

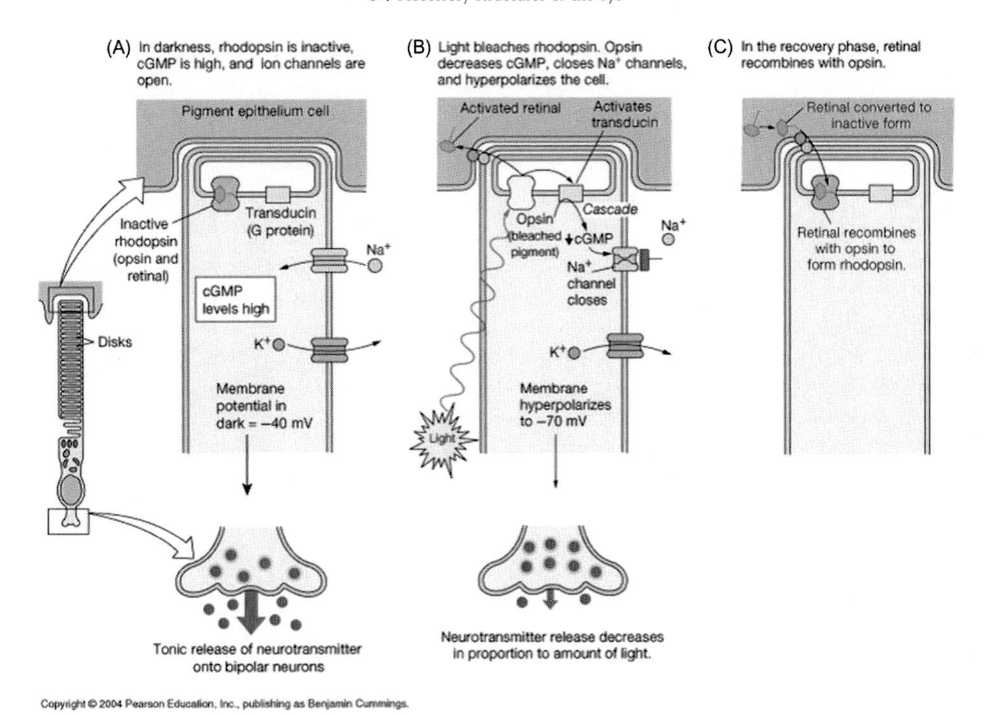

FIGURE 3.40 Light is transduced into electric signals. Important to know is that light bleaches rhodopsin resulting in the closure of sodium channels and hyperpolarization of the cell and a reduction of neurotransmitter release in proportion to the light intensity [141].

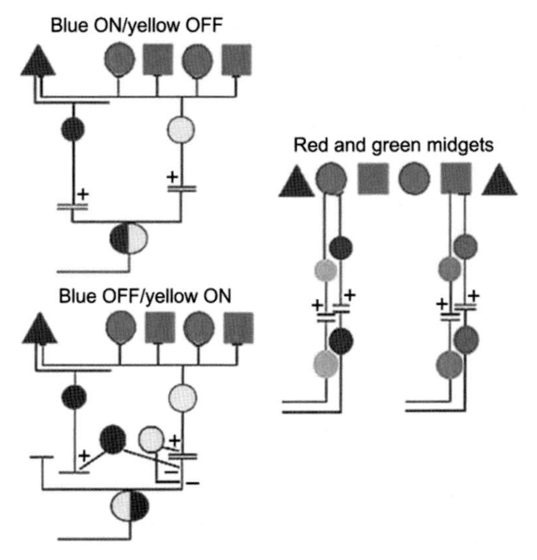

FIGURE 3.41 Hypothetical wiring diagrams of the S-cone pathways [142].

3.7 Accessory structures of the eye

These include the eyelid or palpebrae, a fold of skin with eyelashes on the border, raising and closing under the action of devoted muscles; the caruncle, the small, pinkish portion of the innermost corner of the eye that contains oil and sweat (sebaceous) glands and conjunctival tissue; the lacrimal glands (tear glands) are the almond-shaped body located inside the orbit at the upper, outer corner of each eye, able to make tears to help keep the surface of the eye and lining of the eyelids moist and lubricated. Tears help reduce friction and remove dust and debris from the eye to prevent infection. Small lacrimal ducts drain tears from the lacrimal gland through very tiny openings inside the inner corner of each eyelid. The iris is a highly

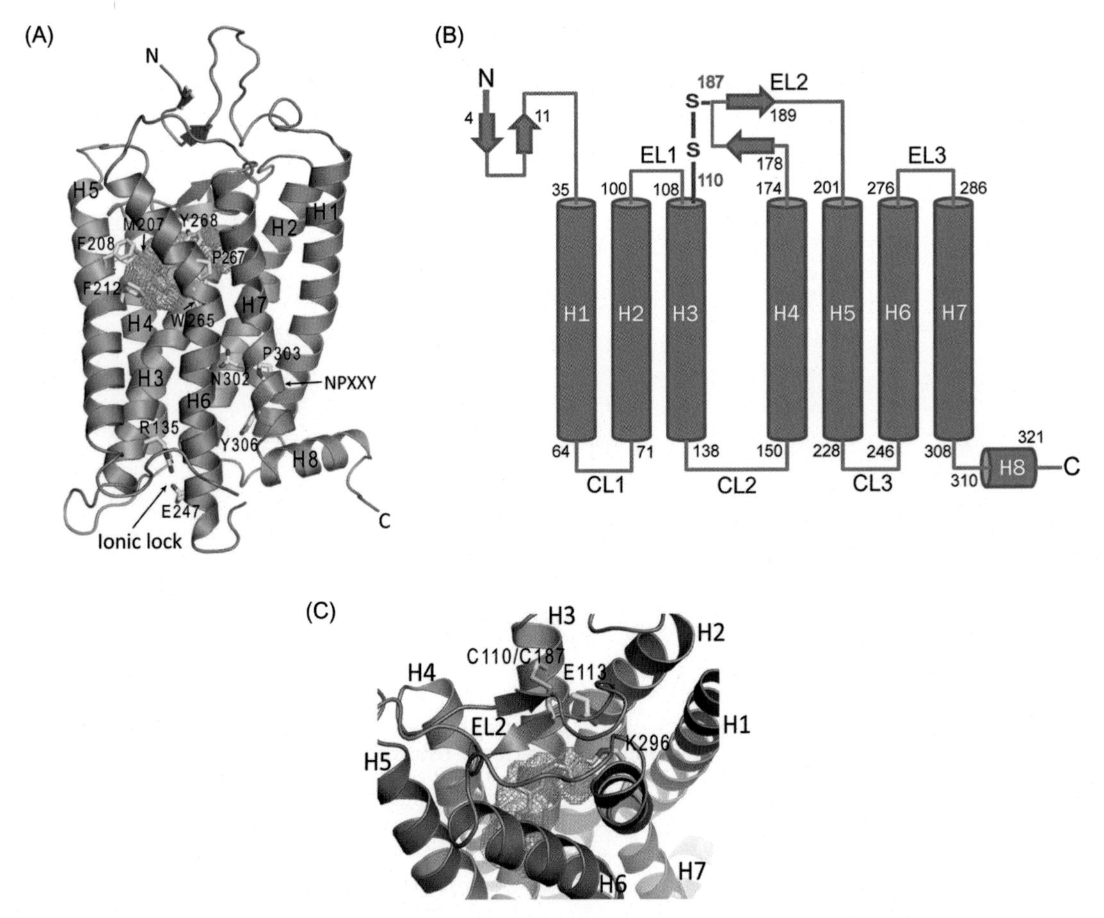

FIGURE 3.42 Overall structure of ground sate bovine rhodopsin and its key features (PDB: 1F88). (A) The seven-transmembrane helix domain with the retinal in gray stick and the ligand binding pocket shown as a pink mesh. Major ligand binding residues around the ligand binding pocket are shown as yellow sticks and are labeled. Other features include the ionic lock (yellow sticks) and the NPXXY motif (orange). (B) Two-dimensional sequence of bovine rhodopsin with the starting and ending residues of secondary structural elements indicated. The disulfide bond connecting EL2 to helix 3 is shown in orange. *N*, amino terminus; *C*, carboxyl terminus; *EL*, extracellular loop; *CL*, cytoplasmic loop. (C) The ligand binding pocket (pink mesh) of rhodopsin with EL2 (the lid of the pocket) shown in dark brown. The disulfide bond between C110 and C187 is labeled [143] (http://faculty.pasadena.edu/dkwon/chap10_C/chap%2010%20part%20C_files/textmostly/slide19.html)

complex structure, indeed the most complex in the human body, and on this ground is increasingly used for personal identification in banks, border controls, etc.

However, the color perception of iris is not "correct." What is perceived as a brown iris is actually due to an abundance of melanocytes and melanin in the outer layer and stroma, while a blue iris contains very little melanin. Different shades of blue depend on the amount of melanin. In irises with a small amount of melanin, different shades of gray, green, and hazel, are determined by the thickness and density of the iris itself and the extent of accumulation of white collagen fibers, as well as patches of tissue loss in the anterior border layer and stroma. As light traverses these relatively melanin-free layers, collagen fibrils of the iris scatter the short blue wavelengths to the surface, thus a blue iris is a consequence of structure not of major differences in chemical composition and the iris pigmented epithelium (IPE). The IPE is always pigmented in the examination of the iris in all eye colors, except in individuals with albinism, and contributes little to the impression of eye color. Where the above lying stroma is thin this layer can have some influence on patterning. For example, the dark coloration of the IPE can absorb the penetrating light in eyes with a very thin stroma, which will give the white collagen fibers in the deeper cell layers of the stroma a gray tinge (Fig. 3.44) [145].

The main aspects of the iris color classification are reported in Fig. 3.45.

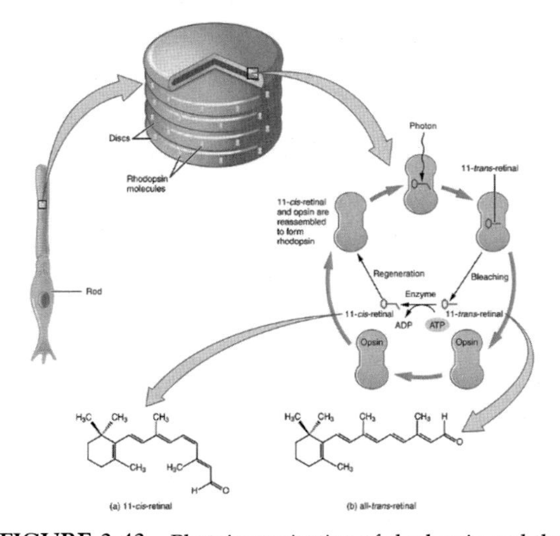

FIGURE 3.43 Photoisomerization of rhodopsin and the vision process. The *retinal* molecule inside an *opsin* protein absorbs a photon of light. Absorption of the photon causes *retinal* to change from its *11-cis-retinal* isomer into its *all-trans-retinal* isomer. This change in shape of *retinal* pushes against the outer *opsin* protein to begin a signal cascade, which may eventually result in chemical signaling being sent to the brain as visual perception. The *retinal* is reloaded by the body so that signaling can happen again [144]. Source: *Wikipedia*.

3.8 Damage caused by light

Given its function, the eye is necessary exposed to damage by light. This may occur either at a slow or fast process. As the chemicals involved in vision are highly hydrophobic, and are located in a hydrophobic medium, it is to be expected that radical chain mechanisms have a significant role in degradation. The fast degradation first noticed by Boll when he discovered the pigment [103, 146].

Thermodynamic data support that 2 + 2 photocycloaddition is repaired with a counterion-independent efficiency in DNA-DNA and NA-RNA duplex [147]. The damage caused by short

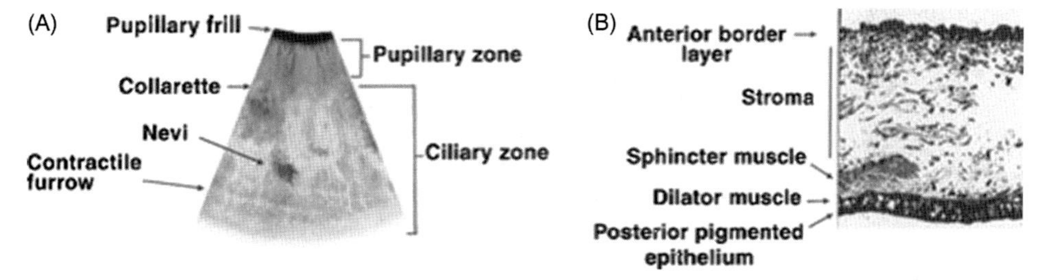

FIGURE 3.44 Human iris structure. (A) Frontal sector of the iris is shown with the collarette separating the papillary and ciliary zones. (B) A transverse view illustrating the five layers found within the iris [145].

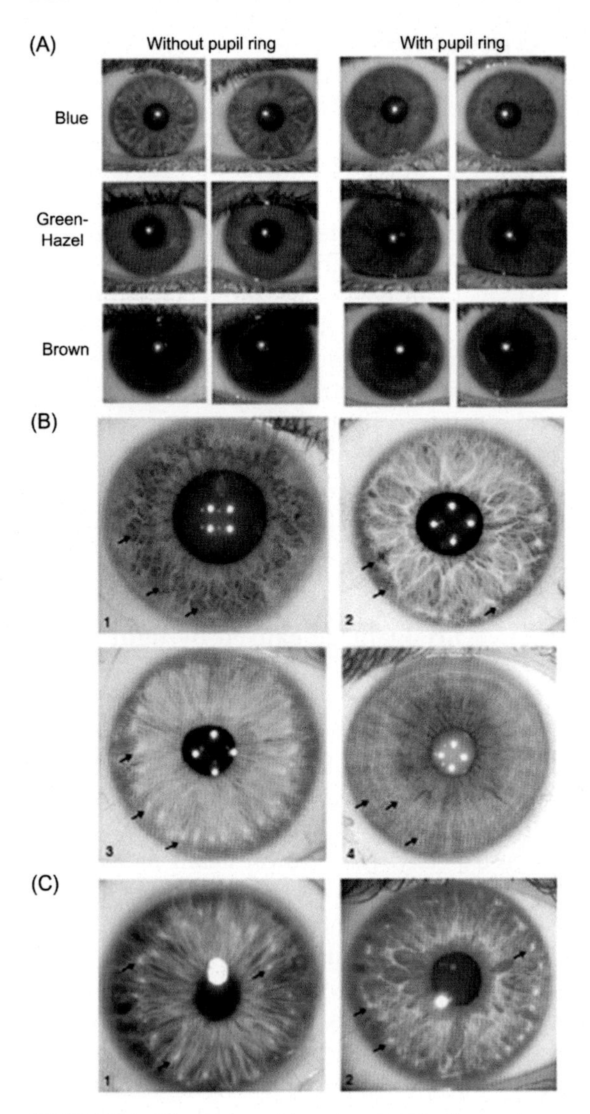

FIGURE 3.45 Human iris color classification and patterns. (A) Three major classes of eye color are shown as blue, green-hazel, brown with and without a brown peripapillary ring. (B) Patterns found within the iris highlighted by arrows: 1. Fuchs' crypts, mild stroma atrophy; 2. nevi, dots of accumulated melanin; 3. Wolfflin nodules, dots of accumulated collagen fibrils; 4. contraction furrows, fold marks in thicker irises due to iris contraction and dilation. (C) Patterns found within the iris highlighted by arrows: 1. Brushfield spots, observed in Down's syndrome; 2. Wolfflin nodules, observed in normal controls [145].

wavelength light may either involve the formation of radicals or cycloaddition involving the DNA bases. The former effect is indicated inter alia by the fact that molecules easily forming radicals, such as linalol, prevent carcinogenesis. Thus it was found that UV-B-irradiation (180 mJ/cm^2) caused both acute (hyperplasia, edema formation, lipid peroxidation, antioxidant depletion) and chronic effects [overexpression of cyclooxygenase-2 (COX-2) and ornithine decarboxylase upon periodical exposure to UV-B] in mouse skin, but both of these effects were suppressed by topical or intraperitoneal treatment with linalol prevented the acute effects as well as overexpression of COX-2 and the decarboxylase. 2 + 2 Photocycloaddition affords a useful method for DNA conversion [148]. Selective photocrosslinking reaction between 3-cyanovinylcarbazole nucleoside and 5-methylcytosine (mC), which is known as epigenetic modification in genomic DNA, was developed. The reaction was completely finished within 5 s of 366 nm irradiation, and the rate of this photocrosslinking reaction was c.30-fold higher than that in the case of unmodified normal cytosine. There were no significant differences in the thermodynamic parameters and the kinetics of hybrid formation of oligonucleotides (ODN), through the in situ DNA manipulation. Thus cyanocarbazole nucleotide gave crosslinking upon irradiation at 36 nm for 1 s, and reverted by irradiation at 312 nm for 1 min [149], containing the fast cross-linked cyanovinylcarbazole phosphoramidite and its complementary short single-stranded synthetic DNA (ODN) containing C or 5-mC at the photocrosslinking site, and suggesting that the quick and selective photoreaction has potential for the selective detection of mC in the DNA strand via the photocrosslinking reaction [150].

Micronutrients in many cellular processes, such as riboflavin (vitamin B_2), FMN, and FAD are photosensitive to UV and visible light to generate ROS. The photochemical treatment with UV light has been applied for the inactivation of microorganisms to serve as an effective and safe technology. UV or high-intensity radiation is, however, considered as a highly risky practice. A study was devoted to the application of visible LED lights to riboflavin photochemical reactions to develop an effective antimicrobial treatment. The photosensitization of bacterial genome with riboflavin was investigated in vitro and in vivo by light quality and irradiation dosage. The riboflavin photochemical treatment with blue LED light was proved to be able to inactivate *Escherichia coli* by damaging nucleic acids with generated ROS [151]. Riboflavin is capable of intercalating between the bases of bacterial DNA or RNA and absorbs lights in the visible regions. LED light illumination could be a more accessible and safe practice for riboflavin photochemical treatments to achieve hygienic requirements in vitro. PUVA has been clinically used for healing various skin dermatosis, including psoriasis; however, the mechanism involved required several investigations. In a recent study, it was demonstrated that PUVA differentially regulates miRNA expression profile with a significant upregulation of hsa-miR-4516 [152]. Microarray analysis revealed 1932 differentially expressed gene and their in silico analysis revealed retinoic acid inducible gene-I (RIG-1) signaling, apoptosis, and p53 pathway to be associated with PUVA-induced effects. miR-4516-mediated downregulation of UBE2N promotes p53 nuclear translocation and proapoptotic activity of PUVA is independent of IRF3 but is mediated by the RIG-I in a p53 and NFκB-dependent manner. In addition, PUVA inactivated the AKT/mTOR pathway in concert with inhibition of autophagy and reviles the mechanism of action of PUVA [153]. A new photocaged nucleoside was synthesized and incorporated into DNA with the use of standard synthesis conditions. This approach enabled the disruption of specific H-bonds and all wed for the analysis of their contribution to the activity of a DNAzyme. Brief irradiation with nonphotodamaging UV light led to rapid decaging and almost quantitative restoration of DNAzyme activity. The developed strategy has the potential to find widespread application in the light-induced regulation of ODN function [154]. Thymine-Hg(II)-thymine base pairs have been incorporated in an ODN duplex to study their effect on DNA-mediated charge transport. In the experiment, the introduction of a formally charged Hg atom inside the DNA base core does not significantly alter the charge hopping and trapping properties, and Hg(II) replaces the protons normally found on thymines within the complex and acts like a "big proton" in terms of its role in DNA charge transport. The widespread of G-quadruplex-forming sequences in genomic DNA and their role in regulating gene expression has made G-quadruplex structures attractive therapeutic targets against a variety of diseases, in particular cancer. Information on the structure of G-quadruplexes has been found a requirement for understanding their physiological roles and designing effective drugs against them. However, resolving the structures of G-quadruplexes, has been a challenge as yet, specially for those in double-stranded DNA. A photocleavage footprinting technique has been developed to determine the folding orientation of each individual G-tract in intramolecular G-quadruplex formed in both single- and double-stranded nucleic acids [155].

From the therapeutic point of view, primary care avoids the use of allergens and rather using cold compresses and preservative-free tears. Secondary care involves the use of topical antihistamines, such as levocabastine, emedastine, pheniramine, pyrilamine, and antazoline as well as decongestants, such as phenylephrine, as naphazoline or topical antihistamine/decongestant combination, as well as oral antihistamines, containing sedating antihistamines, or nonsedating antihistamines. Topical mast cell stabilizers (cromolyn, lodoxamide, pemirolast), nonsteroidal antiinflammatory agents (ketorolac, diclofenac, flurbiprofen, nedocromil), and antihistamine/mast cell stabilizer, such as olopatadine, azelastine, ketotifen have a role. Tertiary care involves topical corticosteroids, such as loteprednol, rimexolone. As for immunotherapy, new topical agents are under development, including tacrolimus, cyclosporine, anti-IgE [156].

As mentioned above, the solar UV light induces human skin cancers. First, epidemiological studies have demonstrated a negative correlation between the latitude of residence and incidence and mortality rates of both melanoma and nonmelanoma skin cancers in homogeneous populations. Second, skin cancer can be produced in mice by UV irradiation; the action spectrum of Ph deficiencies in repairing UV-induced DNA damage are prone to develop cancers in sun-exposed areas of the skin. Photocarcinogenesis is a multistage process that involves initiation, promotion, and progression. In addition, UV-induced immunosuppression is closely involved in photocarcinogenesis. Accumulation of DNA lesions caused by UV in several cancer-related genes plays a crucial role in carcinogenesis. Indeed, even in actinic keratosis, precancerous lesions, genetic alterations can be observed. A conventional knowledge demonstrated that UV-B-induced DNA lesion causes genetic mutation (initiation) and UV-B inflammation (sunburn) induces promotion. However, recent findings revealed that the photocarcinogenesis pathway is more complex consequences where each of these processes, mediated by various cellular, biochemical, and molecular changes, is closely related to each other. The pyrimidine photoproducts that result from direct DNA damage induced by UV are involved in developing skin cancer through mutations that lead to the upregulation or downregulation of signal transduction pathways, cell cycle dysregulation, and depletion of antioxidant defenses. In addition, pyrimidine dimers have been shown to trigger UV-induced immunosuppression, which also plays an important role in photocarcinogenesis, partly by upregulation of IL-10, an immunosuppressive cytokine. UV also produces oxidative stress and oxidative DNA damage in skin cells, which cause alteration of the genes involved in the cell cycle, apoptosis, and modification of cell signaling by redox regulation, resulting in inflammation. It has been shown that in Ogg1 knockout mice which are deficient in repairing 8-oxo-7, 8-dihydroguanine (UV-B irradiation upregulates the inflammatory gene, implying that 8-oxoG is involved in triggering inflammation). Thus the key aspects regarding photocarcinogenesis and implication for clinical viewpoints are shown in Fig. 3.46 [157].

To this end, a search of the literature was accomplished by using Medline/Index Medicus, EMBASE/Excerpta Medica, and Chemical Abstracts; most of the relevant citations were studied and summarized. In order to better understand the nature of oxidative stress, the principles of free radical production and the body's normal defense system are discussed. The pesticides are categorized and discussed according to their ability to produce lipid peroxidation or alter body antioxidant status. It is concluded that stimulation of free radical production, induction of lipid peroxidation, and disturbance of the total antioxidant capability of the body are mechanisms of toxicity in most pesticides, including organophosphates, bipyridyl herbicides, and organochlorines [158]. The World Health Organization has clearly identified prevention and early detection as major objectives in the control of the oral cancer burden worldwide. Screening of oral cancer and its preinvasive intraepithelial stages is still based on visual examination of the mouth. However, evidence suggests while visual inspection of the oral mucosa is effective in reducing mortality from oral cancer in individuals exposed to risk factors. Simple visual examination, however, is by subjective interpretation and by the potential

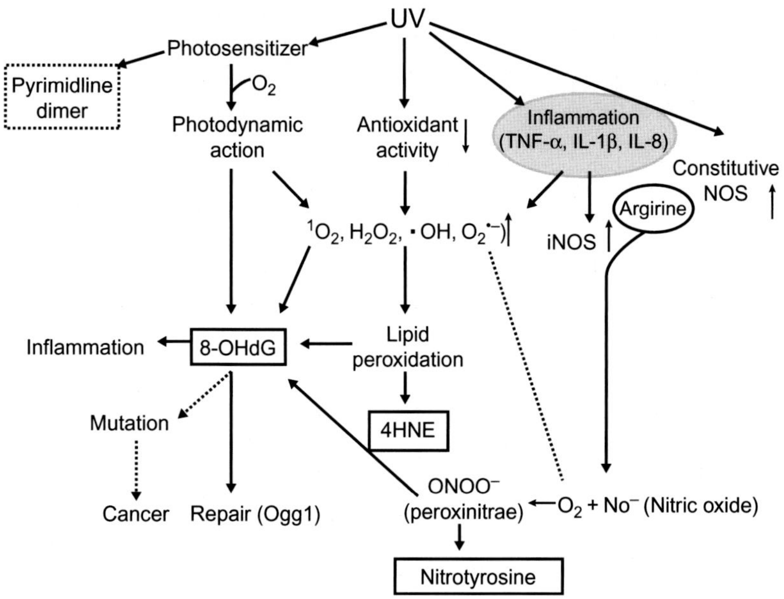

FIGURE 3.46 UV light and cancer [157].

occurrence of dysplasia in normal looking oral mucosa. Therefore adjunctive techniques have been considered to increase the ability to differentiate between benign abnormalities and dysplastic/malignant changes, as well as to identify areas of dysplasia/early oral squamous cell carcinoma (OSCC) that are not visible to naked eye. Examples are the use of toluidine blue, brush biopsy, chemiluminescence, and tissue autofluorescence, with promising results, but not firm evidence [159]. Occult cancerous changes the biological potential of clinically abnormal mucosal lesions can benefit from modern techniques, such as aids or adjuncts such as toluidine blue, brush cytology, tissue reflectance, and autofluorescence. The increased public awareness of oral cancer fostered the marketing of recently introduced screening adjuncts is commendable, the tantalizing implication that such technologies may improve detection of oral cancers and precancers beyond conventional oral examination alone has yet to be rigorously confirmed [160].

3.9 Physical aspects

Improving the visual performance means improving the signal-to-noise ratio. A useful operative concept is that of detective quantum efficiency, defined as a suitably selected ratio of the actual detecting ability of the eye to the maximum conceivable detecting ability, meant as the performance that could be achieved by an ideal device that makes optimum use of all of the photons that enter the pupil of the eye. Recent publications report an average short-term quantum efficiency of about 0.11, a value that is close to that previously reported (0.1) for the fraction of blue–green light incident on the cornea which is absorbed by the rods [161,162]. In fact, three types of ambient radiation can be defined (Figs. 3.47 and 3.48).

In order to understand the light behavior in the environment, (1) the signal radiation has to be detected and usually varies with time; (2) the blackbody radiation field produced by the detector and its environment; (3) other steady

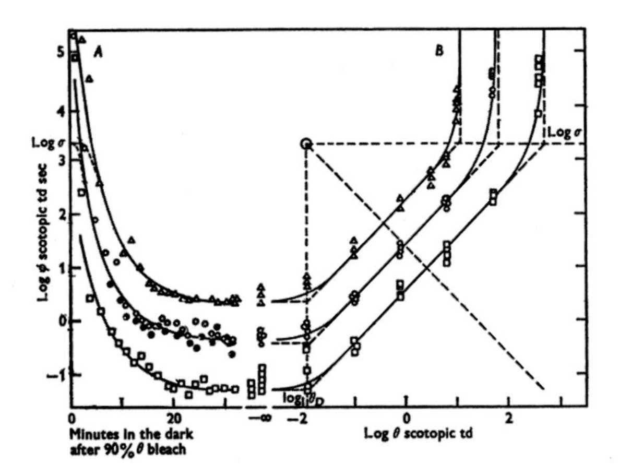

FIGURE 3.47 Dark adaptation of ϕ measured by the criterion of the production of the fixed signal N that just inhibits the fixed test [161].

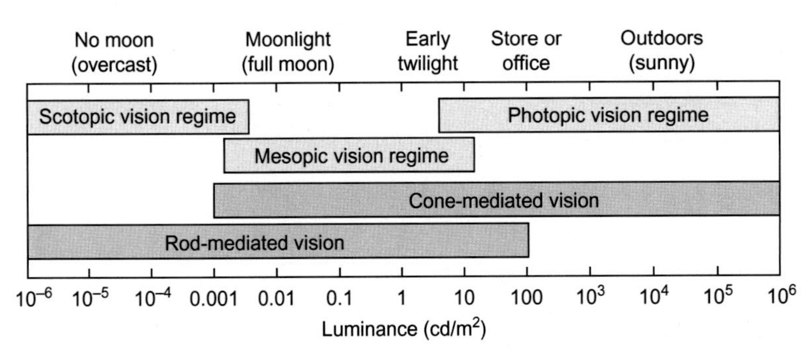

FIGURE 3.48 Scotopic, mesopic, and photopic regimes in absorption of light [162,163].

radiation, such as daylight, moonlight, or steady manmade illumination. The combination of (2) and (3) is called "ambient radiation," and is the steady radiation that falls on the detector [164]. Following three months of monocular deprivation, opening the eye for up to 5 years produces only a very limited recovery in the cortical physiology, and no obvious recovery of the geniculate atrophy, even though behaviorally there is some return of vision in the deprived eye. Closing the normal eye, though necessary for behavioral recovery, has no detectable effect on the cortical physiology. The amount of possible recovery in the striate cortex is probably no greater if the period of eye closure is limited to weeks, but after a five weeks closure there is a definite enhancement of the recovery, even though it is far from complete. The present high values are not surprising, because an important source of variation which can affect frequency of seeing curves has been eliminated. It seems that little of the information absorbed from the external world is lost [165]. A quantitative assessment may be guessed by an extensive application of previous analyses. The distribution of the number of photons in successive periods is a Poisson distribution. Then, the root-mean-square fluctuation N of the number of photons is given by

$$N - (Mb - Mb)2)Al = Mb$$

where the symbol N is intended a physical detector, the detective quantum efficiency Q is defined as the square of the ratio of the signal-to-noise ratio in the output, to the signal-to-noise ratio in the incident radiation. But with the detection process of human vision, the output of the observer is not a signal-to-noise ratio but is rather a probability of detecting a given radiation signal. The output is thus the result of a decision-making process. In that the reference noise is radiation noise (also called photon noise), rather than a Johnson-Nyquist noise, that is the noise involving the noise generated by the motions of charge carriers. Then the shaded area to the right of the vertical line at T is the probability f that the device falsely concludes that a signal is present when it is not Type I error, and the shaded area to the left of the vertical line is the probability $1-P$ that the device concludes a signal is not present when it is Type II error. Thus, P is the probability that the signal is detected when it is present. We call f the false alarm fraction, and p we call the (unnormalized) reliability of detection. In fact, the unnormalized reliability p just defined is not what one would logically define as the reliability. One notes that to achieve an unnormalized reliability equal to the false alarm fraction, the required signal-to-noise ratio k is zero. Accordingly, it is customary in situations of this kind to introduce a renormalized reliability q that is zero when the probability of recognizing a signal is equal to the false alarm fraction. A suitable definition of q is

$$q\_[p\text{-}PA/(\text{-}f)]$$

(see Fig. 3.49).

A graphical relation between the signal-to-noise ratio k, the false alarm fraction f, and the reliability q can be established as Eq. (3.1) [166]. That may be used to obtain the corresponding analytic relatio, the Gauss error function n x = erf(y).

$$x = \text{erf y} \tag{3.1}$$

as the relation that is inverse to Eq. (3.2)

$$y = \text{erf}\, x - (2r) - \text{if}\, \exp(-u2/2)\text{d}u \tag{3.2}$$

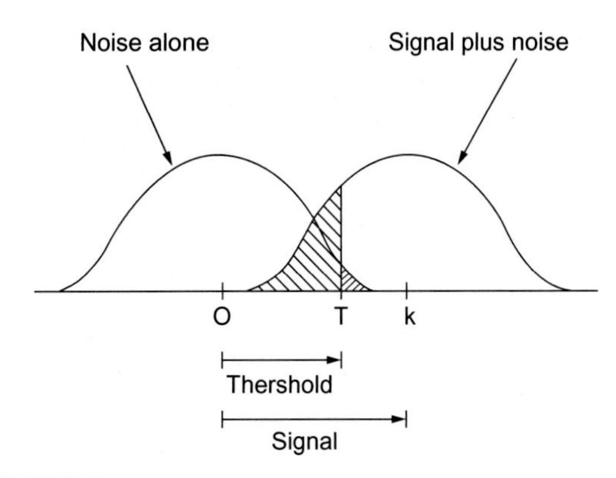

FIGURE 3.49 The Gaussian bell on the left represents the distribution amplitudes when the signal is absent and the similar bell to the right is the distribution of the signal-to-noise amplitudes when the signal is present. If the decision threshold at the position T, the shaded area to the left on the line represents the probability that the device fails to detect the signal that is present, and the shaded area to the right represent the probability that the device judges the signal present when it is not (a false alarm) [163].

$$y = \text{erf}\, x - (2r) - \text{if}\, \exp(-u2/2)\text{d}u \tag{3.3}$$

This issue was confronted again in the post-genetic era [167]. A group of observers (16 young males) viewed an annular stimulus field (i.d. 3, o.d. 11) with an alternation of red primary (690 nm) and green primary (546 nm) light and a 600 nm comparison light (the intensity of which he could vary). On each trial, he had to choose the best answer among the three: mixed light indistinguishable from comparison light, too much red or green added. In parallel, genomic DNA was isolated from peripheral leukocytes and appropriately purified and used to amplify exon 3 of the X-linked visual pigment genes. In the polymerase chain reaction both M and L pigment genes were hybridized, and the reaction of the 5′ primer is specific to the 3′ end of intron 2 and its sequence is 5′ TTCCCCTTTGCTTTGGCTCAAAGC 3′. The 3′ primer is homologous of the 5′ end of intron 3 and is 5′ GACCCTGCCCACTCCATCTTGC 3′

[168]. Furthermore, substitution of Ser^{180} was effected and corresponded to a base change of thymine for guanine at the nucleotide position 1032 in the genes. DNA sequencing on agarose and radiogram allowed to measure the band intensity ratio G^{1032}/T^{1032}. That was taken as an indication of the number of x-linked pigment gene with G^{1032} to the number with T^{1032}, in turn converted to the relative proportion of genes with G at 1032, $G/G+T$. These values were compared with the proportions of red light in the mixture required to match the comparison light in Rayleigh matches. The results showed a high correspondence ($R^2=0.90$) between the individual difference at position 1032 and the Rayleigh matching differences.

The wavelength of light that appears to be yellow is surprisingly consistent across people even though the ratio of middle (M) to long (L) wavelength-sensitive cones is strikingly variable. This observation has been explained by normalization to the mean spectral distribution of our shared environment. On this basis, the possibility of reconciling the nearly perfect alignment of everyone's unique yellow through a normalization process with the striking variability in unique green, which varies by as much as 60 nm between individuals. The spectral location of unique green was measured in a group of volunteers, whose cone ratios were estimated with a technique that combined genetics and flicker photometric electroretinograms. In contrast to unique yellow, unique green was highly dependent upon relative cone numerosity. The difference in neural architecture of the blue–yellow and red–green opponent systems in the presence of a normalization process creates the surprising dependence of unique green on cone ratio. Comparison of the predictions of different theories of color vision processing that incorporate L and M cone ratio and a normalization process. The results of this analysis reveal that—contrary to prevailing notions—postretinal contributions may not be required to explain the phenomena of unique hues.

Most modern theories of color vision circuitry propose that the small bistratified ganglion cell is the retinal basis for blue–yellow color vision. The separation of achromatic and chromatic sensations is a major challenge for the visual system. In the retina, the majority of midget ganglion cells carry both RG chromatic and achromatic spatial information. The prevailing theories of color vision propose that the midget system, with its intermingled chromatic and achromatic signals, performs "double duty," serving both high-resolution spatial vision and chromatic vision. These theories then presume that the two channels are demultiplexed in cortex; although the actual circuitry are not highly defined, it appears that the predictions from standard theories are sufficient to explain the data, while making clear the complexity of the vision phenomenon that has really no connection with a "natural" rendering of colors as we imagine them [169,170]

Color vision depends on the activity of cone photoreceptors that is compared with postreceptoral circuitry. In a recent work, the eye's aberrations were corrected with adaptive optics and retinal position was precisely tracked in real-time to compensate for natural movement. Subjects reported the color appearance of each spot. A majority of L- and M-cones consistently gave rise to the sensation of white, while a smaller group repeatedly elicited hue sensation. As for which hue, while blue sensations were reported they were more likely mediated by M- than L-cones, blue sensations were elicited from M-cones against a short-wavelength light that preferentially elevated the quantal catch in surrounding S-cones, while stimulation of the same cones against a white background elicited green sensations (see Figs. 3.50, and 3.51). In a recent study on two subjects, proximity to S-cones increased the probability of blue reports when M-cones were probed. Thus M-cone increments excited both green and blue opponent

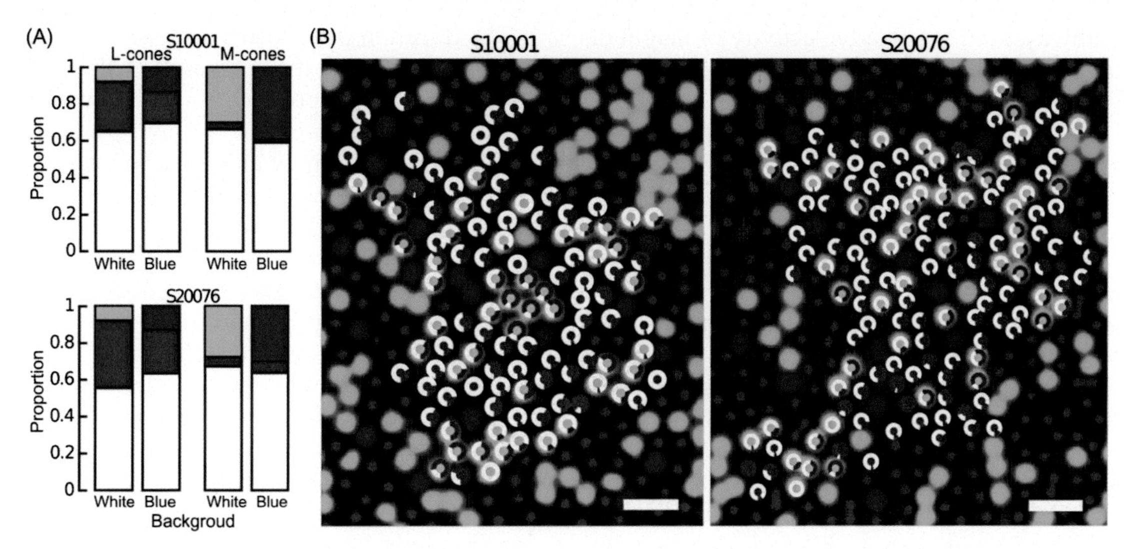

FIGURE 3.50 Color appearance of cone-targeted stimuli on a blue background. Color naming distributions from S10001 to S20076. (A) Proportion of responses after stimulation of L- and M-cones on a white and blue background. (B) The response to stimulation with a 543-nm light against a blue background is represented for each cone by a donut plot. The center of the donut indicates the type of cone targeted (L = red, M = green). Colors correspond to reported percepts. Scale bar = 2.5 arcmin [171].

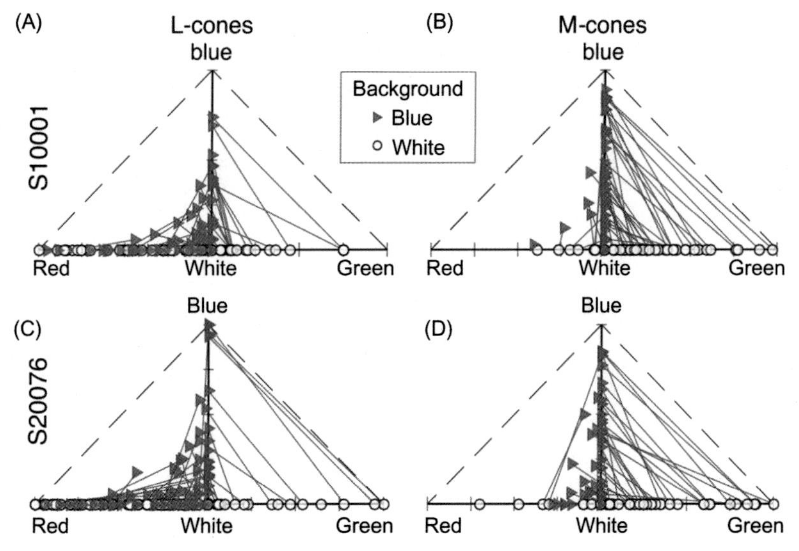

FIGURE 3.51 Opponent color responses from a population of L- and M-cones. Color distributions for each cone were transformed into a red–green$(g - r/T)$ and a blue–yellow $(y - b/T)$dimension, where r, g, b, y are the number of trials that produced red, green, blue, and yellow responses, respectively, and T equals the total number of trials. Each marker represents a cone that was targeted on both a white (circles) and blue (triangles) background. Left column = L-cones, right column = M-cones. Each cone is plotted twice and connected by a solid line to represent its color distribution in both contexts. (A and B) Display the data from S10001 and (C and D) represent S20076. Yellow dimension is not shown because subjects never used yellow in these experiments [171].

pathways, but the relative activity of neighboring cones favored one pathway over the other [169,171].

3.10 Vision impairment

Some idea about how vision works has been introduced in the previous sections, but the topic would require a detailed treatment, out of question in the present case. Obviously, while in general protection of tissues from light is achieved by absorbing a part of it, for example, by tanning, such a method cannot be simply chosen for protecting eyes, the job of which is exactly to monitor and elaborate light signals. Nature has evolved a way to cope with this limitation by filtering away the shortest wavelength part of the radiation (blue light) and involves the generation of polymers from polyenes (typically zeaxanthin) that are in fact deeply yellow.

Such damage may in turn lead to infections [172]. Although it is frequently claimed that the retina is burned by looking at the sun, retinal damage appears to occur primarily due to photochemical injury rather than thermal injury. The temperature rises from looking at the sun with a 3-mm pupil only causes a 4°C increase in temperature, insufficient to photocoagulate. The energy is still phototoxic: since light promotes oxidation, chemical reactions occur in the exposed tissues with unbonded oxygen molecules. It also appears that central serous retinopathy can be a result of a depression in a treated sun damaged eye.

Under these conditions, the absicic acid (ABA) level increases rapidly. Such processes may in turn lead to infections [173]. A damage to the retina is relatively unimportant, although it is frequently claimed that the retina is burned by looking at the sun, retinal damage appears to occur primarily due to photochemical injury rather than thermal injury [174].

The duration of exposure necessary to cause injury varies with the intensity of light, and also affects the possibility and length of recovery.

A polymerization or any other photochemical reaction causing a damage may in turn lead to infections [175,176].

In summary, a certain extent blue light can promote a development of refraction in the human eye and regulate circadian rhythm, but harmful blue light-induced effects on human eyes should not be ignored blue light can also produce different degree of damage to corneal, crystal lens, and retina to take appropriate protective measures when using blue light-related products, especially at night the appearing of a dazzling light has been often taken as a divine manifestation [177], for example, Dante Alighieri depicts the elevation of the poet to the sky in Paradise, with the well-known verses:

> se non che la mia mente fu percossa
> da un fulgore in che sua voglia venne
> (however, My mind was hit by a light that as it pleased came, Paradise)

Another way is opalizing when occurring on the visual axe, this is indicated as cataract, and involves a significant vision impairment [178].

Cataract involves a hardening of the lens and thus it comes "natural" to the physician to proceed surgically and eliminate the no more absorbing part, and indeed cataract surgery is the most frequently quoted surgical intervention in the antiquity. Cataracts were undoubtedly very common as well as diagnosed and operated, with some success in antiquity. The current word cataract, which means both an opacity of the lens and a torrent of water, comes from the Greek word υπόχυσις (kataráktēs) meaning the fall of water. The Latins called it *suffusio*, an extravasation and coagulation of humors behind the iris; and the Arabs, white water [179]and Ebbel translated the original term into cataract. Although other distinguished linguistics interpreted it as a discharge or accumulation of water in the eyes. According to Ebers Papyrus, the old Egyptians tried to treat cases of cataract by eye ointments and magic spells, to some difference with what a contemporary physician would prescribe. However, no solution is possible if the opalized lens remained in their place. In the liquor aqueous humor of the eyes of patients suffering of cataract there is a significant amount of hydrogen peroxide, resulting from an irreversible process that can, however, been prevented by the presence of glutathione, and is accompanied by extensive damage to both DNA and the pump system and loss of the viability of the cell and death of them by necrosis or apoptosis. It has been conclude that when passing to an oxidizing environment the lens undergoes a rapid elongation process, and the fibers formed are unable to replace the normal macromolecules. In babies, a large number of this process occurs, to allow or a uniform refractive index, and is followed by a slower phenomenon that occurs during all of the life. The detoxification of hydrogen peroxide occurs in the presence of metal ions, according to the usual Fenton mechanism.

Most of the chemical products generated are formed in vitro by reaction with hydrogen peroxide and catalysts, via hydroxy radicals. The opacity in which consists cataract may be limited to a part or be extended to the entire organ [180].

This is an irreversible process that can, however, been prevented by the presence of glutathione, and is accompanied by extensive damage to both DNA and the pump system and loss of the viability of the cell and death of the cell by necrosis or apoptosis. This support that when passing to an oxidizing environment the lens these undergo a rapid elongation process, and the fibers formed are unable to replace the normal macromolecules. In babies, a large number of this process occurs, to allow or a uniform refractive index, and is followed by a slower phenomenon that occurs during all of the life.

Most of the chemical products formed are formed in vitro by reaction with hydrogen peroxide and catalysts, via hydroxy radicals.

Whether the sequence of events delineated with the photochemical model applies to other initiating conditions remains to be determined [181].

Drusen, small roundish yellow particles are randomly distributed in the macula likely to form in the perifoveal retina, where the concentration of rod photoreceptors is comparatively high. Nondegraded material is extruded from the RPE cells basal laminar deposit is located between the RPE plasma membrane and the basal lamina. Basal linear deposit accumulates between the RPE and the lower basal lamina collagenous layer of Bruch's membrane. The basal linear deposit is the ultrastructural correlate to soft drusen. The thickening of Bruch's membrane with crosslinking between the collagen fibers and increasing lipid incorporation render a decreased hydraulic conductivity and the membrane becomes more hydrophobic, impeding fluid transport. The main resistance develops in the inner collagenous layer of Bruch's membrane, most likely due to a progressive accumulation of lipid deposits in this layer [182].

These lipids contain phospholipids and neutral fats and show a continuous increase with age. The peroxidized lipids seem to be derived, at least partly, from polyunsaturated fatty acids in the photoreceptor outer segments [180].

There are three primary types of age-related cataracts: nuclear sclerotic, cortical, and posterior subcapsular. As a person ages, any one type, or a combination of any of these three types, can develop over time [183].

In prospective studies of incident cataract, a person may have developed two or all three of cortical, nuclear sclerotic and posterior subcapsular cataract at the time when clinically significant cataract is apparent. One can compare the impact of risk factors on different types of cataract with methods of competing risk survival analysis that account for tied events [184].

Nuclear sclerotic cataract is the most common type of age-related cataract, caused primarily by the hardening and yellowing of the lens over time. "Nuclear" refers to the gradual clouding of the central portion of the lens, called the nucleus; "sclerotic" refers to the hardening, or sclerosis, of the lens nucleus. As this type of cataract progresses, it changes the eye's ability to focus, and close-up vision (for reading or other types of close work) may temporarily improve. This symptom is referred to as "second sight," but the vision improvement it produces is not permanent [185].

A second type is called cortical cataracts, where "cortical" refers to white opacities, or cloudy areas, that develop in the lens cortex, which is the peripheral (outside) edge of the lens. Changes in the water content of the lens fibers create clefts, or fissures, that look like the spokes of a wheel pointing from the outside edge of the lens in toward the center. These fissures can cause the light that enters the eye to scatter, creating problems with blurred vision, glare, contrast, and depth perception [186]. People with diabetes are at risk for developing cortical cataracts. The third group involves subcapsular cataract, but here treatment with vitamin D is often successful [187].

References

[1] R. Jung, Sensory research in historical perspective: some philosophical foundations of perception, in: I. Darian-Smith (Ed.), Handbook of Physiology, American Physiological Society, Bethesda, MD, 1984, pp. 1–74.

[2] M. Kalloniatis, C. Luu, The Organization of the Retina and Visual System, University of Yuta, Salt Lake City, 2013.

[3] J.L. Benson, The Inner Nature of Color: Studies on the Philosophy of the Four Elements, Steiner Books, MA, 2004.

[4] H.P. Howard, B.J. Rogers, Binocular Vision and Stereopsis, Oxford Press, Oxford, 1993. 1995.

[5] J.L. Benson, Greek Color Theory and the Four Elements, University of Massachusetts Amherst, 2000. Available from: <https://scholarworks.umass.edu/art_jbgc/1>.

[6] G. Sidoli, G. van Brummelen (Eds.), From Alexandria through Bagdad, Survey and Studies in Ancient Greek and Medieval Islamic Mathematical Sciences, in Honor of J.L. Berggen, Springer, Heidelberg, 2000.

[7] R.A. Crone, A History of Color: The Evolution of Theories of Light and Color, Dordrecht, Kluver Academic Publisher, Springer Science & Business Media, NY, 2012.

[8] F. Aguilonius, Opticorum Libri Sex, Ex officina Plantiniana, Antwerpen, 1613.

[9] D.C. Lindberg, Theories of Vision from Al-Kindi to Kepler, The University of Chicago Press, Chicago/London, 1996.

[10] P. Artal, Image formation in the living human eye, Ann. Rev. Vis. Sci. 2015 (2015) 1–17.

[11] G. Deutscher, Through the Language Glass, Why the World Looks Different in Other Languages, Metropolitan Books. Henry Holt and Co, NY, 2011.

[12] P. Kay, L. Maffi, Color appearance and the emergence and evolution of basic color lexicons, Am. Anthropol 101 (1999) 743–760.

[13] A. Byrne, D.R. Hilbert, Color relationalism and relativism, Top. Cognit. Sci. 2016 (2016) 1–21.

[14] O. Beran, Basic color categories' in the language-game perspective, Organon F 19 (4) (2012) 423–443.

[15] M. Schaller, A. Norenz yan, J.H. Steven, T. Yamagishi, T. Kameda (Eds.), Evolution, Culture, and the Human Mind, Psycol Press, Taylor & Francis Group, New York, 2010.

[16] A.F. Franklin, I.R.L. Davies, New evidence for infant categories, Br. J. Dev. Psychol. 22 (2004) 349–377.

[17] M.H. Bornstein, M.E. Arterberry, The development of object categorization in young children hierarchical inclusiveness, age, perceptual attribute, and group versus individual analyses, Dev. Psychol. 46 (2010) 350–365.

[18] M. Bornstein, W. Kessen, S. Weiskopf, The categories of hue in infancy, Science 1971 (4223) (1976) 201–202.

[19] S.C. Levinson, Yeli Dnye and the theory of basic terms, J. Linguisti. Anthropol 10 (2000) 3–55.

[20] B. Berlin, Ethnobiological Classification: Principles of Categorization of Plants and Animals in Traditional Societies, Princeton University Press, Princeton, NJ, 1992.

[21] A. Wierzbicka, The semantics of color: a new paradigm, in: N.J. Pitchford, C.P. Biggam (Eds.), Progress

in Colour Studies: Volume I. Language and Culture, John Benjamins Publishing Company, Philadelphia, 2006, pp. 1–24.

[22] D.M. Powerd, Applications and explanation of Zipf's law, Assoc. Comput. Linguist. (1998) 151–160.

[23] I. Galili, M. Azan, The influence of an historically oriented course on students' content knowledge in optics evaluated by means of facets-schemes analysis, Phys. Educ. Res., Am. J. Phys. 68-71 (Suppl.) (2000) S3–S14.

[24] R. Ferreri Cancho, R.V. Solé, Least effort and the origins of scaling in human language, PNAS 100 (2003) 788–791.

[25] A.H. Munsell, Munsell Book of Colors, ed. Pantone, Carlstadt, NJ, 2015.

[26] J.C. Maxwell, Addition to a paper by Francis Deas on spectra formed by the passage of polarized light through double refracting crystals', Trans. R. Soc. Edinb. XXVI (1872) 185–188 [Refers to: F. Deas, On spectra formed by the passage of polarised light through double refracting crystals), *Trans. Roy. Soc. Edinb.*, XXVI (1872) 177-185].

[27] C.W.F. Everitt, James Clerk Maxwell: Physicist and Natural Philosopher, Scribner, New York, 1976.

[28] T. Birch, The History of the Royal Society of London for Improving of Natural Knowledge from its First Rise in which the Most Considerable of Those Papers Communicated to the Society, Which Have Hitherto Not been Published, are Inserted as a Supplement to the Philosophical Transaction, vol. 3, Millar, London, 1757.

[29] I. Newton, Opticks: Or a Treatise of the Reflexions, Refractions, Inflexions and Colours of Light Questions 13, 14, 16, 17, Smith and Wallford, London, 1704.

[30] T. Young, The Bakerian lecture. on the theory of light and colors, Philos. Trans. R. Soc. Lond., 92 (1802) 121-148.

[31] J.S. Dalton, Extraordinary facts relating to the vision of colors, Mem. Lit. Philos. Soc. Manch. 5 (Pt. 1) (1798) 28–45.

[32] C. Wheatstone, Contributions to the physiology of vision.—Part the first. On some remarkable, and hitherto unobserved, phenomena of binocular vision, Philos. Trans. R. Soc. 128 (1838) 371–394.

[33] K.J. Gaston, J. Bennie, T.W. Davies, J. Hopkins, The ecological impacts of nighttime light pollution: a mechanistic appraisal, Biol. Rev. 88 (2013) 912–927.

[34] R.W. Thimijan, D. Royal, Photometric, radiometric, and quantum light units of measure: a review of procedures for interconversion, Hort. Sci. 18 (1982) 818–822.

[35] M.D. Feit, A. Fleck jr, A. Steiger, Solution of the Schrödinger equation by a spectral method, J. Comput. Phys. 47 (1982) 412–433.

[36] B.B. Lee, The evolution of concepts of color vision, Neurociencias 4 (2008) 209–224.

[37] H.F. von Helmoltz, Handbuch der physiologischen Optick, Leopold Voss, Leipzig, 1867.

[38] J.W. Goethe, Zur Farbenlehre, Cotta, Tübingen, 1810.

[39] J.G. Krüger, F. Hoffmann, Naturlehre, Hemmerde, Halle, 1740.

[40] D.S. Schier, Louis Bertrand Castel, anti-Newtonian scientist, The Torch Press, Cedar Rapids, Iowa, 1941.

[41] L.B. Castel, L.Optique des couleurs: fondée sur les simples observations & tournée sur-tout à la pratique de la peinture, de la teinture, & des autres arts coloristes, Briasson, Paris, 1740.

[42] D.L. Sepper, Goethe contra Newton: Polemics and the Project of a New Science of Colors, Cambridge University Press, Cambridge, 1988.

[43] Z. Vender, Goethe, Wittengstein and the essence of color, Monist. 78 (1995) (1995) 391–410.

[44] M.W. Rowe, Goethe and Wittengstein, Philosophy 66 (257) (1991) 283–303.

[45] J.W. von Goethe, Schiller, in: W.L. Döring (Ed.), Temperamentenrose, Bekennitse, Die Königin der Blumen, 1798.

[46] J.W. von Goethe, Allegorischer, symbolischer, mystischer Gebrauch der Farbe, in: E. Beutler (Ed.), Naturwissenschaftlighe Werke, Sechste Abteilung: Sinnlich-sittliche Wirkung der Farbe, Artemis, Zurich, 1949.

[47] R. Hirsch, Exploring Photography, from Films to Pixel, sixth ed., The Focal Press, NY, 2015.

[48] I. Fatt, B.A. Weissman, Physiology of the Eye: An Introduction to the Vegetative Functions, Butterworth-Heinemann, Boston, 1992.

[49] G.D. Hildebrand, A.F. Fielder, Anatomy and physiology of the retina, in: J. Reynolds, S. Olitsky (Eds.), Pediatric Retina, Springer-Verlag, Berlin Heidelberg, 2011, pp. 39–65.

[50] R.H. Masland, The neuronal organization of the retina, Neuron 76 (2012) 1791–1800.

[51] H. Kolb, Simple anatomy of the retina, in: H. Kolb, E. Fernandez, R. Nelson (Eds.), Webvision (2005): The Organization of the Retina and Visual System [Internet], University of Utah Health Sciences Center, Salt Lake City (UT), 1995. <https://www.ncbi.nlm.nih.gov/books/NBK11518/>.

[52] D. Purves, G.J. Augustine, D. Fitzpatrick, et al. (Eds.), Cones and vision, in: Dale Purves, George J Augustine, David Fitzpatrick, Lawrence C Katz, Anthony-Samuel LaMantia, James O McNamara, and S Mark Williams (Eds.), Neuroscience, second edition, Sinauer Associates, Sunderland (MA), 2001.

[53] <http://thebrain.mcgill.ca/flash/d/d_02/d_02_m/d_02_m_vis/d_02_m_vis.html>

[54] F. Boll, Zur Anatomie und Physiologie der Retina, Monatsberichteder Koniglichen Preussischen Akademieder Wissenschaften zu Berl, Berlin, pp. 783–787, 1876.
[55] K. Palczewski, T. Kusamaka, T. Hori, C.A. Behnke, H. Motoshima, B.A. Fox, et al., Crystal structure of rhodopsin: a G protein-coupled receptor, Science 289 (5480) (2000) 739–745.
[56] P.A. Hargrave, J.H. McDowell, Rhodopsin and phototransduction: a model system for G protein-linked receptors, FASEB J. 1992 (1992) 2323.
[57] P.A. Hargrave, Rhodopsin structure, function, and topography. The Friedenwald lecture, Invest. Ophthalmol. Vis. Sci. 42 (2001) 3–9.
[58] A. Wiechmann, D. Sherry, Role of melatonin and its receptors in the vertebrate retina, Int. Rev. Cell Mol. Biol. 300C (2013) 211–242.
[59] R.H. Kramer, C.M. Davenport, Lateral inhibitionin the vertebrate retina: the case of the missing neurotransmitter, Plos Biol. 13 (2015) e1002322.
[60] R.H. Masland, The neuronal organization of the retina, Retina Neuron 76 (2012) 266–280.
[61] J.A. Liu, The anatomy and physiology of direction-selective retinal ganglion cells. In Webvision: The Organization of the Retina and Visual System, 2015.
[62] H. Kolb, E. Fernandez, R. Nelson (Eds.), Salt Lake City (UT): University of Utah Health Sciences Center, 1995. <https://www.ncbi.nlm.nih.gov/books/NBK321299/>.
[63] H. Kolb, Glial cells of the retina, in: H. Kolb, E. Fernandez, R. Nelson (Eds.), Webvision (1995): The Organization of the Retina and Visual System [Internet], University of Utah Health Sciences Center, Salt Lake City (UT), 2007. <https://www.ncbi.nlm.nih.gov/books/NBK11516/>.
[64] A. Reichenbach, S.R. Robinson, The involvement of Müller cells in the outer retina, in: M.B.A. Djamgoz, S.N. Archer, S. Vallerga (Eds.), Neurobiology and Clinical Aspects of the Outer Retina, Chapman & Hall, London, 1995, pp. 395–416.
[65] R.P. Kraig, M. Chesler, Astrocytic acidosis in hyperglycemic and complete ischemia, J. Cereb. Blood Flow Metab. 10 (1990) 104–114.
[66] A. Trivino, J.M. Ramirez, J.J. Salazar, A.I. Ramirez, J. Garcia-Sanchez, Immunohistochemical study of human optic nerve head astroglia, Vis. Res. 36 (1996) 2015–2028.
[67] E. Strettoi, E. Novelli, F. Mazzoni, I. Barone, D. Damiani, Complexity of retinal cone bipolar cells, Prog. Retin. Eye Res. 29 (2010) 272–283.
[68] H. Wässle, B.B. Boycott, Functional architecture of the mammalian retina, Physiol. Rev. 7 (1991) 447–480.
[69] H. Wässle, C. Puller, F. Müller, S. Haverkamp, Cone contacts, mosaics, and territories of bipolar cells in the mouse retina, J. Neurosci. 29 (2009) 106–117.
[70] R.W. Young, The daily rhythm of shedding and degradation of cone outer segment membranes in the lizard retina, J. Ultrastruct. Res. 61 (1977) 172–185.
[71] M.M. Jablonski, J. Tombran-Tink, D.A. Mrazek, A. Iannaccone, Pigment epithelium-derived factor supports normal development of photoreceptor neurons and opsin expression after retinal pigment epithelium removal, J. Neurosci. 20 (2000) 7149–7157.
[72] P.I. Song, J.I. Matsui, J.E. Dowling, Morphological types and connectivity of horizontal cells found in the adult zebrafish (*Danio rerio*) retina, J. Comp. Neurol. 506 (2008) 328–338.
[73] D. Jiang, G. Xiong, H. Feng, Z. Zhang, P. Chen, B. Yan, et al., Research paper donation of mitochondria by iPSC-derived mesenchymal stem cells protects retinal ganglion cells against mitochondrial complex I defect-induced degeneration, Theranostics 9 (2019) 2395–2410.
[74] Y. Zhang, Y. Yang, C. Trujillo, W. Zhong, Y.F. Leung, The expression of *irx7* in the inner nuclear layer of zebrafish retina is essential for a proper retinal development and lamination, PLoS One 7 (2012) e36145.
[75] B. Mcguire, J.K. Stevens, P. Sterling, Microcircuitry of bipolar cells in cat retina, J. Neurosci. 4 (1985) 2920–2938.
[76] K.A. Martemyanov, A.P. Sampath, The transduction cascade in retinal ON bipolar cells: signal processing and disease, Annu. Rev. Vis. Sci. 3 (2017) 25–51.
[77] J. Snellman, T. Kaur, Y. Shem, S. Nawy, Regulation of ON bipolar cell activity, Prog. Retin. Eye Res. 27 (2008) 450–463.
[78] V.C. Pailun, A.C. Schütz, M.M. Michel, W.S. Geisler, K.R. Gegenfurtner, Visual search under scotopic lighting conditions, Vis. Res. 113 (pt, B) (2015) 155–168.
[79] J.C.A. Read, The place of human psychophysics in modern neuroscience, Neuroscience 296 (2015) 116–127.
[80] M. Kalloniatis, C. Luu, The perception of space, in: H. Kolb, E. Fernandez, R. Nelson (Eds.), Webvision (1995): The Organization of the Retina and Visual System [Internet], University of Utah Health Sciences Center, Salt Lake City (UT), 2005. <https://www.ncbi.nlm.nih.gov/books/NBK11545/>.
[81] M. Kalloniatis, C. Luu, The perception of depth, in: H. Kolb, E. Fernandez, R. Nelson (Eds.), Webvision: The Organization of the Retina and Visual System [Internet], University of Utah Health Sciences Center, Salt Lake City (UT), 2007. 1995 <https://www.ncbi.nlm.nih.gov/books/NBK11512/>.

[82] D. Purves, G.J. Augustine, D. Fitzpatrick, W. Hall, A. S. Lamanthia, J.O. Mcnamara, M.S. Williams (Eds.), Neuroscience, second ed., Sinauer Associates, Sunderland (MA), 2001.

[83] G.B. Awatramani, M.M. Slaughter, Origin of transient and sustained responses in ganglion cells of the retina, J. Neurosci. 20 (18) (2000) 7087–7095.

[84] S.M. Wu, F. Gao, B.R. Maple, Integration and segregation of visual signals by bipolar cells in the tiger salamander retina, Prog. Brain Res. 131 (2001) 125–143.

[85] Y. Cao, J. Pahlberg, I. Sarria, N. Kamasawa, A.P. Sampath, K.A. Martemyanov, Regulators of G protein signaling RGS7 and RGS11 determine the onset of the light response in ON bipolar neurons, Proc. Natl. Acad. Sci. USA. 109 (2012) 7905–7910.

[86] D. Engel, P. Jonas, Presynaptic action potential amplification by voltage-gated Na^+ channels in hippocampal mossy fiber boutons, Neuron 45 (2005) 405–417.

[87] H. Kandori, Y. Shichida, T. Yoshizawa, Photoisomerization in rhodopsin, Biochem. (Mosc.) 66 (2001) 1197–1209.

[88] S.O. Smith, Structure and activation of the visual pigment rhodopsin, Annu. Rev. Biophys. 39 (2010) 309–328.

[89] M. Murakami, T. Kouyama, Crystallographic study of the LUMI intermediate of squid rhodopsin, PLoS One 10 (2015) e0126970/1–e0126970/12.

[90] J.H. Park, P. Scheerer, K.P. Hofmann, H.W. Choe, O. P. Ernst, Crystal structure of the ligand-free G-protein-coupled receptor opsin, Nature 454 (2008) 183–188.

[91] R. Vogel, M. Mahalingam, S. Lüdeke, T. Huber, F. Siebert, T.P. Sakmar, Functional role of the "ionic lock"—An interhelical hydrogen-bond network in family. A heptahelical receptors, J. Mol. Biol. 380 (2008) 648–655.

[92] M. Han, V.V. Gurevich, S.A. Vishnivetskiy, P.B. Sigler, C. Schubert, Crystal structure of β-Arrestin at 1.9 Å: possible mechanism of receptor binding and membrane translocation, Structure 9 (2001) 869–880.

[93] Arrestins - Pharmacology and therapeutic potential, in: V.V. Gurevich (Ed.), Handbook of Experimental Pharmacology, 219, Springer, Berlin, 2014.

[94] Y. Xue, S. Sato, D. Razafsky, B. Sahu, S.Q. Shen, C. Potter, et al., The role of retinol dehydrogenase 10 in the cone visual cycle, Sci. Rep. 7 (2017) 2390.

[95] C. Grimm, C.E. Remé, P.O. Rol, T.P. Williams, Blue light's effects on rhodopsin: photoreversal of bleaching in living rat eyes, Invest. Ophthalmol. Vis. Sci. 41 (2000) 3984–3990.

[96] T.P. Williams, B.N. Baker, The Effects of Constant Light on Visual Processes, Springer-Verlag, Berlin, 1980.

[97] M.E. Burns, Deactivation mechanisms of rod phototransduction: the Cogan lecture, Invest. Ophthalmol. Vis. Sci. 51 (3) (2010) 1283–1288.

[98] C.M. Krispel, D. Chen, N. Melling, Y.J. Chen, K.A. Martemyanov, N. Quillinan, et al., RGS expression rate-limits recovery of rod photoresponses, Neuron. 51 (2006) 409–416.

[99] M.E. Burns, E.N. Pugh, RGS9 concentration matters in rod phototransduction, Biophys. J. 97 (2009) 1538–1547.

[100] J.K. Angleson, T.G. Wensel, A GTPase-accelerating factor for transducin, distinct from its effector cGMP phosphodiesterase, in rod outer segment membranes, Neuron. 11 (1993) 939–949.

[101] E.N. Pugh Jr., Variability in single photon responses: a cut in the Gordian knot of rod phototransduction? Neuron 23 (1999) 205–208.

[102] B. Rayer, M. Naynert, H. Stieve, Phototransduction: different mechanisms in vertebrates and invertebrates, J. Photochem. Photobiol. B. 7 (1990) 107–148.

[103] F. Boll, Zur Physologie und Phisiologie der Retina, Ark. f Physiol (1877). Abt 4-35.

[104] W.F. Kühne, Chemische Vorgänge in der Netxhaut, handbuch der Physiologie. Dritte Part, Erstes Theil. Physiologie der Sinnenorgan, Erstes Theil. Sinneorgan, FCW Vogel ed, Leipzig, 1888.

[105] G. Bassolino, T. Sovdat, Soares, S.A. Duarte, J.M. Lim, C. Schnedermann, et al., Barrierless photoisomerization of 11-cis retinal protonated Schiff base in solution, J. Am. Chem. Soc. 137 (2015) 12434–124347.

[106] M. Kono, R. Crouch, In vitro assays of rod and cone opsin activity: retinoid analogs as agonists and inverse agonist, Methods Mol. Biol. 2010 (652) (2010) 85–94.

[107] M. Eilers, J.A. Goncalves, S. Ahuja, C. Kirkup, A. Hirshfeld, C. Simmerling, Structural transitions of transmembrane helix 6 in the formation of metarhodopsin I, J. Phys. Chem. B 116 (2012) 10477–10489.

[108] J.E. Kim, M.J. Tauber, R.A. Mathies, Analysis of the mode-specific excited-state energy distribution and wavelength-dependent photoreaction quantum yield in rhodopsin, Biophys. J. 84 (2003) 2492–2501.

[109] P.S.H. Park, D.T. Lodowski, K. Palczewski, Activation of G protein–coupled receptors: beyond two-state models and tertiary conformational changes, Annu. Rev. Pharmacol. Toxicol. 48 (2008) 107–141.

[110] N.J. Coughlan, B.D. Adamson, L. Gamon, K. Catani, E.J. Bieske, Retinal shows its true colours: photoisomerization action spectra of mobility-selected isomers of the retinal protonated Schiff base, Phys. Chem. Chem. Phys. 17 (35) (2015) 22623–22631.

[111] P. Kukura, D.W. McCamant, S. Yoon, R.A. Wandschneider, Structural observation of the

primary isomerization in Vision with femtosecond-stimulated Raman, Science 310 (2005) 1006–1009.
[112] A.J. Zele, D. Cao, Vision under mesopic and scotopic illumination, Front. Psychol. 5 (1594) (2015) 1–15.
[113] C.R. McCudden, M.D. Hains, R.J. Kimple, D.P. Siderovski, F.S. Willard, G-protein signaling: back to the future, Cell Mol. Life Sci. 62 (2005) 551–577.
[114] Visual Phototransduction <https://en.wikipedia.org/wiki/Visual_phototransduction>.
[115] I.B. Leskov, V.A. Klenchin, J.V. Handy, G.G. Whitlock, V. Govardovskii, B.D. Bownds, et al., The gain of rod phototransduction: reconciliation of biochemical and electrophysiological measurements, Neuron 27 (2000) 525–537.
[116] D. Larhammar, K. Nordström, T.A. Larsson, Evolution of vertebrate rod and cone phototransduction genes, Phil. Trans. R. Soc. B 364 (2009) 2867–2880.
[117] D. Takeshita, L. Smeds, P. Ala-Laurila, Processing of single-photon responses in the mammalian On and Off retinal pathways at the sensitivity limit of vision, Phil. Trans. R. Soc. B 372 (2017) 20160073.
[118] D.M. Dacey, B.B. Lee, The 'blue-on' opponent pathway in primate retina originates from a distinct bistratified ganglion cell type, Nature 367 (1994) 731–735.
[119] S. Johnsen, A. Kelber, E. Warrant, A.M. Sweeney, E. Widder, R.L. Lee, et al., The color of night: twilight and nocturnal illumination and its effects on color perception, Integr. Comp. Biol. 1021 (2005).
[120] H. Koshitaka, M. Kinoshita, M. Vorobyev, K. Arikawa, Tetrachromacy in a that has eight varieties of spectral receptors, Proc. Biol. Sci. 275 (1637) (2008) 947–954.
[121] A. Kelber, L.S.V. Roth, Nocturnal colour vision - not as rare as we might think, J. Exp. Biol. 209 (2006) 781–788.
[122] A. Kelber, D. Osorio, From spectral information to animal color vision: experiments and concepts, Proc. R. Soc. Lond. B. 277 (2010) 1617–1625.
[123] B.C. Verrelli, S.A. Tishkoff, Signatures of selection and gene conversion associated with human vision variation, Am. J. Hum. Genet. 75 (2004) 363–375.
[124] D. Corney, J. Dylan-Haynes, G. Rees, R.B. Lotto, The brightness of color, PLoS One 4 (2009) e5091.
[125] H.H. Thoen, M.G. How, T.H. Chiou, J. Marshall, A different form of color vision in mantis shrimp, Science 343 (169) (2014) 411–413.
[126] J. Marshall, K. Arikawa, Unconventional colour vision, Curr. Biol. 24 (2014) R1150–R1154.
[127] C.J. Kennedy, P.E. Rakoczy, I.J. Constable, Lipofuscin of the retinal pigment epithelium: a review, Eye (Lond) 9 (1995) 763–771.
[128] O. Hisatomi, F. Tokunaga, Molecular evolution of proteins involved in vertebrate phototransduction, Comp. Biochem. Physiol. B 133 (2002) 509–522.
[129] S.P. Collin, A.E.O. Trezise, The origins of color vision in vertebrates, Clin. Exptl. Optom. 87 (2004) 217–233.
[130] J. Nathan, T.P. Piantanida, T.L. Eddy, T.B. Shows, D. Hogness, Molecular genetics of inherited variation in human color vision, Science 232 (1986) 203–210.
[131] A. Kelber, C. Yovanovich, P. Olsson, Threshold and limitations of color vision in dim light, Philos. Trans. R. Soc. Lond. B Biol. Sci. 372 (2017) 20160065.
[132] J. Nathans, The evolution and physiology of human vision: insights from molecular genetic studies of visual pigments, Neuron 24 (1999) 299–312.
[133] J.K. Bowmaker, Evolution of vertebrate visual pigments, Vis. Res. 48 (2008) 2022–2041.
[134] S. Hecht, S. Shlaer, M.H. Pirenne, Energy, quanta and vision, J. Gen. Physiol. 25 (1942) 819–840.
[135] <http://math.ucr.edu/home/baez/physics/Quantum/see_a_photon.html>.
[136] A. Morshedian, G.L. Fain, The evolution of rod photoreceptors, Phil. Trans. R. Soc. B 372 (2017) 20160074.
[137] V.V. Maximov, Environmental factors which may have led to the appearance of color vision, Philos. Trans. R. Soc. Lond. B 355 (1401) (2000) 1239–1242.
[138] D. Osorio, M. Vorobyev, Color vision as an adaptation to frugivory in primates, Proc. R. Soc. Lond. B 263 (1996) 593–599.
[139] E.P. Hornstein, J. Verweij, J.L. Schnapf, Electrical coupling between red and green cones in primate retina, Nat. Neurosci. 7 (2004) 745–750.
[140] G. Palczewska, F. Vinberg, P. Stremplewski, M.P. Bircher, D. Salom, K. Komar, E. Palczewski, Human infrared vision is triggered by two-photon chromophore isomerization, Proc Natl Acad Sci 111 (2014) E5445–E5545.
[141] <http://faculty.pasadena.edu/dkwon/chap10_C/chap%2010%20part%20C_files/textmostly/slide19.html>.
[142] H. Kolb, S-Cone Pathways, Webvision: The Organization of the Retina and Visual System [web], 2005.
[143] E. Zhou, M. Karsten, E. Xu, Structure and activation of rhodopsin, Acta Pharmacol. Sin. 33 (2012) 291–299.
[144] Y. Schichida, T. Matsuyama, Evolution of opsins and phototransduction, Philos. Trans. R. Soc. Lond. B Biol. Sci. 364 (2009) 2881–2895.
[145] R.A. Sturm, M. Larsson, Genetics of human iris color and patterns, Pigment Cell Melanoma Res. 2009 (2009) 544–562.

[146] W.F. Kühne, The Photochemistry of Vision and Retinal Purple, MacMillan, London, 1879.
[147] E.A. Lesnik, S.M. Freier, Relative thermodynamic stability of DNA, RNA, and DNA:RNA hybrid duplexes: Relationship with base composition and structure, Biochemistry 34 (1995) 10807–10815.
[148] S. Gunaseelan, A. Balupillai, K. Govindasamy, G. Muthusamy, K. Ramasamy, Shanmugam, et al., The preventive effect of linalol on acute and chronic UV-B-mediated skin carcinogenesis in Swiss albino mice, Photochem. Photobiol. Sci. 15 (2017) 851–860.
[149] C. Gstrein, P. Walde, A.D. Schlüter, T. Nauser, Shielding effects in spacious macromolecules: a case study with dendronized polymers, Photochem. Photobiol. Sci. 15 (2016) 964–968.
[150] T.K. Wong, C.W. Cheng, Z.J. Hsieh, Y.J. Liang, Blue light induced free radicals from riboflavin on *E. coli* DNA damage, J. Photochem. Photobiol. B Biol. 119 (2013) 60–64.
[151] J.Y. Liang, J.M. Yuann, C.W. Cheng, H.L. Jian, C.C. Lin, L.Y. Chen, Blue light induced free radicals from riboflavin on *E. coli* DNA damage, J. Photochem. Photobiol. B. 119 (2013) 60–64.
[152] S. Chowdhari, N. Saini, Gene expression profiling reveals the role of RIG1 like receptor signaling in p53 dependent apoptosis induced by PUVA in keratinocytes, Cell Signal. 28 (2016) (2016) 25–33.
[153] H. Lusic, D.D. Young, M.O. Lively, A. Deiters, Photochemical DNA activation, Org. Lett. 9 (2007) 1903–1907.
[154] J. Joshy, G.B. Schuster, Long-distance radical cation hopping in DNA: the Effect of thymine – Hg(II) – thymine base pairs, Org. Lett 9 (2007) 1843–1846.
[155] D. Zhang, L.X. Zhang, Y.W. Hao, S.X. Zhou, Z. Tan, K.W. Zheng, Dissecting the strand folding orientation and formation of G-quadruplexes in single- and double-stranded nucleic acids by ligand-induced photocleavage footprinting, J. Am. Chem. Soc. 133 (2011) 1475–1483.
[156] L. Bielory, Allergic and imunumologic disorder of the eye. Part II. Ocular allergy, J. Allergy Clin. Immunol 106 (2000) 1019–1032.
[157] C. Nishisgori, Current concept of photocarcinogenesis, Photochem. Photobiol. Sci 14 (2015) 1713.
[158] R.C. Jones, Quantum efficiency of human vision, J. Opt. Soc. Am. 49 (1959) 645–653.
[159] P.W. Kämmerer, T. Rahimi-nedjat, T. Ziebart, C. Bemsch, B. Walter, B. Al-nawas, et al., A chemiluminescent light system in combination with toluidine blue to assess suspicious oral lesions—clinical evaluation and review of the literature, Clin. Oral Invest. (2014). Available from: https://doi.org/10.1007/s00784-014-1252-z.
[160] M. Schiffman, P.E. Castle, J. Jeronimo, A.C. Rodriguez, S. Wacholder, Human papillomavirus and cervical cancer, Lancet 370 (9590) (2007) 890–907.
[161] M. Alpern, W.A. Rushton, S. Torii, The attenuation of rod signals by bleachings, J. Physiol. 207 (2) (1970) 449–461.
[162] M. Alpern, W.A.H. Rushton, S. Torii, The size of rod signals, J. Physiol. 206 (1970) 193–208.
[163] E.F. Schubert, Light-emitting Diodes, Cambridge University Press, Cambridge UK, 2003, pp. 276, Fig 16.2.
[164] IAEA Safety Standards Series, Assessment of Occupational Exposure Due to External Sources of Radiation, IAEA, Vienna, pp. 1-98, 1999.
[165] D.H. Hubel, T.N. Wiesel, The period of susceptibility to the physiological effects of unilateral eye closure in kittens, J. Physiol. 206 (1970) 419–436.
[166] R.C. Jones, Human vision, Adv. Electron. Electron. Phys. IX (1959) 154–170.
[167] D. Cao, J. Pokorny, V.C. Smith, Matching rod percepts with cone stimuli, Vis. Res. 45 (16) (2005) 2119–2128.
[168] J. Neitz, M. Neitz, C.H. Jacobs, More than three different cone pigments among people with normal color vision, Vis. Res. 33 (1993) 117–122.
[169] B.P. Schmidt, P. Touch, M. Neitz, J. Neitz, Circuitry to explain how the relative number of L and M cones shapes color experience, J. Vis. 16 (2016) 18.
[170] R. Shapley, M.J. Hawken, Color in the cortex: single- and double-opponent cells, Vis. Res. 51 (2011) 701–717.
[171] B.P. Schmidt, R. Sabestian, W.S. Tuten, I. Neitz, S. Roorda, Sensations from a single M-cone depend on the activity of surrounding S-cones, Sci. Rep. 8 (2018) 8561.
[172] Q.X. Zheng, Y.P. Ren, P.S. Reinach, B. Xiao, H.H. Lu, Y.R. Zhu, et al., Reactive oxygen species activated NLRP inflammasomes initiate inflammation in hyperosmolarity stressed human corneal epithelial cells and environment-induced dry eye patients, Exp. Eye Res. 134 (2015) 133–140.
[173] E. Profoyeva, A. Wegener, E. Zrenner, Cataract prevalence and prevention in Europe: a literature review, Acta Ophthalmol. 91 (5) (2013) 395–405.
[174] I.I. Geneva, Photobiomodulation for the treatment of retinal diseases: a review, Int. J. Ophthalmol. 9 (2016) 145–152.
[175] R. Kline, B. Kline, K.L.P. Linton, Prevalence of age-related maculopathy: the beaver dam eye study, Ophthalmol 99 (6) (1992) 933–943.
[176] M.C. Callegan, M. Engelbert, D.W. Parke, B.D. Jett, M.S. Gilmore, Bacterial endophthalmitis: epidemiology, therapeutics, and bacterium-host interactions, Clin. Microbiol. Rev. 15 (2002) 111–124.

[177] N. Drake, Our nights are getting brighter, and Earth is paying the price, Natl. Geographic (2019). April 3.
[178] P.G. Montan, G. Koranyi, H.E. Settequist, A. Stridh, B.T. Philipson, K. Wiklund, Endophthalmitis after cataract surgery: risk factors relating to technique and events of the operation and patient history: a retrospective case-control study, Ophthalmology 105 (12) (1998) 2171–2177.
[179] F.J. Ascaso, J.A. Cristobal, The oldest cataract in the Nile valley, J. Cataract Refract. Surg. 27 (2001) 1714–1715.
[180] P.V. Algvere, J. Marshall, S. Seregard, Age-related maculopathy and, blue light hazard, Acta Ophthalmol. Scand. 84 (2006) 4–15.
[181] C.A. Curcio, C.L. Millican, Basal linear deposit and large drusen are specific for early age-related maculopathy, Arch. Ophthalmol. 117 (1999) 329–339.
[182] F.G. Holz, D. Pauleikhoff, R.F. Spaide, A.C. Bird, Age-related Macular Degeneration, Springer Verlag, Berlin, 2004.
[183] R.J. Gynn, B. Rosner, W.G. Christen, Evaluation risk factors cataract types a competing risks framework, Ophthalmic. Epidemiol. 16 (2) (2009) 98–106.
[184] W. Schakel, C. Bode, P.A. van der Aa, C.T.J. Hulshof, G.H.M.B. van Rens, R.M.A. van Nispen, The economic burden of visual impairment and comorbid fatigue: a cost-of-Illness study (from a societal perspective), Invest. Ophthalmol. Vis. Sci. 59 (2018) 1916–1923.
[185] Areds, Risk factors associated with age-related nuclear and cortical cataract a case-control study in the age-related eye disease study, AREDS Report No. 5, Ophthalmology 108 (2001) 1400–1408.
[186] A. Dominguez-Vincent, U. Birkeldh, C.G. Laurwell, M. Nilson, R. Brautaset, Objective assessment of nuclear and cortical cataracts through scheimpflug images: agreement with the LOCS III scale, PLoS One 11 (2016) e0152953.
[187] C.J. Brown, F. Akaichi, Vitamin D deficiency and posterior subcapsular cataract, Clin. Ophthmol. 9 (2015) 1093–1098.

Further reading

B. Berlin, P. Kay, Basic Terms: Their Universality and Evolution, University of California Press, Berkeley & Los Angeles, 1969.
W. He, C.W. Cowan, T.G. Wensel, RGS9, a GTPase accelerator for phototransduction, Neuron. 20 (1998) 95–102.

CHAPTER

4

Circadian system

4.1 Principles of chronobiology

The term circadian rhythm (Latin circa diem, along the day) refers to an endogenous rhythm in some biological functions. Circadian rhythms have been first reported in 1729 by the French scientist J.J. D'Ortous De Mairan,[1] who noted that the leaves of the mimosa plant moved with a periodicity of 24 h, even when the plant was brought down to a dark basement. This meant that biological clock involved was "endogenous," that is, the clock was still correctly working when it received no further light signal. Moreover, the rhythms in every organism had to adapted to changing solar cycles throughout the seasons, that is, it had to be "trainable" by environmental cues [1,2]. Initially all of the parameters of the organisms oscillate at a slightly longer pace than 24 h and are brought to an exact 24 h value by interaction with the day/night cycle that thus function as Zeitgeber (German for time-givers). Circadian rhythmicity is present in the sleeping and feeding patterns of animals, including human beings, whether they are interested in the environment around them or not.

From such initial observations, a new discipline, chronobiology, developed that considers the dimension time, and is thus concerned with biological rhythms and how they adapt at the solar and lunar cycles and include both well-apparent patterns of core body temperature, brain wave activity, hormone production, cell regeneration, and other biological activities. In addition, photoperiodism, the physiological reaction of organisms to the length of day or night, is vital to both plants and animals, and the circadian system plays a role in the measurement and interpretation of day length. Timely prediction of seasonal periods of weather conditions, food availability, or predator activity is crucial for survival of many species. Although not the only parameter, the changing length of the photoperiod ("day length") is the most predictive environmental cue. Besides the large day adaptation, further less strong but longer cyclic patterns are detected, in correspondence to the change of the day length with the seasons, the tides, and any cyclic manifestation, as well as short ones, typically those that follow the plot of any day, from the heart beating to the alternance between aggressive/remissive, or optimistic/pessimistic, the seasonal timing of physiology and behavior, most notably for timing of migration, hibernation, and reproduction. Some scientists feel that the confidence to be

[1] On fçait que la Sensitive

DOI: https://doi.org/10.1016/B978-0-12-811659-3.00004-9

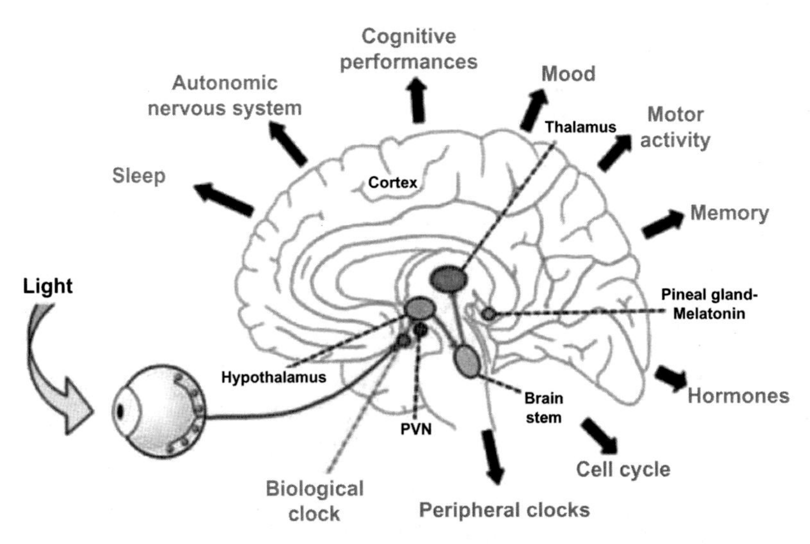

FIGURE 4.1 Diagram of the biological functions controlled by the circadian biological clock (nonexhaustive list). The structures indicated are, respectively, the suprachiasmatic nucleus, the pineal gland, the hypothalamus (containing the ventrolateral preoptic area, known as the *sleep switch*), the brain stem (containing the ascending activator cortical pathway and the slow wave/paradoxical *sleep switch*), the thalamus (responsible for cortical activation and synchronization of the EEG). Source: *Modified from E. Mignot, S. Taheri, S. Nishino, Sleeping with the hypothalamus: emerging therapeutic targets for sleep disorders, Nat. Neurosci. 5 Suppl. (2002) 1071; C. Caramelo Gomes, S. Preto, Blue light: a blessing or a curse? Procedia Manuf. 3 (2015) 4472–4479.*

placed of such parameters as derived from circadian cycles—rather than from societal impulses—has long been, and in part still is, a matter of discussion and polemics. Earth resonates to the daily recurring environmental cycles, a deeply exerts influence on multiple levels, ranging from behavior to physiology, all of the way to gene expression rhythms in organs and individual cells. Light-sensitive cells are also present in the eyes that do not participate to image-forming processes and thus differ from those devoted to sight, but still are sensitive to light. In mammals, this system operates through sensitive cells and through specialized ganglion cells, which are found in the retina along with cones and rods, but, differently from the former, excite directly the suprachiasmatic nucleus (SCN). A diagram showing the anatomy and the most important physiological functions is reported in Fig. 4.1 [3].

Studies on cyanobacteria, where it has been proved that upon cell division the daughter cell maintains the original phase of the mother cell, and thus the results are not complicated by cell division, may be used to illustrate the situation, as reported in Fig. 4.2.

Circadian rhythms of chromosomal compaction as visualized by a fluorescent DNA-binding dye (green; red indicates chlorophyll autofluorescence). The chromosome is more compacted in the subjective night (hours 12–20). Chromosomal topology shows a circadian rhythm as assayed by supercoiling of an endogenous plasmid. Topoisomers of the plasmid are more relaxed in the subjective night and are more supercoiled in the subjective day. The phosphorylation of the cyanobacterial clock protein KaiC leads to the oscillating in vitro reaction. In the figure, upper bands are hyperphosphorylated KaiC, lower bands are hypophosphorylated KaiC. The predominant

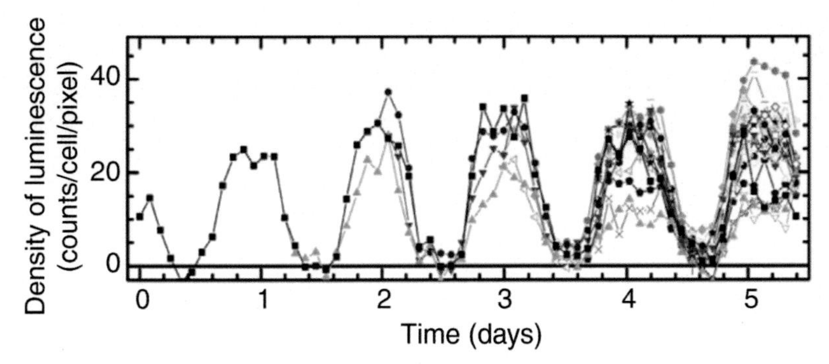

FIGURE 4.2 Quantification of bioluminescence from a single cell as it divides into multiple cells as a function of time under unchanging illumination. Starting at day 1.5, there are two differently colored traces as a result of cell division; the next division occurs at day 2.0 and so on [4].

species of complexes of KaiA, KaiB, and KaiC that form during the in vitro oscillation are hypophosphorylated and uncomplexed KaiC hexamer (lowest row) is present at all phases, but KaiC in complexes with KaiA and/or KaiB form rhythmically in concert with changes in KaiC's phosphorylation status (upper line, only the predominant complex is shown). Light blue KaiC hexamers are in the phosphorylating phase prior to monomer exchange, while dark blue KaiC hexamers are those undergoing dephosphorylation and monomer exchange.

All of the pigments present in the cells contain the pigment melanin as the prosthetic group and an opsin. Among systems operated through specific effects, a variety of important functions are tuned by the circadian system. These include the regulation of the overall sensitivity of vision and the level of activity of the organism. Related to it the wakefulness/sleep equilibrium is helping in establishing the primary circadian clock, through a specific organ, located in mammals in the SCN, as well as a regular sleep–wake rhythm and an overall sensitivity to the light flux crossing the pupil. The SCN takes the information on the lengths of the day and night from the retina, interprets it, and passes it on to the pineal gland, a tiny structure shaped like a pine cone and located on the epithalamus. In response, the pineal secretes the hormone melatonin. Secretion of melatonin peaks at night and ebbs during the day. Its presence provides information about night length. The SCN receives information about illumination through the eyes. If cells from the SCN are removed and cultured, they maintain their own rhythm in the absence of external cues. Several studies have indicated that melatonin feeds back on SCN rhythmicity to modulate circadian patterns of activity and other processes. However, the nature and system-level significance of this feedback are unknown.

The retina of the eye, besides "classical" photoreceptors such as rods and cones, which are used for "conventional" (perhaps better indicated as "image-forming") vision and operates through the above discussed mechanism of phototransduction also contains specialized ganglion cells that are a small number of the light-sensitive cells ($\sim 5\%$), are directly photosensitive and melanopsin-expressing. These are labeled intrinsically photosensitive retinal ganglion cells (ipRGCs), and project directly to the SCN, where they help in the entrainment (synchronization) of this master circadian clock with the lifestyle of each individual. The night alternation is disturbed by the fact that present day "dark" is not comparable with what had been previously experienced. When Thomas Alva Edison installed his lamps, he probably did not anticipate how deeply would this invention change the everyday life of people. Concern about the largely changing conditions human beings are forced to live under is growing. The day–night alternation is disturbed by the fact that present day "dark" is not comparable with what had been previously thought and a fully dark space is available only far enough from the multiple sources of artificial light that make our night lively. A nice representation of what the life was, and indeed has been until the begin of the 20th century can be found in the novel Das Parfum by Patrick Susskind[2] [4–6]. Studies of the effects of light on the circadian system of insects, plants, and animals carried out from the late 1950s through the 1970s demonstrated that the timing of a light stimulus has an important influence on the direction and magnitude of response to that stimulus. Those studies indicated that the circadian system of both nocturnal and diurnal organisms is most sensitive to light during the biological night. Because humans sleep throughout most of their biological night, testing the influence of light on the human circadian system therefore required that in the sleep–wake cycle be shifted in order to deliver the light stimulus at the time of highest expected sensitivity [7–11]. That manipulation of sleep–wake timing was a concern in the earliest human light studies, because of prior reports suggesting that social cues-influenced human circadian rhythms. The human circadian system is quite similar to that of other organisms, and phase angle of entrainment in humans is strongly influenced by light. Actually, the circadian rhythm must be distinguished from the behavior that accompanies going to sleep, such as lying on a bed, avoiding cold, putting down lights, and so on. The information obtained has implications for understanding and developing treatments for circadian rhythm sleep disorders. These studies have allowed for the design of light treatment regimens for night workers, jet travelers, and patients with circadian rhythm sleep disorders [12]. One of the most intriguing questions in chronobiology is how the transmission of rhythmic signals from the core oscillator to clock output functions is implemented across the many regulatory layers of gene expression. Circadian clock-dependent and -independent posttranscriptional regulation underlies temporal mRNA accumulation in mouse liver. Significant fractions, up to 20%, of expressed transcripts are found to oscillate with a 24-h period in various organs. In mouse

[2] est heliotrope, c'est-a-dire que fes rameaux & ses feuilles fe dirige toujours vers le côté d'ou vient la plus grand lumiére & l'on fçait de plus qu'à cette proprieté que lui eft commune ave d'autre Plantes, elle en joint une que lui eft plus particulière, elle est Sensive à l'egard du Soleil ou du jour, les feuilles est leur pedicules fe replient & fe contactent vers le coucher du Soleil, de la même manière dont cela fe fait quand on touche la plante ou qu'on l'agite. Mais M de Mairan a obfervé qu'il n'eft point néceffaire pour ce phénoméne qu'elle foit au Soleil ou au grand air, il eft feulement un peu moins marqué lorfqu'on la tient toûjours enfermée dans un lieu obfcur elle s'épanouit encore très fenfiblement pendant le jour, & fe replie ou fe refferre régulierement le foir pour toute la nuit. (English) it is known that the sensitive is heliotrope, that is, it branches and leaves point always toward the strongest light. It is also known that besides this property that it has in common with other plants, it has one more, that is, characteristic, it is sensitive toward the sun or the light, its leaves ply down when the sunset approaches in the same way as it happens when one touches the plant or shakes it. But M de Marain a observé that in order to observe this phenomenon it is not required that the plant is exposed to the sun or at any rate to full light, and rather it occurs only slightly less intense when it is taken in the dark. It still unfolds in the morning and folds up regularly.

A nicely expressed contrast can be found in the novel by PatrikSuskind, Das Parfum—Die Geschichte einesMörders, where walking around in London after sunset is depicted as a dangerous adventure, where vision was of no use at all as compared with a nocturnal walk in a region in that no light-emitting feature for 25 km around is present, which of course is becoming less and less easy to find at present time.

liver, genome-wide profiling of transcription and RNA accumulation over the 24 h day showed that mRNA abundance can oscillate without rhythmic transcription, suggesting that a posttranscriptional control of a significant fraction of the rhythmic transcriptome. Whether the mRNA degradation rate of a transcript, or equivalently its half-life, is constant or rhythmic predicts quantitatively distinct temporal profiles of mRNA accumulation. Specifically, if a mRNA is rhythmically transcribed but degraded at a constant rate, the peak of mRNA abundance (phase) will be delayed between 0 h and maximally 6 h after synthesis, with a damped oscillation [13]. However, if degradation rate is also rhythmic, then the phase delay can flexibly vary between 0 and 24 h, and the relative amplitudes could be either damped or magnified. Thus the combined effects of rhythmic synthesis and rhythmic degradation can, in principle, lead to temporal gene expression profiles with diverse amplitudes and phases. The rhythm is linked to the light–dark cycle. Animals, including humans, kept in total darkness for extended periods eventually function with a free-running rhythm. Their sleep cycle is pushed back or forward each "day," depending on whether their "day," their endogenous period, is shorter or longer than 24 h (in a recent study, in average 24 h and 11 min) or shorter. Free-running organisms that normally have one or two consolidated sleep episodes will still have them when in an environment shielded from external cues, but the rhythm is not entrained to the 24-h light–dark cycle in nature. The sleep–wake rhythm may, in these circumstances, be shifted out of phase with other circadian or ultradian (Latin, more than a day) rhythms, such as metabolic, hormonal, CNS electrical, or neurotransmitter rhythms [14]. The environmental cues that reset the rhythms each day are called Zeitgebers (from the German, "time-givers") [15]. Notably for this purpose is the defect that blinds subterranean mammals is able to maintain their endogenous clocks in the apparent absence of external stimuli. Although they lack of image-forming eyes, their photoreceptors (which detect light) are still functional, and indeed they surface periodically as well [16]. The role of periodical changes in the vital functions of humans is difficult to disentangle from interfering effects [17,18]. The hypothesis has been formulated that this might be because that some humans indeed are seasonally photoresponsive, but others are not, and that individual variation may be the cause of the inconsistencies that have plagued the study of responsiveness to photoperiod in the past. This hypothesis has been examined in relation to seasonal changes in the reproductive activity of humans, and it is developed by reviewing and combining five bodies of knowledge. These are the correlations of human birth rates with photoperiod; seasonal changes in the activity of the neuroendocrine pathway that could link photoperiod to gonadal steroid secretion in humans; what is known about photoperiod, latitude, and reproduction of nonhuman primates; documentation of individual variation in photoresponsiveness in rodents and humans; and what is known about the evolutionary ecology of humans [19]. Mechanisms regulating the degradation of transcripts include the recruitment of RNA-binding proteins to the 3′-untranslated region of transcripts, as well as targeting of transcripts by miRNAs. Biochemical studies of individual genes have shown that half-lives of mRNAs can fluctuate during the circadian cycle. Indeed, studies on mammalian core clock genes, such as *Per1*, *Per2*, *Per3*, and *Cry*, have been found to fluctuate mRNA half-lives governed by RNA-binding regulators. mRNA degradation also regulates systemically driven rhythmic transcripts, such as *Tfrc*, *Fus*, and *Cirbp*. However, the temporal regulation of physiology and gene expression in a complex organ such as the liver is still a challenging topic at a genome-wide scale. Transcription during the diurnal cycle in tissues can be estimated in vivo through Pol II loading on genes or approximated with nascent RNA or pre-mRNA, direct measurements of mRNA degradation rates, which may also vary over the course of the day, poses challenges. Experimental approaches using inhibitors of transcription as well as metabolic pulse labeling of nascent RNA can yield genome-wide insights in mRNA production and degradation in eukaryotic cells. However, these techniques may complicate analyses due to potential biases. For example, antibiotics that block transcription can arrest growth, and metabolic labeling of RNAs can inhibit rRNA synthesis. Such methods still require to be adapted to measure dynamics of synthesis and degradation of mammalian mRNAs in vivo, such as in the intact liver. Dual-color labeling of introns and exons by single-molecule fluorescence in situ hybridization is a noninvasive technique that could infer transcription and degradation rates of individual genes in mouse liver, although this approach relied on other quantities that are also challenging to measure, such as transcription elongation rates [20]. Recently, a promising avenue to identify regulatory control points has been found in gene expression is to integrate measurements on multiple omics levels with predictions from kinetic production degradation models [21]. As an example, it has been clarified a model selection framework to systematically identify the contributions of transcriptional and posttranscriptional regulation from times series pre-mRNA and mRNA profiles in mouse liver, without additional external input such as mRNA half-lives.

It was found that rhythmic transcription drove a majority of rhythmic mRNAs (~65%), with constant mRNA degradation, while a lesser fraction of rhythmic mRNA degradation (~35%) was regulated by a constant or rhythmic transcription. It was further observed that rhythmic mRNA degradation is exploited not only to generate rhythms but also to flexibly fine-tune oscillatory amplitudes and peak timings of mRNA rhythms depending on the mRNA half-life. Finally, the transcriptomes of liver from *Bmal1*$^{-/-}$ mice were analyzed and showed that rhythmic mRNA degradation was often independent of functional *Bmal1*$^{-/}$, but originated most likely from systemic signals driven by feeding–fasting or sleep–wake cycles [13,22].

Later studies demonstrated that the bases of biological time keeping, that is, the gears of the clock, consisted in a few important genes ("core clock genes"). A model called the transcription–translation feedback loop (TTFL) became the dominant paradigm for conceptualizing circadian oscillation in booth plants and animals. TTFLs are further able to remodel DNA. DNA is normally packaged into dense chromosomes by special proteins called histones, which can be modified chemically to compact and decompress DNA so that it is inactive or active, respectively. CLOCK, for example, is known to function as a histone acetyltransferase and this acetylation is required for the rhythmic expression of other core clock and output genes, including its partner BMAL, which undergoes rhythmic acetylation in the liver. The Sirtuin (SIRT) family of proteins can function as nicotinamide adenine dinucleotide (NAD)-dependent deacetylases, directly linking the energy state of the cell to histone lysine residue deacetylation. One family member, SIRT1, has been shown to interact with CLOCK and BMAL and is also involved in the rhythmic deacetylation of the PER2 protein. NAD is a central metabolite and is required for the oxidation of glucose in respiration reactions such as glycolysis, a fundamental energy production pathway. These results provide the first insights into how metabolic state and the circadian clock may be linked. Further functions of the TTFL framework of circadian rhythms have appeared in a variety of organisms, from bacteria to flies to mammals, and the redox states are likely to be an oscillation that feeds back upon the TTFL, whereby a cell's redox state may alter clock gene expression and the clock genes, in turn, regulate redox states. The redox state of a cell is a careful balance between the generation of oxidants through metabolic processes, and the amount of reducing agents available. Oxidants such as reactive oxygen species (ROS) are damaging to cellular components. In order to compensate, all organisms have evolved mechanisms to buffer oxidants, such as superoxide dismutase (SOD) and catalase, which decomposes H_2O_2 into water and oxygen. Glutathione (GSH) maintains the reduced state of protein sulfhydryl groups that are necessary for DNA repair and prevents oxidative damage to cell membranes by reducing lipid peroxides. When oxidized, GSH forms a dimer (GSSG) that is reduced by the glutathione reductases (GR) via an NADPH-dependent reaction. Otherwise, GSSG is potentially toxic to the cell. Glutathione peroxidases (GPx) are also antioxidant enzymes known to have an affinity for lipid peroxides and may be involved in intracellular signaling. In other organisms both fruit flies and mammals, daily rhythms in the expression of GSH, and in the activity of SOD, GPx, and GR have been observed (Fig. 4.3) [23,25].

The effect of circadian rhythms may be important also on associated diseases, as obesity, or diabetes [24,26,27]. The effect on the heart rate 24 h variability was also revered significant circadian rhythm of normalized units of both high and low frequency power of the variability in healthy subjects, with higher values during sleep (as well as an increase of the low frequency component when awakening into the supine position), while no effect was evidenced with patients with uncomplicated arterial diseases [25].

A disruptor of circadian rhythms can lead to far-reaching consequences. For example, such a disruptor is a risk factor for several human gastrointestinal diseases, ranging from diarrhea to ulcers to cancer. This has been determinate on bioluminescent jejunal explants from mice, where it has been demonstrated that robust rhythms develop for about 72 h [28]. Another pathology where the circadian rhythms are important is the Parkinson disease. In fact, mice have been adopted for a satisfactory model for the Parkinson disease by using 6-hydroxydopamine (6-OHDA). In similar cases, exposure to an intense flux of blue light has been found to cause significant improvements. A study on implanted rats demonstrated a significant alteration of the heart rate, but not of temperature and locomotor activity [29,30]. Data presented in a recent paper suggest that the decrease in exposure to bright light may be associated with declines in sleep quality that accompany normal and pathological aging. This study suggests that increased exposure to bright light may have a therapeutic value in patients with senile dementia of the Alzheimer's type.

While there are little quantitative data regarding the exact nature of chronobiologic sleep disturbance in this disease, the numerous anecdotal accounts of night wandering and excessive day time napping in these individuals are consistent with the notion that a circadian rhythm disturbance may be the source of these symptoms [31].

The regulation of many homeostatic functions including energy regulation is of obvious important and can be reached by appropriate circadian regulation. In fact, multiple genes are involved in nutrient metabolism display rhythmic oscillation. This holds for metabolically related hormones such as glucagon, insulin, ghrelin, leptin, and corticosterone, which are released in a circadian manner. Mice harboring mutations in circadian clock genes have been studied and found to alter feeding

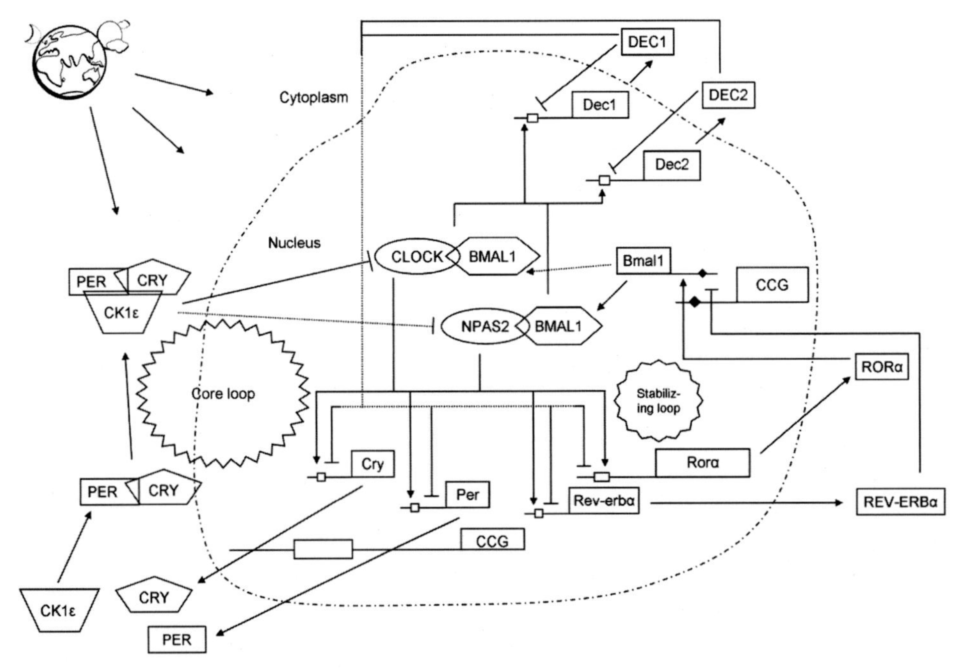

FIGURE 4.3 A simplified depiction of the mammalian molecular circadian clock machinery. Light perceived by the retina is the most potent synchronizer of the SCN clock. The circadian clock consists of positive and negative autoregulatory feedback loops. The oscillator is composed of interlocking transcription/translational feedback loops, controlling circadian timing. The CLOCK:BMAL1 or CLOCK:NPAS2 heterodimer (positive elements) is the "core loop" and induces E-box-mediated transcription of *Per*, *Cry*, and *Dec*; their products are cyclically released in the cytoplasm. When PER and CRY proteins reach a critical concentration, they form heterodimers PER:CRY (negative elements), phosphorylate and translocate into the nucleus, where they inactivate the BMAL1:CLOCK or BMAL1:NPAS2 E-box-mediated transcription, including transcription of their own genes, which reduces their levels sufficiently to allow for the new transcription cycle. In addition, DECs bind to the E-box element of their promoter and inhibit their own transcription directly. CLOCK: BMAL1 also controls the levels of the nuclear receptors retinoid-related orphan receptor α and Rev-erbα [known as nuclear receptor subfamily 1, group D, member 1 (NR1D1)], which constitute the "stabilizing/auxiliary loop" by repressing BMAL1 concentration via competitive actions on the retinoic acid–related orphan receptor response element (black diamond shape) in the *Bmal1* promoter. Cycling of clock components by the core and stabilizing/auxiliary loops also promotes cyclic accumulations of clock-controlled gene mRNA species, thus achieving an oscillating pattern and generating rhythmic physiological outputs in a cell type–specific manner (steroid biosynthesis, cell cycle progression/arrest, cell proliferation, apoptotic pathways, immune function, hormonal oscillations, body temperature, metabolism, DNA repair, response to anticancer drugs, and so on). E-boxes (white rectangle shape): regulatory enhancer sequences present in the promoter regions of the genes to which CLOCK:BMAL1 heterodimer binds. Casein kinase isoforms phosphorylate PER and CRY proteins modulating the nucleocytoplasmic translocation of core clock elements and thereby their transcriptional activity [23,24]. *DEC*, differentially expressed in chondrocytes.

behavior, endocrine signaling, and dietary fat absorption [32,33].

In Fig. 4.4 a schematic view of the mammalian molecular circadian clock machinery is shown. The potential relation between disruptions of normal circadian rhythms with the genetic driving machinery of cancer has been discussed [34].

The upregulation by light involves an important component of the retinal clock network, thus contributing to the circadian physiology. During postnatal development the

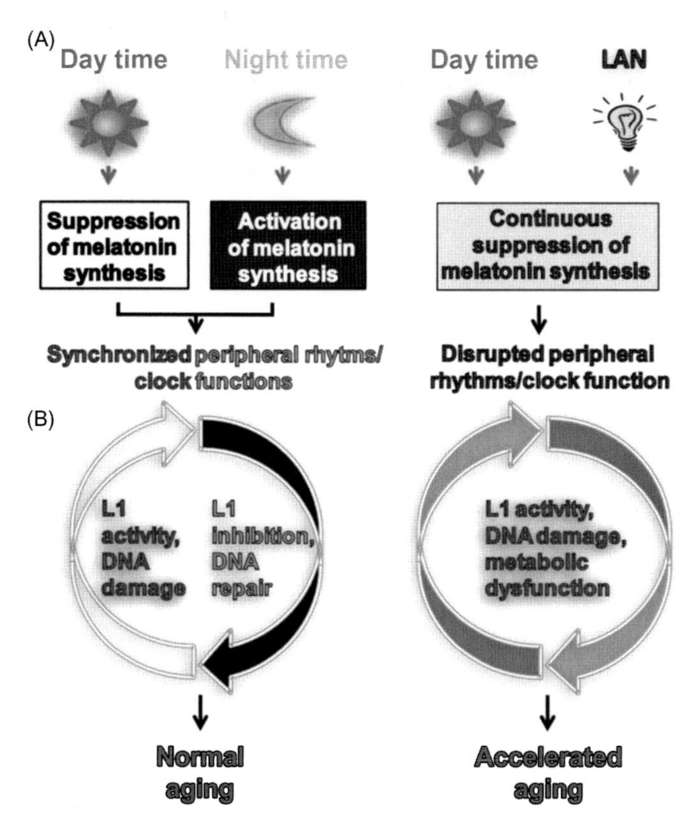

FIGURE 4.4 Light exposure at night accelerates aging by impeding or enhancing processes associated with aging. (A) Usually aging involves normal light exposure that is characterized by alternating intervals of light and dark over a 24-h period, which result in circadian production of nocturnal melatonin. This leads to the synchronization of peripheral clock function controlling many biochemical processes in cells including L1 expression and activity and the DNA damage response (DDR). (B) Exposure to light at night (LAN) is reported to accelerate aging. LAN blocks nocturnal melatonin production which prevents synchronization of PCs, leading to the disruption of timely function of many biochemical processes in cells including L1 expression and activity, DDR, and metabolism [34].

dimensions of the eye undergo various fluctuations in dimensions, and a role of the ambient lighting has been postulated. Experimental myopia may be induced by wearing a concave spectacle lens as well as by the genes encoding a melatonin receptor and the photopigment melanopsin [35,36]. As hinted above in contemporary society there never is a full dark, so that weak emissions to which our progenitors were accustomed, such as a Milky Way in the sky, are no more visible. The benefits of electric light are counterbalanced by detriments such as wasted energy, harm to animals and plants, and increase in severe human maladies such as cancers of breast and prostate [36]. Neurodegenerative diseases and psychiatric disorders find their origin in the oxidative stress combined with proinflammatory mechanisms. In fact, melatonin is a hallmark of circadian rhythm functionality, is a powerful natural antioxidant ($E = +0.65$ V vs Ag/AgCl [37]) with a circadian secretion pattern. Under similar conditions 6-OHDA has been found to control the Parkinson disease [38] and melatonin is consumed by undergoing a series of oxygenation reactions [39]. This appears to involve either singlet oxygen oxidation products (path *a* predominating under oxygen) or products of radical oxygenation (path *b* predominating under nitrogen) (Scheme 4.1).

In contemporary 24 h/7 days society, a shift of work time over three periods is largely present. This causes a disruption of the human circadian time organization, resulting in symptoms similar to those of jet lag in the short

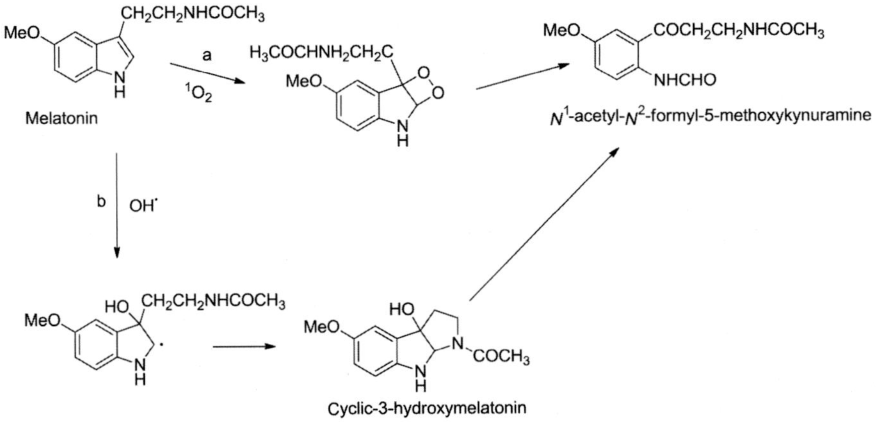

Melatonin

a $^1O_2$

b $OH^{\cdot}$

N^1-acetyl-N^2-formyl-5-methoxykynuramine

Cyclic-3-hydroxymelatonin

SCHEME 4.1 Main reaction paths from irradiation of melatonin.

term, while in the long term it may contribute to a range of syndromes such as weight gain/obesity, metabolic syndrome/type II diabetes, cardiovascular disease. The generation of proinflammatory ROS has also an important role. It has been also suggested that female workers in rotating shifts are exposed to an increased cancer risk, especially for breast cancer. As a matter of fact, the carcinogenic effect of night and shift work constitutes additional serious medical, economic, and social problems for a substantial proportion of the working population [40,41]. Actually, LAN is a true syndrome that affects the integrity of the cellular genome and metabolic function resulting in the suppression of circadian melatonin production. Aging can be described as a progressive decline in the stability, continuity, and synchronization of multifrequency oscillations in biological processes to a temporally disorganized state.

Even worst when shift work is being exposed passively to continuous light. A typical example is the negative impact on human health is observed in critically ill patients confined in the intensive care units (ICU), where environmental constancy throughout both day and night (continuous light, noise, caring activities, medications, etc.) is imposed. This condition induces a syndrome known as circadian misalignment, circadian disruption, or chronodisruption [42]. A way out is using minimally invasive devices (up to 12 h differences in individual circadian phase), has enabled continuous circadian timing system (CTS) assessment in nonhospitalized cancer patients. The data support the personalization of chronotherapy, a method that aims at the adjustment of cancer treatment delivery according to circadian rhythms of each patient. This can be adjusted by using programmable-in-time pumps or by introduced novel release formulations, in order to increase both efficacy and tolerability [41].

4.2 Human medicine aspects

The human mood and societal engagement are strongly dependent on the attitude of each person. Both disruptions in sleep and circadian rhythms are observed in individuals with bipolar disorders (BDs), during acute mood episodes as well as remission. It has been highlighted that sleep and circadian rhythm disturbances are closely linked to the sensitivity to BDs and susceptibility to bad mood recurrence. Lithium is known to act as a

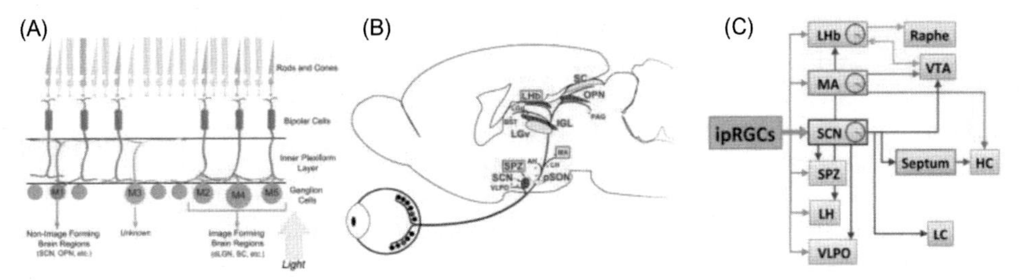

FIGURE 4.5 (A) Schematic view of the retina showing the organization of different neuronal populations and their synaptic connections. Rods and cones are confined to the photoreceptor layer. Light detected by rods and cones is processed and signaled to retinal ganglion cells (RGCs) through horizontal, amacrine and bipolar cells. RGCs are the only output neurons from the retina to the brain. A subset of RGCs (2%–5% of the total number of RCGs) are intrinsically photosensitive RGCs (ipRGCs). There are at least five subtypes of ipRGCs (M1–M5) with different morphological and electrophysiological properties, which show widespread projection patterns throughout the brain. (B) ipRGCs project to numerous brain regions, including many that have a role in driving light-mediated behaviors, including circadian photoentrainment and sleep. In addition, ipRGCs also innervate nuclei involved in depression and/or anxiety, such as the MA, LHb, and SPZ (M1 arrow) indicating a possible direct role of light on mood.(C) Several of the ipRGCs targets (M3 Arrow), including the SPZ, VLPO, LH, and LHb also receive innervation from the SCN, raising the possibility that in addition to its pacemaker function, the SCN can also act as a conduit for light information. Interestingly, the MA and the LHb are also brain peripheral clocks that receive direct retinal innervation. Areas involved in mood regulation (VTA and raphe) and cognition (LH) can be influenced by light either via the SCN or in parallel via the MA and LHb. *HC*, Hippocampus; *LC*, locus coeruleus; *LH*, lateral hypothalamus; *LHb*, Lateral habenula; *MA*, medial amygdala; *SCN*, suprachiasmatic nucleus; *SPZ*, subparaventricular zone; *VLPO*, ventrolateral preoptic area; *VTA*, ventral tegmental area, *RCG*, retinal ganglion cells [45].

synchronizer and stabilizer of circadian rhythms and pharmacogenetic studies testing circadian gene polymorphisms and prophylactic response to lithium have been carried out. Sleep deprivation, light therapy, and psychological therapies have been compared for treatment and prevention of bipolar depression [43].

Evidence of artificial light at night (ALAN) has been increasingly reported with different types of malignant tumors, particularly breast cancer. According to several epidemiological screening studies, exposure to ALAN increases breast cancer risk by abolishing the characteristic nighttime production of melatonin by the pineal gland. Melatonin has been recognized to be antioncogenic in breast and prostate cancers and the involvement of melatonin in regulation of epigenetic responses through DNA modifications has been suggested. Melatonin can regulate epigenetic modifications in cancer cells by both DNA methylation and histone protein remodeling. The connection between chemical pathways of both ALAN and cancer-related epigenetic reactions, in particular, global DNA methylation, is presented in Fig. 4.5 [44,45].

Melatonin is also produced in the peripheral reproductive organs, including granulosa cells, the cumulus oophorus, and the oocyte. Melatonin is a powerful free radical scavenger and protects the oocyte from oxidative stress, especially at the time of ovulation [46,47] as well as aging [48].

The change in the light effect is connected with the substitution of natural by artificial lamps, at the time of the large diffusion of Edison's incandescent lamps. More recently, these lamps have in turn be substituted by further light sources, at present the compact fluorescent light and the light-emitting diode.

The emission spectrum of such lamps contains significant blue component with the warmer incandescent globes has been the cause of emerging health concerns. The blue light bandwidth in the visible spectrum has a significant influence on the disruption of the internal body clock and suppression of melatonin secretion at night. This effect has been reconducted to various illnesses, including breast cancer, prostate cancer, heart disease, obesity, and diabetes. In contrast positive effects have been observed too, such as resetting the body clock to the required sleep pattern, boosting mood, alertness, cognitive performance, and alleviating seasonal affective depression (SAD) [47]. SAD is prevalent when vitamin D stores are typically low. Broad-spectrum light therapy includes wavelengths between 280 and 320 nm (UV-B) which allow the skin to produce vitamin D from dehydrocholesterol [49].

In mammals, daily physiology and metabolism are regulated by circadian clocks [50], and disruption of circadian rhythmicity and has a strong influence on lipid homeostasis and pathologies such as fatty liver and obesity. The interaction between circadian clocks and lipid homeostasis has been confirmed from recent advances in lipidomics methodologies and their application in chronobiology studies [48].

Recent findings show that the disruption of circadian clocks leads to metabolic disorders. In particular, nutrition studies have accumulated that consider molecular relationships between circadian clocks and nutrition (chrononutrition) [51,52].

The role of retina is peculiar, and several studies have shown that multiple sites within this membrane are capable of generating circadian oscillations. In mammals, clock activity is most robust in the inner nuclear layer, where melatonin and dopamine serve as signaling molecules to entrain circadian rhythms. γ-Aminobutyric acid too is an important component of the system that regulates retinal circadian rhythms and the melatonin–dopamine system makes possible the reconfiguration of the retinal circadian clock, so that the retinal circuits are enhanced by light-adapted cone-mediated visual function during the day and dark-adapted rod-mediated visual signaling at night [53]. Circadian rhythms are entrained by light, and their 24-h oscillation is maintained by a core molecular feedback loop composed of canonical circadian ("clock") genes and proteins. Different modulators are involved in the maintaining of the proper rhythmicity of these genes and proteins through such a modulator, and the key role appears to be that of dopamine [54]. Dopamine has been shown to have circadian-like activities in the retina, olfactory bulb, striatum, midbrain, and hypothalamus, where it regulates, and is regulated by, clock genes in some of these areas. Thus it is likely that dopamine is essential to mechanisms that maintain proper rhythmicity of these five brain areas [55]. The circadian clock is also implicated in the pathogenesis of eye disease, disk shedding, and phagocytosis [56,57] (Fig. 4.6) and its contribution to the stability of the system is particularly important in the peripartum phase.

The circadian clock system in the mammalian retina controls several functions. Many studies have shown that rhythms in the eye are under direct control of the retinal circadian clock system. Recent studies have also indicated that the many different cell types within the eye contain circadian clocks that interact to modulate many ocular functions [58,59], as well as accompanying the aging of body and mind, or the enhanced likelihood of the development of cancer [50,60] and/or dementia [61].

Transgenic animals with mutations in certain clock genes display remarkable mood phenotypes. The ClockΔ19 history is exemplary. This is a dominant negative mutation of *clock*. In homozygosity, mice have a lengthened circadian period of 27 h, that is, lost to arrhythmicity under prolonged constant dark conditions. During the day, the ClockΔ19 mouse exhibits marked hyperactivity, increased reward-seeking behavior, reduced immobility in the forced swim test, and decreased anxiety.

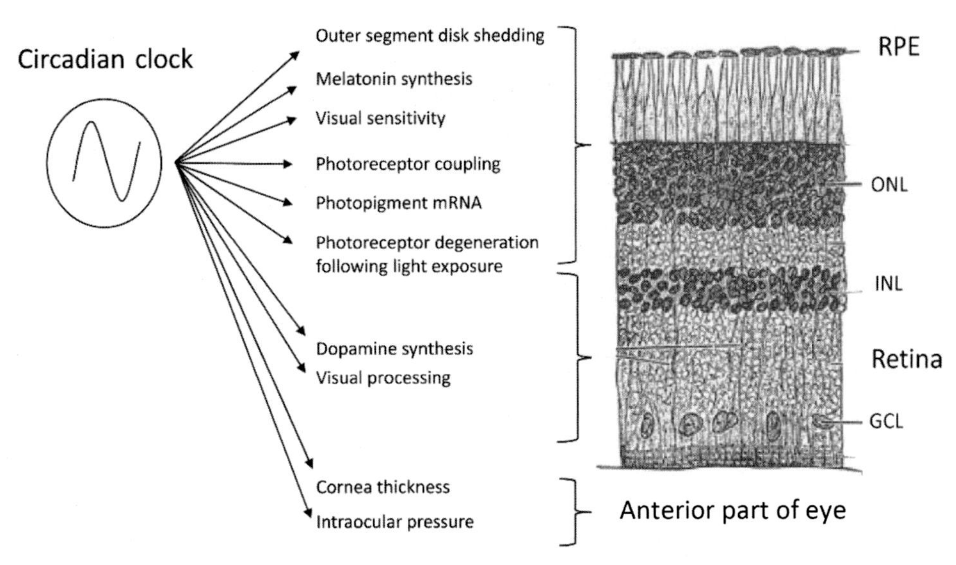

FIGURE 4.6 Circadian processes in the retina and the retinal pigment epithelium, with their approximate location identified by retinal layer. *RPE*, retinal pigment epithelium; *ONL*, outer nuclear layer; *INL*, inner nuclear layer; *GCL*, ganglion cell layer [56,57].

Indeed, the characteristics of these mice are a summary of the most salient features of mania in humans and become euthymic a night. This mouse has an increased expression of tyrosine hydroxylase (TH) and increased ventral tegmental area (VTA) dopaminergic tone. Normally, the VTA harbors a subset of dopaminergic neurons that display, in vivo, circadian activity [62]. This rhythm ceases ex vivo, suggesting a dependence on the SCN, which is connected to the VTA indirectly by way of the medial preoptic nucleus. Therefore since the creation of the ClockΔ19 mouse, researchers have speculated as to the causal relationship between its mood and circadian phenotypes. The denouement is instructive in that it epitomizes how clock genes have pleiotropic effects: *clock* also acts as a negative regulator of TH expression, and pharmacological inhibition of TH during the day reverses the manic phenotype, whereas optogenetic chronic stimulation that increases tonic VTA activity triggers manic behavior. Furthermore, Clock/Bmal1 also activate transcription of cholecystokinin (CCK), which is a digestive hormone but also a neuropeptide that negatively regulates dopaminergic signaling, and the ClockΔ19 phenotype shows reduced CCK levels [62]. This pleiotropy provides a model to explain epiphenomenal circadian alterations. In other words, the same factor can produce a circadian phenotype and an affective phenotype without there being any causal link between them. Sometimes pleiotropic noncircadian effects can be unmasked. To reprise the example above, the manic phenotype of ClockΔ19 is ameliorated by CK1δ/ε inhibitors and CK1δ/ε participates in the phosphorylation-mediated degradation of Per1, 2, and 3. However, CK1δ/ε inhibition causes period lengthening, therefore aggravating ClockΔ19's circadian phenotype.

Pleiotropic effects are by no means exclusive to *clock*. Monoaminoxidase-A, an enzyme that metabolizes 5-HT, norepinephrine, and dopamine, is under direct transcriptional (circadian) regulation by Npas2, Bmal1, and

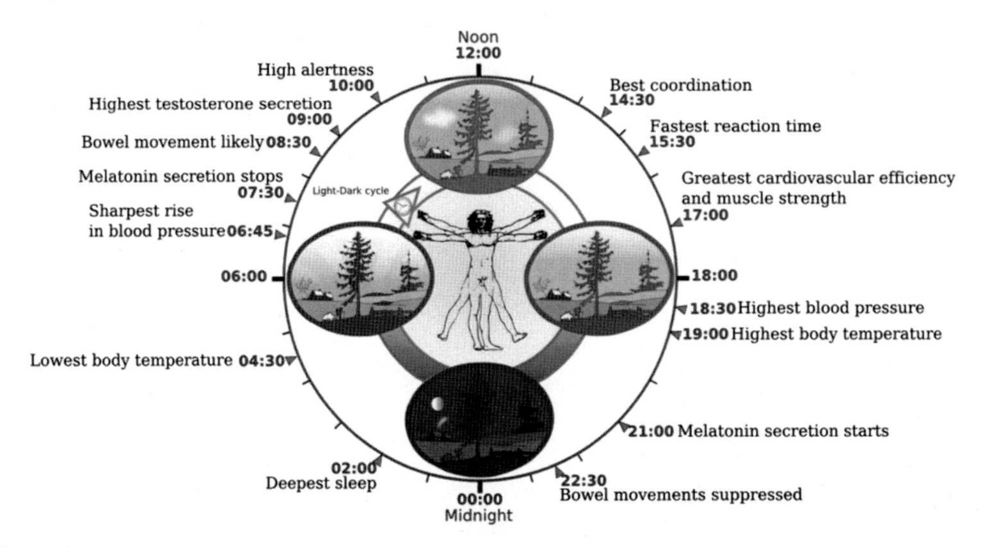

FIGURE 4.7 Changes of important parameters during night and day [65].

Per2 in the striatum, and these proteins are regulated not only by circadian signals from the SCN (with no evidence that their rhythmic expression is important for mood regulation) but also by extrinsic factors, such as chronic stress.

The fact that the cellular circadian machinery is so evolutionarily ancient and so intertwined with cell metabolism, as well as the fact that so much of the genome is under circadian regulation (up to 10% in mice), suggests that pleiotropy could be pervasive [56,63].

It has been observed that rapid eye movement though sleep behavior disorder may be an early marker of neurodegeneration and that this disorder is a possible marker for incoming dementia and for a general worst prognosis. Immune-related diseases too are associated with disruption of circadian homeostasis [62].

Jeffery C. Hall, Michael Rosbash, and Michael W. Young were assigned the Nobel Prize in physiology or medicine in 2014 in recognition of their studies on clock-like ups and downs of daily life [64]. Hall and Rosbash discovered the first molecular gear of the circadian clockworks: further proteins are involved in circadian rhythm (see Fig. 4.7 [65]).

A misalignment in the timing of clocks throughout the body may be, as hinted about, at the root of the problem for many people and contributes to a range of serious conditions including diabetes (type 2), cardiovascular disease, and cancer. Exposure to environmental light (light therapy) and lifestyle changes (scheduled meals, exercise, and sleep) is used to help fix a "broken clock" [66]. The CTS constitutes a novel therapeutic target. Interventions that normalize CTS dysfunction may affect quality of life and survival in cancer patients [67].

For many thousands of years, the sun was our only source of light, and human behavior followed a natural day–night cycle. This milieu began to change approximately 150 years ago with the invention of incandescent lighting. Electric lighting disrupted man's behavioral dependence on the day–night cycles of the sun, and facilitated alterations in our circadian sleep–wake cycles. Recent research has begun to identify the physiologic consequences of unnatural light exposure and subsequently altered circadian rhythms, and correspondingly the potential of light therapy. Thus a number of connections between disrupted circadian rhythms and clinical diseases has been established, as well as the relation with the sleep hygiene, while the concept of daylight as a therapy to restore disrupted circadian rhythms and to improve clinical outcomes has been explored [60].

For nearly 4 billion years, life on Earth, outside of the poles and deep oceans, has evolved under a consistent pattern of alternating bright days and dark nights. As a result, most organisms on our planet synchronize "daily life" to a 24-h cycle. Such circadian rhythm has been the same while in the same environment modern humans evolved over the past 200,000 years [60]. Today we know that the circadian rhythm does not simply regulate sleep–wake cycles, but also influences the molecular biology of individual cells and organ systems. The molecular mechanism of the circadian rhythm itself was discovered around 1970 in the common fruit fly, *Drosophila melanogaster* [68,69] and shortly thereafter described in humans. Genes including Clock and Period were identified as important regulators of the circadian rhythm, through patterns of protein expression that oscillate approximately every 24 h. The expression of these proteins reflects the circadian rhythm on a molecular level in all mammals. The molecular mechanism of the circadian rhythm involves a very complex and autonomous transcriptional–translational feedback loop which consists of a core set of oscillating, ubiquitously expressed genes, including Clock, Bmal1, Period homologues 1 and 2 (Per1 and Per2), and cryptochromes 1 and 2 (Cry1 and Cry2). This transcriptional–translational feedback loop takes approximately 24 h to complete. In addition to this classical transcriptional and posttranslational mechanism, many interacting pathways have been described, but the complete regulatory system is not yet fully understood [50] (Fig. 4.8).

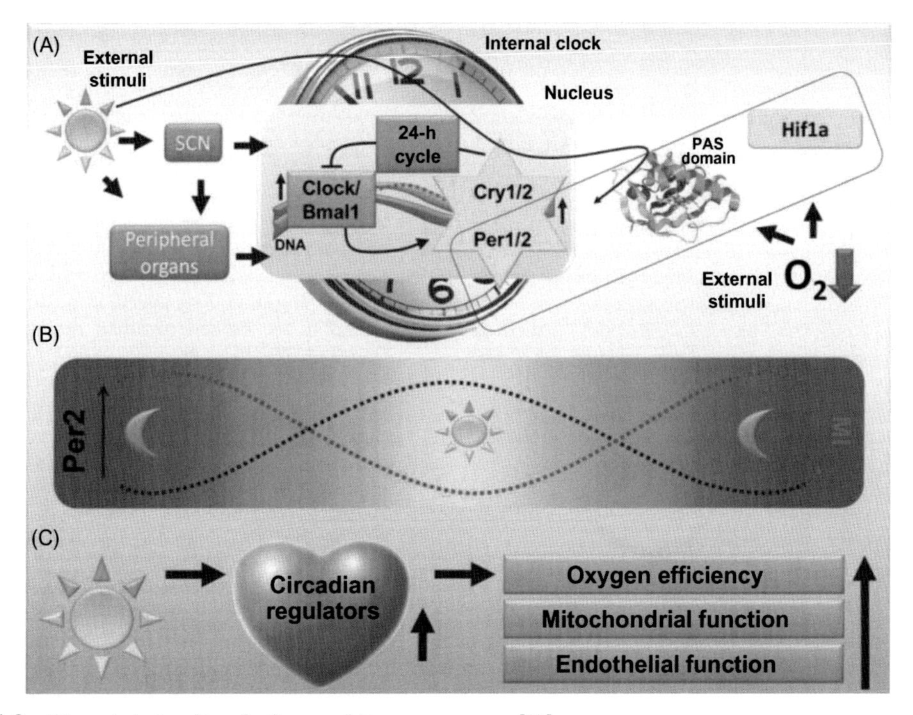

FIGURE 4.8 Disrupted circadian rhythms and its consequences [60].

Many factors in a clinical setting lead to a disrupted circadian rhythm. As proper sleep is a reflection of a functional circadian rhythm, in particular sleep-deprived patients are at risk (e.g., ICU). In addition, many common clinical scenarios disrupt our circadian rhythm, such as severe illness, stress, noise, surgery, sepsis, drugs, and LAN. Based on the current literature, this could increase the risk for myocardial infarction, stroke, sepsis, and obesity. The use of intense daylight (at least 4000 Lux, reflecting a sunny day outside) in conjunction with quiet and dark nights in hospitals could represent a future strategy to restore circadian rhythms and to benefit the overall health of inpatients; the circadian clock is composed of a primary negative-feedback loop involving the genes *Clock*, *Smal1*, Period homologue 1 (*Per1*), *Per2*, cryptochrome 1 (*Cry1*) and *Cry2*; disruption of these genes in animals models leads to the diseases indicated [60].

References

[1] T. Roenneberg, M. Merrow, Circadian clocks—the fall and rise of physiology, Nat. Rev. Mol. Cell Biol. 6 (2005) 965–971.

[2] E. Mignot, S. Taheri, S. Nishino, Sleeping with the hypothalamus: emerging therapeutic targets for sleep disorders, Nat. Neurosci. 5 (Suppl.) (2002) 1071.

[3] C. Caramelo Gomes, S. Preto, Blue light: A blessing or a curse? Procedia Manuf. 3 (2015) 4472–4479.

[4] C.H. Johnson, M. Egli, P.L. Steward, Structural insights into a circadian oscillator, Science 322 (2008) 697–701.

[5] Schlessinger, D.I. and Schlessinger, J., (2019), Biochemistry, Melanin, StatPearls Pub, Treasure Island, FL.

[6] S. Kitsinelis, S. Kitsinelis, Light Sources: Basics of Lighting Technologies and Applications, second ed., Boca Raton, NY, 2015.

[7] D.M. Graham, K.J. Wong, Melanopsin-expressing, intrinsically photosensitive retinal ganglion cells (ipRGCs), in: H. Kolb, E. Fernandez, R. Nelson (Eds.), Webvision: The Organization of the Retina and Visual System, University of Utah Health Sciences Center, Salt Lake City, UT, 2016.

[8] J.J. De Mairan, D'ortous, Observation botanique, Histoire de l'Academie Royaledes Science (1729) 35–36.

[9] Y. Cho, S.H. Ryu, B.R. Lee, K.H. Kim, E. Lee, Effects of artificial light at night on human health: a literature review of observational and experimental studies applied to exposure assessment, J. Chronobiol. Int 32 (2015) 1294–1310.

[10] R.L. Sack, R.W. Brandes, B.S. Kendall, A.J. Lewy, Entrainment of free-running circadian rhythms by melatonin in totally blind people, N. Engl. J. Med. 343 (2000) 1070–1077.

[11] T.M. Schmidt, K. Taniguchi, P. Kofuji, Intrinsic and extrinsic light responses in melanopsin-expressing ganglion cells during mouse development, J. Neurophysiol. 100 (2008) 371–384.

[12] T.M. Schmidt, P. Kofuji, Structure and function of bistratified intrinsically photosensitive retinal ganglion cells in the mouse, J. Comp. Neurol. 519 (2011) 1492–1504.

[13] D.M. Dominoni, J.C. Borniger, R.J. Nelson, Light at night and health: from humans to wild organisms, Biol. Lett. 12 (2016) 20160015. /1200160015/4.

[14] M. Pfeffer, H. Korf, H. Wicht, Synchronizing effects of melatonin on diurnal and circadian rhythms, Gen. Comp. Endocrinol 258 (2017) 215.

[15] J.F. Duffy, C.A. Czeisler, Effect of light on human circadian physiology, Sleep Med. Clin. 4 (2009) 165–177.

[16] J. Wang, L. Symul, Cédric Gobet, J. Yeung, C. Gobet, J. Sobel, et al., Circadian clock-dependent and -independent posttranscriptional regulation underlies temporal mRNA accumulation in mouse liver, PNAS 115 (2018) E1916–E1925.

[17] W. Cromie, Human biological clock set back an hour. Harvard Gazette, 2015 (retrieved 07.04.15).

[18] F.A. Scheer, K.P. Wright, R.E. Kronauer, C.A. Czeisler, K.P. Jr Wright, C. Kronauer, Plasticity of the intrinsic period of the human circadian timing system, PLoS One. 2 (2007) e721.

[19] J.F. Duffy, K.P. Wright, K.P. Wright Jr, Entrainment of the human circadian system by light, J. Biol. Rhythms 20 (2005) 326–338.

[20] A. Quetelet, Recherches Sur La Reproduction Et La Mortalité De L'homme Aux Différens Ages, L. Hauman et cie, Bruxelles, 1832.

[21] F.H. Bronson, Are humans seasonally photoperiodic? J. Biol. Rhythms,. 19 (2004) 180–192.

[22] A. Wehr, Photoperiodism in humans and other primates. Evidence and implications, J. Biol. Rhythms 16 (2001) 348–364.

[23] K. Nozue, M.F. Covington, P.D. Duek, S. Lorrain, C. Fankhauser, S.L. Harmer, et al., Rhythmic growth

explained by coincidence between internal and external cues, Nature 448 (2007) 458–463.
[24] J.W.D. Mauvoisin, E. Martin, F. Atger, A.N. Galindo, L. Dayon, et al., Nuclear proteomics uncovers diurnal regulatory landscapes in mouse liver, Cell Metab. 25 (2007) 102–117.
[25] C. Savvidis, M. Koutsilieris, Circadian rhythm disruption in cancer biology in, Mol. Med 18 (2012) 1248–1260.
[26] T. Roenneberg, T. Kantermann, M. Juda, C. Vetter, K. V. Allebrandt, Light and the human circadian clock, Handb. Exp. Pharmacol. 217 (2013) 311–331.
[27] L. Wulund, A.B. Reddy, A brief history of circadian time: the emergence of redox oscillations as a novel component of biological rhythms, Perspect. Sci. 6 (2015) 27–37.
[28] M. Akashi, A. Okamoto, Y. Tsuchiya, T. Todo, E. Nischida, K. Node, A positive role for PERIOD in mammalian circadian gene expression, Cell Rep. 7 (2014) 1056–1064.
[29] D.E. Blask, S.M. Hill, R.T. Dauchy, S. Xiang, L. Yuan, T. Duplessis, et al., Circadian regulation of molecular, dietary, and metabolic signaling mechanisms of human breast cancer growth by the nocturnal melatonin signal and the consequences of its disruption by light at night, J. Pineal Res. 51 (2011) 259–269.
[30] J.A. Desotelle, M.J. Wilking, N. Ahmad, The control of skin and cutaneous photodamage circadian, Photochem. Photobiol. 88 (2012) 1037–1047.
[31] H.V. Huikuri, M.J. Nimela, S. Ojala, A. Rantala, M.J. Ikaheimo, J. Airaksinem, Circadian rhythms of frequency domain measures of heart rate variability in healthy subjects and patients with coronary artery disease effects of arousal and upright posture, Circulation 90 (1994) 121–126.
[32] S.R. Moore, J. Pruszka, J. Vallance, E. Aihara, T. Matsuura, M.H. Montrose, et al., Robust circadian rhythms in organoid cultures from PERIOD2: LUCIFERASE mouse small intestine, Dis. Model Mech. 7 (2014) 1123.
[33] V. Ben, B. Bruguerolle, Effects of bilateral striatal 6-OHDA lesions on circadian rhythms in the rat: a radiotelemetric study, Life Sci. 67 (2000) 1549–1558.
[34] P.V. Belancio, D.E. Blask, P. Deiniger, S.M. Hill, S.M. Jazwinski, The aging clock and circadian control of metabolism and genome stability, Front. Genet. 5 (2015) 455.
[35] S.S. Campbell, D.F. Kripke, J.C. Gillin, J.C. Hrubovcak, Exposure to light in healthy elderly subjects 789 and Alzheimer's patients, Physiol. Behav. 42 (1988) 141–144.
[36] E. Poggiogalle, H. Jamshed, C.M. Peterson, Circadian regulation of glucose, lipid, and energy metabolism in humans, Metabolism 84 (2018) 11–27.
[37] L.K. Fonken, R.J. Nelson, The effects of light at night on circadian clocks and metabolism, Endocr. Rev. 35 (2014) 648–670.
[38] N. Simola, M. Morelli, A.R. Carta, The 6-hydroxydopamine model of Parkinson's disease, Neurotox. Res. 11 (2007) 151–167.
[39] R.A. Stone, M.T. Pardue, P.M. Iuvone, T.S. Khurana, Pharmacology of myopia and potential role for intrinsic retinal circadian rhythms, Exp. Eye Res. 114 (2013) 35–47.
[40] R.G. Stevens, Electric light causes cancer? Surely you're joking, Mr. Stevens, Rev. Mutat. Res. 682 (2009) 1–6.
[41] A.L. Colin-Gonzalez, G. Aguilera, I.N. Serratos, B.M. Escribano, A. Santamaria, I. Tunez, On the relationship between the light/dark cycle, melatonin and oxidative stress, Curr. Pharm. Des. 21 (2015) 3477–3488.
[42] X.P. Wu, L. Zhang, W.R. Liao, J.P. Duan, H.Q. Chen, G.N. Chen, Study on the electrochemical behavior of melatonin with an activated electrode, Electroanalysis 14 (2002) 1654–1660.
[43] D. Zetner, A.J. Rosenberg, Melatonin as protection against radiation injury: a systematic, review, Drug Res. (Stuttg) 66 (2016) 281–296.
[44] E.L. Haus, M.H. Smolensky, Shift work and cancer risk: potential mechanistic roles of circadian disruption, light at night, and sleep deprivation, Sleep Med. Rev. 17 (2013) 2737–2784.
[45] T.A. LeGates, D.C. Fernandez, S. Hattar, Light as a central modulator of circadian rhythms, sleep and affect, Nat. Rev. Neurosci. 15 (2014) 443–454.
[46] V.P. Belancio, D.E. Blask, P. Deininger, S.M. Hill, M. Jazwinski, The aging clock and circadian control of metabolism and genome stability, Front. Genet. 14 (5) (2015) 455.
[47] T.M. Burke, The Influence of Melatonin, Caffeine and Bright Light on Human Circadian Physiology (Thesis), University of Colorado at Boulder, 2009.
[48] P.F. Innominato, V.P. Roche, O.G. Palesh, A. Ulusakarya, D. Spiegel, F.A. Levi, The circadian timing system, Ann. Med. (London, UK) 46 (2014) 191–207.
[49] C.J. Madrid-Navarro, R. Sanchez-Galvez, A. Martinez-Nicolas, R. Marina, J.A. Garcia, J.A. Madrid, et al., Disruption of circadian rhythms and delirium, sleep impairment and sepsis in critically ill patients. Potential therapeutic implications for increased light-dark contrast and melatonin therapy in an ICU environment., Curr. Pharm. Des. 21 (2015) 3453–3468.
[50] E.D. Buhr, J.S. Takahaschi, Molecular components of the mammalian circadian clock, Handb. Exp. Pharmacol. 217 (2013) 3–27.
[51] F. Bellivier, P.A. Geoffroy, B. Etain, J. Scott, Sleep- and circadian rhythm-associated pathways as therapeutic

targets in bipolar disorder, Expert. Opin. Ther. Targets 19 (2015) 747–763.
[52] S.A. Kunst, Die tageszeitliche Regulation der Genexpression in Retina und Fotorezeptorzelle, Thesis, Johannes Gutenberg-Universität Mainz, 2016 (Fig. 4.2).
[53] A. Agarwal, S. Gupta, R.K. Sharma, Role of oxidative stress in female reproduction, Reprod. Biol. Endocrinol. 3 (2005) 28.
[54] F.H. Abdel-Rahman, Texas Southern University Department of Biology, PhD Thesis, Sept, 1990.
[55] S. Melrose, Seasonal affective disorder: an overview of assessment and treatment approaches, Depress. Res. Treat 2015 (2015) 178564.
[56] R. Chakraborty, L.A. Ostrin, D.L. Nickla, P.M. Iuvone, M.T. Pardue, R.A. Stone, Circadian rhythms, refractive development, and myopia, Ophthalmic Physiol. Opt. 38 (2018) 217–245.
[57] D.G. McMahon, P.M. Iuvone, G. Tosini, Circadian organization of the mammalian circadian: from gene regulation to physiology and diseases, Prog. Retinal Eye Res. 39 (2014) 58–76.
[58] M.P. Giannoccaro, E. Antelmi, G. Plazzi, Sleep and movement disorders, Curr. Opin. Neurol. 26 (2013) 428–434.
[59] A. Arjona, A.C. Silver, W.E. Walker, E. Fikrig, Immunity's fourth dimension: approaching the circadian-immune connection, Trends Immunol. 33 (2012) 607–612.
[60] J. Brainard, M. Gobel, B. Scott, M. Koeppen, T. Eckle, Health implications of disrupted circadian rhythms and the potential for daylight as therapy, Anesthesiology 122 (2015) 1170–1175.
[61] K. Uchida, N. Okamoto, K. Ohara, Y. Morita, Daily rhythm of serum melatonin in patients with dementia of the degenerate type, Brain. Res. 717 (1996) 154–159.
[62] R. Arey, M.A. McClung, An inhibitor of casein kinase 1 ε/δ partially normalizes the manic-like behaviors of the ClockΔ19 mouse, Behav. Pharmacol. 23 (2012) 392–396.
[63] R. Arey, C.A. MacClung, An inhibitor of casein kinase 1 ε/δ partially normalizes the maniac-like behaviors of the ClockΔ19 mouse, Behav. Pharmacol. 23 (2012) 392–396.
[64] Official notice of the Nobel Assembly at Karolinska Institutet, The Nobel Prize in Physiology or Medicine, 2017.
[65] Circadian rhythm, Wikipedia, on October 27 2019, 5 pm.
[66] E. Vieira, P.B. Burris, I. Quesada, Clock genes, pancreatic function, and diabetes, Trends Mol. Med. 20 (2014) 685–693.
[67] M. Sant, T. Aereleid, F. Berrino, M. Bielska Lasota, P.M. Carli, Faivre, et al., EUROCARE-3: survival of cancer patients diagnosed 1990–94—results and commentary, Ann. Oncol. 14 (2003) v61–v118.
[68] C. Dubowy, A. Sehgal, Circadian rhythms and sleep in *Drosophila melanogaster*, Genetics 205 (2017) 01373–01397.
[69] V. Chintapalli, J. Wang, J.A.T. Dow, Using FlyAtlas to identify better *Drosophila melanogaster* models of human disease, Nat. Genet. 39 (2007) 715.

Further reading

C.T. Karlsson, B. Malmer, F. Wiklund, H. Grönberg, Breast cancer as a second primary in patients with prostate cancer--estrogen treatment or association with family history of cancer? J. Urol. 176 (2006) 538–543.

CHAPTER

5

(Photo)chemotherapeutic

In this section, photochemistry will wear the candid robes of the truth and exploit what has been learned for developing useful methodologies. These are two main issues, the use of a sensitizer that absorbs strongly in the ultraviolet A (UVA) and sensitizes a compound that is present and carry out a sensitized process (and here about 50 years of experience suggest psoralens in a process that has come to be called the psoralen UVA, or the PUVA, process), or direct irradiation is chosen by using NB-UVB. Although the activity spectrum of NB-UVB has been measured several times [1], no attempted rationalization is present in the literature. More importantly for the applications is the fact that the two methods are comparable, in the sense that 20 sessions are required for clearance, although the cumulative dose for achieving clearance is much lower with NB (ca.26 J/cm^2) than with PUVA (c.191 J/cm^2).[1] Patients in the narrowband UVB (NB-UVB) group were treated three times weekly with a starting dose of 500 mJ/cm^2. Next dose was increased by 20% of the previous, to a maximum of 2000 mJ/cm^2. Patients underwent up to 25 sessions. The dosage was adjusted if patients developed erythema. Patients in the PUVA group were treated three times weekly with a starting dose of 2 J/cm^2, increasing by 20% each session, to a maximum of 15 J/cm^2 per dose. Patients underwent up to 25 sessions and the dose was adjusted correspondingly (Fig. 5.1).

Thus in principle NB-UVB is to be preferred, although in some cases PUVA is to be preferred, as examined in some details below.

Irradiation in the presence of psoralens (furocumarines) is a large ly used system for remediating disorders such as psoriasis, vitiligo, and many other related chemotherapeutics. The choice of psoralens may appear reckless, as these compounds are expected to have relatively long-lived triplet states of radical nature and are exactly what nature carefully avoids, particularly in view of the radical chain amplification caused by such species, particularly effective in a nonhydrogen donating solvent as water. More precisely, α,β-unsaturated esters and ketones, the moiety present in such compounds are both quite active as abstracting agents, thus efficiently generating radicals, and excellent radical traps, and are in fact the most used class of compounds in preparative radical chemistry (Scheme 5.1). Clearly, the more

[1] In the considered paper, two UV machines were used: (1) Medisun 2800, whole body exposure units fitted with fluorescent lamps (Philips TL100W/01), the intensity (3.1 mW/cm^2) was checked monthly. (2) Houva II, whole body exposure units fitted with 24 UVA lamps, intensity (11.1 mW/cm^2) was measured monthly.

Light, Molecules, Reaction and Health
DOI: https://doi.org/10.1016/B978-0-12-811659-3.00005-0

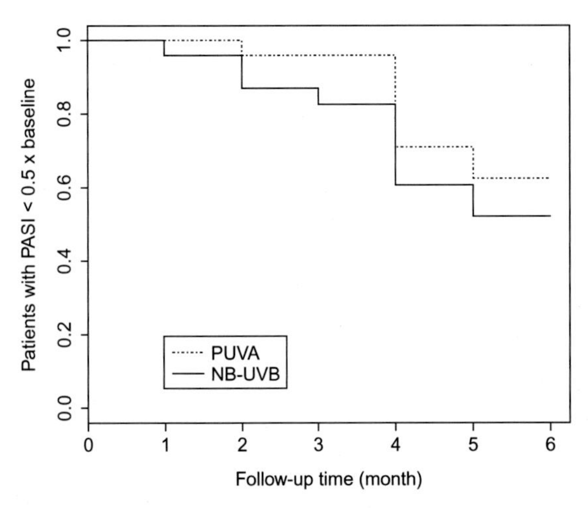

FIGURE 5.1 Kaplan-Meier's curve of relapse after therapy [2]. *NB-UVB*, Narrowband ultraviolet B; *PUVA*, psoralen ultraviolet A.

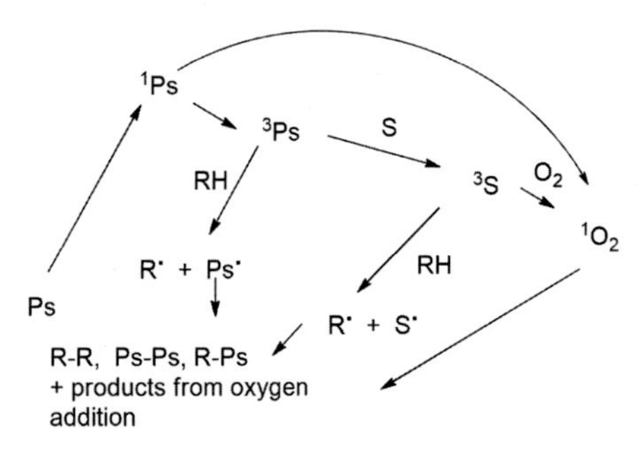

SCHEME 5.1 Psoralen (Ps) forms the triplet states on irradiation. This has radical character and initiates radical chain decomposition. Alternatively, in the first step energy transfer to some other chromophore (S) may occur.

precise is the excitation, the lesser is the risk that potentially absorbing compounds present do absorb and the course of the reaction is less controlled. In fact, the availability of new narrow band lamps emitting mainly in the UVA1 has considerably ameliorated the results [3,4]. In the same direction goes the development of easy to handle excimer lasers (in the usual configuration emitting at 308 nm) [5].

In the meantime, the action spectrum of PUVA was defined in the UVA range (320–400 nm) [6]. Special UVA light boxes were constructed for total body irradiation. At first, the 8-methoxypsoralen (8-MOP) was applied topically. Within a few years, high-intensity UVA bulbs became available, which permitted the 8-MOP to be administered orally. Photochemotherapy or PUVA treatment as we know it today had begun. The reaction occurring may involve both the hydrogen transfer to the carboxyl oxygen, or initial energy transfer, in both cases thus leading to the classic radical chemistry. In this case oxygen (or air) equilibrated solutions are used, energy transfer to that gas becomes a further possibility, leading to hydroperoxides.

Hydrogen transfer to triplet psoralens (^{3}Ps) from a substrate and then a series of radical steps, or the initial energy transfer to the substrate from it and the ensuing steps as above [6–8].

The formation of plaques and alternate paths of pigmented/depigmented areas on the skin have been described (psoriasis, vitiligo, in Sanskrit *shvitra*). Known from ancient times, vitiligo is an autoimmune disease directed against melanocytes characterized by depigmented/hypopigmented patches. Modern science confronts this disorder by treating with topical corticosteroids, topical immunomodulators, phototherapy including PUVA, and by surgical options including autologous mini punch grafting, blister roof grafting, and epidermal cell transplantation, all of them are not cost-effective methods [9].

In order to have always an efficient instrument at their disposal, operators are recommended to abide to recommendations (Table 5.1).

The exposure to UV radiations and visible light, or phototherapy, is a well-known therapeutic tool available for the treatment of many dermatological disorders. The continuous medical and technological progresses, of the last 50 years, have involved the field of phototherapy, which evolved from UVA and PUVA

TABLE 5.1 Recommendation for operator of UV lamps [9].

1. Calibrate the unit with a radiometer every 2–4 weeks;
2. Change all bulbs at the same time every 8–10 months for regularly used machines or when the UV meter shows that the power has reached 3 or 4 mW/cm^2;
3. Avoid phototherapy sessions on consecutive days to prevent burn on burn;
4. In event of a burn, reduce the last PUVA or UVB dose by 50%, reinitiating when erythema has fully disappeared, which might take 2 days to 2 weeks;
5. Nausea from oral psoralens can be avoided by taking it with a fixed amount of milk or food, taking an antiemetic with meal prior to dosing, taking five ginger tablets 15 min before dosing, or dividing the dosage;
6. To achieve 0.03% concentration of 8-methoxypsoralen for hand and foot soak PUVA, dissolve a 10-mg tablet or 1.0 cc of Oxsoralen solution 1% in 3 L of water;
7. To achieve 0.000075% concentration of 8-methoxypsoralen for full-body soak PUVA, dissolve 60-mg 8-methoxypsoralen in 80 L of water (i.e., a bath tub).

in its various forms, to the development of NB-UVB and NB-UVB micro-focused phototherapies. Further advances in technology have now permitted the introduction of new devices emitting UV-A1 radiations, both lamps and excimer lasers, in the most used assembling, 308 nm [10].

Phototherapy is now also used successfully with biological agents as combination therapy to treat recalcitrant psoriasis. Therefore though one of the oldest therapeutic modalities for psoriasis, phototherapy remains a mainstay treatment with promise for further advancement [11]. In the same direction, targeted phototherapy by using the excimer lasers holds potential for more aggressive and effective treatment and long-lasting remission of psoriasis [12].

5.1 Psoralen ultraviolet A process versus narrowband ultraviolet B other treatments

NB-UVB phototherapy and PUVA photochemotherapy are widely used phototherapeutic modalities for a range of skin diseases. The main indication for NB-UVB and PUVA therapies is psoriasis, and other key diagnoses include atopic eczema, vitiligo, cutaneous T-cell lymphoma (CTCL), and the photodermatoses. The decision on choice of phototherapy is important and NB-UVB is usually the primary choice. NB-UVB phototherapy is a safe and effective therapy which is usually considered when topical agents have failed. PUVA requires prior psoralen sensitization but remains a highly effective mainstay therapy, often used when NB-UVB fails, there is rapid relapse following NB-UVB or in specific indications, such as pustular or erythrodermic psoriasis [13].

NB-UVB phototherapy and PUVA photochemotherapy are widely used light-based treatments for a range of diverse skin diseases and can be highly effective, well-tolerated, safe, cost-saving, and reduce the need for topical therapies [14]. The main indication for NB-UVB or PUVA is psoriasis [15] but other mainstay indications include atopic dermatitis or dermatitis of other cause, vitiligo, CTCL, and a range of other conditions, including the photodermatoses, pityriasis rubra pilaris, urticaria, aquagenic pruritus, urticaria pigmentosum, pityriasis lichenoides, lichen planus, granuloma annulare, alopecia areata, and graft versus host disease (Table 5.2).

The decision on choice of phototherapy is important and NB-UVB is usually the primary choice [16]. NB-UVB phototherapy is a safe and effective therapy which is usually considered

TABLE 5.2 Primary and secondary disorders to which psoralen ultraviolet A (marked with an asterisk) or ultraviolet B photochemotherapy applied [16].

Psoriasis
Pustular or erythrodermic*
Eczema--atopic or other type
Vitiligo
Cutaneous T-cell lymphoma
Patch
Plaque*
Photodermatoses
Polymorphic light eruption, actinic prurigo, solar urticaria, hydrozoa vacciniforme
Erythropoietic protoporphyria
Chronic actinic dermatitis*
Urticaria
Urticaria pigmentosa
Aquagenic pruritus
Mastocytosis
Generalized pruritus
Secondary to cholestasis or uremia
Pityriasis lichenoides chronica
Lichen planus
Granuloma annulare
Graft versus host disease
Alopecia areata*
Pityriasis rubra pilaris*
Hand and foot eczema*
Palmoplantar pustulosis

when topical agents have failed. PUVA requires prior psoralen sensitization but remains a highly effective mainstay therapy, often used when NB-UVB fails, there is rapid relapse following NB-UVB or in specific indications, such as pustular or erythrodermic psoriasis. The use of NB-UVB and PUVA therapies and a comparative information on these important dermatological treatments are briefly summarized below [16,17].

As indicated previously, the use of psoralenes or the direct irradiation at about 310 nm is intrinsically dangerous because of the large absorption of DNA in that region and thus of the possible cancer induction. Oral 8-MOP-UVA (PUVA) and narrowband are effective and widely used treatments for chronic plaque psoriasis. Although the role of PUVA therapy in skin carcinogenesis in humans with psoriasis has been clearly demonstrated, there is still controversy regarding the risk of skin cancer with NB-UVB. In practical terms, what is sought is how many doses can be administered, of which amount, with which frequency. A systematic literature search was carried out in Medline, Embase, and Cochrane Library databases from 1980 to December 2010 in English and French, with the keywords "psoriasis" AND "UVB therapy" AND "UVA therapy" AND "cancer" AND "skin" OR "neoplasm" OR "cutaneous carcinoma" OR "melanoma." Of the resulting 243 identified references [18] assessed the risk of nonmelanoma skin cancers (NMSC) following PUVA. All publications referring to the US prospective PUVA follow-up study revealed an increased risk of NMSC with the following characteristics: risk most pronounced for squamous cell carcinomas developing even with low exposures and increasing linearly with the number of sessions, tumors occurring also on nonexposed skin including invasive penile tumors, a risk persisting after cessation of treatment. An increased risk of basal cell carcinomas was observed in patients receiving more than 100 PUVA sessions. The four prospective European studies selected in a recent review and most of the pre-1990 European and US retrospective studies failed to find a link between exposure to PUVA and skin cancer. Only the most recent cohorts, including three large long-term retrospective European studies comparing records with their respective national cancer registries reported on an independent increased risk of NMSC with PUVA. The risk was lower as compared to the US prospective PUVA follow-up study. Six studies assessed the risk of melanoma following PUVA therapy: two of the three US publications coming from the same PUVA prospective follow-up study revealed an increased risk with more than doubled incidence of both invasive and in situ melanoma among patients exposed to at least 200 PUVA treatments compared with patients exposed to lower doses, whereas the three retrospectives European studies, comparing the incidence of melanoma in PUVA users with national cancer registers, did not find any increased risk of melanoma [14,19].

5.1.1 Eczema

Whilst any light-based treatment approach is less straightforward for eczema than psoriasis, not least for the reason of flaring of eczema in the early stages of treatment mainly due to the heat load of therapy, both NB-UVB and PUVA can be highly effective for the treatment of atopic eczema and other forms of eczema. However, the evidence-based is relatively weak and there are no prospective studies comparing head-to-head systemic PUVA with NB-UVB [20]. Systemic 5-MOP PUVA was shown to be superior to medium dose UVA1 for atopic eczema in an intraindividual randomized controlled comparison study [21,22]. Further methods successfully used have been bath PUVA, using 8-MOP and NB-UVB. It appears that NB-UVB is usually be the first line of choice for atopic eczema, including the use for children [20,23].

5.1.2 Vitiligo

For the treatment of vitiligo, NB-UVB has been shown to be superior to PUVA with respect to rates of repigmentation, particularly for unstable extensive vitiligo, and in achieving more cosmetically acceptable even return of the normal color [6,24]. Thus NB-UVB would be the phototherapy of choice for vitiligo, although PUVA may be considered in certain cases, particularly if there is lack of response to NB-UVB.

5.1.3 Cutaneous T-cell lymphoma

Both NB-UVB and PUVA have been found to be effective for early stage CTCL, reaching 81% of patients with early-stage CTCL achieved complete remission with the former one and the 71% with the latter [25], with complete remission in approximately three quarters of patients being achievable, often within 6 months [26]. Further studies have supported the efficacy of both methods in achieving remission of early stage CTCL and thus NB-UVB should be the phototherapy of choice for early patch stage CTCL disease. In another study, it was found that more doses of romidepsin were helpful [27]. Most patients (71%) had advanced stage disease (disease, although not worst than the other group.≤ IIB). The response rate was 34% (primary end point), including six patients with complete response (CR). Twenty-six of 68 patients (38%) with advanced disease achieved a response, including five CRs. The median time to response was 2 months, and the median duration of response was 15 months. A clinically meaningful improvement in pruritus was observed in 28 (43%) of 65 patients, including patients who did not achieve an objective response [27] (Scheme 5.2).

Whether initially formed radicals results from a thermal or photochemical processes, the reaction occurs via chain processes. For thicker plaque stage CTCL, the increased depth of penetration of PUVA makes this the method of choice, although for tumor stage disease, PUVA as monotherapy would not suffice and combination therapy is likely to be required [28]. Adjunctive agents should be considered and combination with retinoids, rexinoids, or interferon may be required in later stages, as well as the use of radiotherapy for localized tumor stage disease or total skin electron beam treatment for more extensive involvement, up to photopheresis for Sezary syndrome [29] (Scheme 5.3).

5.1.4 The photodermatoses

Trial evidence supports the use of NB-UVB and PUVA for the abnormal photosensitivity conditions, with similar efficacy, also for desensitization of Polymorphous Light Eruption (PLE), generally administered as a regular annual desensitization course may be required from a relatively young age. NB-UVB is preferred for PLE as the phototherapy of choice, although PUVA should be considered for treatment failures and when reported its use may be

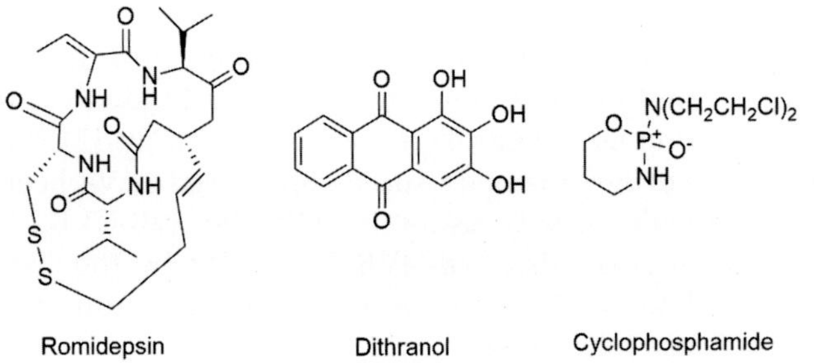

Romidepsin Dithranol Cyclophosphamide

SCHEME 5.2 Radical traps used in PUVA reactions (see the weak S–S bond in romidepsin, the activate phenolic bond in ditranol).

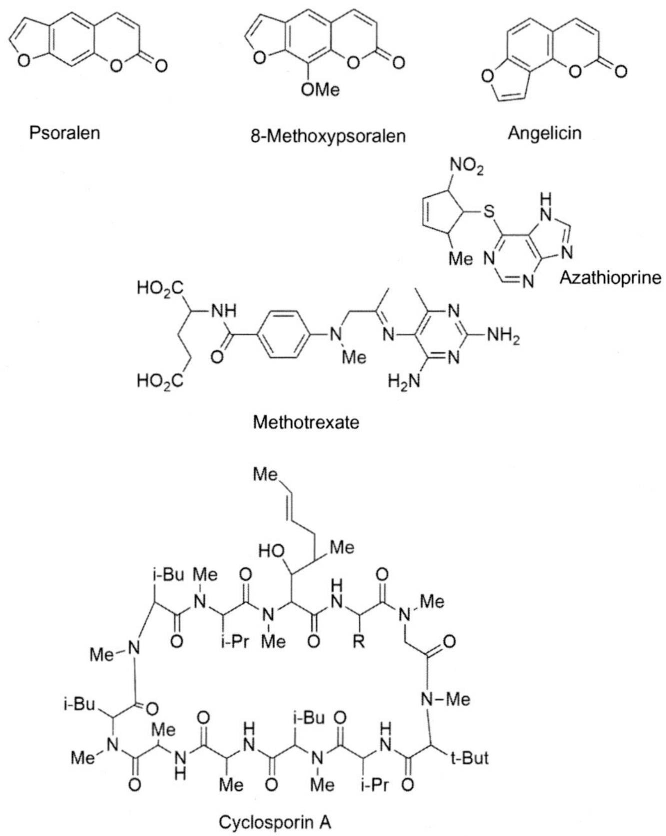

Psoralen 8-Methoxypsoralen Angelicin

Azathioprine

Methotrexate

Cyclosporin A

SCHEME 5.3 Some agents that promote radical decomposition. Psoralen and relative compound, such as angelicin, react photochemically as shown in Scheme 5.1. The highly toxic cyclosporin A would be the structure of choice for a water soluble radical trap due to the ionic groups and the many hydrogen donating groups.

for more severe PLE [30]. Induction of PLE during treatment is common and to be expected but usually not to the point that an early termination of the course is required, and a reduction of dose increments and a reduction of the dose is sufficient [31]. Desensitization phototherapies with either NB-UVB or PUVA are to be considered and appropriate with other phototherapies, but will depend on the action spectrum for induction of abnormal photosensitivity and thus which light-based treatment approach can be tolerated. In general, these patients should be investigated and managed through a specialist in photodermatology unit as there may be additional needs, such as inpatient requirements for suppression and light-protected care and advice regarding subsequent natural sunlight top up exposure. In chronic actinic dermatitis, the action spectrum for induction of abnormal photosensitivity is usually maximal in the UVB region and therefore NB-UVB phototherapy cannot often be tolerated. In this setting PUVA may need to be considered, sometimes in combination with topical superpotent or systemic corticosteroids in order to reduce the risk of disease flare, particularly in the early stages of treatment [32].

NB-UVB and PUVA may also be useful therapeutic approaches for the other photodermatoses, such as erythropoietic protoporphyria (EPP), hydroa vacciniforme, actinic prurigo, and idiopathic solar urticaria [33]. Patients with EPP will usually require annual treatment courses from a young age, NB-UVB is advised and the recourse to PUVA is rarely justified [34,35].

5.1.5 Localized hand, foot, and mouth disease

Hand and foot dermatoses make up a varied group of conditions, which include hyperkalemia. There is no reason to consider that one approach will suit all conditions, but undoubtedly NB-UVB and PUVA photochemotherapy may be useful for localized hand and foot dermatoses, as an example for eczema of the palms and soles, or oral disorders. The depth of penetration of 8-MOP systemic PUVA may be desirable for recalcitrant hand and foot dermatitis and other studies found satisfactory results with oral PUVA for hand and foot eczema. Such a disease is prone to epidemic diffusion as is happened in Singapore in 2000 that lasted 2 months and required a large effort, but inspired a detailed plan to cope with such events [36].

In contrast, topical PUVA has not been shown to be superior to placebo or any other active treatment. For palmoplantar pustulosis, again oral PUVA either as monotherapy or combined with retinoids, may be highly effective, although wide scope reports are not yet available. Both NB-UVB and PUVA may be effective for urticaria and indeed randomized controlled trial evidence demonstrates the superior efficacy of NB-UVB plus antihistamine compared with antihistamine alone [37]. More recently, superiority of NB-UVB compared with PUVA has been shown for urticaria [38], and thus NB-UVB should be considered as a treatment option if antihistamines and other pharmacological therapies fail and may provide useful disease remission. Other conditions effectively treated by NB-UVB and PUVA include pityriasis lichenoides [39], granuloma annulare [40], urticaria pigmentosa, and cutaneous mastocytosis [41], aquagenic pruritus [18], lichen planus [42,43], alopecia areata [44], generalized pruritus, such as secondary to uremia or cholestasis [45], and graft versus host disease [6] that at any rate would be difficult to treat otherways. For conditions such as pityriasis rubra pilaris, which may be aggravated and flared by the use of NB-UVB, 8-MOP systemic PUVA should be considered.

A look to what presented above confirms the important possibilities both direct irradiation and PUVA have, in view of their easy administration, good tolerability and with few adverse effects, with excellent disease remission. It should be noted, however, that although erythema is always a risk, this may be decreased by the concomitant use of phototoxic drugs [46].

Psoralen is used either systemic or topically (usually now 8-MOP as bath, soak, gel, cream, or lotion). The mechanism of action of PUVA is quite distinct from that of UVB or of UVA alone, with PUVA inducing a delayed erythemal reaction peaking around 96 h after irradiation of psoralen-sensitized skin [47,48]. This contrasts with the peak time for development of erythema after NB-UVB exposure of 12–24 h. With systemic PUVA, appropriate skin and eye protection must be used for 24 h after psoralen ingestion. Oral 8-MOP may cause some gastrointestinal upset, although switching to 5-MOP minimizes this adverse effect and of course this is not an issue with topical PUVA. However, PUVA treatment can be highly effective and very safely administered in any dermatology department with a significantly sized phototherapy unit.

With the exception of less common adverse effects such as PUVA pain, treatment is otherwise usually well tolerated [6,15]. Undoubtedly, there is a longer term risk of skin carcinogenesis with high numbers of PUVA exposures, but the risks can be minimized by vigilance, limitation of lifetime numbers of PUVA exposures, and avoidance of the use of maintenance PUVA where possible.

It is essential that adequate governance is ensured for the safe delivery of both NB-UVB and PUVA during therapies to children, young people, and women. In Scotland, a National

Managed Clinical Network for phototherapy has been established (Photonet; www.photonet.scot.nhs.uk), which employs a central database (Photolysys), enabling standardization of treatment protocols, recording of treatment parameters, and outcomes and facilitating linkage studies to ascertain longer term risks of treatment, notably skin cancer risk [49], also for children and childbearing women.

5.1.6 Psoriasis

In an initial controlled comparative half-body study in 10 patients with widespread psoriasis, no significant difference in efficacy was seen between twice-weekly NB-UVB or systemic PUVA and the same result was obtained, observation was also reported in a separate intraindividual open nonrandomized controlled paired comparison study of three times weekly NB-UVB and PUVA. However, there was a trend to superior efficacy with PUVA and this was particularly evident for patients with a higher baseline European Medicines Agency (PASI) score [50], possibly suggestive of a role for PUVA in more severe psoriasis or relapsing psoriasis, although given the convenience of NB-UVB this would generally be the preferred initial approach. In a separate interindividual study of 100 patients with psoriasis, twice-weekly PUVA was superior in efficacy to twice-weekly NB-UVB, with 35% of patients still being clear at 6 months after completion of PUVA, compared with only 12% after NB-UVB [6].

Phototherapy was successfully applied with patients living in remote areas far from their nearest hospital phototherapy unit [51], despite the fact that dermatologists were of the opinion that home phototherapy should be used with caution, and that general, nonevidence-based opinions were widespread about this form of therapy.

For patients undergoing home therapy for the first time, the phototherapy nurse should clearly explain the characteristics and the risks and provide a regular, close supervision. The nurse should contact the patient regularly during the course of treatment, to ensure that progress is satisfactory and that they understand how to carry out the self-treatment. This contact also permits the nurse to advise the patient on the next three dosages of UVB. Such close follow-up of patients minimizes the risk of incorrect usage, and may also allow any adverse side effects to be detected earlier. Each phototherapy session is recorded in the equipment's memory so that the nurse can verify at any time whether the recommended protocol is being followed correctly [52].

The NB-UVB is considered the first-line treatment for extensive plaque-type psoriasis, and advocated the use of home phototherapy to encourage patient access. Other studies have also concluded that home phototherapy is an effective and safe form of therapy for photoresponsive diseases when used with appropriate guidelines and patient education [6].

Clinical trial evidence for home phototherapy is robust, showing noninferiority for efficacy and safety compared with hospital-based phototherapy within optimized clinical governance systems. Despite this, uptake and implementation of home phototherapy in NHS dermatology departments remains poor. Contemporary clinical guidelines for home phototherapy are now needed to establish a clinical governance framework for the UK within which home phototherapy can be extended more widely. There is growing recognition that investment in phototherapy services to improve patient access is likely to reduce the number of patients progressing along the pathway of systemic therapies (including the use of biologics). When considered within a whole system, excellent phototherapy services, including home phototherapy, provide choice for patients, and may be less costly than whole systems that neglect phototherapy as an important treatment option [53].

5.1.7 Treatment for palmoplantar

Palmoplantar psoriasis and palmoplantar pustulosis are chronic skin diseases with a large impact on patient quality of life. They are frequently refractory to treatment, being generally described as a therapeutic challenge. Classical management of mild-to-moderate palmoplantar pustulosis and palmoplantar psoriasis relies on use of potent topical corticosteroids, phototherapy, and/or acitretin. Nevertheless, these drugs have proven to be insufficient in long-term control of extensive disease. Biological therapy-namely antiinterleukin-17 agents, and phosphodiesterase type 4 inhibitors-has recently shown promising results in the treatment of palmoplantar psoriasis [54].

Apitherapy is the medical use of honey bee products, honey, propolis, royal jelly, bee wax, and bee venom to relieve human ailments. Propolis is one of the most well-documented products derived from the honey bee and has always played an important role in traditional folk medicine [55].

5.2 Theranostics

As several times mentioned above, excitation of suitable molecules often causes the liberation of a compound active as a drug. In this way, the "best" amount of a medicine is established avoiding too low and too high (and thus toxic doses). As an example, the medical speciality of theranostic nuclear oncology has taken more than the 70 years to learn to avoid single-center reports and rather give weight to the quantitative and personal aspects of medicine. Theranostics is the epitome of personalized medicine. The specific tumor biomarker is quantitatively imaged on positron emission tomography/computed tomography (CT) or single photon emission. If it is clearly demonstrated that a tumoricidal radiation absorbed dose can be delivered, the theranostic beta or alpha-emitting radionuclide pair, coupled to the same targeted molecule, is then administered, to control advanced metastatic cancer in that individual patient. This prior selection of patients who may benefit from theranostic treatment is in direct contrast to the evolving oncological indirect treatments using immune-check point inhibitors, where there is an urgent need to define biomarkers which can reliably predict response, and thus avoid the high cost and toxicity of these agents in patients who are unlikely to benefit. The immune and molecular treatment approaches of oncology are a recent phenomenon and the efficacy and safety of immune-check point blockade and chimeric antigen receptor T-cell therapies are currently under evaluation in personalize controlled trials. Such objective evaluation is compromised by the inadequacy of conventional response evaluation criteria in solid tumor CT/MR anatomical/functional imaging to define tumor response, in both immuneoncology and theranostic nuclear oncology. This introduction to the clinical practice of theranostics explores ways in which nuclear physicians can learn from the lessons of history, and join with their medical, surgical, and radiation oncology colleagues to establish a symbiotic collaboration to realize the potential of personalized molecular medicine to control advanced cancer and actually enhance quality of life whilst prolonging survival [56,57].

5.3 Laser for medicine

The following table has been prepared by using Sci-finder with the search terms "laser in medicine/surgery", on August 2, 2019.

1. Main factors affecting the structure and properties of titanium and cobalt alloys manufactured by the 3D printing

By Kazantseva, N.

From Jourmal of Physics: Conference Series (2018), 1115 (6th International Congress on Energy Fluxes and Radiation Effects, 2018), 042008/1–042008/6.

2. Architectures and Mechanical Properties of Drugs and Complezes of Surface-Active Compounds at Air-Water and Oil-Water Interfaces

By Sarker, Dipak K.

From Current Drug Discovey Technologies (2019), 16(1), 11–29.

3. Laser ablation of polymers: a review

By Ravi-Kumar, Sandeep; Lies, Benjamin; Zhang, Xiao; Lyu, Hao; Qin, Hantang

From Polymer International (2019), 68(8), 1391–1401.

4. Microfluidic Model for Evaluation of Immune Checkpoint Inhibitors in Human Tumors

By Beckwith, Ashley L.; Velasquez-Garcia, Luis F.; Borenstein, Jeffrey T.

From Advanced Healthcare Materials (2019), 8(11), n/a.

5. 3D Printing Methods for Pharmaceutical Manufacturing: Opportunity and Challenges

By Warsi, Musarrat H.; Yusuf, Mohammad; Al Robaian, Majed; Khan, Maria; Muheem, Abdul; Khan, Saba

From Current Pharmaceutical Design (2018), 24(42), 4949–4956.

6. Three-Dimensional (3-D) Printing Technology Exploited for the Fabrication of Drug Delivery Systems

By Zeeshan, Farrukh; Madheswaran, Thiagarajan; Pandey, Manisha; Gorain, Bapi

From Current Pharmaceutical Design (2018), 24(42), 5019–5028.

7. 3D Printing Technology in Customized Drug Delivery System: Current State of the Art, Prospective and the Challenges

By Khan, Farooq A.; Narasimhan, Kaushik; Swathi, C. S. V.; Mustak, Sayyad; Mustafa, Gulam; Ahmad, Mohammad Zaki; Akhter, Sohail

From Current Pharmaceutical Design (2018), 24(42), 5049–5061.

8. Molecular mechanisms of skin photoaging and plant inhibitors

By Garg, Chanchal; Khurana, Priyanka; Garg, Munish

From International Journal of Green Pharmacy (2017), 11 (2Suppl.), S217–S232.

9. Hybrid architectures made of nonlinear-active and metal nanostructures for plasmon-enhanced harmonic generation

By Falamas, Alexandra; Tosa, Valer; Farcau, Cosmin

From Optical Materials (Amsterdam, Netherlands) (2019), 88, 653–666.

10. Method of preparation and evaluation of green synthesis silver nanoparticle: an updated review

By Chintamani, Ravindra B.; Salunkhe, Kishor S.; Chavan, Machindra J.

From International Journal of Pharma and Bio Sciences (2018), 9(2), 69–84.

11. Review of the recent advances and applications of LIBS-based imaging

By Jolivet, L.; Leprince, M.; Moncayo, S.; Sorbier, L.; Lienemann, C.-P.; Motto-Ros, V.

From Spectrochimica Acta, Part B: Atomic Spectroscopy (2019), 151, 41–53.

12. An Overview of 3D Printing Technologies for Soft Materials and Potential Opportunities for Lipid-based Drug Delivery Systems

By Vithani, Kapilkumar; Goyanes, Alvaro; Jannin, Vincent; Basit, Abdul W.; Gaisford, Simon; Boyd, Ben J.

From Pharmaceutical Research (2019), 36(1), 1–20.

13. Diabetic retinopathy: role of traditional medicinal plants in its management and their molecular mechanism

By Gond, Anand Kumar; Gupta, S. K.

From International Journal of Pharmaceutical Science Invention (2017), 6(6), 1−14.

14. Biomedical applications of mid-infrared quantum cascade lasers—a review

By Isensee, Katharina; Kroeger-Lui, Niels; Petrich, Wolfgang

From Analyst (Cambridge, United Kingdom) (2018), 143 (24), 5888−5911.

15. Pharmacokinetic considerations concerning the use of bronchodilators in the treatment of chronic obstructive pulmonary disease

By Matera, Maria Gabriella; Rinaldi, Barbara; Page, Clive; Rogliani, Paola; Cazzola, Mario

From Expert Opinion on Drug Metabolism & Toxicology (2018), 14(10), 1101−1111.

16. Laser-plasma particle sources for biology and medicine

By Giulietti, Antonio; Bussolino, Giancarlo; Fulgentini, Lorenzo; Koester, Petra; Labate, Luca; Gizzi, Leonida A.

From Springer Series in Chemical Physics (2015), 112 (Progress in Ultrafast Intense Laser Science, Volume 12), 151−178.

17. Applications of single plasmonic nanoparticles in biochemical analysis and bioimaging

By Lei, Gang; He, Yan

From Wuli Huaxue Xuebao (2018), 34(1), 11−21.

18. 3D printing: basic role in pharmacy

By Bala, Rajni; Madaan, Reecha; Kaur, Amandeep; Mahajan, Kriti

From European Journal of Biomedical and Pharmaceutical Sciences (2017), 4(4), 242−247.

19. Fabrication technologies of the sintered materials including materials for medical and dental application

By Dobrzanski, Leszek A.; Dobrzanska-Danikiewicz, Anna D.; Achtelik-Franczak, Anna; Dobrzanski, Lech B.; Hajduczek, Eugeniusz; Matula, Grzegorz

Edited by:Dobrza5ski, Leszek A

From Powder Metallurgy: Fundamentals and Case Studies (2017), 17−52.

20. MALDI-TOF MS for determination of resistance to antibiotics

By Hrabak, Jaroslav; Dolejska, Monika; Papagiannitsis, Costas C.

Edited by: Kostrzewa, Markus; Schubert, Soeren

From MALDI-TOF Mass Spectrometry in Microbiology (2016), 93−108.

21. Single-cell metabolomics

By Emara, Samy; Amer, Sara; Ali, Ahmed; Abouleila, Yasmine; Oga, April; Masujima, Tsutomu

From Advances in Experimental Medicine and Biology (2017), 965(Metabolomics: From Fundamentals to Clinical Applications), 323−343.

22. Laser applications in surgery

By Azadgoli, Beina; Baker, Regina Y.

From Annals of Translational Medicine (2016), 4(23), 452/1−452/7.

23. Current status and domestic trends of laser treatment techniques

By Ishii, Katsunori; Awazu, Kunio

From Reza Kenkyu (2016), 44(3), 147−1513

24. Laser Irradiation of Metal Oxide Films and Nanostructures: Applications and Advances

25. Elemental imaging using laser-induced breakdown spectroscopy: A new and promising approach for biological and medical applications

By Busser, Benoit; Moncayo, Samuel; Coll, Jean-Luc; Sancey, Lucie; Motto-Ros, Vincent

From Coordination Chemistry Reviews (2018), 358, 70−79.

26. Applications of Raman Spectroscopy in Biopharmaceutical Manufacturing: A Short Review

By Buckley, Kevin; Ryder, Alan G.

From Applied Spectroscopy (2017), 71(6), 1085−1116.

27. Hierarchical Multicomponent Inorganic Metamaterials: Intrinsically Driven Self-Assembly at the Nanoscale

By Levchenko, Igor; Bazaka, Kateryna; Keidar, Michael; Xu, Shuyan; Fang, Jinghua

From Advanced Materials (Weinheim, Germany) (2018), 30 (2), n/a.

28. Bio-interfaces engineering using laser-based methods for controlled regulation of mesenchymal stem cell response in vitro

By Dinca, Valentina; Sima, Livia Elena; Rusen, Laurentiu; Bonciu, Anca; Lippert, Thomas; Dinescu, Maria; Farsari, Maria

Edited by: Perveen, Farzana Khan

From Recent Advances in Biopolymers (2016), 221–251.

29. Efficiency of an Electric-Discharge N2 Laser

By Apollonov, V. V.

From Springer Series in Optical Sciences (2016), 201 (High-Energy Molecular Lasers), 325–332.

30. Overview of PAT process analysers applicable in monitoring of film coating unit operations for manufacturing of solid oral dosage forms

By Korasa, Klemen; Vrecer, Franc

From European Journal of Pharmaceutical Sciences (2018), 111, 278–292

31. Design and functionalization of the NIR-responsive photothermal semiconductor nanomaterials for cancer theranostics

By Huang, Xiaojuan; Zhang, Wenlong; Guan, Guoqiang; Song, Guosheng; Zou, Rujia; Hu, Junqing

From Accounts of Chemical Research (2017), 50(10), 2529–2538

32. The mechanisms and research progress of laser fabrication technologies beyond diffraction limit

By Zhang, Xin-zheng; Xia, Feng; Xu, Jing-jun

From Wuli Xuebao (2017), 66(14), 144207/1–144207/16

33. Craniofacial wound healing with photobiomodulation therapy: new insights and current challenges

By Arany, P. R.

From Journal of Dental Research (2016), 95(9), 977–984

34. Cell processing engineering for regenerative medicine

By Takagi, Mutsumi

From Advances in Biochemical Engineering/Biotechnology (2016), 152 (Bioreactor Engineering Research and Industrial Applications II), 53–74

35. Heteroepitaxy, an amazing contribution of crystal growth to the world of optics and electronics

By Tassev, Vladimir L.

From Crystals (2017), 7(6), 178/1–178/38.

36. Application of high-energy chemistry methods to the modification of the structure and properties of polylactide (a review)

By Demina, T. S.; Gilman, A. B.; Zelenetskii, A. N.

From High Energy Chemistry (2017), 51(4), 302–314.

37. Update of Ablative Fractionated Lasers to Enhance Cutaneous Topical Drug Delivery

By Waibel, Jill S.; Rudnick, Ashley; Shagalov, Deborah R.; Nicolazzo, Danielle M.

From Advances in Therapy (2017), 34(8), 1840–1849.

38. Topical treatment of glaucoma: established and emerging pharmacology

By Dikopf, Mark S.; Vajaranant, Thasarat S.; Edward, Deepak P.

From Expert Opinion on Pharmacotherapy (2017), 18(9), 885–898.

39. 3D bioprinting for drug discovery and development in pharmaceutics

By Peng, Weijie; Datta, Pallab; Ayan, Bugra; Ozbolat, Veli; Sosnoski, Donna; Ozbolat, Ibrahim T.

From Acta Biomaterialia (2017), 57, 26–46.

40. Laser induced breakdown spectroscopy in ecology, biology and medicine (review)

By Agrafenin, A. V.; Bezrukova, P. V.

From Mikroelementy v Meditsine (Moscow, Russian Federation) (2014), 15(4), 8–22.

41. Commentary on "Analyses of human dentine and tooth enamel by laser ablation-inductively coupled plasma-mass spectrometry (LA-ICP-MS) to study the diet of medieval Muslim individuals from Tauste (Spain)" by Guede et al., 2017, Microchemical Journal 130, 287–294

By Lugli, Federico; Cipriani, Anna

From Microchemical Journal (2017), 133, 67–69.

42. Recent update on the role of Chinese material medical and formulations in diabetic retinopathy

By More, Sandeep Vasant; Kim, In-Su; Choi, Dong-Kug

From Molecules (2017), 22(1), 76/1–76/20.

43. Remote Control and Modulation of Cellular Events by Plasmonic Gold Nanoparticles: Implications and Opportunities for Biomedical Applications

By Li, Jiayang; Liu, Jing; Chen, Chunying

From ACS Nano (2017), 11(3), 2403–2409.

44. Combination therapy for melisma

By Zhu, Liping; Liu, Haiyang; Pang, Qin; He, Li

From Zhonghua Pifuke Zazhi (2016), 49(2), 147–150.

45. Peripheral nerve repair: theory and technology application

By He, Xin-ze; Wang, Wei; Hu, Tie-min; Ma, Jian-jun; Yu, Chang-yu; Gao, Yun-feng; Cheng, Xing-long; Wang, Pei

From Zhongguo Zuzhi Gongcheng Yanjiu (2016), 20(7), 1044–1050.

46. An overview of recent developments in metabolomics and proteomics—phytotherapic research perspectives

By Mumtaz, Muhammad Waseem; Abdul Hamid, Azizah; Akhtar, Muhammad Tayyab; Anwar, Farooq; Rashid, Umer; Al-Zuaidy, Mizher Hezam

From Frontiers in Life Science (2017), 10(1), 1–37.

47. Therapeutic Options in Refractory Diabetic Macular Oedema

By Shah, Sanket U.; Maturi, Raj K.

From Drugs (2017), 77(5), 481–492.

48. 3D printing & pharmaceutical manufacturing: opportunities and challenges

By Bhusnure, O. G.; Gholve, S. V.; Dongre, R. C.; Munde, B. S.; Tidke, P. M.

From International Journal of Pharmacy and Pharmaceutical Research (2015), 5(1), 136–175.

49. Long-acting slow effective release antiretroviral therapy

By Edagwa, Benson; McMillan, JoEllyn; Sillman, Brady; Gendelman, Howard E.

From Expert Opinion on Drug Delivery (2017), 14(11), 1281–1291.

50. Bioprinting for vascular and vascularized tissue biofabrication

By Datta, Pallab; Ayan, Bugra; Ozbolat, Ibrahim T.

From Acta Biomaterialia (2017), 51, 1–20.

51. A review on biogenic synthesis of ZnO nanoparticles using plant extracts and microbes: A prospect towards green chemistry

By Ahmed, Shakeel; Annu; Chaudhry, Saif Ali; Ikram, Saiqa

From Journal of Photochemistry and Photobiology, B: Biology (2017), 166, 272–284.

52. Laser-assisted biofabrication in tissue engineering and regenerative medicine

By Koo, Sangmo; Santoni, Samantha M.; Gao, Bruce Z.; Grigoropoulos, Costas P.; Ma, Zhen

From Journal of Materials Research (2017), 32(1), 128–142.

53. Gut flora connects obesity with pathological angiogenesis in the eye

By Scholz, Rebecca; Langmann, Thomas

From EMBO Molecular Medicine (2016), 8(12), 1361–1363.

54. MALDI mass spectrometry in medical research and diagnostic routine laboratories

By Fuh, Manka M.; Heikaus, Laura; Schlueter, Hartmut

From International Journal of Mass Spectrometry (2017), 416, 96–109.

55. Welcoming a new age for gene therapy in hematology

By Weiss, Mitchell J.; Mullighan, Charles G.

From Blood (2016), 127(21), 2523–2524.

56. Nuclear science and applications with the next generation of high-power lasers and brilliant low-energy gamma beams at ELI-NP

By Gales, S.

Edited by: Cherepanov, Evgeni

From Exotic Nuclei, Proceedings of the International African Symposium on Exotic Nuclei, 1st, Cape Town, South Africa, Dec. 2–6, 2013 (2015), 63–77.

57. Application of spectroscopic techniques for monitoring microbial diversity and bioremediation

By Chakraborty, Jaya; Das, Surajit

From Applied Spectroscopy Reviews (2017), 52(1), 1–38.

58. Generation of gaseous singlet oxygen for interaction experiments

By Censky, M.; Jirasek, V.; Schmiedberger, J.; Kodymova, J.

From Chemicke Listy (2016), 110(1), 11–17.

59. Recent development in drug treatment of keloid

By Li, Zhou-na; Zhu, Lian-hua; Fang, Yu-hui; Jin, Zhe-hu

From Zhongguo Pifu Xingbingxue Zazhi (2016), 30(2), 196–199.

60. 3D bioprinting technology in biochemical engineering

By Eom, Tae Yoon

From Hwahak Konghak (2016), 54(3), 285–292.

61. Cancer and electromagnetic radiation therapy: Quo Vadis?

By Makropoulou, Mersini

From arXiv.org, e-Print Archive, Physics (2016), 1–22.

62. Emergence of 3D Printed Dosage Forms: Opportunities and Challenges

By Alhnan, Mohamed A.; Okwuosa, Tochukwu C.; Sadia, Muzna; Wan, Ka-Wai; Ahmed, Waqar; Arafat, Basel

From Pharmaceutical Research (2016), 33(8), 1817–1832.

63. Contributed Review: A review of the investigation of rare-earth dopant profiles in optical fibers

By Sidiroglou, F.; Roberts, A.; Baxter, G.

From Review of Scientific Instruments (2016), 87(4), 041501/1–041501/11. Language: English, Database: CAPLUS, DOI:10.1063/1.4947066.

64. Recent progress of tunable terahertz sources based on difference frequency generation

By Chai, Lu; Niu, Yue; Li, Yan-feng; Hu, Ming-lie; Wang, Qing-yue

From Wuli Xuebao (2016), 65(7), 070702/1–070702/15.

65. Lasers in periodontics

By Singh, Harmandeep; Kaur, Navkiran; Verma, Ashish

From Elixir International Journal (2015), (Oct.), 35664–35669.

66. Application of imaging mass spectrometry for drug discovery

By Hayasaka, Takahiro

From Yakugaku Zasshi (2016), 136(2), 163–170.

67. Review on VUV to MIR absorption spectroscopy of atmospheric pressure plasma jets

By Reuter, Stephan; Sousa, Joao Santos; Stancu, Gabi Daniel; van Helden, Jean-Pierre Hubertus

From Plasma Sources Science & Technology (2015), 24(5), 054001/1–054001/41.

68. Graphene scaffolds in progressive nanotechnology/stem cell-based tissue engineering of the nervous system

By Akhavan, Omid

From Journal of Materials Chemistry B: Materials for Biology and Medicine (2016), 4(19), 3169–3190.

69. Characterization of tissue engineered cartilage products: Recent developments in advanced therapy

By Maciulaitis, Justinas; Rekstyte, Sima; Usas, Arvydas; Jankauskaite, Virginija; Gudas, Rimtautas; Malinauskas, Mangirdas; Maciulaitis, Romaldas

From Pharmacological Research (2016), 113(Part_B), 823–832.

70. Rapid microbiological diagnostics in medicine using electromigration techniques

By Buszewski, Boguslaw; Klodzinska, Ewa

From TrAC, Trends in Analytical Chemistry (2016), 78, 95–108.

71. Novel advances in shotgun lipidomics for biology and medicine

By Wang, Miao; Wang, Chunyan; Han, Rowland H.; Han, Xianlin

From Progress in Lipid Research (2016), 61, 83–108.

72. Direct analysis of traditional Chinese medicines by mass spectrometry

By Wong, Melody Yee-Man; So, Pui-Kin; Yao, Zhong-Ping

From Journal of Chromatography B: Analytical Technologies in the Biomedical and Life Sciences (2016), 1026, 2–14.

73. 1064 nm Q-switched Nd:YAG laser for the treatment of Argyria: a systematic review

By Griffith, R. D.; Simmons, B. J.; Bray, F. N.; Falto-Aizpurua, L. A.; Yazdani Abyaneh, M.-A.; Nouri, K.

From Journal of the European Academy of Dermatology and Venereology (2015), 29(11), 2100–2103.

74. Insight of the application and progress of synchrotron radiation X-ray source in biology

By Zhao, Yan; Liu, Jiao; Xie, Binghe; Feng, Wei; Zhang, Zhen; Yan, Peng; Zuo, Kaili; Li, Li; Wang, Hongfei

From Jiguang Shengwu Xuebao (2015), 24(2), 111–118.

75. Non-invasive subcutaneous fat reduction: a review

By Kennedy, J.; Verne, S.; Griffith, R.; Falto-Aizpurua, L.; Nouri, K.

From Journal of the European Academy of Dermatology and Venereology (2015), 29(9), 1679–1688.

76. Pharmacotherapy and Adherence Issues in Treating Elderly Patients with Glaucoma

By Broadway, David C.; Cate, Heidi

From Drugs & Aging (2015), 32(7), 569–581.

77. New research progress on epidemiology of age-related macular degeneration

By Wu, Ming-xing; Zheng, Zheng; Zhou, Xi-yuan

From Guoji Yanke Zazhi (2015), 15(2), 223–227.

78. Laser acceleration of ions: recent results and prospects for applications

By Bychenkov, V. Yu.; Brantov, A. V.; Govras, E. A.; Kovalev, V. F.

From Physics-Uspekhi (2015), 58(1), 71–102.

79. Current research and future development of organic laser materials and devices

By Zhang, Qi; Zeng, Wen-jin; Xia, Rui-dong

From Wuli Xuebao (2015), 64(9), 094202/1–094202/22.

80. Recent progress in preparation and application of microfluidic chip electrophoresis

By Cong, Hailin; Xu, Xiaodan; Yu, Bing; Yuan, Hua; Peng, Qiaohong; Tian, Chao

From Journal of Micromechanics and Microengineering (2015), 25(5), 1–11.

81. The role of polymers in random lasing

By Sznitko, Lech; Mysliwiec, Jaroslaw; Miniewicz, Andrzej

From Journal of Polymer Science, Part B: Polymer Physics (2015), 53(14), 951–974.

82. MALDI Profiling and Applications in Medicine

By Dudley, Ed

From Advances in Experimental Medicine and Biology (2014), 806(Advancements of Mass Spectrometry in Biomedical Research), 33–58.

83. Recent advances in 3D printing of biomaterials

By Chia, Helena N.; Wu, Benjamin M.

From Journal of Biological Engineering (2015), 9, 1–33.

84. Terahertz spectroscopy and imaging

By Angeluts, Andrei A.; Balakin, Aleksei V.; Borodin, Aleksandr V.; Evdokimov, Maksim G.; Esaulkov, Mikhail N.; Nazarov, Maksim M.; Ozheredov, Il'ya A.; Sapozhnikov, Dmitrii A.; Solyankin, Petr M.; Shkurinov, Aleksandr P.; et al.

From Vestnik RFFI (2014), (3), 21–36.

85. Additive manufacturing techniques for the production of tissue engineering constructs

By Mota, Carlos; Puppi, Dario; Chiellini, Federica; Chiellini, Emo

From Journal of Tissue Engineering and Regenerative Medicine (2015), 9(3), 174–190.

86. Core-shell quantum dots: Properties and applications

By Vasudevan, D.; Gaddam, Rohit Ranganathan; Trinchi, Adrian; Cole, Ivan

From Journal of Alloys and Compounds (2015), 636, 395–404.

87. Research advance of treatment on androgenetic alopecia

By Gan, Chao-nan; Yang, Ding-quan

From Zhongri Youhao Yiyuan Xuebao (2014), 28(1), 53–55.

88. Commentary on "Efficacy and safety of photodynamic therapy for recurrent, high grade nonmuscle invasive bladder cancer refractory or intolerant to bacille Calmette-Guerin immunotherapy." Lee JY, Diaz RR, Cho KS, Lim MS, Chung JS, Kim WT, Ham WS, Choi YD, Department of Urology, Yonsei University College of Medicine, Seoul, Republic of Korea; Severance Hospital and Urological Science Institute, Yonsei University College of Medicine, Seoul, Republic of Korea.

By Kamat, Ashish

From Urologic Oncology: Seminars and Original Investigations (2015), 33(1), 46.

89. Fiber lasers and their applications [Invited]

By Shi, Wei; Fang, Qiang; Zhu, Xiushan; Norwood, R. A.; Peyghambarian, N.

From Applied Optics (2014), 53(28), 6554–6568, 15 pp.

90. Wound healing in urology

By Ninan, Neethu; Thomas, Sabu; Grohens, Yves

From Advanced Drug Delivery Reviews (2015), 82–83, 93–105.

91. Beam by design: Laser manipulation of electrons in modern accelerators

By Hemsing, Erik; Stupakov, Gennady; Xiang, Dao; Zholents, Alexander

From Reviews of Modern Physics (2014), 86(3), 897–941.

92. All solid-state Raman lasers

By Fan, Ya-xian; Wang, Hui-tian

From Liangzi Dianzi Xuebao (2014), 31(4), 394–402.

93. Review: Lanthanide coordination chemistry: from old concepts to coordination polymers

By Bunzli, Jean-Claude G.

From Journal of Coordination Chemistry (2014), 67(23–24), 3706–3733.

94. QD ultrafast and continuous wavelength laser diodes for applications in biology and medicine

By Loza-Alvarez, Pablo; Aviles-Espinosa, Rodrigo; Matcher, Steve J.; Childs, D.; Sokolovski, Sergei G.

Edited by: Rafailov, Edik U

From Physics and Engineering of Compact Quantum Dot-Based Lasers for Biophotonics (2014), 171–229.

95. Field of researches and applications domains for compact and large- scale DPF devices: Current assets, problems and essentials

By Gribkov, V. A.

From International Journal of Modern Physics: Conference Series (2014), 32, 1460314/1–1460314/12.

96. Techniques for microscale patterning of zeolite-based thin films

By Mandal, Swarnasri; Williams, Heather L.; Hunt, Heather K.

From Microporous and Mesoporous Materials (2015), 203, 245–258.

97. Spectroscopic enhancement in nanoparticles embedded glasses

By Sahar, M. R.; Ghoshal, S. K.

From AIP Conference Proceedings (2014), 1617(1, 3rd International Conference on Theoretical and Applied Physics 2013), 12–15.

98. Optical Technologies in Biophysics and Medicine XV; and Laser Physics and Photonics XV. (Saratov Fall Meeting 2013 held in Saratov, Russian Federation 24–28 September 2013). [In: Proc. SPIE, 2014; 9031]

By Genina, Elina A.; Derbov, Vladimir L.; Meglinski, Igor; Tuchin, Valery V.; Editors

From No Corporate Source data available (2014), No pp. given.

99. Research progresses for tumor theranostics based on gold nanomaterials

By Kou, Yu; Wang, Jing; Chen, Chunying

From Zhongguo Zhongliu Linchuang (2014), 41(1), 51–55.

100. Photothermal spectroscopy

By Proskurnin, M. A.

From Woodhead Publishing Series in Electronic and Optical Materials (2014), 43(Laser Spectroscopy for Sensing), 313–361.

101. Carbon nanotubes: a promising tool in drug delivery

By Basu, B.; Mehta, Gunjan Kumar

From International Journal of Pharma and Bio Sciences (2014), 5(1), 533–555.

102. The laser technology: new trends in biology and medicine

By Legres, Luc G.; Chamot, Christophe; Varna, Mariana; Janin, Anne

From Journal of Modern Physics (2014), 5(5), 267–279, 13 pp.

103. Advances in 2-μm Tm-doped mode-locked fiber lasers

By Rudy, Charles W.; Digonnet, Michel J. F.; Byer, Robert L.

From Optical Fiber Technology (2014), 20(6), 642–649.

104. Nano-theranostics with plasmonic nanobubbles

By Lukianova-Hleb, Ekaterina Y.; Lapotko, Dmitri O.

From IEEE Journal of Selected Topics in Quantum Electronics (2014), 20(3), 7300412/1–7300412/12.

105. Design and establishment of urologic laser surgical procedures on basis of physical characteristic of lasers

By Xia, Shujie; Wang, Kui

From Linchuang Miniao Waike Zazhi (2014), 29(1), 1–4.

106. Extreme Light Infrastructure—Nuclear Physics a new Research Infrastructure at the interface of laser and subatomic physics

By Ursescu, Daniel; Tesileanu, Ovidiu; Cernaianu, Mihail O.; Gales, Sydney; Zamfir, Nicolae V.

From Reza Kenkyu (2014), 42(2), 123–126.

107. Recent advances of medicine treatment research in retinopathy of prematurity

By Zheng, Ling; Feng, Guang-qiang

From Guoji Yanke Zazhi (2014), 14(3), 460–463.

108. Carbon nanotube—a burgeoning field in nanotechnology

By Jayronia, Sonali; Hardenia, Anu; Jain, Sanjay

From World Journal of Pharmaceutical Research (2014), 3 (1), 295–310, 16 pp.

109. Rapid prototyping technologies for tissue regeneration

By Tran, V.; Wen, X.

From Woodhead Publishing Series in Biomaterials (2014), 70(Rapid Prototyping of Biomaterials), 97–155.

110. Beam by design: laser manipulation of electrons in modern accelerators

By Hemsing, Erik; Stupakov, Gennady; Xiang, Dao

From arXiv.org, e-Print Archive, Physics (2014), 1–51, arXiv:1404.2579v1 [physics.acc-ph].

111. Formation of the Titanium Oxide Surface by the Laser Ablation Method

By Ermakov, V. V.; Leitman, A. G.; Reimer, I. V.; Sheikin, V. V.; Chuchalin, V. S.; Osipov, A. N.; Vusovich, O. V.; Gol'tsova, P. A.

From Russian Physics Journal (2014), 56(11), 1292–1296.

112. Striae distensae: a comprehensive review and evidence-based evaluation of prophylaxis and treatment

By Al-Himdani, S.; Ud-Din, S.; Gilmore, S.; Bayat, A.

From British Journal of Dermatology (2014), 170(3), 527–547.

113. Two-photon fluorescence microscopy-a prospective tool in the research of traditional medicine

By Wang, Changming; Yu, Guang; Jian, Tunyu; Yang, Niuniu; Tang, Zongxiang

From Research Journal of Biotechnology (2014), 9(1), 90–94.

114. The optical tweezer. Micro world in light grasp

By Alpmann, Christina; Kruse, Annika; Denz, Cornelia

From Physik in Unserer Zeit (2014), 45(1), 36–42.

115. Identification of fungal microorganisms by MALDI-TOF mass spectrometry

By Chalupova, Jana; Raus, Martin; Sedlarova, Michaela; Sebela, Marek

From Biotechnology Advances (2014), 32(1), 230–241.

116. Cavity ringdown spectroscopy of plasma species

By Wang, Chuji

Edited by: Chu, Paul K.; Lu, XinPei

From Low Temperature Plasma Technology (2014), 207–260.

117. Next-generation protein analysis in the pathology department

By Ahmed Melek; Broeckx Glenn; Pauwels Patrick; Dendooven Amelie; Baggerman Geert; Baggerman Geert; Schildermans Karin; Schildermans Karin; Pauwels Patrick; Dendooven Amelie; et al.

From Journal of clinical pathology (2019)

118. Laser Acupuncture: A Concise Review

By Chon Tony Y; Mallory Molly J; Yang Juan; Bublitz Sara E; Do Alexander; Yang Juan; Dorsher Peter T

From Medical acupuncture (2019), 31(3), 164–168.

119. Dispelling Myths about Antenatal TAPS: A Call for Action for Routine MCA-PSV Doppler Screening in the United States

By Nicholas Lauren; Fischbein Rebecca; Aultman Julie; Ernst-Milner Stephanie

From Journal of clinical medicine (2019), 8(7)

120. Dermatologic care of sexual and gender minority/ LGBTQIA youth, Part 2: Recognition and management of the unique dermatologic needs of SGM adolescents

By Kosche Cory; Mansh Matthew; Luskus Mark; Yeung Howa; Nguyen Andy; Martinez-Diaz Gabriel; Inwards-Breland David; Boos Markus D

From Pediatric dermatology (2019)

121. Placebo and nocebo response magnitude on temporomandibular disorders related-pain: a systematic review and meta-analysis

By Porporatti Andre Luis; De Luca Canto Graziela; Costa Yuri Martins; Reus Jessica Conti; Stuginski-Barbosa Juliana; Conti Paulo Cesar Rodrigues; Velly Ana Miriam

From Journal of oral rehabilitation (2019)

122. Pathophysiology of Fibrosis in the Vocal Fold: Current Research, Future Treatment Strategies, and Obstacles to Restoring Vocal Fold Pliability

By Kumai Yoshihiko

From International journal of molecular sciences (2019), 20(10)

123. Radiocarbon Tracers in Toxicology and Medicine: Recent Advances in Technology and Science

By Malfatti Michael A; Enright Heather A; Stewart Benjamin J; Loots Gabriela G; Turteltaub Kenneth W; Buchholz Bruce A; Ognibene Ted J; McCartt A Daniel; Bench Graham; Zimmermann Maike; et al.

From Toxics (2019), 7(2)

124. Concomitant Use of Hyaluronic Acid and Laser in Facial Rejuvenation

By Urdiales-Galvez Fernando; Martin-Sanchez Sandra; Maiz-Jimenez Monica; Castellano-Miralla Antonio; Lionetti-Leone Leonardo

From Aesthetic plastic surgery (2019)

125. Hair Testing for Drugs in the Field of Forensics

By Shima Noriaki; Sasaki Keiko; Kamata Tohru; Miki Akihiro; Katagi Munehiro

From Yakugaku zasshi: Journal of the Pharmaceutical Society of Japan (2019), 139(5), 705–713

126. Additive manufacturing of biodegradable metals: Current research status and future perspectives

By Qin Yu; Wen Peng; Guo Hui; Xia Dandan; Zheng Yufeng; Jauer Lucas; Poprawe Reinhart; Voshage Maximilian; Schleifenbaum Johannes Henrich

From Acta biomaterialia (2019)

127. Advances on Photonic Crystal Fiber Sensors and Applications

By Portosi Vincenza; Laneve Dario; Falconi Mario Christian; Prudenzano Francesco

From Sensors (Basel, Switzerland) (2019), 19(8)

128. Molecular impacts of photobiomodulation on bone regeneration: A systematic review

By Hosseinpour Sepanta; Fekrazad Reza; Arany Praveen R; Ye Qingsong

From Progress in biophysics and molecular biology (2019)

129. CO2 laser application in stomatology

By Sutter Eveline; Giacomelli-Hiestand Barbara; Rucker Martin; Valdec Silvio

From Swiss dental journal (2019), 129(3), 214–215.

130. Collagen-based bioinks for hard tissue engineering applications: a comprehensive review

By Marques C F; Diogo G S; Pina S; Oliveira J M; Silva T H; Reis R L; Marques C F; Diogo G S; Pina S; Oliveira J M; et al.

From Journal of materials science. Materials in medicine (2019), 30(3), 32.

131. Methods for mammalian single cell research—a review

By Jiang Wenqian; Zuo Rui; Lin Jun; Jiang Wenqian; Zuo Rui; Lin Jun; Tian Yarong; Lin Jun

From Sheng wu gong cheng xue bao = Chinese journal of biotechnology (2019), 35(1), 27–39.

132. Platelet-rich plasma and its utility in the treatment of acne scars: A systematic review

By Hesseler Michael J; Shyam Nikhil

From Journal of the American Academy of Dermatology (2019), 80(6), 1730–1745.

133. Mechanisms of repigmentation induced by photobiomodulation therapy in vitiligo

By Yu Sebastian; Lan Cheng-Che E; Yu Hsin-Su; Yu Sebastian; Yu Hsin-Su; Yu Sebastian; Lan Cheng-Che E; Yu Sebastian; Yu Hsin-Su

From Experimental dermatology (2019), 28 Suppl 110–14.

134. Confocal laser endomicroscopy in the diagnostics of gastrointestinal lesions literary review and personal experience

By Kollar M; Krajciova J; Hus ak R; Maluskova J; Kment M; Vackova Z; Spicak J; Martinek J

From Rozhledy v chirurgii: mesicnik Ceskoslovenske chirurgicke spolecnosti (2018), 97(12), 531–538.

135. Somatic mutations and immune checkpoint biomarkers

By Parris Brielle A; Shaw Eloise; Fong Kwun; Parris Brielle A; Shaw Eloise; Fong Kwun; Pang Brendan; Soong Richie; Soo Ross A

From Respirology (Carlton, Vic.) (2019), 24(3), 215–226.

136. Urologic Oncology

By Burgess Kristine Elaine; DeRegis Carol J

From The Veterinary clinics of North America. Small animal practice (2019), 49(2), 311–323.

137. Three-dimensional bio-printing equipment technologies for tissue engineering and regenerative medicine

By Ahn Sang Hyun; Lee Junhee; Park Su A; Kim Wan Doo; Ahn Sang Hyun; Kim Wan Doo

From Tissue engineering and regenerative medicine (2016), 13(6), 663–676.

138. The Evidence for Adjunctive Facelift Procedures

By Urso-Baiarda Fulvio; Edmondson Sarah-Jayne; Grover Rajiv

From Facial plastic surgery: FPS (2018), 34(6), 624–630.

139. The Current Status of Glaucoma and Glaucoma Care in Sub-Saharan Africa

By Kyari Fatima; Adekoya Bola; Abdull Mohammed Mahdi; Mohammed Abdullahi Sadiq; Garba Farouk

From Asia-Pacific journal of ophthalmology (Philadelphia, Pa.) (2018), 7(6), 375–386.

140. New Concepts, Concerns, and Creations in Acne

By Marson Justin W; Baldwin Hilary E

From Dermatologic clinics (2019), 37(1), 1–9.

141. Regenerative Medicine Therapies for Equine Wound Management

By Dahlgren Linda A

From The Veterinary clinics of North America. Equine practice (2018), 34(3), 605–620.

142. Biomedical applications of mid-infrared quantum cascade lasers—a review

By Isensee Katharina; Kroger-Lui Niels; Petrich Wolfgang

From The Analyst (2018), 143(24), 5888–5911.

143. Diabetic retinopathy: a complex pathophysiology requiring novel therapeutic strategies

By Whitehead Michael; Osborne Andrew; Martin Keith R; Wickremasinghe Sanjeewa; Van Wijngaarden Peter; Wickremasinghe Sanjeewa; Van Wijngaarden Peter; Martin Keith R; Martin Keith R; Martin Keith R

From Expert opinion on biological therapy (2018), 18(12), 1257–1270.

144. An Overview of 3D Printing Technologies for Soft Materials and Potential Opportunities for Lipid-based Drug Delivery Systems

By Vithani Kapilkumar; Boyd Ben J; Goyanes Alvaro; Basit Abdul W; Gaisford Simon; Jannin Vincent; Basit Abdul W; Gaisford Simon; Boyd Ben J

From Pharmaceutical research (2018), 36(1), 4.

145. An Introduction to 3D Bioprinting: Possibilities, Challenges and Future Aspects

By Kacarevic Zeljka P; Rider Patrick M; Barbeck Mike; Alkildani Said; Retnasingh Sujith; Smeets Ralf; Barbeck Mike; Smeets Ralf; Jung Ole; Ivanisevic Zrinka; et al.

From Materials (Basel, Switzerland) (2018), 11(11).

146. The Aging Face

By Shah Anil R; Kennedy Paige M

From The Medical clinics of North America (2018), 102(6), 1041–1054.

147. Approaches to limit systemic antibiotic use in acne: Systemic alternatives, emerging topical therapies, dietary modification, and laser and light-based treatments

By Barbieri John S; Spaccarelli Natalie; James William D; Margolis David J

From Journal of the American Academy of Dermatology (2019), 80(2), 538–549.

148. Noninvasive Methods for Lower Facial Rejuvenation

By Sieber David A; Kenkel Jeffrey M

From Clinics in plastic surgery (2018), 45(4), 571–584.

149. Pharmacokinetic considertions concerning the use of bronchodiletors in the treatment of chronic obstructive pulmonary disease.

By Matera Maria Gabriella; Rinaldi Barbara; Page Clive; Rogliani Paola; Cazzola Mario

From Expert opinion on drug metabolism & toxicology (2018), 14(10), 1101–1111.

150. Electromagnetic-Acoustic Sensing for Biomedical Applications

By Liu Siyu; Zhang Ruochong; Zheng Zesheng; Zheng Yuanjin

From Sensors (Basel, Switzerland) (2018), 18(10).

151. Photobiomodulation Therapy in Veterinary Medicine: A Review

By Hochman Lindsay

From Topics in companion animal medicine (2018), 33(3), 83–88.

152. Veterinary Neurologic Rehabilitation: The Rationale for a Comprehensive Approach

By Frank Lauren R; Roynard Patrick F P

From Topics in companion animal medicine (2018), 33(2), 49–57.

153. Outcomes of bevacizumab and cidofovir treatment in HPV-associated recurrent respiratory papillomatosis—review of the literature

By Jackowska Joanna; Piersiala Krzysztof; Klimza Hanna; Wierzbicka Malgorzata

From Otolaryngologia polska = The Polish otolaryngology (2018), 72(4), 1–8.

154. Acinetobacter in veterinary medicine, with an emphasis on *Acinetobacter baumannii*

By van der Kolk J H; Endimiani A; Graubner C; Gerber V; Perreten V

From Journal of global antimicrobial resistance (2019), 1659–71.

155. Minimally Invasive Management of Uroliths in Cats and Dogs

By Cleroux Andreanne

From The Veterinary clinics of North America. Small animal practice (2018), 48(5), 875–889.

156. Applications of MALDI-TOF mass spectrometry in clinical proteomics

By Greco Viviana; Urbani Andrea; Greco Viviana; Urbani Andrea; Piras Cristian; Pieroni Luisa; Ronci Maurizio; Ronci Maurizio; Putignani Lorenza; Putignani Lorenza; et al.

From Expert review of proteomics (2018), 15(8), 683–696.

157. A mixed-reality surgical trainer with comprehensive sensing for fetal laser minimally invasive surgery

By Javaux Allan; Gruijthuijsen Caspar; Denis Kathleen; Vander Poorten Emmanuel; Bouget David; Stoyanov Danail; Vercauteren Tom; Ourselin Sebastien; Deprest Jan; Vercauteren Tom; et al.

From International journal of computer assisted radiology and surgery (2018), 13(12), 1949–1957.

158. Musculoskeletal Therapies: Adjunctive Physical Therapy

By Beutler Anthony

From FP essentials (2018)

159. Study on the Design and Application of Combining Low-Frequency Ultrasound with Laser Radiation in Surgery and Therapy

By Zharov V P; Menyaev Yu A; Kabisov R K; Al'kov S V; Nesterov A V; Savrasov G V

From Critical reviews in biomedical engineering (2017), 45 (1–6), 153–170.

160. Diagnostic Molecular Microbiology: A 2018 Snapshot

By Fairfax Marilynn Ransom; Bluth Martin H; Salimnia Hossein

From Clinics in laboratory medicine (2018), 38(2), 253–276.

161. Three-Dimensional Printing and Cell Therapy for Wound Repair

By van Kogelenberg Sylvia; Yue Zhilian; Dinoro Jeremy N; Wallace Gordon G; van Kogelenberg Sylvia; Baker Christopher S; Baker Christopher S

From Advances in wound care (2018), 7(5), 145–155.

162. Application of laser scanning confocal microscopy in the soft tissue exquisite structure for 3D scan

By Zhang Zhaoqiang; Ibrahim Mohamed; Fu Yang; Wu Xujia; Ren Fei; Chen Lei

From International journal of burns and trauma (2018), 8 (2), 17–25.

163. Bevacizumab for Epistaxis in Hereditary Hemorrhagic Telangiectasia: An Evidence-based Review

164. 5-ALA in the management of malignant glioma

By Stepp Herbert; Stummer Walter

From Lasers in surgery and medicine (2018), 50(5), 399–419.

165. Near infrared low-level laser therapy and cell proliferation: The emerging role of redox sensitive signal transduction pathways

By Migliario Mario; Mortellaro Carmen; Sabbatini Maurizio; Reno Filippo

From Journal of biophotonics (2018), 11(11), e201800025

166. How iMALDI can improve clinical diagnostics

By Popp R; Basik M; Spatz A; Batist G; Zahedi R P; Borchers C H

From The Analyst (2018), 143(10), 2197–2203.

167. Use of Nitrous Oxide in Dermatology: A Systematic Review

By Brotzman Erica A; Sandoval Laura F; Crane Jonathan; Sandoval Laura F; Crane Jonathan; Crane Jonathan

From Dermatologic surgery: official publication for American Society for Dermatologic Surgery [et al.] (2018), 44(5), 661–669.

168. New technologies and techniques for prostate cancer focal therapy

By Linares-Espinos Estefania; Linares-Espinos Estefania; Martinez-Salamanca Juan I; Bianco Fernando; Carneiro Arie; Martinez-Salamanca Juan I; Bianco Fernando; Castro-Alfaro Adalberto; Cathelineau Xavier; Valerio Massimo; et al.

From Minerva urologica e nefrologica = The Italian journal of urology and nephrology (2018), 70(3), 252–263.

169. Emerging Therapies for Acne Vulgaris

By Trivedi Megha K; Bosanac Suzana S; Sivamani Raja K; Larsen Larissa N; Sivamani Raja K

From American journal of clinical dermatology (2018), 19 (4), 505–516.

170. Not all that glitters is gold: A guide to surgical trials in epilepsy

By Jehi Lara; Jette Nathalie; Jette Nathalie

From Epilepsia open (2016), 1(1–2), 22–36.

171. Perspective on Broad-Acting Clinical Physiological Effects of Photobiomodulation

By Shanks Steven; Leisman Gerry; Leisman Gerry

From Advances in experimental medicine and biology (2018), 109641–52.

172. Review and clinical experience exploring evidence, clinical efficacy, and safety regarding nonsurgical treatment of feminine rejuvenation

By Gold Michael; Gold Michael; Gold Michael; Gold Michael; Andriessen Anneke; Andriessen Anneke; Bader Alexandros; Alinsod Red; French Elizabeth Shane; Guerette Nathan; et al.

From Journal of cosmetic dermatology (2018), 17(3), 289–297.

173. Inducible nitric oxide synthase as a target for osteoarthritis treatment

By Leonidou Andreas; Lepetsos Panagiotis; Mintzas Michalis; Kenanidis Eustathios; Tsiridis Eleftherios; Leonidou Andreas; Tzetis Maria; Lepetsos Panagiotis; Macheras George; Potoupnis Michael; et al.

From Expert opinion on therapeutic targets (2018), 22(4), 299–318.

174. Acupuncture for depression

By Smith Caroline A; Armour Mike; Lee Myeong Soo; Wang Li-Qiong; Hay Phillipa J

From The Cochrane database of systematic reviews (2018), 3CD004046

175. Therapeutics for Adult Nail Psoriasis and Nail Lichen Planus: A Guide for Clinicians

By McClanahan Danielle R; English Joseph C 3rd

From American journal of clinical dermatology (2018), 19 (4), 559–584.

176. Vitiligo in Children: What's New in Treatment?

By Gianfaldoni Serena; Lotti Torello; Tchernev Georgi; Tchernev Georgi; Wollina Uwe; Lotti Jacopo; Rovesti Miriam; Satolli Francesca; Franca Katlein

From Open access Macedonian journal of medical sciences (2018), 6(1), 221–225.

177. The efficacy and safety of non-pharmacological therapies for the treatment of acne vulgaris: A systematic review and best-evidence synthesis

By de Vries F M C; Meulendijks A M; van Dooren A A; Tjin E P M; de Vries F M C; Driessen R J B; van de Kerkhof P C M

From Journal of the European Academy of Dermatology and Venereology: JEADV (2018), 32(7), 1195–1203.

178. Pediatric Interventional Pulmonology

By Donato Leonardo; Mai Hong Tran Thi; Ghori Uzair K; Musani Ali I

From Clinics in chest medicine (2018), 39(1), 229–238.

179. The Role of Diet in Glaucoma: A Review of the Current Evidence

By Al Owaifeer Adi M; Al Taisan Abdulaziz A

From Ophthalmology and therapy (2018), 7(1), 19–31.

180. Laser Irradiation of Metal Oxide Films and Nanostructures: Applications and Advances

By Palneedi Haribabu; Peddigari Mahesh; Hwang Geon-Tae; Kim Jong-Woo; Choi Jong-Jin; Hahn Byung-Dong; Ryu Jungho; Park Jung Hwan; Lee Keon Jae; Maurya Deepam; et al.

From Advanced materials (Deerfield Beach, Fla.) (2018), 30 (14), e1705148.

181. Laser Application in Iran Urology: A Narrative Review

By Razzaghi Mohammad Reza; Fallah Karkan Morteza; Ghiasy Saleh; Javanmard Babak

From Journal of lasers in medical sciences (2018), 9(1), 1–6.

182. The clinical effectiveness and cost-effectiveness of fractional CO2 laser in acne scars and skin rejuvenation: A meta-analysis and economic evaluation

By Ansari Fereshteh; Sadeghi-Ghyassi Fatemeh; Yaaghoobian Barmak

From Journal of cosmetic and laser therapy: official publication of the European Society for Laser Dermatology (2018), 20(4), 248–251.

183. Biofabrication: new approaches for tissue regeneration

By Horch Raymund E; Weigand Annika; Arkudas Andreas; Wajant Harald; Groll Jurgen; Boccaccini Aldo R

From Handchirurgie, Mikrochirurgie, plastische Chirurgie: Organ der Deutschsprachigen Arbeitsgemeinschaft fur Handchirurgie: Organ der Deutschsprachigen Arbeitsgemeinschaft fur Mikrochirurgie der Peripheren Nerven und Gefasse: Organ der V. (2018), 50(2), 93–100.

184. Laser therapy for mens infertility. Part 2. Systematic review of clinical trials

By Apolikhin O I; Moskvin S V; Apolikhin O I; Moskvin S V

From Urologiia (Moscow, Russia: 1999) (2017), (6), 164–171.

185. Treatment of Carious Lesions Using Self-Assembling Peptides

By Alkilzy M; Santamaria R M; Schmoeckel J; Splieth C H

From Advances in dental research (2018), 29(1), 42–47.

186. Interventional bronchoscopy in adults

By Yu Diana H; Feller-Kopman David

From Expert review of respiratory medicine (2018), 12(3), 239–248.

187. Pathology, proteomics and the pathway to personalised medicine

By Jin Ping; Lan Jiang; Huang Canhua; Lan Jiang; Wang Kui; Huang Canhua; Baker Mark S; Nice Edouard C

From Expert review of proteomics (2018), 15(3), 231–243.

188. Systematic review and meta-analysis of randomized controlled trials evaluating long-term outcomes of endovenous management of lower extremity varicose veins

By Kheirelseid Elrasheid A H; Crowe Gillian; Sehgal Rishabh; Liakopoulos Dimitrios; Bela Hafiz; Mulkern Edward; McDonnell Ciaran; O'Donohoe Martin

From Journal of vascular surgery. Venous and lymphatic disorders (2018), 6(2)

189. When Fluorescent Proteins Meet White Light-Emitting Diodes

By Fernandez-Luna Veronica; Costa Ruben D; Coto Pedro B; Coto Pedro B

From Angewandte Chemie (International ed. In English) (2018), 57(29), 8826–8836.

190. Prostate cancer diagnosis and characterization with mass spectrometry imaging

By Kurreck Annika; Vandergrift Lindsey A; Fuss Taylor L; Cheng Leo L; Kurreck Annika; Vandergrift Lindsey A; Fuss Taylor L; Cheng Leo L; Kurreck Annika; Habbel Piet; et al.

From Prostate cancer and prostatic diseases (2018), 21(3), 297–305.

191. Generation and Role of Reactive Oxygen and Nitrogen Species Induced by Plasma, Lasers, Chemical Agents, and Other Systems in Dentistry

By Jha Nayansi; Ryu Jae Jun; Jha Nayansi; Choi Eun Ha; Kaushik Nagendra Kumar

From Oxidative medicine and cellular longevity (2017), 20177542540.

192. Interventions for Old World cutaneous leishmaniasis

By Heras-Mosteiro Julio; Monge-Maillo Begona; Pinart Mariona; Lopez Pereira Patricia; Reveiz Ludovic; Garcia-Carrasco Emely; Campuzano Cuadrado Pedro; Royuela Ana; Mendez Roman Irene; Lopez-Velez Rogelio

From The Cochrane database of systematic reviews (2017), 12CD005067

193. Could adverse effects and complications of selective laser trabeculoplasty be decreased by low-power laser therapy?

By Paiva Alexandre de Carvalho Mendes; da Fonseca Adenilson de Souza; da Fonseca Adenilson de Souza; da Fonseca Adenilson de Souza

From International ophthalmology (2019), 39(1), 243–257

194. Biomarkers of ocular surface disease using impression cytology

By Hagan Suzanne

From Biomarkers in medicine (2017), 11(12), 1135–1147.

195. Hierarchical Multicomponent Inorganic Metamaterials: Intrinsically Driven Self-Assembly at the Nanoscale

By Levchenko Igor; Xu Shuyan; Levchenko Igor; Bazaka Kateryna; Keidar Michael; Fang Jinghua

From Advanced materials (Deerfield Beach, Fla.) (2018), 30 (2)

196. Laser treatment of cutaneous angiokeratomas: A systematic review

By Nguyen Jannett; Chapman Lance W; Korta Dorota Z; Zachary Christopher B

From Dermatologic therapy (2017), 30(6)

197. Interventions for Old World cutaneous leishmaniasis

By Heras-Mosteiro Julio; Monge-Maillo Begona; Pinart Mariona; Lopez Pereira Patricia; Reveiz Ludovic; Garcia-Carrasco Emely; Campuzano Cuadrado Pedro; Royuela Ana; Mendez Roman Irene; Lopez-Velez Rogelio

From The Cochrane database of systematic reviews (2017), 11CD005067

198. Laser therapy for male infertility. Part 1. Etiology and pathogenesis. Experimental studies

By Apolikhin O I; Moskvin S V; Apolikhin O I; Moskvin S V

From Urologiia (Moscow, Russia: 1999) (2017), (5), 115–123.

199. A systematic review and meta-analysis of the effect of low-level laser therapy (LLLT) on chemotherapy-induced oral mucositis in pediatric and young patients

By He Mengxue; Zhang Binghua; Shen Nanping; Sun Jiwen; Wu Na

From European journal of pediatrics (2018), 177(1), 7–17.

200. Comparison of the in vitro effects of low-level laser therapy and low-intensity pulsed ultrasound therapy on bony cells and stem cells

By Bayat Mohammad; Virdi Amarjit; Rezaei Fatemehalsadat; Chien Sufan

From Progress in biophysics and molecular biology (2018), 13336–48.

201. X-ray free electron laser: opportunities for drug discovery

By Cheng Robert K Y; Abela Rafael; Hennig Michael

From Essays in biochemistry (2017), 61(5), 529–542.

202. Combination Therapy for Male Cosmetic Patients

By Gold Michael H

From Dermatologic clinics (2018), 36(1), 69–73.

203. Photobiomodulation in Neuroscience: A Summary of Personal Experience

By Rochkind Shimon

From Photomedicine and laser surgery (2017), 35(11), 604–615.

204. The Future of Non-Invasive Rejuvenation Technology: Devices

By Gold Michael H

From Journal of drugs in dermatology: JDD (2017), 16(6), s104–s107.

205. Overview of PAT process analysers applicable in monitoring of film coating unit operations for manufacturing of solid oral dosage forms

By Korasa Klemen; Vrecer Franc

From European journal of pharmaceutical sciences: official journal of the European Federation for Pharmaceutical Sciences (2018), 111278−292.

206. 3D bioprinting using stem cells

By Ong Chin Siang; Yesantharao Pooja; Mattson Gunnar; Boktor Joseph; Fukunishi Takuma; Zhang Huaitao; Hibino Narutoshi; Huang Chen Yu

From Pediatric research (2018), 83(1−2), 223−231.

207. Nonablative Fractional Laser Resurfacing in Skin of Color: Evidence-based Review

By Kaushik Shivani B; Alexis Andrew F

From The Journal of clinical and aesthetic dermatology (2017), 10(6), 51−67.

208. Conventional and novel stem cell based therapies for androgenic alopecia

By Talavera-Adame Dodanim; Newman Nathan; Newman Daniella

From Stem cells and cloning: advances and applications (2017), 1011−19.

209. A Phytochemical-Sensing Strategy Based on Mass Spectrometry Imaging and Metabolic Profiling for Understanding the Functionality of the Medicinal Herb Green Tea

By Fujimura Yoshinori; Miura Daisuke; Tachibana Hirofumi

From Molecules (Basel, Switzerland) (2017), 22(10).

210. Personalised 3D Printed Medicines: Which Techniques and Polymers Are More Successful?

By Konta Andrea Alice; Garcia-Pina Marta; Serrano Dolores R; Serrano Dolores R

From Bioengineering (Basel, Switzerland) (2017), 4(4).

211. Sickle cell retinopathy: improving care with a multidisciplinary approach

By Menaa Farid; Menaa Farid; Menaa Abder; Khan Barkat Ali; Uzair Bushra

From Journal of multidisciplinary healthcare (2017), 10335−346.

212. Pediatric Burn Reconstruction: Focus on Evidence

By Fisher Mark

From Clinics in plastic surgery (2017), 44(4), 865−873.

213. Management of trauma and burn scars: the dermatologist's role in expanding patient access to care

By Miletta Nathanial R; Hivnor Chad M; Donelan Matthias B

From Cutis (2017), 100(1), 18−20.

214. Concise Review: Bioprinting of Stem Cells for Transplantable Tissue Fabrication

By Leberfinger Ashley N; Ravnic Dino J; Dhawan Aman; Ozbolat Ibrahim T; Ozbolat Ibrahim T; Ozbolat Ibrahim T; Ozbolat Ibrahim T

From Stem cells translational medicine (2017), 6(10), 1940−1948.

215. Update of Ablative Fractionated Lasers to Enhance Cutaneous Topical Drug Delivery

By Waibel Jill S; Rudnick Ashley; Shagalov Deborah R; Nicolazzo Danielle M

From Advances in therapy (2017), 34(8), 1840−1849.

216. Beneficial effects of phytochemicals in diabetic retinopathy: experimental and clinical evidence

By Ojha S; Balaji V; Sadek B; Rajesh M

From European review for medical and pharmacological sciences (2017), 21(11), 2769−2783.

217. Low-Level Laser Therapy in Russia: History, Science and Practice

By Moskvin Sergey Vladimirovich

From Journal of lasers in medical sciences (2017), 8(2), 56−65.

218. Mass spectrometry imaging for clinical research—latest developments, applications, and current limitations

By Vaysse Pierre-Maxence; Heeren Ron M A; Porta Tiffany; Balluff Benjamin

From The Analyst (2017), 142(15), 2690−2712.

219. Presenting a Method to Improve Bone Quality Through Stimulation of Osteoporotic Mesenchymal Stem Cells by Low-Level Laser Therapy

By Bayat Mohammad; Jalalifirouzkouhi Ali

From Photomedicine and laser surgery (2017), 35(11), 622−628.

220. Recent medical techniques for peripheral nerve repair: nerve guidance conduits update

By Pyatin V F; Kolsanov A V; Shirolapov I V

From Advances in gerontology = Uspekhi gerontologii (2016), 29(5), 742–750.

221. Stem Cell-Induced Biobridges as Possible Tools to Aid Neuroreconstruction after CNS Injury

By Lee Jea Y; Xu Kaya; Nguyen Hung; Guedes Vivian A; Borlongan Cesar V; Acosta Sandra A

From Frontiers in cell and developmental biology (2017), 551.

222. Applications of Raman Spectroscopy in Biopharmaceutical Manufacturing: A Short Review

By Buckley Kevin; Ryder Alan G

From Applied spectroscopy (2017), 71(6), 1085–1116.

223. Molluscum Contagiosum: An Update

By Leung Alexander K C; Barankin Benjamin; Hon Kam L E

From Recent patents on inflammation & allergy drug discovery (2017), 11(1), 22–31.

224. 3D bioprinting for drug discovery and development in pharmaceutics

By Peng Weijie; Datta Pallab; Ayan Bugra; Sosnoski Donna; Ozbolat Veli; Ozbolat Ibrahim T

From Acta biomaterialia (2017), 5726–46.

225. Physiotherapy, occupational therapy and physical therapy in fibromyalgia syndrome: Updated guidelines 2017 and overview of systematic review articles

By Winkelmann A; Bork H; Bruckle W; Dexl C; Heldmann P; Henningsen P; Hauser W; Krumbein L; Pullwitt V; Schiltenwolf M; et al.

From Schmerz (Berlin, Germany) (2017), 31(3), 255–265.

226. New lasers and light sources—old and new risks?

By Paasch Uwe; Grunewald Sonja; Schwandt Antje; Seeber Nikolaus; Kautz Gerd; Haedersdal Merete

From Journal der Deutschen Dermatologischen Gesellschaft = Journal of the German Society of Dermatology: JDDG (2017), 15(5), 487–496.

227. Topical treatment of glaucoma: established and emerging pharmacology

By Dikopf Mark S; Vajaranant Thasarat S; Edward Deepak P

From Expert opinion on pharmacotherapy (2017), 18(9), 885–898.

228. Low level laser therapy: A narrative literature review on the efficacy in the treatment of rheumatic orthopaedic conditions

By Baltzer A W A; Stosch D; Seidel F; Ostapczuk M S; Ostapczuk M S

From Zeitschrift fur Rheumatologie (2017), 76(9), 806–812.

229. Precision toxicology based on single cell sequencing: an evolving trend in toxicological evaluations and mechanism exploration

By Zhang Boyang; Huang Kunlun; Zhu Liye; Luo Yunbo; Xu Wentao; Zhang Boyang; Huang Kunlun; Zhu Liye; Luo Yunbo; Xu Wentao

From Archives of toxicology (2017), 91(7), 2539–2549.

230. Review on diagnosis and management of urolithiasis in pregnancy: an ESUT practical guide for urologists

By Somani Bhaskar K; Dellis Athanasios; Liatsikos Evangellos; Skolarikos Andreas

From World journal of urology (2017), 35(11), 1637–1649.

231. Recent Advances in Bioink Design for 3D Bioprinting of Tissues and Organs

By Ji Shen; Guvendiren Murat

From Frontiers in bioengineering and biotechnology (2017), 523.

232. New Uses of AbobotulinumtoxinA in Aesthetics

By Schlessinger Joel; Gilbert Erin; Cohen Joel L; Kaufman Joely

From Aesthetic surgery journal (2017), 37(suppl_1), S45–S58.

233. Sustained attractiveness and natural youthful appearance by upper lip rejuvenation: Minimally invasive procedures to combat facial aging

By Wollina Uwe; Goldman Alberto

From Wiener medizinische Wochenschrift (1946) (2018), 168(13–14), 361–366.

234. Microendoscopic minimally invasive techniques in lacrimal surgery

By Emmerich K-H; Ungerechts R; Amin S; Meyer-Rusenberg H-W

From Der Ophthalmologe: Zeitschrift der Deutschen Ophthalmologischen Gesellschaft (2017), 114(5), 409–415.

235. New approaches for antifungal susceptibility testing

By Sanguinetti M; Posteraro B

From Clinical microbiology and infection: the official publication of the European Society of Clinical Microbiology and Infectious Diseases (2017), 23(12), 931–934.

236. Of Cytometry, Stem Cells and Fountain of Youth

By Galkowski Dariusz; Ratajczak Mariusz Z; Kocki Janusz; Darzynkiewicz Zbigniew

From Stem cell reviews (2017), 13(4), 465–481.

237. Role of Laser Resection in Pulmonary Metastasectomy

By Macherey S; Doerr F; Wahlers T; Hekmat K

From Pneumologie (Stuttgart, Germany) (2017), 71(7), 475–479.

238. Photobiomodulation with non-thermal lasers: Mechanisms of action and therapeutic uses in dermatology and aesthetic medicine

By Nestor Mark; Berman Brian; Nestor Mark; Kirsner Robert S; Andriessen Anneke; Andriessen Anneke; Berman Brian; Katz Bruce E; Gilbert Dore; Gilbert Dore; et al.

From Journal of cosmetic and laser therapy: official publication of the European Society for Laser Dermatology (2017), 19(4), 190–198.

239. Remote Control and Modulation of Cellular Events by Plasmonic Gold Nanoparticles: Implications and Opportunities for Biomedical Applications

By Li Jiayang; Liu Jing; Chen Chunying

From ACS nano (2017), 11(3), 2403–2409.

240. Interventions for the management of radiotherapy-induced xerostomia and hyposalivation: A systematic review and meta-analysis

By Mercadante Valeria; Al Hamad Arwa; Lodi Giovanni; Porter Stephen; Fedele Stefano

From Oral oncology (2017), 6664–74.

241. Minimally invasive bypass surgery for nasolacrimal duct obstruction: Transcanalicular laser-assisted dacryocystorhinostomy

By Koch K R; Cursiefen C; Heindl L M

From Der Ophthalmologe: Zeitschrift der Deutschen Ophthalmologischen Gesellschaft (2017), 114(5), 416–423.

242. Effect of laser-assisted scaling and root planing on the expression of pro-inflammatory cytokines in the gingival crevicular fluid of patients with chronic periodontitis: A systematic review

By Kellesarian Sergio Varela; Malignaggi Vanessa Ros; Majoka Hasham Abdullah; Javed Fawad; Al-Kheraif Abdulaziz A; Kellesarian Tammy Varela; Romanos Georgios E

From Photodiagnosis and photodynamic therapy (2017), 1863–77.

243. Light and energy based therapeutics for genitourinary syndrome of menopause: Consensus and controversies

By Tadir Yona; Nelson John Stuart; Gaspar Adrian; Lev-Sagie Ahinoam; Alexiades Macrene; Alinsod Red; Bader Alex; Calligaro Alberto; Elias Jorge A; Gambaciani Marco; et al.

From Lasers in surgery and medicine (2017), 49(2), 137–159.

244. Optical coherence tomography and confocal laser endomicroscopy in pulmonary diseases

By Wijmans Lizzy; d'Hooghe Julia N S; Bonta Peter I; Annema Jouke T

From Current opinion in pulmonary medicine (2017), 23 (3), 275–283.

245. Therapeutic Options in Refractory Diabetic Macular Oedema

By Shah Sanket U; Maturi Raj K

From Drugs (2017), 77(5), 481–492.

246. Rehabilitation of hamstring muscle injuries: a literature review

By Ramos Gabriel Amorim; Arliani Gustavo Goncalves; Astur Diego Costa; Pochini Alberto de Castro; Ejnisman Benno; Cohen Moises

From Revista brasileira de ortopedia (2017), 52(1), 11–16.

247. Clinical efficacy of photodynamic therapy adjunctive to scaling and root planing in the treatment of chronic periodontitis: A systematic review and meta-analysis

By Xue Dong; Tang Lu; Bai Yuhao; Ding Qian; Wang Pengcheng; Zhao Ying

From Photodiagnosis and photodynamic therapy (2017), 18119–127.

248. Stereotactic laser thermocoagulation in epilepsy surgery

By Hoppe C; Witt J-A; Helmstaedter C; Elger C E; Gasser T; Vatter H

From Der Nervenarzt (2017), 88(4), 397–407.

249. Effects of low-level laser therapy on pain in patients with musculoskeletal disorders: a systematic review and meta-analysis

By Clijsen Ron; Clijsen Ron; Clijsen Ron; Brunner Anina; Barbero Marco; Clarys Peter; Taeymans Jan; Taeymans Jan

From European journal of physical and rehabilitation medicine (2017), 53(4), 603–610.

250. Limitations of SMILE (Small Incision Lenticule Extraction)

By Seiler T; Koller T; Wittwer V V

From Klinische Monatsblatter fur Augenheilkunde (2017), 234(1), 125–129.

251. Historical Overview of the Clinical Development of the Small Incision Lenticule Extraction Surgery (SMILE)

By Blum M; Kunert K S; Sekundo W

From Klinische Monatsblatter fur Augenheilkunde (2017), 234(1), 117–122.

252. SMILE: Re-Treatment Options—Techniques and Results

By Meyer B; Kunert K S

From Klinische Monatsblatter fur Augenheilkunde (2017), 234(1), 98–101.

253. Single-Cell Metabolomics

By Emara Samy; Amer Sara; Ali Ahmed; Abouleila Yasmine; Amer Sara; Ali Ahmed; Abouleila Yasmine; Oga April; Masujima Tsutomu

From Advances in experimental medicine and biology (2017), 965323–343.

254. Need for Evidence and Consensus on Laser Treatment for Management of Select Primary Penile Tumors

By Leone Andrew; Inman Brant; Spiess Philippe E

From European urology (2017), 72(1), 4–6.

255. Long-acting slow effective release antiretroviral therapy

By Edagwa Benson; McMillan JoEllyn; Sillman Brady; Gendelman Howard E; Gendelman Howard E

From Expert opinion on drug delivery (2017), 14(11), 1281–1291.

256. Antimicrobial photodynamic therapy as an adjunct for treatment of deep carious lesions-A systematic review

By Cieplik Fabian; Buchalla Wolfgang; Hiller Karl-Anton; Hellwig Elmar; Al-Ahmad Ali; Karygianni Lamprini; Maisch Tim

From Photodiagnosis and photodynamic therapy (2017), 1854–62.

257. The application of lasers in otorhinolaryngology

By Kryukov A I; Tsarapkin G Yu; Arzamasov S G; Panasov S A

From Vestnik otorinolaringologii (2016), 81(6), 62–66.

258. Laser applications in surgery

By Azadgoli Beina; Baker Regina Y

From Annals of translational medicine (2016), 4(23), 452.

259. Bioprinting for vascular and vascularized tissue biofabrication

By Datta Pallab; Ayan Bugra; Ozbolat Ibrahim T

From Acta biomaterialia (2017), 511–20.

260. Melasma: An update on the clinical picture, treatment, and prevention

By Becker S; Schiekofer C; Vogt T; Reichrath J

From Der Hautarzt; Zeitschrift fur Dermatologie, Venerologie, und verwandte Gebiete (2017), 68(2), 120–126.

261. New Treatments for Hair Loss

By Vano-Galvan S; Camacho F

From Actas dermo-sifiliograficas (2017), 108(3), 221–228.

262. Evidence-Based Treatment of Diabetic Macular Edema

By Barham Rasha; El Rami Hala; Sun Jennifer K; Silva Paolo S; Sun Jennifer K; Silva Paolo S

From Seminars in ophthalmology (2017), 32(1), 56–66.

263. Recent Update on the Role of Chinese Material Medica and Formulations in Diabetic Retinopathy

By Vasant More Sandeep; Kim In-Su; Choi Dong-Kug

From Molecules (Basel, Switzerland) (2017), 22(1).

264. Primary angle closure glaucoma: What we know and what we don't know

By Sun Xinghuai; Dai Yi; Chen Yuhong; Chen Junyi; Kong Xiangmei; Wang Xiaolei; Jiang Chunhui; Yu Dao-Yi; Cringle Stephen J

From Progress in retinal and eye research (2017), 5726–45.

265. Evidence-Based Medicine: A Graded Approach to Lower Lid Blepharoplasty

By Hashem Ahmed M; Couto Rafael A; Waltzman Joshua T; Drake Richard L; Zins James E

From Plastic and reconstructive surgery (2017), 139(1), 139e–150e.

266. A review on biogenic synthesis of ZnO nanoparticles using plant extracts and microbes: A prospect towards green chemistry

By Ahmed Shakeel; Ikram Saiqa; Annu; Chaudhry Saif Ali

From Journal of photochemistry and photobiology. B, Biology (2017), 166272–284.

267. Refractive Lenticule Implantation for Correction of Ametropia: Case Reports and Literature Review

By Lazaridis A; Messerschmidt-Roth A; Sekundo W; Schulze S

From Klinische Monatsblatter fur Augenheilkunde (2017), 234(1), 77–89.

268. The Use of Low-Level Energy Laser Radiation in Basic and Clinical Research

By Rola Piotr; Doroszko Adrian; Derkacz Arkadiusz

From Advances in clinical and experimental medicine: official organ Wroclaw Medical University (2014), 23(5), 835–842.

269. EUS for pancreatic cystic neoplasms: The roadmap to the future is much more than just a few shades of gray

By Kirtane Tejas; Bhutani Manoop S

From Asian Pacific journal of tropical medicine (2016), 9 (12), 1218–1221.

270. Pathophysiology, diagnosis and treatment of Zenker's diverticulum

By Hussain T; Lang S; Stuck B A; Maurer J T

From HNO (2017), 65(2), 167–176.

271. Recommendations for Potassium-Titanyl-Phosphate Laser in the Treatment of Cholesteatoma

By le Nobel Gavin John; James Adrian Lewis

From The journal of international advanced otology (2016), 12(3), 332–336.

272. Laser flow cytometry as a tool for the advancement of clinical medicine

By Aebisher David; Bartusik Dorota; Tabarkiewicz Jacek

From Biomedicine & pharmacotherapy = Biomedecine & pharmacotherapie (2017), 85434–443.

273. Stem cell bioprinting for applications in regenerative medicine

By Tricomi Brad J; Corr David T; Dias Andrew D

From Annals of the New York Academy of Sciences (2016), 1383(1), 115–124.

274. Physical modalities for the treatment of rosacea

By Hofmann Maja A; Lehmann Percy

From Journal der Deutschen Dermatologischen Gesellschaft = Journal of the German Society of Dermatology: JDDG (2016), 14 Suppl 638–43.

275. Physikalische Methoden zur Behandlung der Rosazea

By Hofmann Maja A; Lehmann Percy

From Journal der Deutschen Dermatologischen Gesellschaft = Journal of the German Society of Dermatology: JDDG (2016), 14 Suppl 638–44.

276. The Role of MRgLITT in Overcoming the Challenges in Managing Infield Recurrence After Radiation for Brain Metastasis

By Patel Purvee D; Patel Nitesh V; Davidson Christian; Danish Shabbar F

From Neurosurgery (2016), 79 Suppl 1S40–S58, Language: English

277. Attributes Associated with Adherence to Glaucoma Medical Therapy and its Effects on Glaucoma Outcomes: An Evidence-Based Review and Potential Strategies to Improve Adherence

By Joseph Arun; Pasquale Louis R; Pasquale Louis R

From Seminars in ophthalmology (2017), 32(1), 86–90.

278. Update on Image-Guided Percutaneous Ablation of Breast Cancer

By Fleming Margaret M; Holbrook Anna I; Newell Mary S

From AJR. American journal of roentgenology (2017), 208 (2), 267–274.

279. Twin pregnancy complicated by selective growth restriction

By Townsend Rosemary; Khalil Asma

From Current opinion in obstetrics & gynecology (2016), 28(6), 485–491.

280. Evidence-Based Treatment of Diabetic Retinopathy

By El Rami Hala; Barham Rasha; Sun Jennifer K; Silva Paolo S; Sun Jennifer K; Silva Paolo S

From Seminars in ophthalmology (2017), 32(1), 67–74.

281. New treatment strategies for benign prostatic hyperplasia in the frail elderly population: a systematic review

By Albisinni Simone; Aoun Fouad; Roumeguere Thierry; Porpiglia Francesco; Tubaro Andrea; DE Nunzio Cosimo

From Minerva urologica e nefrologica = The Italian journal of urology and nephrology (2017), 69(2), 119–132.

282. Mass Spectrometry-based Proteomics in Acute Respiratory Distress Syndrome: A Powerful Modality for Pulmonary Precision Medicine

By Xu Xue-Feng; Dai Hua-Ping; Wang Chen; Li Yan-Ming; Xiao Fei

From Chinese medical journal (2016), 129(19), 2357–64.

283. Patient safety in otolaryngology: a descriptive review

By Danino Julian; Muzaffar Jameel; Metcalfe Chris; Coulson Chris

From European archives of oto-rhino-laryngology: official journal of the European Federation of Oto-Rhino-Laryngological Societies (EUFOS): affiliated with the German Society for Oto-Rhino-Laryngology—Head and Neck Surgery (2017), 274(3), 1317–1326.

284. Why acupuncture in pain treatment?

By Ondrejkovicova Alena; Petrovics Gabriel; Svitkova Katarina; Bajtekova Bibiana; Bangha Ondrej

From Neuro endocrinology letters (2016), 37(3), 163–168.

285. A Critical Assessment of the Evidence for Low-Level Laser Therapy in the Treatment of Hair Loss

By Gupta Aditya K; Foley Kelly A

From Dermatologic surgery: official publication for American Society for Dermatologic Surgery [et al.] (2017), 43(2), 188–197.

286. Possibilities and limitations of eye drops for glaucoma therapy

By Lanzl I M; Lanzl I M; Poimenidou M; Spaeth G L

From Der Ophthalmologe: Zeitschrift der Deutschen Ophthalmologischen Gesellschaft (2016), 113(10), 824–832.

287. Laser therapy for the treatment of pearly penile papules

By Maranda Eric L; Moore Kevin J; Jimenez Joaquin J; Akintilo Lisa; Hundley Kelsey; Nguyen Austin H; Zullo Joseph

From Lasers in medical science (2017), 32(1), 243–248.

288. Evidence-Based Scar Management: How to Improve Results with Technique and Technology

By Khansa Ibrahim; Harrison Bridget; Janis Jeffrey E

From Plastic and reconstructive surgery (2016), 138(3 Suppl), 165S–78S.

289. How to achieve safe, high-quality clinical studies with non-Medicinal Investigational Products? A practical guideline by using intra-bronchial carbon nanoparticles as case study

By Berger M; van der Zee J S; Sterk P J; Kooyman P J; Makkee M; van der Zee J S; van Dijk J; Kemper E M

From Respiratory research (2016), 17(1), 102.

290. The application of laser in endodontics

By He W X; Liu N N; Wang X L; He X Y

From Zhonghua kou qiang yi xue za zhi = Zhonghua kouqiang yixue zazhi = Chinese journal of stomatology (2016), 51(8), 470–4.

291. Femtosecond Laser-Assisted Cataract Surgery: Current Status and Outlook

By Dick H B; Schultz T

From Klinische Monatsblatter fur Augenheilkunde (2016), 233(8), 967–86.

292. Physiotherapy of cancer patients

By Gomez Izabella; Szekanecz Eva; Szekanecz Zoltan; Bender Tamas

From Orvosi hetilap (2016), 157(31), 1224–31.

293. Acupuncture for Pediatric Pain

By Golianu Brenda; Brooks Meredith; Yeh Ann Ming

From Children (Basel, Switzerland) (2014), 1(2), 134–48.

294. ThermiVa: The Revolutionary Technology for Vulvovaginal Rejuvenation and Noninvasive Management of Female SUI

By Magon Navneet; Alinsod Red

From Journal of obstetrics and gynaecology of India (2016), 66(4), 300–2.

295. Incorporating Minimally Invasive Procedures into an Aesthetic Surgery Practice

By Matarasso Alan; Nikfarjam Jeremy; Abramowitz Lauren

From Clinics in plastic surgery (2016), 43(3), 449–57.

296. The application of electrolaser myostimulation and laserophoresis of biologically active compounds in sports medicine (a review)

By Khadartsev A A; Fudin N A; Moskvin S V

From Voprosy kurortologii, fizioterapii, I lechebnoi fizicheskoi kultury (2016), 93(2), 59–67.

297. Photodynamic Therapy: A Clinical Consensus Guide

By Ozog David M; Rkein Ali M; Fabi Sabrina G; Gold Michael H; Goldman Mitchel P; Lowe Nicholas J; Martin George M; Munavalli Girish S

From Dermatologic surgery: official publication for American Society for Dermatologic Surgery [et al.] (2016), 42(7), 804–27.

298. Tumor Heterogeneity, Single-Cell Sequencing, and Drug Resistance

By Schmidt Felix; Efferth Thomas

From Pharmaceuticals (Basel, Switzerland) (2016), 9(2).

299. Cerebral Perfusion Enhancing Interventions: A New Strategy for the Prevention of Alzheimer Dementia

By de la Torre Jack C

From Brain pathology (Zurich, Switzerland) (2016), 26(5), 618–31.

300. PKP for Keratoconus—From Hand/Motor Trephine to Excimer Laser and Back to Femtosecond Laser

By Seitz B; Szentmary N; Hager T; Viestenz A; El-Husseiny M; Langenbucher A; Janunts E

From Klinische Monatsblatter fur Augenheilkunde (2016), 233(6), 727–36.

301. Clinical Application of Analytical and Medical Instruments Mainly Using MS Techniques

By Tanaka Koichi

From Rinsho byori. The Japanese journal of clinical pathology (2016), 64(2), 193–201.

302. Low-Level Laser Therapy to the Bone Marrow Ameliorates Neurodegenerative Disease Progression in a Mouse Model of Alzheimer's Disease: A Minireview

By Oron Amir; Oron Uri

From Photomedicine and laser surgery (2016), 34(12), 627–630.

303. Treatment of warts and molluscum: what does the evidence show?

By Sterling Jane

From Current opinion in pediatrics (2016), 28(4), 490–9.

304. The Future of Fractional Lasers

By Paasch Uwe

From Facial plastic surgery: FPS (2016), 32(3), 261–8.

305. Therapy of basal cell carcinoma

By Schmitz L; Schmitz L; Dirschka T; Dirschka T

From Der Hautarzt; Zeitschrift fur Dermatologie, Venerologie, und verwandte Gebiete (2016), 67(6), 483–99.

306. Lasers in tattoo and pigmentation control: role of the PicoSure(®) laser system

By Torbeck Richard; Saedi Nazanin; Bankowski Richard; Henize Sarah

From Medical devices (Auckland, N.Z.) (2016), 963–7.

307. Emergence of 3D Printed Dosage Forms: Opportunities and Challenges

By Alhnan Mohamed A; Okwuosa Tochukwu C; Sadia Muzna; Wan Ka-Wai; Arafat Basel; Ahmed Waqar

From Pharmaceutical research (2016), 33(8), 1817–32.

308. Strategies for the noninvasive diagnosis of melanoma

By Fink C; Haenssle H A

From Der Hautarzt; Zeitschrift fur Dermatologie, Venerologie, und verwandte Gebiete (2016), 67(7), 519–28.

309. Multiparametric Magnetic Resonance Imaging of Recurrent Prostate Cancer

By Mertan Francesca V; Greer Matthew D; Borofsky Sam; Kabakus Ismail M; Merino Maria J; Wood Bradford J; Pinto Peter A; Choyke Peter L; Turkbey Baris

From Topics in magnetic resonance imaging: TMRI (2016), 25(3), 139–47.

310. Craniofacial Wound Healing with Photobiomodulation Therapy: New Insights and Current Challenges

By Arany P R

From Journal of dental research (2016), 95(9), 977–84.

311. Contributed Review: A review of the investigation of rare-earth dopant profiles in optical fibers

By Sidiroglou F; Baxter G; Roberts A

From The Review of scientific instruments (2016), 87(4), 041501.

312. Laser Treatment of Nongenital Verrucae: A Systematic Review

By Nguyen Jannett; Korta Dorota Z; Chapman Lance W; Kelly Kristen M

From JAMA dermatology (2016), 152(9), 1025–34.

313. Evidence-Based Management of Vascular Malformations

By Morgan Patrick; Keller Robert; Patel Krishna

From Facial plastic surgery: FPS (2016), 32(2), 162–76.

314. Rapid prototyping for patient-specific surgical orthopaedics guides: A systematic literature review

By Popescu Diana; Laptoiu Dan

From Proceedings of the Institution of Mechanical Engineers. Part H, Journal of engineering in medicine (2016), 230(6), 495–515.

315. Liquid-Phase Laser Induced Forward Transfer for Complex Organic Inks and Tissue Engineering

By Nguyen Alexander K; Narayan Roger J

From Annals of biomedical engineering (2017), 45(1), 84–99.

316. Photodynamic therapy for the treatment of different severity of acne: A systematic review

By Keyal Uma; Bhatta Anil Kumar; Wang Xiu Li

From Photodiagnosis and photodynamic therapy (2016), 14191–9.

317. Welcoming a new age for gene therapy in hematology

By Weiss Mitchell J; Mullighan Charles G

From Blood (2016), 127(21), 2523–4.

318. Diagnosis, treatment and monitoring of diabetic eye control

By Radda Stephan; Radda Stephan; Bolz Matthias; Egger Stefan; Gasser-Steiner Vanessa; Kralinger Martina; Mennel Stefan; Scholda Christoph; Stolba Ulrike; Wedrich Andreas; et al.

From Wiener klinische Wochenschrift (2016), 128 Suppl 2S97–102.

319. Diagnosis and Management of Onychomycosis Perspectives from a Joint Podiatric Medicine-Dermatology Roundtable

By Vlahovic Tracey C; Joseph Warren S; Scher Richard K; Tosti Antonella; Plasencia Jesse; Pariser David M; Markinson Bryan C

From Journal of the American Podiatric Medical Association (2016), 106(2), 155–62.

320. Treatment of primary cutaneous amyloidosis with laser: a review of the literature

By Al Yahya Rand S

From Lasers in medical science (2016), 31(5), 1027–35.

321. Current Evidence of Minimally Invasive Spine Surgery in the Treatment of Lumbar Disc Herniations

By Quirno Martin; Vira Shaleen; Errico Thomas J

From Bulletin of the Hospital for Joint Disease (2013) (2016), 74(1), 88–97.

322. Central Retinal Vein Occlusion: A Review of Current Evidence-based Treatment Options

By Patel Amy; Nguyen Christine; Lu Stephanie

From Middle East African journal of ophthalmology (2016), 23(1), 44–8.

323. The use of laser in hysteroscopic surgery

By Nappi Luigi; Sorrentino Felice; Angioni Stefano; Pontis Alessandro; Greco Pantaleo

From Minerva ginecologica (2016), 68(6), 722–6.

324. Evidence-based review of diabetic macular edema management: Consensus statement on Indian treatment guidelines

By Das Taraprasad; Aurora Ajay; Chhablani Jay; Giridhar Anantharaman; Kumar Atul; Raman Rajiv; Nagpal Manish; Narayanan Raja; Natarajan Sundaram; Ramasamay Kim; et al.

From Indian journal of ophthalmology (2016), 64(1), 14–25.

325. Changing Paradigms in Cranio-Facial Regeneration: Current and New Strategies for the Activation of Endogenous Stem Cells

By Mele Luigi; Tirino Virginia; Paino Francesca; Liccardo Davide; Papaccio Gianpaolo; Desiderio Vincenzo; Vitiello Pietro Paolo; De Rosa Alfredo

From Frontiers in physiology (2016), 762.

326. Features and Role of Minimally Invasive Palliative Procedures for Pain Management in Malignant Pelvic Diseases: A Review

By Cascella Marco; Viscardi Daniela; Cuomo Arturo; Muzio Maria Rosaria

From The American journal of hospice & palliative care (2017), 34(6), 524–531.

327. Translational medicine in the field of ablative fractional laser (AFXL)-assisted drug delivery: A critical review from basics to current clinical status

By Haedersdal Merete; Erlendsson Andres M; Paasch Uwe; Anderson R Rox

From Journal of the American Academy of Dermatology (2016), 74(5), 981–1004.

328. Optical technologies for intraoperative neurosurgical guidance

By Valdes Pablo A; Golby Alexandra; Valdes Pablo A; Roberts David W; Lu Fa-Ke; Golby Alexandra; Golby Alexandra

From Neurosurgical focus (2016), 40(3), E8.

329. Characterization of tissue engineered cartilage products: Recent developments in advanced therapy

By Maciulaitis Justinas; Rekstyte Sima; Malinauskas Mangirdas; Usas Arvydas; Maciulaitis Romaldas; Jankauskaite Virginija; Gudas Rimtautas

From Pharmacological research (2016), 113(Pt B), 823–832.

330. Oral leukoplakia treatment with the carbon dioxide laser: A systematic review of the literature

By Mogedas-Vegara Alfonso; Hueto-Madrid Juan-Antonio; Bescos-Atin Coro; Chimenos-Kustner Eduardo

From Journal of cranio-maxillo-facial surgery: official publication of the European Association for Cranio-Maxillo-Facial Surgery (2016), 44(4), 331–6.

331. Recent advances in treatment for Benign Prostatic Hyperplasia

By van Rij Simon; Gilling Peter

From F1000Research (2015), 4.

332. Practical Rehabilitation and Physical Therapy for the General Equine Practitioner

By Kaneps Andris J

From The Veterinary clinics of North America. Equine practice (2016), 32(1), 167–80.

333. Current diagnosis and treatment of basal cell carcinoma

By Alter Mareike; Hillen Uwe; Leiter Ulrike; Sachse Michael; Gutzmer Ralf

From Journal der Deutschen Dermatologischen Gesellschaft = Journal of the German Society of Dermatology: JDDG (2015), 13(9), 863–74; quiz 875

334. Minimally Invasive Techniques to Accelerate the Orthodontic Tooth Movement: A Systematic Review of Animal Studies

By Qamruddin Irfan; Alam Mohammad Khursheed; Khamis Mohd Fadhli; Husein Adam

From BioMed research international (2015), 2015608530.

335. Application of Imaging Mass Spectrometry for Drug Discovery

By Hayasaka Takahiro

From Yakugaku zasshi: Journal of the Pharmaceutical Society of Japan (2016), 136(2), 163–70.

336. Efficacy and Safety of Topical Timolol Eye Drops in the Treatment of Myopic Regression after Laser In Situ Keratomileusis: A Systematic Review and Meta-Analysis

By Wang Xiaochen; Zhao Guiqiu; Lin Jing; Jiang Nan; Wang Qian; Xu Qiang

From Journal of ophthalmology (2015), 2015985071.

337. Does surgical sympathectomy improve clinical outcomes in patients with refractory angina pectoris?

By Holland Luke C; Navaratnarajah Manoraj; Taggart David P

From Interactive cardiovascular and thoracic surgery (2016), 22(4), 488–92.

338. A systematic review of low-level light therapy for treatment of diabetic foot ulcer

By Tchanque-Fossuo Catherine N; Ho Derek; Koo Eugene; Isseroff R Rivkah; Jagdeo Jared; Tchanque-Fossuo Catherine N; Dahle Sara E; Isseroff R Rivkah; Jagdeo Jared; Dahle Sara E; et al.

From Wound repair and regeneration: official publication of the Wound Healing Society [and] the European Tissue Repair Society (2016), 24(2), 418–26.

339. Novel advances in shotgun lipidomics for biology and medicine

By Wang Miao; Wang Chunyan; Han Rowland H; Han Xianlin

From Progress in lipid research (2016), 6183–108.

340. In quest of optimal drug-supported and targeted bone regeneration in the cranio facial area: a review of techniques and methods

By Lucaciu Ondine; Crisan Bogdan; Crisan Liana; Baciut Mihaela; Bran Simion; Hurubeanu Lucia; Hedesiu Mihaela; Vacaras Sergiu; Campian Radu Septimiu; Baciut Grigore; et al.

From Drug metabolism reviews (2015), 47(4), 455–69.

341. Low level laser therapy and hair regrowth: an evidence-based review

By Zarei Mina; Wikramanayake Tongyu C; Falto-Aizpurua Leyre; Schachner Lawrence A; Jimenez Joaquin J

From Lasers in medical science (2016), 31(2), 363–71.

342. Application of high-frequency ultrasonography in closing small blood vessels

By Mlosek Robert Krzysztof; Malinowska Sylwia

From Journal of ultrasonography (2014), 14(58), 320–7.

343. Reported concepts for the treatment modalities and pain management of temporomandibular disorders

By Wieckiewicz Mieszko; Boening Klaus; Wiland Piotr; Shiau Yuh-Yuan; Paradowska-Stolarz Anna

From The journal of headache and pain (2015), 16106.

344. Complications of corneal lamellar refractive surgery

By Kohnen T; Remy M

From Der Ophthalmologe: Zeitschrift der Deutschen Ophthalmologischen Gesellschaft (2015), 112(12), 982–9.

345. Effectiveness of Practices To Increase Timeliness of Providing Targeted Therapy for Inpatients with Bloodstream Infections: a Laboratory Medicine Best Practices Systematic Review and Meta-analysis

By Buehler Stephanie S; Madison Bereneice; Snyder Susan R; Derzon James H; Liebow Edward B; Cornish Nancy E; Saubolle Michael A; Weissfeld Alice S; Weinstein Melvin P; Wolk Donna M

From Clinical microbiology reviews (2016), 29(1), 59–103.

346. Screening and Treatment in Retinopathy of Prematurity

By Stahl Andreas; Gopel Wolfgang

From Deutsches Arzteblatt international (2015), 112(43), 730–5.

347. Clinical Personnel Training in Laboratory Medicine in Chiba University Hospital during the Past 15 Years

By Nomura Fumio

From Rinsho byori. The Japanese journal of clinical pathology (2015), 63(4), 507–13.

348. Systematic Review of Endoscopic Obliteration Techniques for Managing Congenital Piriform Fossa Sinus Tracts in Children

By Lachance Sophie; Chadha Neil K

From Otolaryngology—head and neck surgery: official journal of American Academy of Otolaryngology-Head and Neck Surgery (2016), 154(2), 241–6.

349. Oncoproteomics: Trials and tribulations

By Zhou Li; Huang Canhua; Nice Edouard C; Zhou Li; Li Qifu; Wang Jiandong; Nice Edouard C

From Proteomics. Clinical applications (2016), 10(4), 516–31.

350. Coherent Raman Scattering Microscopy in Biology and Medicine

By Zhang Chi; Zhang Delong; Cheng Ji-Xin

From Annual review of biomedical engineering (2015), 17415–45.

351. Laser treatment of granuloma annulare: a review

By Verne Sebastian H; Kennedy Johnathan; Falto-Aizpurua Leyre A; Griffith Robert D; Nouri Keyvan

From International journal of dermatology (2016), 55(4), 376–81.

352. Novel approaches to imaging basal cell carcinoma

By Rossi Anthony M; Sierra Heidy; Rajadhyaksha Milind; Nehal Kiswher

From Future oncology (London, England) (2015), 11(22), 3039–46.

353. Pain and analgesia following onychectomy in cats: a systematic review

By Wilson Deborah V; Pascoe Peter J

From Veterinary anaesthesia and analgesia (2016), 43(1), 5–17.

354. Esthetic dermatology for the elderly

By Wollina U; Goldman A

From Der Hautarzt; Zeitschrift fur Dermatologie, Venerologie, und verwandte Gebiete (2016), 67(2), 148–52.

355. An historical overview of the activities in the field of exposure and risk assessment of non-ionizing radiation in Bulgaria

By Israel Michel

From Electromagnetic biology and medicine (2015), 34(3), 183–9.

356. Best Reconstructive Techniques: Improving the Final Scar

By Balaraman Brundha; Geddes Elizabeth R; Friedman Paul M

From Dermatologic surgery: official publication for American Society for Dermatologic Surgery [et al.] (2015), 41 Suppl 10S265–75.

357. The Three-Dimensional Techniques in the Objective Measurement of Breast Aesthetics

By Yang Jiqiao; Lv Qing; Zhang Run; Shen Jiani; Hu Yuanyuan

From Aesthetic plastic surgery (2015), 39(6), 910–5.

358. Endoscopic Ultrasound-guided Local Therapy of Pancreatic Tumors

By Yoon Won Jae; Seo Dong Wan

From The Korean journal of gastroenterology = Taehan Sohwagi Hakhoe chi (2015), 66(3), 154–8.

359. Chemical peel treatments in dermatology

By Wiest L G; Habig J

From Der Hautarzt; Zeitschrift fur Dermatologie, Venerologie, und verwandte Gebiete (2015), 66(10), 744–7.

360. Treatment of superficial venous incompetence

By Esponda Omar; Sadek Mikel; Kabnick Lowell S

From Seminars in vascular surgery (2015), 28(1), 29–38.

361. Management and prevention of adverse events in esthetic interventions

By Hartmann D; Heppt M; Gauglitz G G

From Der Hautarzt; Zeitschrift fur Dermatologie, Venerologie, und verwandte Gebiete (2015), 66(10), 764–71.

362. Very high frequency ultrasound: New therapeutic method in aesthetic medicine and dermatology

By Kruglikov I

From Der Hautarzt; Zeitschrift fur Dermatologie, Venerologie, und verwandte Gebiete (2015), 66(11), 829–33.

363. Effectiveness of interventions to enhance healing of chronic ulcers of the foot in diabetes: a systematic review

By Game F L; Apelqvist J; Londahl M; Attinger C; Hartemann A; Hinchliffe R J; Price P E; Jeffcoate W J

From Diabetes/metabolism research and reviews (2016), 32 Suppl 1154–68.

364. Efficacy of adjunctive laser in non-surgical periodontal treatment: a systematic review and meta-analysis

By Cheng Y; Chen J W; Ge M K; Zhou Z Y; Yin X; Zou S J

From Lasers in medical science (2016), 31(1), 151–63.

365. Undesirable pigmentation

By Bayerl C

From Der Hautarzt; Zeitschrift fur Dermatologie, Venerologie, und verwandte Gebiete (2015), 66(10), 757–63.

366. A systematic review of non-surgical treatments for lentigo maligna

By Read T; Wagels M; Foote M; Smithers B M; Read T; Wagels M; Smithers B M; Read T; Noonan C; Noonan C; et al.

From Journal of the European Academy of Dermatology and Venereology: JEADV (2016), 30(5), 748–53.

367. Three-Dimensional Printing and Medical Imaging: A Review of the Methods and Applications

By Marro Alessandro; Bandukwala Taha; Mak Walter

From Current problems in diagnostic radiology (2016), 45 (1), 2–9.

368. Low Reactive Level Laser Therapy for Mesenchymal Stromal Cells Therapies

By Kushibiki Toshihiro; Hirasawa Takeshi; Okawa Shinpei; Ishihara Miya

From Stem cells international (2015), 2015974864.

369. Nonablative fractional lasers: Acne scars and other indications

By Degitz K

From Der Hautarzt; Zeitschrift fur Dermatologie, Venerologie, und verwandte Gebiete (2015), 66(10), 753–6.

370. Retinal vein occlusion: Therapy of retinal vein occlusion371. Recent progress in protein profiling of clinical tissues for next-generation molecular diagnostics

By Boellner Stefanie; Becker Karl-Friedrich

From Expert review of molecular diagnostics (2015), 15 (10), 1277–92.

372. Evidence-Based Medicine in the Treatment of Infantile Hemangiomas

By Keller Robert G; Patel Krishna G

From Facial plastic surgery clinics of North America (2015), 23(3), 373–92.

373. Evidence-Based Medicine in Laser Medicine for Facial Plastic Surgery

By Marcus Benjamin C; Hyman David

From Facial plastic surgery clinics of North America (2015), 23(3), 297–302.

374. Drug compound characterization by mass spectrometry imaging in cancer tissue

By Kwon Ho Jeong; Kim Yonghyo; Sugihara Yutaka; Baldetorp Bo; Welinder Charlotte; Watanabe Ken-ichi; Watanabe Ken-ichi; Nishimura Toshihide; Vegvari Akos; Fehniger Thomas E; et al.

From Archives of pharmacal research (2015), 38(9), 1718–27.

375. Electromagnetic fields in medicine—The state of art

By Pasek Jaroslaw; Cieslar Grzegorz; Sieron Aleksander; Pasek Jaroslaw; Pasek Tomasz; Sieron-Stoltny Karolina

From Electromagnetic biology and medicine (2016), 35(2), 170–5.

376. Laser therapy on points of acupuncture: Are there benefits in dentistry?

By de Oliveira Renata Ferreira; da Silva Camila Vieira; Cersosimo Maria Cecilia Pereira; Borsatto Maria Cristina; de Freitas Patricia Moreira

From Journal of photochemistry and photobiology. B, Biology (2015), 15176–82.

377. Anti-VEGF treatment for myopic choroid neovascularization: from molecular characterization to update on clinical application

By Zhang Yan; Ru Yusha; Bo Qiyu; Wei Rui Hua; Han Qian

From Drug design, development and therapy (2015), 93413–21.

378. Pharmacotherapy and Adherence Issues in Treating Elderly Patients with Glaucoma

By Broadway David C; Cate Heidi

From Drugs & aging (2015), 32(7), 569–81.

379. Therapeutic approach in persistent diabetic macular edema

By Branisteanu Daniel; Moraru Andreea

From Oftalmologia (Bucharest, Romania: 1990) (2014), 58 (4), 3–9.

380. Geotrichum capitatum septicaemia in a haematological patient after acute myeloid leukaemia relapse: identification using MALDI-TOF mass spectrometry and review of the literature

By Miglietta Fabio; Vella Antonietta; Faneschi Maria Letizia; Lobreglio Giambattista; Rizzo Adriana; Palumbo Claudio; Palumbo Carla; Di Renzo Nicola; Pizzolante Maria

From Le infezioni in medicina: rivista periodica di eziologia, epidemiologia, diagnostica, clinica e terapia delle patologie infettive (2015), 23(2), 161–7.

381. Role of tele-medicine in retinopathy of prematurity screening in rural outreach centers in India—a report of 20,214 imaging sessions in the KIDROP program

By Vinekar Anand; Jayadev Chaitra; Shetty Bhujang; Mangalesh Shwetha; Vidyasagar Dharmapuri

From Seminars in fetal & neonatal medicine (2015), 20(5), 335–45.

382. Recent Innovations in Medical and Surgical Retina

By Bhagat Neelakshi; Zarbin Marco

From Asia-Pacific journal of ophthalmology (Philadelphia, Pa.) (2015), 4(3), 171–9.

383. Laser-driven electron beam and radiation sources for basic, medical and industrial sciences

By Nakajima Kazuhisa

From Proceedings of the Japan Academy. Series B, Physical and biological sciences (2015), 91(6), 223–45.

384. Fibroids (uterine myomatosis, leiomyomas)

By Lethaby Anne; Vollenhoven Beverley

From BMJ clinical evidence (2015), 2015.

385. Recent developments in vascular imaging techniques in tissue engineering and regenerative medicine

By Upputuri Paul Kumar; Sivasubramanian Kathyayini; Mark Chong Seow Khoon; Pramanik Manojit

From BioMed research international (2015), 2015783983.

386. Anesthesiology in interventional pneumology-endoscopic interventions: part 1

By Kern Michael; Niemeyer Daniel; Kerner Thoralf; Tank Sascha

From Anasthesiologie, Intensivmedizin, Notfallmedizin, Schmerztherapie: AINS (2015), 50(4), 228–36.

387. Complementary and alternative medicine for the treatment of obesity: a critical review

By Esteghamati Alireza; Mazaheri Tina; Vahidi Rad Mona; Noshad Sina

From International journal of endocrinology and metabolism (2015), 13(2), e19678.

388. Modern retinal laser therapy

By Kozak Igor; Luttrull Jeffrey K

From Saudi journal of ophthalmology: official journal of the Saudi Ophthalmological Society (2015), 29(2), 137–46.

389. Aiming for zero blindness

By Nakazawa Toru

From Nippon Ganka Gakkai zasshi (2015), 119(3), 168–93; discussion 194.

390. 1064 nm Q-switched Nd:YAG laser for the treatment of Argyria: a systematic review

By Griffith R D; Simmons B J; Bray F N; Falto-Aizpurua L A; Yazdani Abyaneh M-A; Nouri K

From Journal of the European Academy of Dermatology and Venereology: JEADV (2015), 29(11), 2100–3.

391. Home-use devices in aesthetic dermatology

By Keller Emily C

From Seminars in cutaneous medicine and surgery (2014), 33(4), 198–204.

392. Noninvasive imaging technologies for cutaneous wound assessment: A review

By Paul Dereck W; Prindeze Nicholas J; Moffatt Lauren T; Alkhalil Abdulnaser; Shupp Jeffrey W; Ghassemi Pejhman; Ramella-Roman Jessica C; Shupp Jeffrey W

From Wound repair and regeneration: official publication of the Wound Healing Society [and] the European Tissue Repair Society (2015), 23(2), 149–62.

393. Survey on computer aided decision support for diagnosis of celiac disease

By Hegenbart Sebastian; Uhl Andreas; Vecsei Andreas

From Computers in biology and medicine (2015), 65348–58.

394. Role of human albumin in the management of complications of liver cirrhosis

By Bernardi Mauro; Ricci Carmen S; Zaccherini Giacomo

From Journal of clinical and experimental hepatology (2014), 4(4), 302–11.

395. Photobiomodulation in oral medicine: a review

By Pandeshwar Padma; Das Reshma; Shastry Shilpa P; Kaul Rachna; Srinivasreddy Mahesh B; Roa Mahesh Datta

From Journal of investigative and clinical dentistry (2016), 7(2), 114–26.

396. Non-invasive subcutaneous fat reduction: a review

By Kennedy J; Verne S; Griffith R; Falto-Aizpurua L; Nouri K

From Journal of the European Academy of Dermatology and Venereology: JEADV (2015), 29(9), 1679–88.

397. Tissue Engineering Applications of Three-Dimensional Bioprinting

By Zhang Xiaoying; Zhang Yangde

From Cell biochemistry and biophysics (2015), 72(3), 777–82.

398. New therapeutic modalities to modulate orthodontic tooth movement

By Andrade Ildeu Jr; Sousa Ana Beatriz dos Santos; da Silva Gabriela Goncalves

From Dental press journal of orthodontics (2014), 19(6), 123–33.

399. Pulsed dye laser therapy for molluscum contagiosum: a systematic review

By Griffith Robert Denison; Yazdani Abyaneh Mohammad-Ali; Falto-Aizpurua Leyre; Nouri Keyvan

From Journal of drugs in dermatology: JDD (2014), 13(11), 1349–52.

400. Prospects of nanoscience with nanocrystals

By Kovalenko Maksym V; Manna Liberato; Cabot Andreu; Hens Zeger; Talapin Dmitri V; Kagan Cherie R; Klimov Victor I; Rogach Andrey L; Reiss Peter; Milliron Delia J; et al.

From ACS nano (2015), 9(2), 1012–57.

401. Therapeutic laser in veterinary medicine

By Pryor Brian; Millis Darryl L

From The Veterinary clinics of North America. Small animal practice (2015), 45(1), 45–56.

402. Laser Doppler flowmetry in manual medicine research

By Zegarra-Parodi Rafael; Snider Eric J; Park Peter Yong Soo; Degenhardt Brian F

From The Journal of the American Osteopathic Association (2014), 114(12), 908–17.

403. The Effects of Low-Level Laser Therapy on Pain Associated With Tendinopathy: A Critically Appraised Topic

By Doyle Andrew T; Lauber Christine; Sabine Kendra

From Journal of sport rehabilitation (2016), 25(1), 83–90.

404. Diagnosis of inflammatory bowel disease: Potential role of molecular biometrics

By M'Koma Amosy E

From World journal of gastrointestinal surgery (2014), 6 (11), 208–19.

405. Significance of transcanalicular laser assisted dacryocystorhinostomy in modern lacrimal drainage surgery

By Koch K R; Kuhner H; Cursiefen C; Heindl L M

From Der Ophthalmologe: Zeitschrift der Deutschen Ophthalmologischen Gesellschaft (2015), 112(2), 122–6.

406. Wound healing in urology

By Ninan Neethu; Thomas Sabu; Grohens Yves

From Advanced drug delivery reviews (2015), 82–83, 93–105.

407. Biomaterials for integration with 3-D bioprinting

By Skardal Aleksander; Atala Anthony

From Annals of biomedical engineering (2015), 43(3), 730–46.

408. Proteomic and metabolic prediction of response to therapy in gastric cancer

By Aichler Michaela; Luber Birgit; Lordick Florian; Walch Axel

From World journal of gastroenterology (2014), 20(38), 13648–57.

409. Rigid bronchoscopy

By Alraiyes Abdul Hamid; Machuzak Michael S

From Seminars in respiratory and critical care medicine (2014), 35(6), 671–80.

410. Guidelines for treatment of chronic primary angle-closure glaucoma

By Munoz-Negrete F J; Gonzalez-Martin-Moro J; Casas-Llera P; Urcelay-Segura J L; Rebolleda G; Ussa F; Guerri Monclus N; Pablo L E; Mendez Hernandez C; Garcia-Feijoo J; et al.

From Archivos de la Sociedad Espanola de Oftalmologia (2015), 90(3), 119–38.

411. Prostato-symphyseal fistula after photoselective vaporization of the prostate: case series and literature review of a rare complication

By Sanchez Alejandro; Rodriguez Dayron; Cheng Jed-Sian; McGovern Francis J; Tabatabaei Shahin

From Urology (2015), 85(1), 172−7.

412. Acupuncture (zhen ji)—an emerging adjunct in routine oral care.

By Gupta Devanand; Dalai Deepak Ranjan; Jain Ankita; Swapnadeep; Mehta Parul; Indra B Niranjanaprasad; Rastogi Saurabh; Chaturvedi Mudita; Sharma Saumya; Singh Sanjeev; et al.

From Journal of traditional and complementary medicine (2014), 4(4), 218−23.

413. Treatments for nail psoriasis: a systematic review by the GRAPPA Nail Psoriasis Work Group

By Armstrong April W; Tuong William; Love Thorvardur J; Carneiro Sueli; Grynszpan Rachel; Lee Steve S; Kavanaugh Arthur

From The Journal of rheumatology (2014), 41(11), 2306−14.

414. Acupuncture for schizophrenia

By Shen Xiaohong; Xia Jun; Adams Clive E

From The Cochrane database of systematic reviews (2014), (10), CD005475.

415. Molecular imaging for theranostics in gastroenterology: one stone to kill two birds

By Ko Kwang Hyun; Kown Chang-Il; Park Jong Min; Hahm Ki Baik; Lee Hoo Geun; Han Na Young

From Clinical endoscopy (2014), 47(5), 383−8.

416. Fiber lasers and their applications [Invited]

By Shi Wei; Fang Qiang; Zhu Xiushan; Norwood R A; Peyghambarian N

From Applied optics (2014), 53(28), 6554−68.

417. Scleroderma and evidence based non-pharmaceutical treatment modalities for digital ulcers: a systematic review

By Moran M E

From Journal of wound care (2014), 23(10), 510−6.

418. Assessment of skin blood flow following spinal manual therapy: a systematic review

By Zegarra-Parodi Rafael; Park Peter Yong Soo; Heath Deborah M; Makin Inder Raj S; Degenhardt Brian F; Roustit Matthieu

From Manual therapy (2015), 20(2), 228−49.

419. Management of sickle cell disease: summary of the 2014 evidence-based report by expert panel members

By Yawn Barbara P; Buchanan George R; Afenyi-Annan Araba N; Ballas Samir K; Hassell Kathryn L; James Andra H; Jordan Lanetta; Lanzkron Sophie M; Lottenberg Richard; Savage William J; et al.

From JAMA (2014), 312(10), 1033−48.

420. Mass spectrometry-based serum and plasma peptidome profiling for prediction of treatment outcome in patients with solid malignancies

By Labots Mariette; Schutte Lisette M; van der Mijn Johannes C; Pham Thang V; Jimenez Connie R; Verheul Henk M W

From The oncologist (2014), 19(10), 1028−39.

421. The advent of ultrasound-guided ablation techniques in nodular thyroid disease: towards a patient-tailored approach

By Papini Enrico; Pacella Claudio M; Bizzarri Giancarlo; Misischi Irene; Guglielmi Rinaldo; Dossing Helle; Hegedus Laszlo

From Best practice & research. Clinical endocrinology & metabolism (2014), 28(4), 601−18.

422. Management of scars: updated practical guidelines and use of silicones

By Meaume Sylvie; Le Pillouer-Prost Anne; Richert Bertrand; Roseeuw Diane; Vadoud Javid

From European journal of dermatology: EJD (2014), 24(4), 435−43.

423. Fetal surgery: an overview

By Sala Paolo; Prefumo Federico; Pastorino Daniela; Buffi Davide; De Biasio Pierangela; Gaggero Chiara Roberta; Foppiano Marilena

From Obstetrical & gynecological survey (2014), 69(4), 218−28.

424. Devices and topical agents for rosacea management

By Mansouri Yasaman; Goldenberg Gary

From Cutis (2014), 94(1), 21−5.

425. Nanoparticle counting: towards accurate determination of the molar concentration

By Shang Jing; Gao Xiaohu

From Chemical Society reviews (2014), 43(21), 7267–78.

426. Therapeutic indications for percutaneous laser in patients with vascular malformations and tumors

By Labau D; Cadic P; Ouroussoff G; Ligeron C; Laroche J-P; Quere I; Guillot B; Dereure O; Galanaud J-P

From Journal des maladies vasculaires (2014), 39(6), 363–72.

427. Updated international clinical recommendations on scar management: part 1—evaluating the evidence

By Gold Michael H; Berman Brian; Clementoni Matteo Tretti; Gauglitz Gerd G; Nahai Foad; Murcia Crystal

From Dermatologic surgery: official publication for American Society for Dermatologic Surgery [et al.] (2014), 40(8), 817–24.

428. Interventional treatment of atrial fibrillation—contemporary methods and perspectives

By Zuchowski Bartosz; Kaczmarek Krzysztof; Szumowski Lukasz; Li Yi-Gang; Ptaszynski Pawel

From Expert review of medical devices (2014), 11(6), 595–603.

429. Meta-analysis of randomized controlled trials comparing selective laser trabeculoplasty with prostaglandin analogue in the primary treatment of open-angle glaucoma or ocular hypertention

By Peng Wei; Zhong Xiao; Yu Minbin

From [Zhonghua Yank Ke Za Zhi] Chinese journal of ophthalmology (2014), 50(5), 343–8.

430. A new look at drugs targeting malignant melanoma—an application for mass spectrometry imaging

By Sugihara Yutaka; Vegvari Akos; Welinder Charlotte; Jonsson Goran; Ingvar Christian; Lundgren Lotta; Olsson Hakan; Breslin Thomas; Wieslander Elisabet; Laurell Thomas; et al.

From Proteomics (2014), 14(17–18), 1963–70.

431. Low risk papillary thyroid cancer

By Brito Juan P; Hay Ian D; Morris John C

From BMJ (Clinical research ed.) (2014), 348g3045.

432. The Effect of Low-level Laser Therapy on Trigeminal Neuralgia: A Review of Literature

By Falaki Farnaz; Nejat Amir Hossein; Dalirsani Zohreh

From Journal of dental research, dental clinics, dental prospects (2014), 8(1), 1–5.

433. Analysis of exhaled breath for disease detection

By Amann Anton; Miekisch Wolfram; Schubert Jochen; Buszewski Boguslaw; Ligor Tomasz; Jezierski Tadeusz; Pleil Joachim; Risby Terence

From Annual review of analytical chemistry (Palo Alto, Calif.) (2014), 7455–82.

434. Interventional pulmonology: an update for internal medicine physicians

By Beaudoin E L; Chee A; Stather D R

From Minerva medica (2014), 105(3), 197–209.

435. Treatment for palmoplantar pustular psoriasis: systematic literature review, evidence-based recommendations and expert opinion

By Sevrain M; Richard M-A; Barnetche T; Rouzaud M; Villani A P; Paul C; Beylot-Barry M; Jullien D; Aractingi S; Aubin F; et al.

From Journal of the European Academy of Dermatology and Venereology: JEADV (2014), 28 Suppl 513–6.

436. Pemphigus vulgaris and laser therapy: crucial role of dentists

By Pavlic Verica; Aleksic Vesna Vujic; Zubovic Nina; Veselinovic Valentina

From Medicinski pregled (2014), 67(1–2), 38–42.

437. Low level laser therapy for the treatment of diabetic foot ulcers: a critical survey

By Beckmann Kathrin H; Meyer-Hamme Gesa; Schroder Sven

From Evidence-based complementary and alternative medicine: eCAM (2014), 2014626127.

438. MALDI profiling and applications in medicine

By Dudley Ed

From Advances in experimental medicine and biology (2014), 80633–58.

439. Acupuncture for treating acute ankle sprains in adults

By Kim Tae-Hun; Lee Myeong Soo; Kim Kun Hyung; Kang Jung Won; Choi Tae-Young; Ernst Edzard

From The Cochrane database of systematic reviews (2014), (6), CD009065.

440. Warts (non-genital)

By Loo Steven King-Fan; Tang William Yuk-Ming

From BMJ clinical evidence (2014), 2014.

441. Effects of low-level laser therapy on wound healing

By Andrade Fabiana do Socorro da Silva Dias; Clark Rosana Maria de Oliveira; Ferreira Manoel Luiz

From Revista do Colegio Brasileiro de Cirurgioes (2014), 41 (2), 129–33.

442. Surface-assisted laser desorption ionization mass spectrometry techniques for application in forensics

By Guinan Taryn; Ronci Maurizio; Voelcker Nicolas H; Kirkbride Paul; Pigou Paul E; Kobus Hilton

From Mass spectrometry reviews (2015), 34(6), 627–40.

443. Evidence-based review of photodynamic therapy in the treatment of acne

By Zheng Weifeng; Wu Yan; Xu Xuegang; Gao Xinghua; Chen Hong-Duo; Li Yuanhong

From European journal of dermatology: EJD (2014), 24(4), 444–56.

444. Hirsutism: an evidence-based treatment update

By Somani Najwa; Turvy Diane

From American journal of clinical dermatology (2014), 15 (3), 247–66

445. Destruction of cancer cells by laser-induced shock waves: recent developments in experimental treatments and multiscale computer simulations

By Steinhauser Martin Oliver; Schmidt Mischa

From Soft matter (2014), 10(27), 4778–88.

446. What's new in objective assessment and treatment of facial hyperpigmentation?

By Molinar Vanessa E; Taylor Susan C; Pandya Amit G

From Dermatologic clinics (2014), 32(2), 123–35.

447. Treatment of diabetic macular oedema with the VEGF inhibitors ranibizumab and bevacizumab: conclusions from basic in vitro studies

By Lang G E; Lang G K; Deissler H L

From Klinische Monatsblatter fur Augenheilkunde (2014), 231(5), 527–34.

448. Diode lasers: a magical wand to an orthodontic practice

By Srivastava Vipul Kumar; Mahajan Shally

From Indian journal of dental research: official publication of Indian Society for Dental Research (2014), 25(1), 78–82.

449. Mass spectrometry in plant metabolomics strategies: from analytical platforms to data acquisition and processing

By Ernst Madeleine; Silva Denise Brentan; Silva Ricardo Roberto; Vencio Ricardo Z N; Lopes Norberto Peporine

From Natural product reports (2014), 31(6), 784–806.

450. Tattoo ink-related cutaneous pseudolymphoma: a rare but significant complication. Case report and review of the literature

By Marchesi Andrea; Parodi Pier Camillo; Brioschi Marco; Marchesi Matteo; Bruni Barbara; Cangi Maria Giulia; Vaienti Luca

From Aesthetic plastic surgery (2014), 38(2), 471–8.

451. Proteomics in the search for biomarkers of animal cancer

By Kycko Anna; Reichert Michal

From Current protein & peptide science (2014), 15(1), 36–44.

452. Efficacy of low-level laser therapy for accelerating tooth movement during orthodontic treatment: a systematic review and meta-analysis

By Ge M K; He W L; Chen J; Wen C; Yin X; Hu Z A; Liu Z P; Zou S J

From Lasers in medical science (2015), 30(5), 1609–18.

453. Evidence-based management of primary angle closure glaucoma

By Emanuel Matthew E; Parrish Richard K 2nd; Gedde Steven J

From Current opinion in ophthalmology (2014), 25(2), 89–92.

454. Systematic review and meta-analysis of the effect of various laser wavelengths in the treatment of peri-implantitis

By Kotsakis Georgios A; Konstantinidis Ioannis; Karoussis Ioannis K; Ma Xiaoye; Chu Haitao

From Journal of periodontology (2014), 85(9), 1203–13.

455. Systematic review of the limited evidence base for treatments of Eustachian tube dysfunction: a health technology assessment

By Norman G; Llewellyn A; Harden M; Coatesworth A; Kimberling D; Schilder A; McDaid C

From Clinical otolaryngology: official journal of ENT-UK; official journal of Netherlands Society for Oto-Rhino-Laryngology & Cervico-Facial Surgery (2014), 39 (1), 6–21.

456. New techniques on the horizon: interventional radiology and interventional endoscopy of the urinary tract ('endourology')

By Berent Allyson

From Journal of feline medicine and surgery (2014), 16(1), 51–65.

457. Recent advancements of flow cytometry: new applications in hematology and oncology

By Woo Janghee; Baumann Alexandra; Arguello Vivian

From Expert review of molecular diagnostics (2014), 14(1), 67–81.

458. SIU/ICUD Consultation on Urethral Strictures: Dilation, internal urethrotomy, and stenting of male anterior urethral strictures.

459. A critical evaluation of the current state-of-the-art in quantitative imaging mass spectrometry

By Ellis Shane R; Bruinen Anne L; Heeren Ron M A

From Analytical and bioanalytical chemistry (2014), 406(5), 1275–89.

460. Evidence-based guidelines for the chiropractic treatment of adults with neck pain

By Bryans Roland; Decina Philip; Descarreaux Martin; Duranleau Mireille; Marcoux Henri; Potter Brock; Ruegg Richard P; Shaw Lynn; Watkin Robert; White Eleanor

From Journal of manipulative and physiological therapeutics (2014), 37(1), 42–63.

461. Laser therapy of onychomycosis

By Nenoff Pietro; Grunewald Sonja; Paasch Uwe

From Journal der Deutschen Dermatologischen Gesellschaft = Journal of the German Society of Dermatology: JDDG (2014), 12(1), 33–8.

462. Identification of fungal microorganisms by MALDI-TOF mass spectrometry

By Chalupova Jana; Raus Martin; Sedlarova Michaela; Sebela Marek

From Biotechnology advances (2014), 32(1), 230–41.

463. Striae distensae: a comprehensive review and evidence-based evaluation of prophylaxis and treatment

By Al-Himdani S; Ud-Din S; Gilmore S; Bayat A

From The British journal of dermatology (2014), 170(3), 527–47.

464. Phototherapy approaches in treatment of oral lichen planus

By Pavlic Verica; Vujic-Aleksic Vesna

From Photodermatology, photoimmunology & photomedicine (2014), 30(1), 15–24.

465. Proteomics in veterinary medicine: applications and trends in disease pathogenesis and diagnostics

By Ceciliani F; Eckersall D; Burchmore R; Lecchi C

From Veterinary pathology (2014), 51(2), 351–62.

466. Clinical practice guidelines for treatment of acne vulgaris: a critical appraisal using the AGREE II instrument

By Sanclemente Gloria; Acosta Jorge-Luis; Tamayo Maria-Eulalia; Bonfill Xavier; Alonso-Coello Pablo

From Archives of dermatological research (2014), 306(3), 269–77.

467. The role of focal therapy in the management of localised prostate cancer: a systematic review

By Valerio Massimo; Ahmed Hashim U; Emberton Mark; Lawrentschuk Nathan; Lazzeri Massimo; Montironi Rodolfo; Nguyen Paul L; Trachtenberg John; Polascik Thomas J

From European urology (2014), 66(4), 732–51.

468. Evidence for the effectiveness of electrophysical modalities for treatment of medial and lateral epicondylitis: a systematic review

By Dingemanse Rudi; Randsdorp Manon; Huisstede Bionka M A; Koes Bart W

From British journal of sports medicine (2014), 48(12), 957–65.

469. Additive manufacturing techniques for the production of tissue engineering constructs

By Mota Carlos; Puppi Dario; Chiellini Federica; Chiellini Emo

From Journal of tissue engineering and regenerative medicine (2015), 9(3), 174–90.

470. Current evidence for postoperative monitoring of microvascular free flaps: a systematic review

By Chae Michael P; Rozen Warren Matthew; Whitaker Iain S; Chubb Daniel; Grinsell Damien; Ashton Mark W; Hunter-Smith David J; Lineaweaver William C

From Annals of plastic surgery (2015), 74(5), 621–32.

5.3.1 Photochemical strategies

The effects of irradiation on the skin in the presence of xenobiotics may be quite serious. Such compounds when on the excited state interact with the immune system and make much more severe effect of sunlight. A hypothesis of the mechanism is presented in Fig. 5.2.

As shown in Fig. 5.2, three avenues that may be hypothesized to operate in the initiation of Cutaneous Drug Reactions(CDR) after exposure to reactive metabolites: (1) a direct cytotoxic effect, (2) stimulation of danger signal(s) expression, and/or (3) stimulation of adhesion/costimulatory molecule expression. It has been demonstrated that incubation of

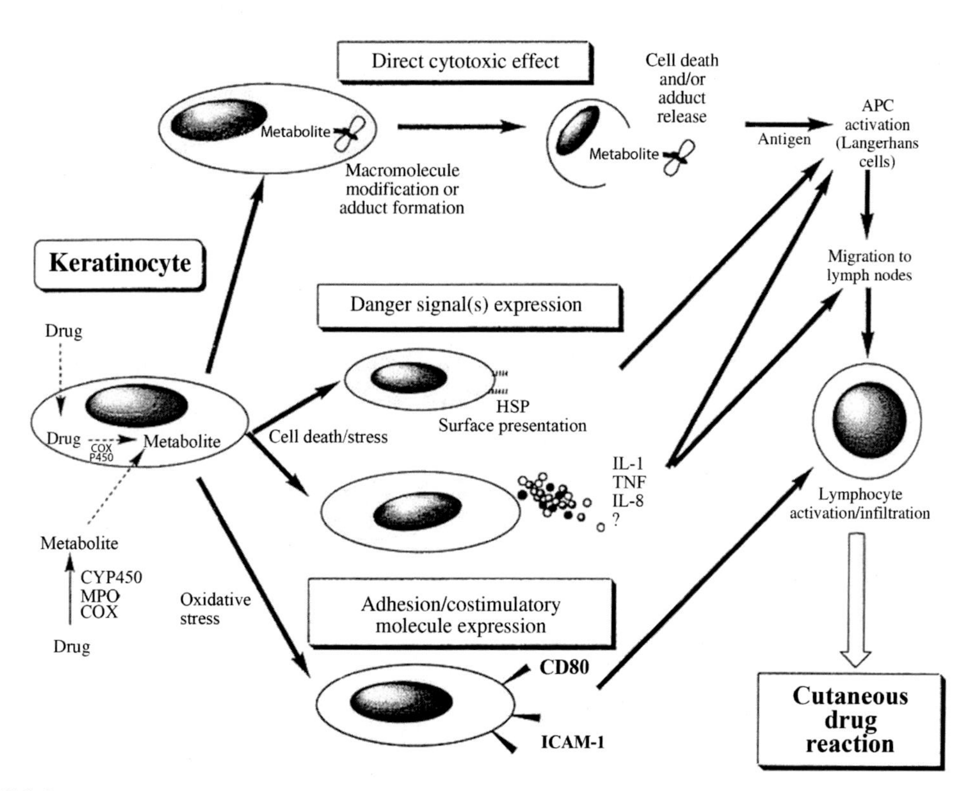

FIGURE 5.2 Working hypothesis for the mechanism of cutaneous drug reactions [58].

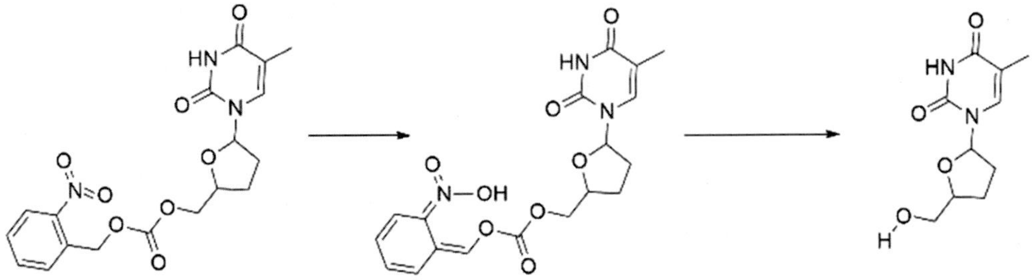

SCHEME 5.4 Intramolecular hydrogen transfer and oxidation at the benzyl position.

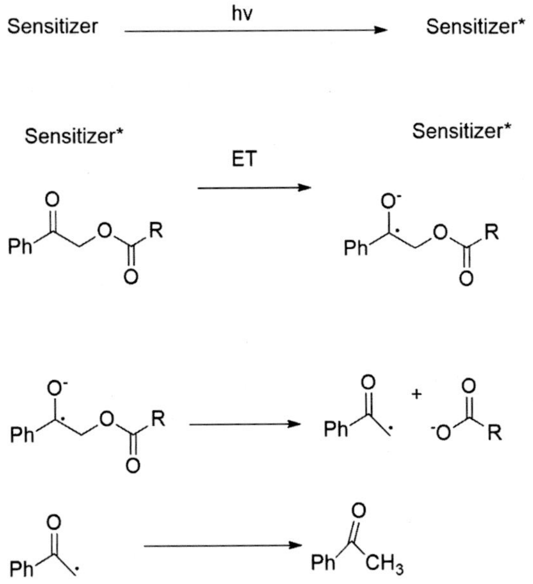

SCHEME 5.5 Intramoleular electron transfer and cleavage of esters [61].

Sulphamethoxazole (SMX), dapsone, or their respective hydroxylamine metabolites, with normal human keratinocytes results in the formation of drug/metabolite-protein adducts, even in the absence of cytotoxicity. Thus antigen presentation may occur as the result of cytotoxicity or in its absence. Since keratinocytes have been shown to express human leukocyte antigen (HLA-DR) at inflammatory sites it is possible that these cells may serve as Antigen-Presenting Cells (APC) in situations that predispose subjects to CDRs. Moreover, keratinocytes have been shown to express accessory molecules that serve as important "second signals" in the activation of T-cells [58].

Apart from the negative effect mentioned above, it is possible to exploit such a preferred localization by exploiting the different physicochemical state in at least three ways. First, through complexation and using photoremovable protecting groups. These have been developed for organic synthesis where it is often useful to quench the reaction of a group, and in a later step, repristinate the reactivity of that group, by treating with a mild reactive, light has been demonstrated to be an effective mild reaction. On the other hand, the same reactions, provided that conditions avoid biological damage [59], may be used for drug generation when it is desired to start from an inactive prodrug and at a desired time or place generate the actual drug [59,60] (see Scheme 5.4).

As it has been often recalled in this book the uncommon high energy of electronic excited states makes also possible to produce a large rearrangement of the bonds, or cleave some of them and give a different high energy intermediate (Scheme 5.5).

This approach may seem too complex, as well understandable in the case of previtamin D, that, however, has had millions of years to check everything, and at any rate stays on the sure side by using only concerted reactions. Otherways using the cell as a photochemical reactor is too risky as a general choice for humans in general, but well suited for ill individuals (see the retrocycloaddition of DNA dimers). Notice further

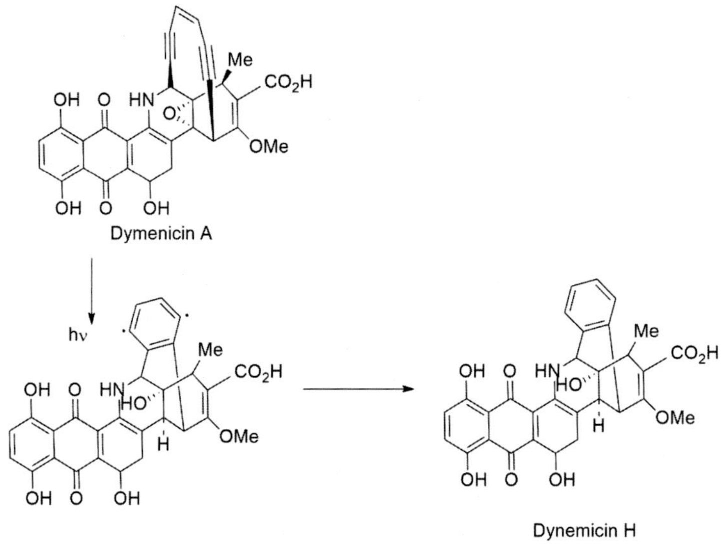

SCHEME 5.6 Reaction of Dhnemicyn H via a diradalic intermediate [62,63].

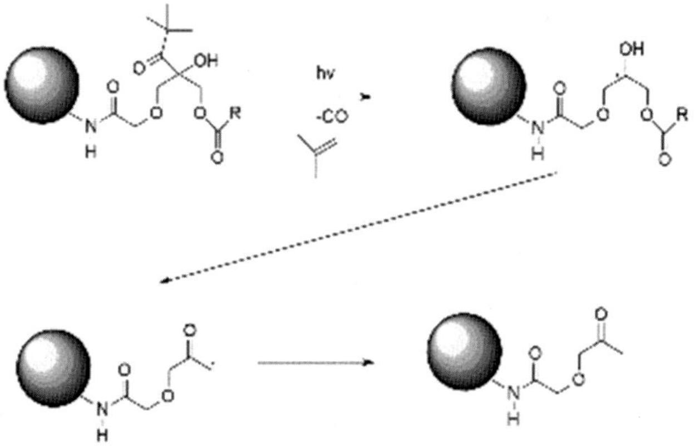

Dymenicin A

Dynemicin H

SCHEME 5.7 Intermolecular electron transfer of tethered esters [64].

that the generation of intermediates is perhaps more large than thought, as indicated in a couple of cases below via radical, ions, or diradicals (Schemes 5.6 and 5.7).

An interesting variation is photochemical internalization, where the drug is taken up in a vesicle, whether of an artificial or a natural origin (in a typical example, a liposome) that is smoothly punctured by a photochemical reaction transfer the cargo to a desired location (Fig. 5.3).

Still another alternative is generating chemicals from inactive precursors by photochemistry, leaving to the hand of the surgeon where he/she exactly make the move. Note in Scheme 5.4, the dye pheoforbide is fixed through a suitable chain to the tip of a glass tube, where a flux of oxygen passes as well as light through a fiber. All of is required is now there, and excitation by light brings the dye to the triplet state, which in turn excites and cleaves the dialkoxyalkene moiety. This result opens the way to a fiberoptic-guided sensitizer delivery for the potential photodynamic therapy of hypoxic structures requiring cytotoxic control [66,67] (Scheme 5.8).

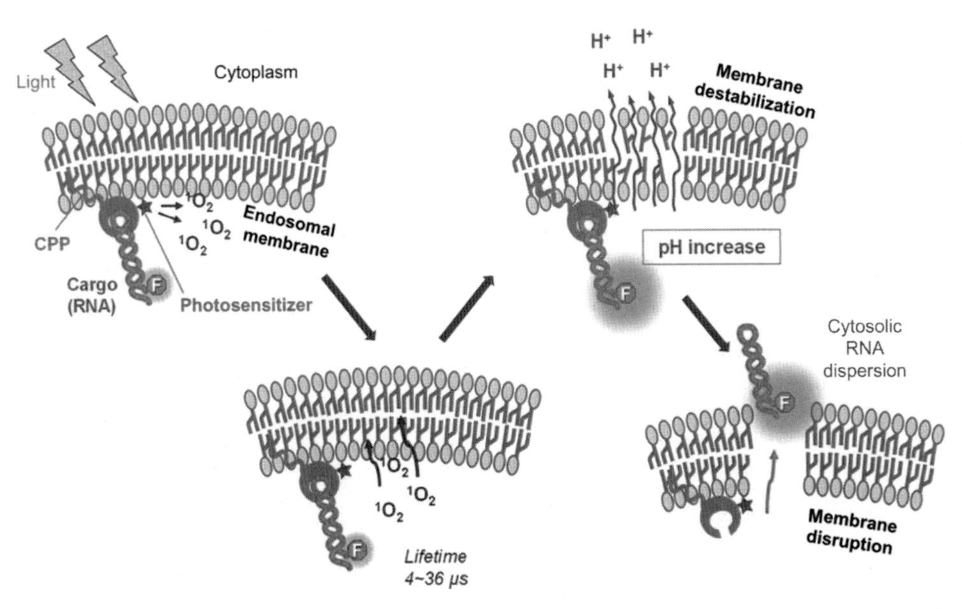

FIGURE 5.3 Photochemical internalization. The drug is ingested or at any rate taken within a membrane that is punctured when appropriate. Schematic illustration of the endosomal disruption process using the cell penetrating peptide (CPP)-cargo-photosensitizer conjugate suggested by this study [65].

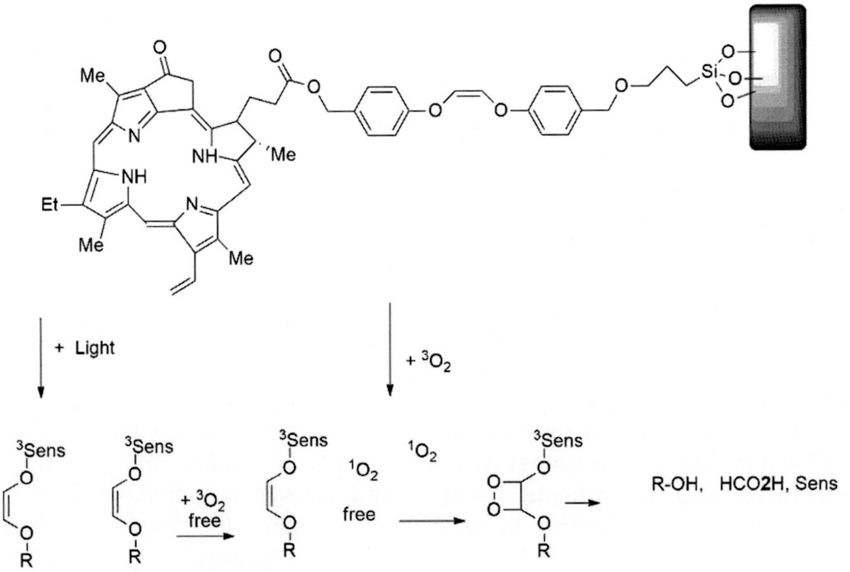

+ Light

+ $^3O_2$

^{3}Sens

+ $^3O_2$ free

$^1O_2$ free

R-OH, HCO2H, Sens

SCHEME 5.8 Oxidative cleavage of alkylatedienol ethers by reaction with singlet oxygen.

References

[1] B.L. Diffey, Solar ultraviolet radiation effects on biological systems, Rev. Phys. Med. Biol. 36 (3) (1991) 299–328.

[2] T.N. Van, T.H. Van, Minh Thi, H.N. Trong, T.G. Van, N.D. Huu, et al., Efficacy of narrow - band UVB phototherapy versus PUVA chemophototherapy for psoriasis in Vietnamese patients, Maced. J. Med. Sci. 7 (2019) 227–230.

[3] B. Lindelöf, Risk of melanoma with psoralen/ultraviolet A therapy for psoriasis. Do the known risks now outweigh the benefits? Drug Saf. 20 (1999) 289–297.

[4] T.M. Lotti, S. Gianfaldoni, Ultraviolet A-1 in dermatological diseases, Adv. Exp. Med. Biol. 996 (2017) 105–110.

[5] S. Mehraban, A. Felly, 308 nm laser in dermatology, J. Lasers Med. Sci. 5 (2014) 8–12.

[6] S.H. Ibbotson, A perspective on the use of NB-UVB phototherapy vs. PUVA photochemotherapy, Front. Med. 5 (2018) 184.

[7] E. Riklis (Ed.), Photobiology: The Science and Its Applications, Springer, New York, 1988.

[8] H. Harada, Irradiation apparatus and irradiation method, H Harada – US Patent 6,984,835, 2006 – Google Patents, 2006.

[9] C.L. By Tsui, J. Levitt, Practical pearlsin phototherapy, Int. J. Dermatol. 52 (2013) 1395–1397.

[10] K. Köllner, M.B. Wimmershoff, C. Hintz, M. Landthaler, U. Hohenleutner, Comparison of the 308-nm excimer laser and a 308-nm excimer lamp with 311-nm narrowband ultraviolet B in the treatment of psoriasis, Br. J. Dermatol. 2005 (152) (2005) 750–754.

[11] M. Galvan-Banqueri, R. Marin Gil, B. Santos Ramos, F.J. Bautista Paloma, Biological treatments for moderate-to-severe psoriasis: indirect comparison, J. Clin. Pharm. Ther. 38 (2013) 121–130.

[12] J. Foerster, K. Boswell, J. West, H. Cameron, C. Fleming, S. Ibbotson, et al., Narrowband UVB treatment is highly effective and causes a strong reduction in the use of steroid and other creams in psoriasis patients in clinical practice, PLoS One. 12 (2017) e0181813.

[13] S.H. Ibbotson, D. Bilsland, N.H. Cox, R.S. Dawe, B. Diffey, C. Edwards, et al., An update and guidance on narrowband ultraviolet B phototherapy: a British photodermatology group workshop report, Br. J. Dermatol. 151 (2004) 283–297.

[14] E. Archier, S. Devaux, E. Castela, A. Gallini, F. Aubin, M. Le Maitre, et al., Carcinogenic risks of psoralen UV-A therapy and narrowband UV-B therapy in chronic plaque psoriasis: a systematic literature review, J. Eur. Acad. Dermatol. Venereol. 26 (Suppl.3) (2012) 22–31.

[15] T.C. Ling, T.H. Clayton, J. Crawley, L.S. Exton, V. Goulden, S. Ibbotson, et al., British association of dermatologists and British photodermatology group guidelines for the safe and effective use of psoralen-ultraviolet a therapy 2015, Br. J. Dermatol. 174 (2016) 24–55.

[16] S.H. Ibbostson, A perspective on the use of NB_UVB phototeraphy vs. PUVA photochemoterapy, Front. Med. 5 (2018) 184.

[17] X. Chen, M. Yang, Y. Cheng, G.J. Liu, M. Zhang, Narrow-band ultraviolet B phototherapy versus broad-band ultraviolet B or psoralen-ultraviolet A photochemotherapy for psoriasis, Cochrane Database Syst. Rev. (2013). CD009481.

[18] D. Morgado-Carrasco, J. Riera-Monroig, X. Fusta-Novell, S. Podlipnik, P. Aguilera, Resolution of aquagenic pruritus with intermittent UVA/NBUVB combined therapy, Photodermatol. Photoimmunol. Photomed. 33 (2017) 291–292.

[19] F.M. Garritsen, M.W.D. Brouwer, J. Limpens, P.I. Spuls, Photo(chemo)therapy in the management of atopic dermatitis: an updated systematic review with implications for practice and research, Br. J. Dermatol. 170 (2014) 501–513.

[20] M.P. Sheehan, D.J. Atherton, P. Norris, J. Hawk, Oral psoralen photochemotherapy in severe childhood atopic eczema: an update, Br. J. Dermatol. 129 (1993) 431–436.

[21] R. Sapam, S. Agrawal, T.K. Dhali, Systemic PUVA vs. narrowband UVB in the treatment of vitiligo: a randomized controlled study, Int. J. Dermatol. 51 (2012) 1107–1115.

[22] S. Tzaneva, H. Kittler, G. Holzer, D. Reljic, M. Weber, H. Honigsmann, et al., 5-Methoxypsoralen plus ultraviolet (UV) A is superior to medium-dose UVA1 in the treatment of severe atopic dermatitis: a randomized crossover trial, Br. J. Dermatol. 162 (2010) 655–660.

[23] T.H. Clayton, S.M. Clark, D. Turner, V. Goulden, The treatment of severe atopic dermatitis in childhood with narrowband ultraviolet B phototherapy, Clin. Exp. Dermatol. 32 (2007) 28–33.

[24] S.S. Yones, D. Der, R.A. Palmer, T.M. Garibaldinos, J. L.M. Hawk, Randomized double-blind trial of treatment of vitiligo, Arch. Dermatol. 143 (2007) 578–584.

[25] P.V.M.M. Diederen, H. Van Weelden, C.J.G. Sanders, J. Toonstra, W.A. Van Vloten, Narrowband UVB and psoralen-UVA in the treatment of early-stage mycosis fungoides: a retrospective study, J. Am. Acad. Dermatol. 48 (2003) 215–219.

[26] A. Hofer, L. Cerroni, H. Kerl, P. Wolf, Narrowband (311-nm) UV-B therapy for small plaque parapsoriasis and early-stage mycosis fungoides, Arch. Dermatol. 135 (1999) 1377–1380.

[27] S.J. Whittaker, M.F. Demierre, E.J. Kim, A.H. Rook, A. Lerner, M. Duvic, et al., Final results from a multicenter, international, pivotal study of romidepsin in refractory cutaneous T-cell lymphoma, J. Clin. Oncol. 28 (2010) 4485–4491.
[28] K. Ahmad, S. Rogers, P.D. Mcnicholas, P. Collins, Narrowband UVB and PUVA in the treatment of mycosis fungoides: a retrospective study, Acta Derm. Venereol. 87 (2007) 413–417.
[29] R. van Doorn, M.S. van Kester, R. Dijkman, M.H. Vermeer, A.A. Mulder, K. Szuhai, et al., Oncogenomic analysis of mycosis fungoides reveals major differences with Sezary syndrome, Blood 113 (2009) 127–136.
[30] F. Libon, A.F. Nikkels, Polymorphous light eruption: photerapy-based desensitization vs instramuscola stereoids. Who is right, who is wrong? Dermatology 234 (2018) 192–193.
[31] S. Lembo, A. Raimondo, Polymorphic light eruption: what's new in pathogenesis and management, Front. Med. (Lausanne) 5 (2018) 252.
[32] L. Misery, Chapter 8, Pruritus in cutaneous T-cell lymphomas, in: E. Carstens, L. Akiyama (Eds.), Itch: Mechanisms and Treatment, Taylor & Francis, Boca Raton, FL, 2014.
[33] S. Goetze, P. Elsner, Solar urticaria, J. Dtsch. Dermatologische Ges. 13 (2015) 1250–1253.
[34] U. Mastalier, H. Kerl, P. Wolf, Clinical, laboratory phototest and phototherapy findings in polymorphic light eruption: a retrospective study of 133 patients, Eur. J. Dermatol. 8 (1998) 554–559.
[35] A. Aslam, L. Fullerton, S.H. Ibbotson, Phototherapy and photochemotherapy for polymorphic light eruption desensitisation: a five year case series review from a university teaching hospital, Photodermatol. Photoimmunol. Photomed. 33 (2017) 225–227.
[36] T. Zhang, X. Fu, S. Ma, G. Xiao, L. Wong, C.K. Kwoh, et al., Evaluating temporal factors in combined interventions of workforce shift and school closure for mitigating the spread of influenza, PLoS One 7 (2012) e32203.
[37] A. Bishnoi, D. Parsad, K. Vinay, M.S. Kumaran, Phototherapy using narrowband ultraviolet B and psoralen plus ultraviolet A is beneficial in steroid-dependent antihistamine-refractory chronic urticaria: a randomized, prospective observer-blinded comparative study, Br. J. Dermatol. 176 (2017) 62–70.
[38] K.P. Chan, K.T. Goh, C.Y. Chong, E.S. Teo, G. Lau, A.E. Ling, Epidemic hand, foot and mouth disease caused by human enterovirus, Emerg. Infect. Dis. 9 (2003) 78–85.
[39] F. Farnaghi, H. Seirafi, A.H. Ehsani, M.E. Agdari, P. Noormohammadpour, Comparison of the therapeutic effects of narrow band UVB vs. PUVA in patients with pityriasis lichenoides, J. Eur. Acad. Dermatol. Venereol. 25 (2011) 913–936.
[40] M. Pavlovsky, L. Samuelov, E. Sprecher, H. Matz, NB-UVB phototherapy for generalized granuloma annulare, Dermatol. Ther. 29 (2016) 152–154.
[41] V. Brazzelli, S. Grassi, S. Merante, V. Grasso, R. Ciccocioppo, G. Bossi, et al., Narrow-band UVB phototherapy and psoralen-ultraviolet A photochemotherapy in the treatment of cutaneous mastocytosis: a study in 20 patients, Photodermatol. Photoimmunol. Photomed. 32 (2016) 238–246.
[42] F. Iraji, G. Faghihi, A. Asilian, A.H. Siadat, F.T. Larijani, M. Akbari, Comparison of the narrow band UVB versus systemic corticosteroids in the treatment of lichen planus: a randomized clinical trial, J. Res. Med. Sci. 16 (2011) 1578–1582.
[43] B. Solak, S.B. Sevimli, B. Dikicier, T. Erdem, Narrow band ultraviolet B for the treatment of generalized lichen planus, Cutan. Ocul. Toxicol. 35 (2016) 190–193.
[44] Z. Mohamed, A. Bhouri, A. Jallouli, B. Fazaa, M.R. Kamoun, I. Mokhtar, Alopecia areata treatment with a phototoxic dose of UVA and topical 8-methoxypsoralen, J. Eur. Acad. Dermatol. Venereol. 19 (2005) 552–55537.
[45] S. Decock, R. Roelandts, W.V. Steenbergen, W. Laleman, D. Cassiman, C. Verslype, et al., Cholestasis-induced pruritus treated with ultraviolet B phototherapy: an observational case series study, J. Hepatol. 57 (2012) 637–641.
[46] P. Hammerness, E. Basch, C. Ulbricht, P. Barrette, I. Foppa, S. Basch, et al., St. John's Wort: a systematic review of adverse effects and drug interactions for the consultation psychiatrist, Psychosomatics 44 (2003) 271–282.
[47] S. Behrens-Williams, C. Gruss, M. Grundmann-Kollmann, R.U. Peter, M. Kerscher, Assessment of minimal phototoxic dose following 8- methoxypsoralen bath: maximal reaction on average after 5 days, Br. J. Dermatol. 142 (2000) 112–115.
[48] R.S. Dawe, J. Ferguson, S.H. Ibbotson, An intraindividual study of the characteristics of erythema induced by bath and oral methoxsalen photochemotherapy and narrowband ultraviolet B, Photochem. Photobiol. 78 (2003) 55–60.
[49] M.D. Njoo, J.D. Bos, W. Westerhof, Treatment of generalized vitiligo in children with narrow-band (TL-01) UVB radiation therapy, J. Am. Acad. Dermatol. 42 (2000) 245–253.
[50] A. Tanew, S. Radakovic-Fijan, M. Schemper, H. Honigsmann, Narrowband UV-B phototherapy vs photochemotherapy in the treatment of chronic plaque-type psoriasis - a paired comparison study, Arch. Dermatol. 135 (1999) 519–524.

[51] M. Nakamura, B. Farahnik, T. Bhutani, Recent advances in phototherapy for psoriasis, Sao Paulo Medical J. 114 (1996) 1134–1140.

[52] R. Hung, S. Ungureanu, C. Eswards, B. Gambles, A.V. Anstey, Home phototherapy for psoriasis: a review and update, Clin. Dermatol. 40 (2015) 827–833.

[53] M.B.G. Koek, E. Buskens, C.A.F. Brjinzeel-Koomen, et al., Home ultraviolet B phototherapy for psoriasis: discrepancy between literature, guidelines, general opinions and actual use. Results of a literature review, a web search and a questionnaire among dermatologists, Br. J. Dermatol. 154 (2006) 701–711.

[54] I. Raposo, T. Torres, Palmoplantar psoriasis and palmoplantar pustulosis: current treatment and future prospects, Am. J. Clin. Dermatol. 17 (2016) 191–197.

[55] E. El-Gammal, V. Di Nardo, F. Daaboul, G. Tchernev, U. Wollina, J. Lotti, et al., Apitherapy as a new approach in treatment of palmoplantar psoriasis, Open Acsess Maced. J. Med. Sci. 6 (2018) 1059–1061.

[56] X.J. Chao, K.N. Wang, L.L. Sun, Q. Cao, Z.F. Ke, D.X. Cao, et al., Cationic organochalcogen with monomer/excimer emissions for dual-color live cell imaging and cell damage diagnosis, ACS Appl. Mater. Interfaces 10 (2018) 13266–13278. Figs. 3–5.

[57] J.H. Turner, An introduction to the clinical practice of theranostics in oncology, Br. J. Radiol. 91 (2018) 2018 0440.

[58] C.K. Svensson, E.W. Cowen, A.A. Gastari, Cutaneous drug reactions, Pharmacol. Rev. 53 (2001) 357–379.

[59] A. Hasan, K.P. Stengele, H. Giegrich, P. Cornwell, K.R. Isham, R.A. Sachleben, et al., Photolabile protecting groups for nucleosides: synthesis and photodeprotection rates, Tetrahedron 53 (1997) 4247–4264.

[60] C.G. Bochet, Photolabile protecting groups and linkers, J. Chem. Soc., Perkin Trans 1 (2002) 125–142.

[61] K. Lee, D.E. Falvey, Photochemically removable protecting groups based on covalently linked electron donor-acceptor systems, J. Am. Chem. Soc. 122 (2000) 9361–9366.

[62] B. Ahlström, E. Kraka, D. Cremer, The Bergman reaction of dynemicin A – a quantum chemical investigation, Chem. Phys. Let. 361 (2002) 129–135.

[63] S. Protti, D. Ravlli, M. Fagnoni, A. Albini, Smooth photogeneration of a,n-didehydrotoluenes (DHTs), Pure Appl. Chem. 85 (2013) 1479–1486.

[64] S. Peukert, B. Giese, The pivaloylglycol anchor group: a new platform for a photolabile linker in solid-phase synthesis, J. Org. Chem. 63 (1998) 9045–9051.

[65] T. Ohtsuki, S. Miki, S. Kobayashi, T. Haraguchi, E. Nakata, K. Hirakawa, et al., The molecular mechanism of photochemical internalization of cell penetrating peptide-cargophotosensitizer conjugates, Sci. Rep. 5 (2015) 18577.

[66] M. Zamadar, G. Ghosh, A. Mahendran, M. Minnis, B.I. Kruft, A. Ghogare, et al., Photosensitizer drug delivery via an optical fiber, J. Am. Chem. Soc. 133 (2011) 7682–7691.

[67] S. Protti, A. Albini, R. Viswanathan, A. Greer, Targeting photochemical scalpels or lancets in the photodynamic therapy field-the photochemist's role, Photochem. Photobiol. 93 (2017) 1139–1153.

CHAPTER

6

Oxidations

6.1 Photodynamic effect

Most of the known photochemical reactions are oxidations, which is perhaps not unexpected, since starting from the high energy location of excited states should remove any hindering to take a path toward high stability, that is, to highly oxidized compounds. In particular, the ubiquitous presence of molecular oxygen makes oxygenation reactions, where O_2 has the double role of electron acceptor and reagent, one of the most frequent photochemical reactions. First of all, molecular oxygen is not as reactive as one may expect. Atomic oxygen and ozone are similar under many properties. Mechanistic studies have been carried out and have revealed basic mechanistic aspects. The first point is that the dioxygen molecule is much less reactive than atomic oxygen (or ozone, which is similar to it under many aspects). The reaction of molecular oxygen with organic molecules became one of the focuses of research, quite interesting under both from the theoretical and practical and from the purposes as an important family of reactions, typical examples are the cooxidation of iron or cerium ions (see the below equations), where electronically excited state might be formed from the thermodynamic point of view, though not necessarily from the kinetic one [1].

$$Fe^{3+} + H_2O_2 \rightarrow Fe^{2+} + H_2O + 1/2\ O_2$$
$$Ce^{4+} + H_2O_2 \rightarrow Ce^{3+} + H_2O + 1/2\ O_2$$
$$SO_3^{2-} + O_2 \rightarrow SO_5^{2-} \rightarrow SO_3^{2-} + O_2$$
$$R_3C - O_2^{\cdot} + R_3CH \rightarrow R_3CO_2H + R_3C$$
$$\left[(CN)_2Co\right]^{-3} + O_2 \rightarrow \left[(CN)_5CoO2\right]^{-3}$$

Hypoxia (the situation in which a tissue is deprived of the adequate amount of oxygen) causes a cascade of activities from the level of the individual down to the regulation and function of the cell nucleus. Prolonged periods of low oxygen tension are a core feature of several disease states. Advances in the study of molecular biology have begun to bridge the gap between the cellular response to hypoxia and physiology. Hyperbaric oxygen therapy (HBOT) is a treatment for hypoxic- and inflammatory-driven conditions, in which patients are treated with 100% oxygen at pressures greater than atmospheric pressure [2,3].

In the United States, hypoxia is a significant component of the pathology of conditions such as stroke, cancer, and heart failure, as the unavailability of oxygen leads to physiological responses that, if not resolved, progress to localized hypoxic responses, cell metabolic inefficiency, organ dysfunction, and finally death, even if oxygen is given to breath to most of these patients in the hospital. HBOT is used to return to normal conditions of an hypoxic state, that is, one where an imbalance of oxygen results from an excessive demand or a reduced supply of it (on an average,

Light, Molecules, Reaction and Health
DOI: https://doi.org/10.1016/B978-0-12-811659-3.00006-2

tissues at rest utilize 5–6 mL of O_2 per deciliter of blood delivered, with large differences between various tissues). Hypoxia could be fairly defined as a scenario when tissue fails to receive this amount of oxygen. However, hypoxia is better understood as a component of the pathology of many disease states, such as ischemia.

Under ischemic conditions, cells are unable to perform their usual functions and nutrients are not consumed or wastes are not removed.

At the level of the cell, 80% of the available oxygen is used by the mitochondria, while only 20% is used by other organelles, indicating the importance of oxygen in metabolism. Mitochondria function as the cell's power station, and ATP is the cell's currency for energy. Oxygen is used as the final electron receptor in the electron transport chain (ETC), and the energy generated by this reaction is used to pump hydrogen ions across an electrochemical gradient outside the mitochondria. The hydrogen then diffuses back into the mitochondria, and the energy generated is used to phosphorylate ADP to form ATP. Interestingly, although the mitochondrion utilizes most of the oxygen in the cell, the partial pressure is very low, only 1–3 mmHg and oxygen's role is that of receiving the electrons at the end of the ETC at complex IV, cytochrome *c* oxidase.

With a given pressure of a gas, it is possible to use Henry's law to determine how much oxygen is dissolved in plasma. Henry's law states, "at a constant temperature, the amount of a given gas that dissolves in a given type and volume of liquid is directly proportional to the partial pressure of that gas in equilibrium with that liquid." This means that the concentration of a given gas into a liquid is in relation to the partial pressure that the gas exerts, as well as its coefficient of solubility. Therefore at an oxygen partial pressure of 100 mmHg with a solubility coefficient of 0.0024 mL O_2/(dL blood per mmHg), the amount of oxygen in the plasma at sea level is 0.24 mL O_2/dL blood (see Table 6.1).

The maximum concentration of oxygen in plasma at sea level is only 0.24 mL O_2/dL of blood, but tissues require 5–6 mL O_2/dL of blood for homeostasis. The delivery of more oxygen than the plasma is capable of carrying is obtained accomplished through hemoglobin, which more than makes up for the low capacity of plasma to carry oxygen. One gram of hemoglobin can carry as much as 1.34 mL O_2 if all four binding sites are occupied on each molecule of hemoglobin. The concentration of hemoglobin can indicate the maximum capacity for carrying oxygen in the blood. Assuming 15 g/dL blood hemoglobin (normal is 11–16 g/dL), a hypothetical total of ~20 mL O_2/dL blood can be calculated for the carrying capacity of hemoglobin. This makes the total oxygen concentration in the blood 20.24 mL O_2/dL blood at sea level, assuming that the hemoglobin-carrying capacity is maximized (see Table 6.1).

The capillaries serve as the locations at which oxygen is transferred to tissue from the plasma, and hemoglobin functions as an oxygen reservoir in this context. A maximum of 20.24 mL O_2/dL of blood enters the capillary bed, and at rest the local tissue consumes 5 mL O_2/dL of

TABLE 6.1

	Ambient air	100% Oxygen	Hyperbaric oxygen (3 atm)
Oxygen partial pressure (mmHg)	150	713	2233
Plasma oxygen content (mL O_2/dL blood)	0.24	1.71	4.8
Oxygen content of blood (mL O_2/dL blood)	20.24	21.71	24.8
Net change in plasma oxygen content (%)	Not applicable	+1.47 (7.26)	+4.56 (22.5)

Reproduced by permission from R. Choudhury, Hypoxia and hyperbaric oxygen therapy: a review, Int. J. Gen. Med. 11 (2018) 431–442.

blood, leaving the postcapillary blood with 15.24 mL O_2/dL of blood. During exercise, tissues can consume up to 15 mL O_2/dL of blood, but may still have higher demands for oxygen.

To meet the increased oxygen demands, the body undergoes physiological processes that involve the lungs, heart, and vasculature. Cardiac output is increased as needed by increase in stroke volume and heart rate, delivering more blood, and hence more oxygen to the capillary beds per unit of time. Pulmonary vessels constrict shunting blood from areas of low oxygen tension in the lungs to areas with higher oxygen tension, thereby maximizing the exchange of oxygen in the hemoglobin and plasma. This allows for the maintenance of the reservoir of oxygen stored by hemoglobin in red blood cells. Systemic vessels dilate to perfuse tissues with higher oxygen demand, which also aids in blood delivery, and hence oxygen delivery.

Local oxygen delivery is based upon the components discussed above and is a direct result of the concentration of oxygen in the blood and the amount of blood delivered. The body has many mechanisms to adjust oxygen delivery and maintain adequate oxygenation, including ventilation rate, cardiac output, stroke volume, hemoglobin concentration, dilation of systemic capillaries with constriction of pulmonary capillaries, and increasing the size of alveoli. A careful balance is maintained by coordinating all of these systems; however, the cellular response plays a major role in how hypoxia is handled at the cellular level, and this is where disease begins.

As the level of oxygen drops in the blood, the body undergoes responses such as increasing respiration and blood flow. Simultaneously, individual cells experiencing hypoxia begin reacting to the decreased oxygen tension. The cells sense low oxygen via cellular signaling that starts with the enzyme class prolyl hydroxylase domain (PHD) proteins. This class of enzymes contain oxygen-sensing hydroxylases that will hydroxylate specific proline residues on the α-subunit of the transcription factor hypoxia-inducible factor (HIF), which signals for the destruction of HIF. HIF is a heterodimer composed of subunits HIF-1α or HIF-2α, which dimerize with a HIF-1β subunit. HIF-1α is a ubiquitous subunit, meaning that it is produced in all cell types, and the HIF-2α subunit is found in myeloid cells, liver parenchyma, vascular endothelia, type II pneumocytes, and renal interstitium. PHD forms part of the oxygen-dependent system that regulates the inhibition of HIF. Another molecule, known as factor-inhibiting HIF, serves as another oxygen-dependent system that regulates HIF.

HIF is found in all nucleated cells of metazoan species (which includes humans). This enzyme is activated under low oxygen tension and leads to the regulation of certain genes. In the presence of low oxygen levels, PHD is unable to cause the degradation of HIF, and HIF enters the nucleus and affects transcription by directly binding to promoter sequences in the DNA, which results in an increased transcription rates of certain hypoxic response genes. This binding results in changes in glucose metabolism from aerobic to anaerobic, increased cytochrome oxidase transcription, inhibition of lipid catabolism, and promotion of lipid storage. Overal this means that through this regulation of gene expression, HIF accomplishes the switch from oxidative metabolism to glycolytic metabolism.

Hyperbaric oxygen was first used in 1937 by Behnke and Shaw to treat decompression sickness. After the success of Behnke and Shaw, multiple uses for HBOT were explored, from leprosy to osteomielite, to CO poisoning. Current indications for the use of HBOT are shown in Ref. [3].

At sea level, atmospheric pressure is 1 bar absolute (1 standard atmosphere = 101 kPa = 1.013 bar). The weight of the atmosphere exerts a pressure which will support a column of water 10 m high; 10 m under water, the pressure on a diver is 200 kPa. The volume of gas in an early diving bell full of air at sea level is halved at 10 m according to Boyle's law; at 20 m pressure is 300 kPa absolute and the gas is compressed into one-third of the volume.

That the oxygenation is enhanced by light was known since the earnest times, but was quantitatively reported during renaissance. As an example, the commentary of the Bologna Academy of Science reports a discussion that took place in 1757 about degradation of flowers. About what may be the cause of the loss of the beautiful colors of flowers, one of the Academician said *dixi satis, constet, neque colorem, neque aerem in labe sactandis coloribus multum valore, proclive est credere, in primo illo, cum luce adjunta tantas labes attulissent, id luci magis, quam ipsis, tribuendum fuisse. Itque etiam illorum opinionem minuit, qui existimat colores rerum idcirco in lumine extenuari, quod illarum superficies extima quodammodo aduratur qua opinione lumen libera, culpa omnis in calorem transferunt, cuius adurere est proprium. Sed hos experimentum, quod modo dixit, satis arguit … As I said, it is apparent that neither heat, nor air has a main role in changing colors, and I think rather that it has to be attributed to light, first of all because of the strong damage occurring when light is added. This is against the opinion of those who think that the color of things is as much weakened in the light, as much as their surface is damaged, an opinion that frees light and trasfers the blame onto heat, the nature of which is to damage. But this can be understood from the experiments…* [4].

That light together with oxygen could originate a characteristic reactions emerged first in Bayern, where at the beginning of the 20th century, a student in medicine, Mr. O Raabe was assigned by his mentor, Prof. von Tappeinera, theme for defending his thesis at the end of the Summer semester. Previous work had shown that preparates of unimolecular organisms (dutifully dyed so that this was visible at the microscope, one of the portents that microbiology offered). As the story (or the legend) goes, he found indeed that there were dyes that caused a similar effect to larger extent, but he was sorry not to find any regular patterns in his experiments. It seems, however, Gott (or perhaps Wotan) manifested his will by sending down a peculiarly capricious Spring, with many, if short, thunderstorms and a possible rationalization was considered that attributed the key action to the lightnings. As a matter of fact, the omission of any of the three reagents, light, dye, or oxygen, quenched the reaction [5,6]. Prof. von Tappeiner was quick in grasping the important application in human medicine, published a few years later. This was one of the bases of medical photochemistry, further developed in Copenhagen, that lost importance, however, when the more practical synthetic antimicrobials entered in the field [7,8]. However, a possible activation of oxygen by light came into the attention of scientist some decade later. Prof. Kautsky was involved in what appeared to be in a quite different field, the activation of molecules in the presence of various agents, for example, light. Indeed, the first hypothesis of oxygen activation was formulated by Kautsky in 1931 by demonstrating that the reagent was a gas (see Scheme 6.1).

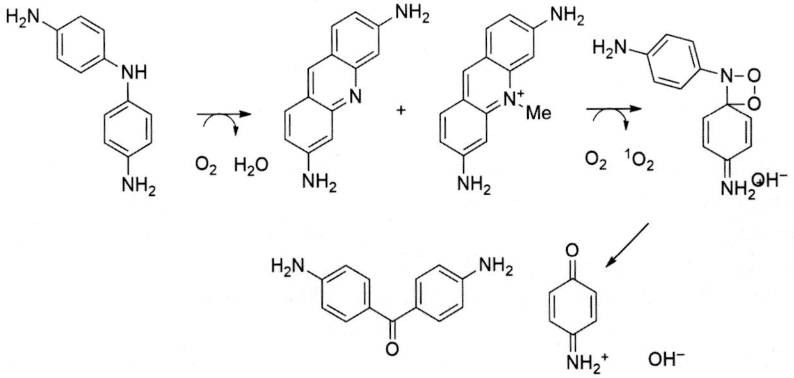

H_2N NH NH_2 O_2 H_2O NH_2 N NH_2 + NH_2 N^+ Me NH_2 O_2 $^1O_2$ H_2N N–O O NH_2 OH^- H_2N NH_2 O O NH_2^+ OH^-

SCHEME 6.1 In the Kautsky three-phase experiment, the dye trypaflavin is generated by reaction of the leuco form. This sensitizes the generation of singlet oxygen. Addition to the leuco dye causes the observed cleavage and in competition a yellow-green florescence from the acridinium. Differentiated previous absorption onto silica gel granules supports that the actual reagent must be a gas and the scientist finds "tempting' to suggest that it is electronically excited oxygen.

Thus, in a three-phase experiment, he intimately mixed silica gel granules that had been separately treated with a dye (tripaflavin) functioning as energy donor and leucomalachite green as an energy acceptor and light emitter [9] (which he considered a good move, since the strong green fluorescence of the triaryl dye revealed immediately the occurring of the oxygenation reaction). Under the conditions he used ($\lambda = 660$ nm), there was no question of cleavage of the oxygen molecule, and thus a role of ozone that is formed via oxygen atom. In view of the dependence of the reaction on oxygen concentration, he rather thought of a photosensitized oxidation involving an excited state of molecular oxygen. Looking for a rationalization of the observed high reactivity and the other properties of this metastable state, Kautsky found in the literature a low-energy species indicated as singlet oxygen, investigated only spectroscopically as yet, with an very weak absorption in the red (37.2 kcal) mol,$^1\Sigma_g\ O_2$) [10].[1]

[1] (Original in German) Die auf den Sauerstoff übertragene Energie wird deshalb kaum den Betrag von 40000 cal übersteigen. An eine Spaltung des Sauerstoff-Molekuls in Atome darf unter diesen Umstanden gar nicht gedacht werden, so daβ auch eine Bildung von Ozon ausgeschlossen ist. Wir denken im Hinblick auf die Druck-Abhangigkeit der, photosensibilisierten Oxydation auf Entfernung" in erster Link an einen Anregungs-Zustand des Sauerstoff-Moleküls. Dieser Anregungs-Zustand muβ metastabil sein. Das besagt eindeutig die nachgewiesene Diffusion des aktivierten Sauerstoffs. Eine weitere Eigenschaft dieses aktivierten Sauer- stoffs ist es, sehr viel starker oxydierend zu wirken als normaler Sauerstoff. Alle Versuche über die photo-sensibilisierte Oxydation sind Beispiele hierfür. Sehen wir uns in der Literatur nach Anregungs-Zustanden geringer Energie des Sauerstoffs um, so finden wir einen, der unseren Forderungen entspricht. Er wurde zuerst von Heurlingers) auf grund rein spektroskopischer Messungen gefunden. Sauerstoff besitzt eine Absorptionsbande in lang-welligen Rot bei 7623 Å = 37.257 cal). Die Absorption des Lichtes dieser Wellenlange durch das Sauerstoff-Molekul ist so gering, daβ schon daraus geschlossen wurde, daβ es sich um einen metastabilen Zustand handeln dürfte. W. Childs und R. Meckeg ist es gelungen, durch quantitative Intensitats-Messungen in der Absorptionsbande die Lebensdauer dieses $^1\Sigma$-Terms des Sauerstoff-Moleküls mit Bestimmtheit auf einige Sekunden festzusetzen. Die Versuchung ist groβ, den durch Photo-sensibilisierung aktivierten Sauerstoff dem $^1\Sigma$-Zustand des Sauerstoff-Moleküls gleichzusetzen. Einen uninittelbaren Beweis dafür haben wir nicht. Die Konzentration des durch direkte Licht-Absorption erhaltenen Sauerstoffs im $^1\Sigma$-Zustand ist so auβer-ordentlich gering, daβ seine chemischen Eigenschaften, seine besonderen oxydierenden Wirkungen nicht untersucht und somit auch nicht mit denen des von uns untersuchten aktiven Zustandes des Sauerstoff-Moleküles ver-glichen werden können. Die Wahrscheinlichkeit aber, daβ die beiden Akti-vierungszustande identisch sind, ist recht groβ in Anbetracht der Ubereinstimmung wichtiger physikalischer Eigenschaften (Anregungs-Energie unter 40 ooo cal und metastabiler Zustand). Ob dieser "aktivierte Sauerstoff" in den weitverbreiteten Vorgängen der Autosydation eine Rolle spielt, wissen wir noch nicht; wir halten es aber für wahrscheinlich, daβ dieser Aktivierungs-Zustand auch bei Über-tragung chemischer Energie auf das Sauerstoff-Molekül erreicht wird. Die Hauptbedeutung dieses aktiven Sauerstoffs liegt wohl auf biologischem Gebiet. Bedenkt man, daβ in jeder grünen, assimilierenden Pflanze die Fluorescenz des Chlorophylls durch Sauerstoff weitgehend getilgt wird und daβ diese Fluorescenz-Tilgung in unmittelbarer Beziehung zu den Energie- Umwandlungen im Assimilations-Vorgang steht , so kann man ermessen, welche Bedeutung der aktivierte Sauerstoff für das biologische Geschehen auf der Erdoberflache hat. (Translated into English) The energy transferred by oxygen thus barely overcomes 40,000 cal. One cannot think that under these conditions a cleavage of the dioxygen molecule occurs, and the same holds for the formation of ozone. We examined the pressure dependence of this "oxidation from a while" considered an excited state of oxygen that has to be metastable. Clearly, this was due to the gradual diffusion of activated oxygen. A further property of such activated oxygen is that it reacts much more strongly as normal O2. All of the experiments of photosensitized oxidation prove this. When looking in the literature for low-energy excited states of oxygen, we find one that corresponds to our search. This has first reported by Hurekingen that found it by purely spectroscopic measurements. Oxygen has an absorption in the long-wavelength red at 7623 Å (= 37,257 cal. The absorption of light of this wavelength by the oxygen molecule is so small that, as soon as closed, it became a metastable state. Childs and Mecke [a] succeeded trough quantitative intensity measurements to evaluate the lifetime of this 1S state of oxygen and this resulted of some seconds.

This proposal was rebuted because the sensitizer was not high enough in energy, but this objection was overcome a few years later, when a second species of singlet oxygen was discovered ($^1\Delta_gO_2$, $^3\Sigma_g$ is he ground state of O_2.) that had ample energy for arriving at this oxygen state [11–16]. The situation remained unsteady and the predominant publications in biological journals did not favor the any final decision [17].

At any rate, nice mechanistic particulars were not in the best moment, due to the breakdown of the Second World War shortly. When Germany came up from the ruins after the destructions, two facts were to be observed in this field one in the large dimension of the industry, where a variety of macromolecule gradualy substituted the natural ones, a change that would not be possible without the rapid development of radical chemistry that had been obtained at the Max Plank (formerly Kaiser Wilhems Institut) [18] for coal chemistry. A fast development of the large scale chemical industry dimension was the result of radical chemistry in this country, that was close to the climax. In particular, Coal Chemistry in Mulheim now led by Prof. K. Ziegler, extending the properties of radicals to diradical, one may find a consistent rationalization of the oxygenation reactions, while preparative interest on oxygenation reaction grew, since, for example, in this way a disinfectant was prepared though the photosensitized oxidation was a powerful disinfectant, was carried out primarily by Prof. G.O. Schenck, a pupil of Ziegler, under conditions well suited to the small scale application in the back yard (ascaridol a powerful, if toxic, disinfectant that was generated by photosensitized oxygenation, using chlorophyll from urtical leaves) [19], for whom an Abteilung (Section) was then opened at Mulheim.

In the following decade, a mechanistic controversy developed between a (mainly) German [20] group of scientists and a (more) US one—that found all important to recognize the role of triplet states [21].

The controversy lasted some years and has been of the utmost importance for the level of the scientists involved, the experimental advancements it achieved, and the theoretical advancements it allowed. A number of highly valuable synthetic processes based on the base of the strong electrophilic properties of singlet oxygen is now available. However, in the biological field it is generally more important to recognize the oxygenation as involving a class of electrophilic or radical intermediates indicated collectively as reactive oxygen species (ROS) that exhibit a similar reactivity. Characteristics are the addition onto electron heterocycles, in particular guanine, the hydrogen abstraction from allylic or bis-allylic bonds, the addition to sulfides to yield sulfoxides or disulfides. In this way, hydroperoxides, peroxides, and further compounds are formed and again show a similar reactivity (Schemes 6.2 and 6.3).

The temptation of comparing such states is great, but we have no definite proof for it. The likelihood that such activated states are identic is great and supported by the identic physical properties (activation energy below 40,000 a metastable state). Whether this "activated oxygen" has a role in the large field of autoxidation we do not know as yet; but we feel it is likely that is involved also in the transfer of chemical energy to the oxygen molecule. The importance of this activated oxygen is fully in biology. Think that in every green, assimilating plant chlorophyll is present and its fluorescence for the largest part is quenched and such quenching is unmistakably involved in the assimilation process then one can understand the significance of activated oxygen for life on the Earth surface. [a] Childs, WHI and Mecke, R, Intensitätsmessungen in der atmosphärischen Sauerstoffbande λ 7600, Zeitshr Phys., 68 (1931) 344–347. [b] By this term we meant the chemical process that reduces carbon dioxide to formaldehyde and polymerizes it to polysaccharides before the chlorophyll machinery was understood.

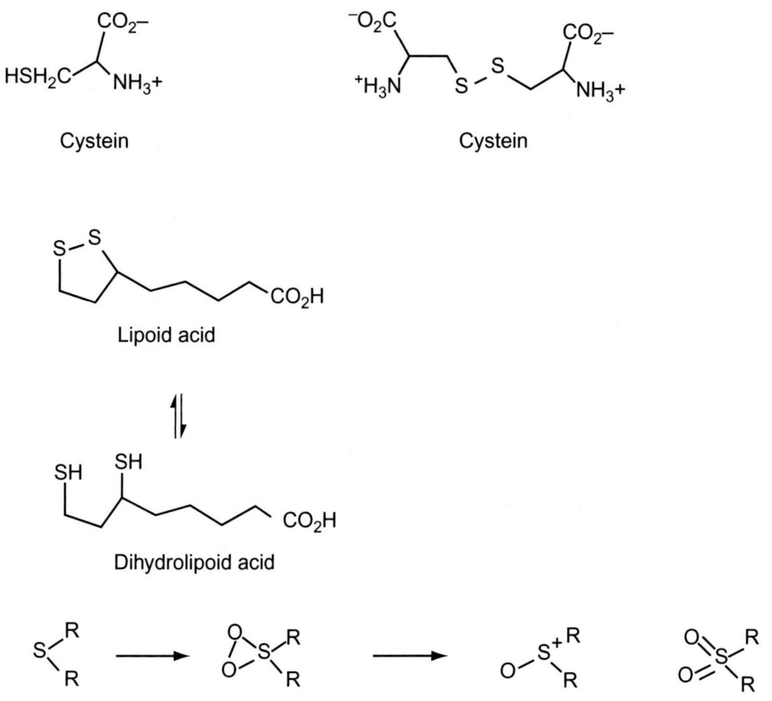

SCHEME 6.2 Examples of the reactivity of ROS viz. cysteine, lipoic acid.

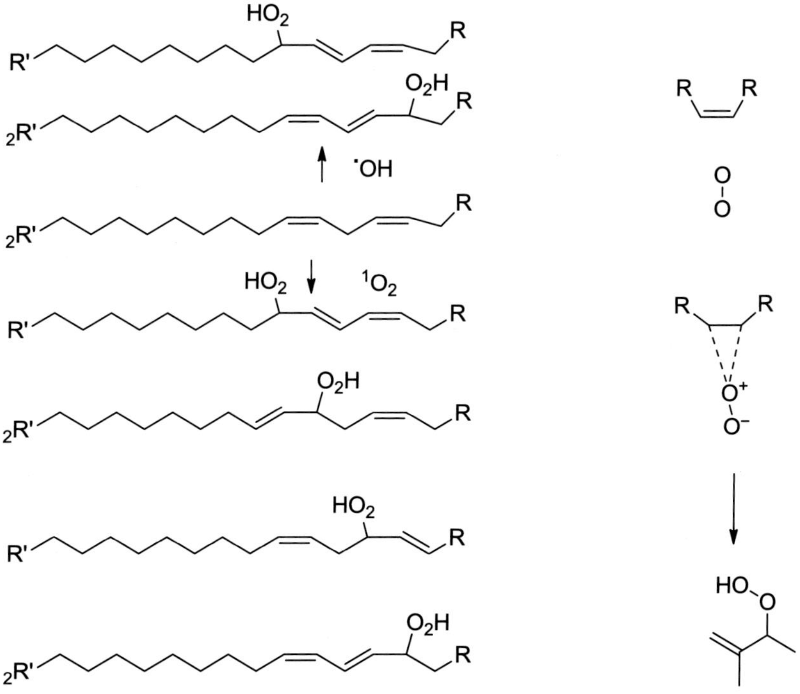

SCHEME 6.3 Mechanism of ene reaction and footprint of singlet oxygen and hydroxy radical.

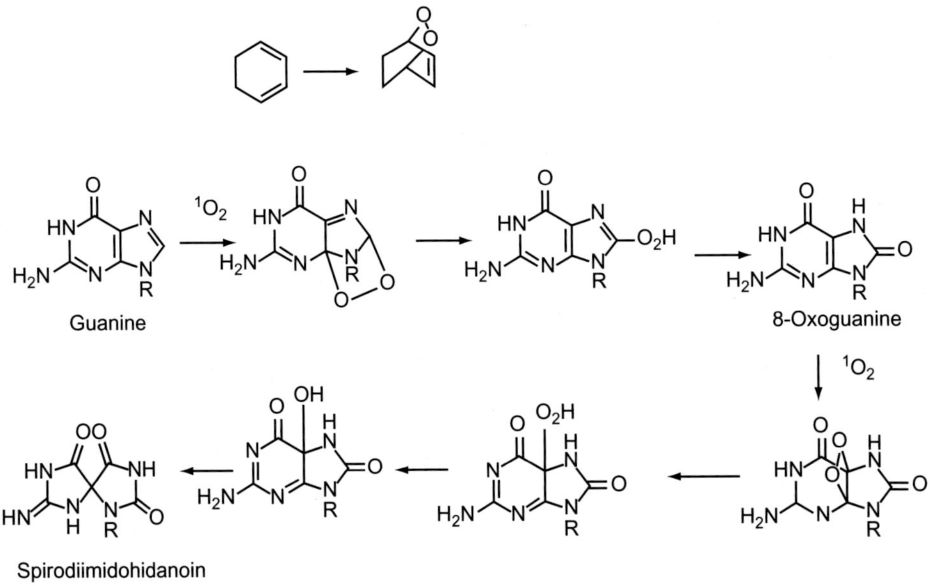

Guanine

8-Oxoguanine

Spirodiimidohidanoin

SCHEME 6.4 4 + 2 Cycloaddition and reaction of guanine with two moles of singlet oxygen.

Singlet oxygen reactions include a variation of the ene reactions, addition to sulfides, phosphines, and other donors.

As for the skin, there is first of all a visual difference, in that the Blaschko lines [22] become visible. These lines are not visible under normal health conditions, but are clearly seen when their development is disturbed, and are believed to trace the migration of embryonic cells. They follow a "V" shape over the back, "S" shaped whirls over the chest and sides, and wavy shapes on the head. The lines are believed to trace the migration of embryonic cells (Scheme 6.4).

The stripes are lines of normal cell development in the skin. These lines are invisible under normal conditions. They become apparent when some diseases of the skin or mucosa manifest themselves according to these patterns. They follow a "V" shape over the back, "S" shaped whirls over the chest and sides, and wavy shapes on the head. The lines are believed to trace the migration of embryonic cells. The stripes are a type of genetic mosaicism. They do not correspond to nervous, muscular, or lymphatic systems. These lines can be observed in other animals such as cats and dogs.

Further, skin diseases caused by sensitized photoreactions include lupus, a chronic inflammatory autoimmune disease with a wide range of clinical presentations resulting from its effect on multiple organ systems. There are four main types of lupus: neonatal, discoid, drug-induced, and systemic lupus erythematosus, the type that affects the majority of patients. Patients with lupus experience a loss of self-tolerance as a result of abnormal immunological function and the production of autoantibodies, which lead to the formation of immune complexes that may adversely affect healthy tissue [23].

On this basis a new science developed, called photodynamic reaction, an appropriate label considering the fact that light caused the killing of the microorganisms.

However, the proposal by Kautsky was rebuted because of the absence of precedents in the literature, and the thermodynamic impossibility of the reaction. This stumbling block was removed when a second excited state of singlet oxygen was discovered. At any rate, it was no good moment in Germany for discussing mechanistic niceties, particularly when a

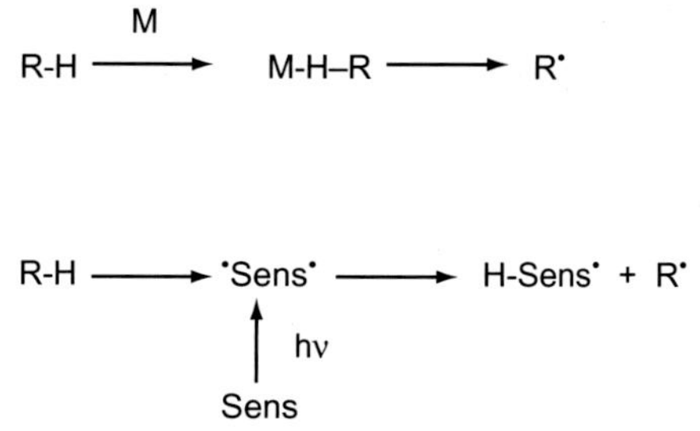

R-H —M→ M-H–R → R$^\bullet$

R-H → $^\bullet$Sens$^\bullet$ → H-Sens$^\bullet$ + R$^\bullet$

↑ hν

Sens

SCHEME 6.5 Activation of alkanes by metals.

non-Arian scientist was involved, and further advancement was to be postponed to after the War when Germany woke up from the debris of the Second War and found two important differences, one a large scale, the other one in the small scale. The large one was the discover that such a dirty ingredient as coal could be transformed into a variety of new materials, the hydrocarbon chain polymers, a new system that involved the activation by metals. In the meantime major achievements were obtained both in quantum chemistry [24] and in chemical biology [25] (Scheme 6.5).

6.2 Physical and chemical decay of singlet oxygen

Solvent	τ (μs)	
	5°	50°
C_6H_6	30.4	30.3
C_6D_6	731 (25°C)	592
$C_6H_5CH_3$	30.3	28.7
$C_6D_5CD_3$	407	205
C_6H_{14}	32.6	32.1
C_6D_{14}	573	608
CH_3OH	8.9	9.7
CH_3OD	32.0	30.3
CD_3OD	282	257
$(CH_3)_2CO$	43.9	47.6
$(CD_3)_2CO$	989	1083
CH_3CN	81.6	79.9
CD_3CN	1604	1662
H_2O	3.57	3.27
D_2O	77.8	58.3

Recompiled from [26] M. Brengnøj, M. Westberg, F. Jensen, P.R. Ogilby, Solvent-dependent singlet oxygen lifetimes: temperature effects implicate tunneling and charge-transfer interactions, Phys. Chem. Chem. Phys. 19 (2016) 22946–22961.

The decay of singlet oxygen occurs through a complex mechanism, as one may expect from the high symmetry of this species, which requires peculiar conditions for electronic transitions. The best model at present involves a term depending on the frequency where absorption take place ($h\nu$) of bimolecular term, and referring both to the solvent and to the solute, as indicated below, where the subscripts q and rxn refer to solute-dependent quenching and reaction channels, respectively, M represents the solvent and k_{nr}, and radiative, k_r, processes.

$$k_\Delta = k_{nr}[M] + k_r[M] + k_q[R] + k_{rxn}[R]$$

These points are of a high relevance, and may be coupled with the footprint of the reaction with lipids (see above) for the unequivocal establishment of the mechanism and the comparison with the thermal generation of singlet oxygen. However, in many cases the structure of the photoproducts is per se not sufficient for the indepth determination of the path followed.

Some representative reactions of singlet oxygen, as applied to the synthesis of complex molecules are listed below [27] (Schemes 6.6–6.9).

6.3 Alternative modes of generation of singlet oxygen

A nonsecondary advantage of working with singlet oxygen is the multiple possibility of arriving at this species. Thus, direct irradiation

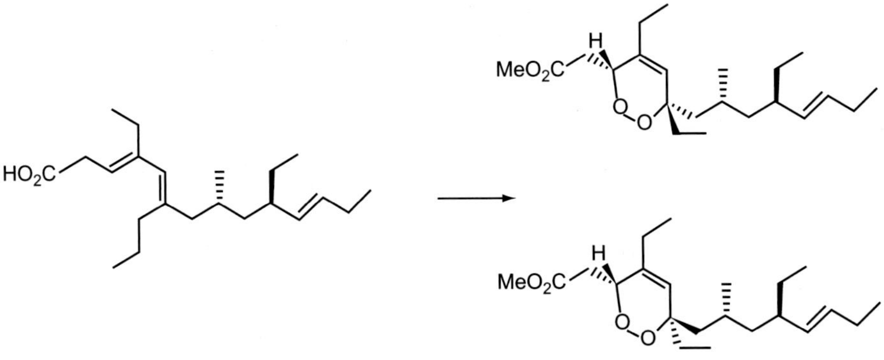

SCHEME 6.6 4 + 2 Cycloaddition to give an endoperoxide [28].

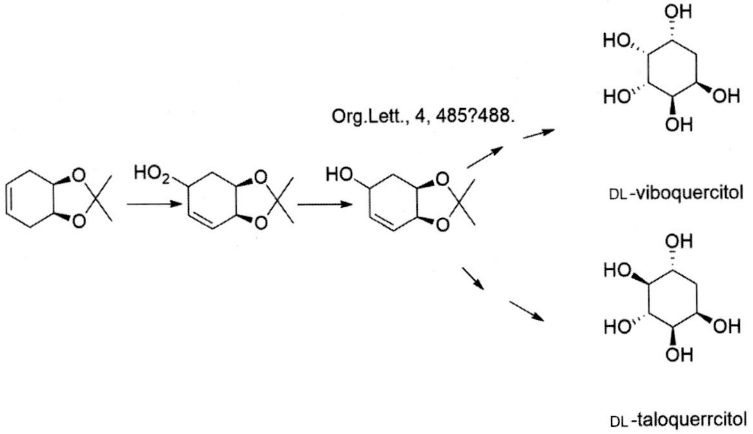

Org.Lett., 4, 485?488.

DL-viboquercitol

DL-taloquerrcitol

SCHEME 6.7 Key role of an allylic alcohol prepared through the ene reaction in the synthesis of some quercitols [29].

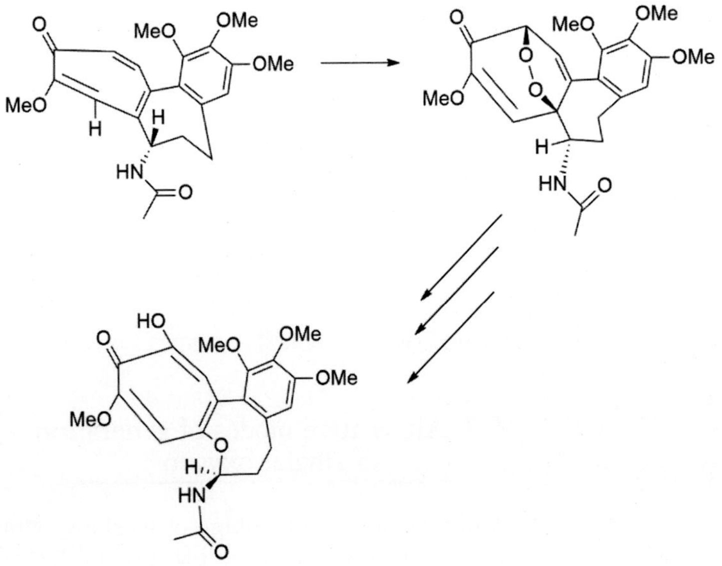

SCHEME 6.8 4 + 2 Cycloaddition and Ru catalized rearrangement to a bisepoxide in the synthesis of elisiapyrones [30].

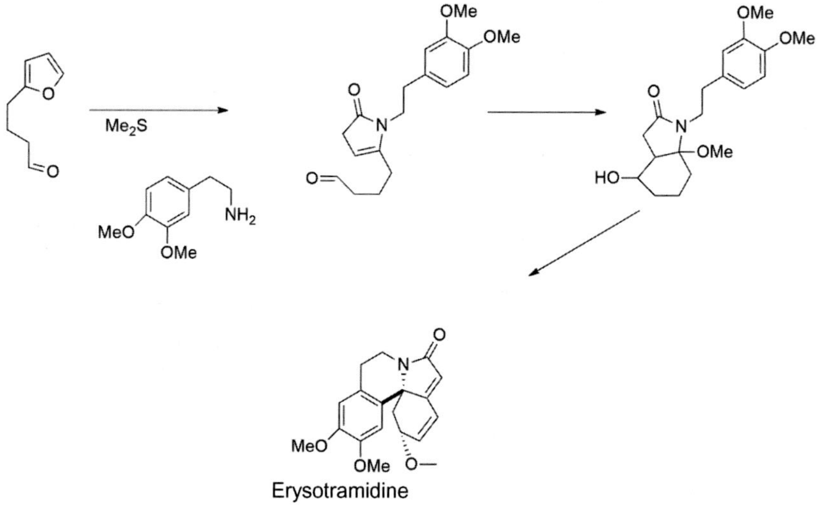

SCHEME 6.9 Singlet oxygen reaction with a furan derivative [31].

is, as previously mentioned, a poor access in view of the quite low molar absorptivity, but sensitization has been successfully also applied on a preparative scale [32,33].

4 + 2 Cycloaddition is one of the most typical reactions of singlet oxygen, in parallel to general electrophile reactions. Thus, from three condensed rings up the adducted easily form, but with naphthalene derivates adduct both form and cleave at low temperatures, so that it can be decided which are the best conditions for the desired experiment [34].

Since there is no charged intermediate, the stability of the endoperoxides is strongly affected by steric strain. This results both in the rate and in the regioselectivity of the cycloaddition.

In mammalian tissues, an ultraweak chemiluminescence arisingfrom biomolecule oxidation has been attributed to the radiative deactivation of singlet molecular oxygen [O_2 ($^1\Delta_g$)]. Thus the generation of O_2 ($^1\Delta_g$) in aqueous solution via energy transfer has been demonstrated from excited triplet acetone, via decomposition of 3,3,4,4-tetramethyl-1,2-dioxetane, a chemical source and horseradish peroxidase-catalyzed oxidation of 2-methylpropanal, as an enzymatic source. Both sources of excited carbonyls showed characteristic light emission at 1270 nm that indicated the monomolecular decay of O_2 ($^1\Delta_g$). Indirect analysis of O_2 ($^1\Delta_g$) by electron paramagnetic resonance using the chemical trap 2,2,6,6-tetramethylpiperidine showed the formation of 2,2,6,6-tetramethylpiperidine-1-oxyl. Using [^{18}O]-labeled triplet, ground state molecular oxygen [$^{18}O_2$ ($^3\Sigma_g^-$)], chemical trapping of $^{18}O_2$ ($^1\Delta_g$) with disodium salt of anthracene-9,10-diyldiethane-2,1-diyl disulfate yielding the corresponding double-[^{18}O]-labeled 9,10-endoperoxide was detected [35] (Scheme 6.10).

As an example, in the anthracene derivative shown in Scheme 6.11 the strong deformation of the benzene rings is enough to allow addition affording the mono- and di-EPO (see Scheme 6.11) [36]. Periinteractions between two neighboring methyl groups bound to a polycyclic aromatic hydrocarbon enhance its reactivity toward $^1O_2$ because the steric strain is somewhat relieved in the transition state. This phenomenon explains why 1,8-dimethylnaphthalene 3 is four times more reactive than the 1,5-isomer (see Scheme 6.12 [37]).

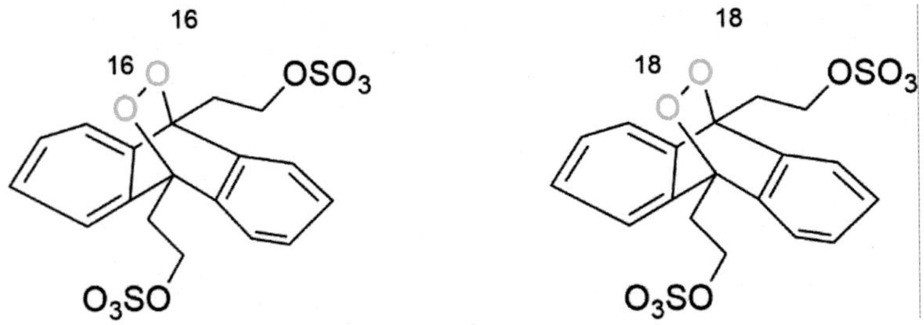

SCHEME 6.10 [35]

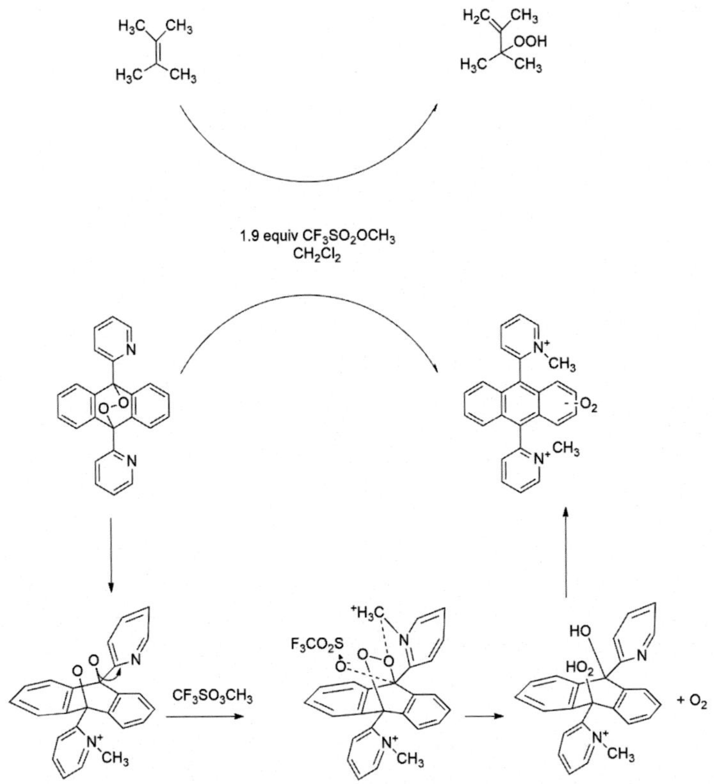

SCHEME 6.11

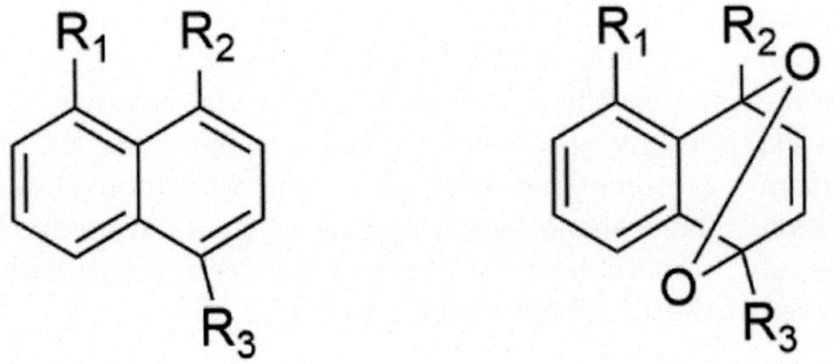

SCHEME 6.12

6.4 Application

Jaundice is the most common condition that requires medical attention and hospital readmission in newborns. The yellow coloration of the skin and sclera in newborns with jaundice is the result of accumulation of unconjugated bilirubin. In most infants, unconjugated hyperbilirubinemia reflects a normal transitional phenomenon. However, in some infants, serum bilirubin levels may rise excessively, which can be the cause for concern because unconjugated bilirubin is neurotoxic and can cause death in newborns or lifelong neurologic sequelae in infants who survive (kernicterus). For these reasons, the presence of neonatal jaundice frequently results in diagnostic evaluation [38,39].

Neonatal jaundice has been first described in detail in a Chinese textbook 1000 years ago. The 18th and 19th centuries contain discussions about the causes and treatment of neonatal jaundice. In 1875 Orth first described yellow staining of the brain, in a pattern later referred to by Schmorl as kernicterus [40].

The excretion of conjugated bilirubin and other substances destined to be excreted in bile are actively transported across the bile canalicular membrane of the hepatocyte (see Scheme 6.13). The concentration gradient is very high and can reach 1:1000. There are at least four known canalicular transporters that participate in excretion of conjugated bilirubin. However, the multidrug resistance-associated protein 2 (MRP2) appears to play a dominant role in the canalicular secretion of conjugated bilirubin.

(Z, Z) bilirubin

(ZE)bilirubin

(E, Z)bilirubin

(EE)bilirubin

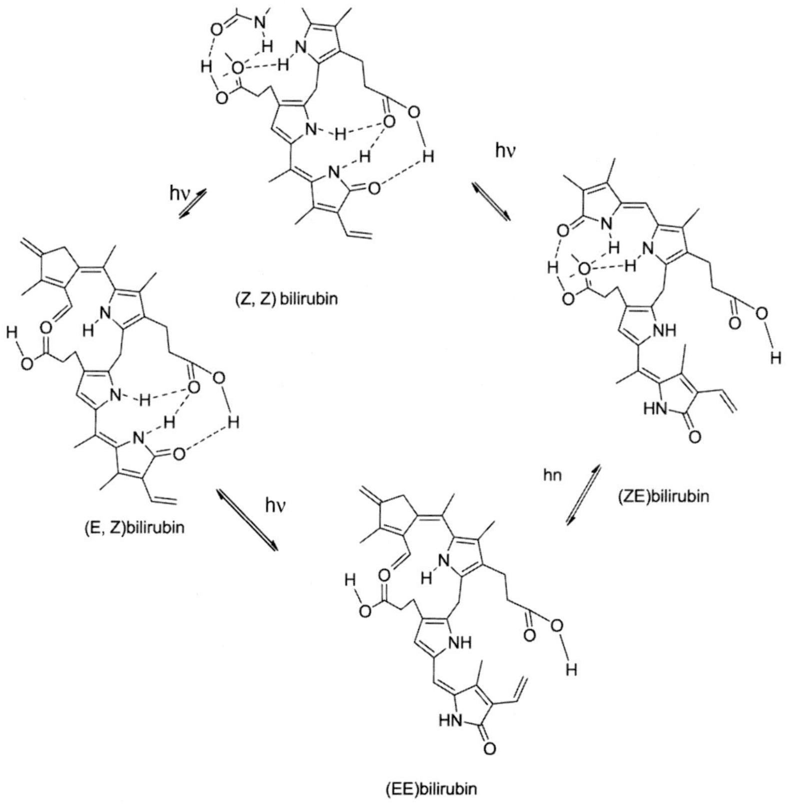

SCHEME 6.13 Steoisomers of bilirubin.

A portion of conjugated bilirubin is transported into the sinusoids and portal circulation by MRP3, which can undergo hepatocyte reuptake via the sinusoidal proteins, organic anion transport protein 1B1 and 1B3 (OATP1B1 and OATP1B3) [41]. Thus some conjugated and unconjugated bilirubin may escape the hepatocyte cytosol into the plasma where it binds to albumin and gets transported around the body. However, only conjugated bilirubin can enter the bile. The conjugated bilirubin is then actively secreted into canalicular bile, and drains into the small intestine. The rate-limiting step in bilirubin throughput is the hepatic excretory capacity of conjugated bilirubin. Part of the conjugated bilirubin may accumulate in serum when the hepatic excretion of the conjugated bilirubin is impaired as in prolonged biliary obstruction or intrahepatic cholestasis. This fraction of conjugated bilirubin gets covalently bound to albumin, and is called delta bilirubin or delta fraction or biliprotein. As the delta bilirubin is bound to albumin, its clearance from serum takes about 12–14 days (equivalent to the half-life of albumin) in contrast to the usual 2–4 h (half-life of bilirubin).

The process of conjugation alters the physiochemical properties of bilirubin giving it many special properties. Most importantly, it makes the molecule water soluble which allows it to be transported in bile without a protein carrier. Conjugation also increases the size of the molecule and prevents bilirubin from being passively reabsorbed by the intestinal mucosa due to its hydrophilicity and large molecular size. Thus conjugation works to promote the elimination of potentially toxic metabolic waste products. Furthermore, conjugation modestly decreases the affinity of bilirubin for albumin.

Although unconjugated bilirubin is always bound to albumin in serum, it cannot be filtered by the glomeruli (in the absence of glomerular disease). Thus unconjugated bilirubin is never found in urine even when there is an elevated level of unconjugated bilirubin in circulation. Jaundice that occurs with unconjugated hyperbilirubinemia is termed acholuric because the urine is not darkened. Dark urine, however, occurs when there is excretion of an excess of water-soluble conjugated bilirubin. This is seen in conjugated hyperbilirubinemia and signifies the presence of either liver or biliary disease. Thus the presence of bilirubin in urine will help identify subtle hepatobiliary dysfunction leading to conjugated hyperbilirubinemia, even when the measured concentration of conjugated bilirubin in serum is only slightly elevated. An exception to this rule is when bilirubinuria is not detected in a patient with prolonged cholestasis and marked jaundice. This is due to the formation of delta bilirubin or conjugated bilirubin that is tightly bound to serum albumin. The absence of bilirubinuria in such patients should not cause any difficulty in diagnosing conjugated hyperbilirubinemia, as the patient is clearly jaundiced and serum conjugated bilirubin is markedly elevated in such cases.

Degradation in the digestive tract: Conjugated bilirubin is not reabsorbed from the proximal intestine as mentioned above; in comparison, unconjugated bilirubin is partially reabsorbed across the lipid membrane of the small intestinal epithelium and undergoes enterohepatic circulation. Within the proximal small intestine, there is no additional metabolism of bilirubin, and very little deconjugation takes place. In stark contrast, when the conjugated bilirubin reaches the distal ileum and colon, it is rapidly reduced and deconjugated by colonic flora to a series of molecules termed urobilinogen. The major urobilinoids seen in stool are known as urobilinogen and stercobilinogen, nature and relative proportion of which will depend on the presence and composition of the gut bacterial flora. These substances are colorless but turn orange-yellow after oxidation to urobilin, giving stool its distinctive color [42].

an important application of selective oxygenation may makes alternative recourse to

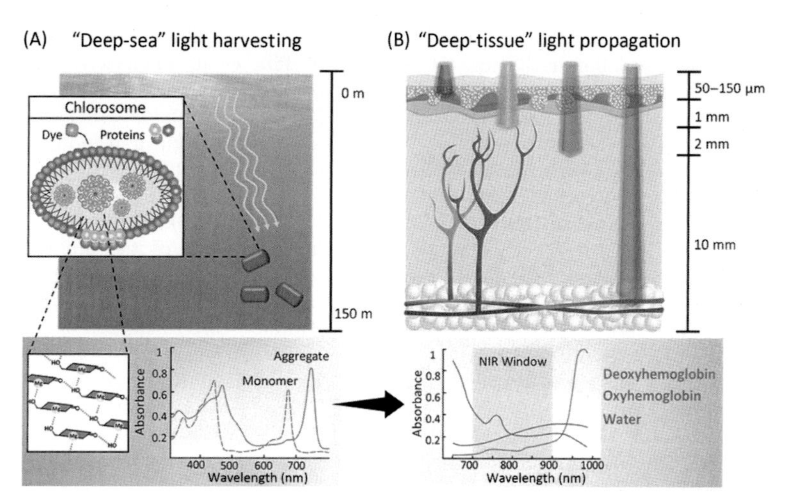

FIGURE 6.1 Natural ordered dye self-assemblies as an inspiration to overcome light-tissue penetration. (A) Schematic representation of chlorosome light-harvesting structures found in green sulfur bacteria that inhabit deep sea waters. Ordered aggregation of bacteriochlorophyll *c* results in exciton coupling and a bathochromic absorbance shift. (B) Propagation of light with different wavelengths through tissue. Major endogenous absorbers prevent shorter wavelengths of light from penetrating into deeper tissue layers, leaving a 700–900 nm window where the intrinsic absorbance is minimal for maximum light penetration [43].

light, rather than oxygen, penetration in tissues. Chlorosomes are ellipsoid structures around 100 nm in length containing >250,000 bacteriochlorophyll, coordinated at a Mg atom and 31–O, hydrogen bonding between the 31–OH with 13–C = O, and $\pi - \pi$ stacking. The efficient charge separation is exploited for a absorption band broadening in synthetic J-aggregates. Overall, the exceptional light-harvesting ability and the lack of a complex protein scaffold are major reasons why chlorosomes are a great design inspiration for artificial light-harvesting systems. Harmatys at al recently outlined the strategies of light use [43] (Fig. 6.1).

References

[1] H. Taube, Mechanism of oxidation with oxygen, J. Gen. Physiol. 49 (1965) 29–50.

[2] R. Choudhury, Hypoxia and hyperbaric oxygen therapy: a review, Int. J. Gen. Med. 11 (2018) 431–442.

[3] C.E. Fife, K.A. Eckert, M.J. Carter, An update on the appropriate role for hyperbaric oxygen: indications and evidence, Plast. Reconstr. Surg. 138 (2016) 107s–116s.

[4] G.B. Beccarius, De Bononiensi Instituto Scientiarum Artium Commentarii, vol. 4, Lili a Vulpe, Bologna, (1757) 74–87.

[5] R.R. Allison, K. Moghissi, Photodynamic therapy (PDT): PDT mechanisms, Clin. Endosc. 46 (2013) 24–29.

[6] O. Raab, Über die Wirkung fluoreszierender Stoffe auf Infusorien, Z Biol. 39 (1900) 524–546.

[7] H. von Tappeiner, A. Jesionek, Therapeutische Versuche mit fluorescierenden Stoffen, Munch Med Wochenschr 47 (1903) 2042–2044.

[8] H. Von Tappeiner, A. Jodlbauer, Die sensibilisierende Wirkung fluorieszierender Substanzer. Gesammte Untersuchungen uber die photodynamische Erscheinung, F. C. W. Vogel, Leipzig, 1907.

[9] H. Kautsky, H. de Bruijn, The explanation of the inhibition of photoluminescence of fluorescent systems by oxygen: the formation of active, diffusing oxygen molecules by sensitization, Naturwissenschaften 19 (1931) 1043.

[10] H. Kautsky, H. de Bruijn, R. Neuwirth, W. Baumeister, Energy transfers at surfaces. VII. Photosensitized oxidation mediated by a reactive, metastable state of the oxygen molecule, Ber. 66B (1933) 1588–1600.

[11] H. Kautsky, A. Hirsch, W. Flesch, Energy transformations on boundary surfaces. VIII. The significance of the metastable state in photosensitized oxidations, Ber. 68B (1935) 152–162.

[12] H. Gaffron, The mechanism of oxygen activation by illuminated dyes. II. Photooxidation in the near infrared, Ber 68B (1935) 1409–1411.

[13] H. Gaffron, Metastable oxygen and carbon dioxide assimilation, Biochem. Z. 287 (1936) 130–139.

[14] A.N. Terenin, The nature of the photochemical action in sensitized oxygen oxidation reactions and hydroperoxide breakdown reactions, Akad. Nauk S.S.S.R. (1955) 85–91.

[15] H. Kautsky, Reciprocal action between sensitizers and oxygen in light, Biochem. Z. 291 (1937) 271–284.

[16] H. Kautsky, Quenching of luminescence by oxygen, Trans. Faraday Soc. 35 (1939) 216–219.

[17] M. Kasha, Singlet oxygen electronic structure and energy transfer, in: A.A. Frimer (Ed.), *Singlet* O_2, Vol. 1, CRC Press, Boca Raton, FL, 1985, pp. 1–12.

[18] G. Wilke, Karl Ziegler – The last alchemist, In Ziegler Catalysts, Springer, Berlin, 1995.

[19] G.O. Schenck, H. Schulze-Bushoff, Synthetic ascaridol: new possibility of specific therapy of ascariasis, Dtsch. Med. Wschr. 73 (1948) 341–344.

[20] G.O. Schenck, M. Cziesla, K. Eppinger, G. Matthias, M. Pape, Isobenzpinakol als ursache des spektralen effekts bei der photoreduktion des benzophenons, Tetrahedron Lett. 8 (3) (1967) 193–198.

[21] R.S.H. Liu, A learning experience through studies in photochemistry, Front. Chem. China 4 (2009) 403.

[22] C. Moss, M. Stacey, Epidermal mosaicism and Blaschko lines, J. Med. Genet. 30 (1993) 752–755.

[23] W. Maidhof, O. Hilas, Lupus: an overview of the disease and management options, Pharm. Ther. 37 (2012) 240–246, 249.

[24] I.N. Levine, Quantum Chemistry, fifth ed., Prentice Hall, Upper Saddle River, NJ, 2000, pp. 402–407.

[25] O. Warburg, W. Christian, The yellow oxidation enzyme, Biochem. Z. 258 (1933) 496–498.

[26] M. Brengnøj, M. Westberg, F. Jensen, P.R. Ogilby, Solvent-dependent singlet oxygen lifetimes: temperature effects implicate tunneling and charge-transfer interactions, Phys. Chem. Chem. Phys. 19 (2016) 22946–22961.

[27] A.A. Goghare, A. Greer, Using singlet oxygen to synthetize natural products and drugs, Chem. Rev. 116 (2016) 9994–10034.

[28] G. Yao, K. Steliou, Synthetic studies toward bioactive cyclic peroxides from the marine sponge *Plakortis angulospiculatus*, Org. Lett. 4 (2002) 485–488.

[29] A. Maraş, H. Seçen, Y. Sütbeyaz, M. Balcı, A convenient synthesis of (±)-talo-quercitol (1-deoxy-neo-inositol) and (±)-viboquercitol (1-Deoxy-myo-inositol) via ene reaction of singlet oxygen, J. Org. Chem. 63 (1998) 2039–2041.

[30] R. Brecht, F. Buttner, M. Bohm, G. Seitz, G. Frenzen, A. Pilz, et al., Photooxygenation of the helimers of (–)-isocolchicine: regio- and facial selectivity of the [4 + 2] cycloaddition with singlet oxygen and surprising endoperoxide transformations, J. Org. Chem. 66 (2001) 2911–2917.

[31] D. Kalaitzakis, T. Montagnon, E. Antonatou, G. Vassilikogiannakis, One-pot synthesis of the tetracyclic framework of the aromatic erythrina alkaloids from simple furans, Org. Lett. 15 (2013) 3714–3717.

[32] J. Baier, T. Maisch, M. Maier, M. Landthaler, W. Bäumler, Direct detection of singlet oxygen generated by UVA irradiation in human cells and skin, J. Invest. Dermatol. 127 (2007) 1498–1506.

[33] Z. Zou, J. Ye, K. Sayama, H. Arakawa, Direct splitting of water under visible light irradiation with an oxide semiconductor photocatalyst, Nature 414 (2001) 625–627.

[34] J.M. Aubry, C. Pierlot, J. Rigaudy, R. Schmidt, Reversible binding of oxygen to aromatic compounds, Acc. Chem. Res. 36 (2003) 668–675.

[35] M.C. Mano, F.M. Predo, J. Massari, G.E. Ronsein, G.R. Martinez, Miyamoto, et al., Excited singlet molecular O_2 ($^1\Delta_g$) is generated enzymatically from excited carbonyls in the dark, Sci. Rep. 4 (2014). Article number: 5938.

[36] W. Fudickar, T. Linker, Release of singlet oxygen from aromatic endoperoxides by chemical triggers, Angew. Chem. 57 (2018) 12971–12975.

[37] L. Martinez-Fernandez, J. Gonzalez-vazquez, L. Gonzalez, I. Corral, Time-resolved insight into the photosensitized generation of singlet oxygen in endoperoxides, J. Chem. Theory Comput. 11 (2015) 406–414.

[38] T.W.R. Hansen, Pioneers in the scientific study of neonatal jaundice and kernicterus, Pediatrics 106 (2000) E15.

[39] S. Ullah, K. Rahman, M. Hedayati, Hyperbilirubinemia in neonates: types, causes, clinical examinations, preventive measures and treatments: a Narrative review article, Iran J. Public Health 45 (5) (2016) 558–568.

[40] S. Onishi, K. Isobe, S. Itoh, Metabolism of bilirubin and its photoisomers in newborn infants during phototherapy, J. Biochem 100 (1986) 789.

[41] K. Aditya, J. Savio, Physiology, Bilirubin, Stat Pearls Publishing, Tresure Island, FL, 2019.

[42] I.J. Beckingham, S.D. Ryder, Investigation of liver and biliary disease, Brit. Med. J. 322 (7277) (2001) 33–36.

[43] K.M. Harmatys, M. Overchuk, G. Zheng, Rational design of photosynthesis-inspired nanomedicines, Acc. Chem. Res. 52 (2019) 1265–1274.

CHAPTER

7

Emission

7.1 Emission for diagnosis in medicine

Emission is intrinsically a more efficient system for revealing the presence of analytes. In fact, in the first case a weak signal is compared with no emission at all, whereas in the latter one an almost undistinguishable absorption is compared with the blank (99% vs 100%, although this advantage is in large part balanced by the fact that emission occurs in every direction here and in single one in absorption). As a matter of fact, however, the most rapidly developing emission techniques in the last two decades have been two nonoptical methods: (1) laser assisted computed tomography (CT or CAT), which involves the use of suitable nuclides (nonzero nuclear spin, ^{11}C, ^{13}N, ^{15}O, and ^{18}F) and obtains 3D images of internal organs and their tumors and essentially is a highly elaborated X-ray method, and (2) positron emission (PET) that reveals alterations of the metabolism, for example, an increased consumption of plasma dissolved oxygen as it generally happens in tumors [1]. On the other hand, optic emission has found diagnostic application, too, for instance in the case of autofluorescence (due to aromatic aminoacids, dyes, etc.) that has been applied through the most interesting characteristics of fluorescence, that is, time resolution [2]. Such procedure has been adapted for instance for obtaining a clear demarcation of malignant tumors from normal tissue and for distinguishing atherosclerotic plaque of different degrees from normal vessel wall. The time-correlated single-photon counting was used in a recent research with an argon puled laser and fluorescence band peaking at 380 nm with a lifetime of $\backsimeq 7$ ns specific for atherosclerotic plaque [3]. A great leap forward has been done with the introduction of confocal microscopy, where the thin-cut "section" of fixed or frozen tissue was substituted to the actual section used in conventional microscopy, which is obviously not suitable for in vivo investigations. In vivo microscopy requires a virtual, rather than a physical, section of the specimen, and indeed confocal microscopy uses optical imaging to create a virtual slice or plane, many micrometers deep, within the tissue. It provides very-high-quality images with fine detail and more contrast than conventional microscopy. In addition, the imaging technique allows for reconstruction of virtual 3D images of the tissue when multiple sections are combined [4]. In a recent application, algorithms to diagnose basal cell carcinomas (BCCs) and melanomas (MMs) were reported based on

Light, Molecules, Reaction and Health
DOI: https://doi.org/10.1016/B978-0-12-811659-3.00007-4

in vivo reflectance confocal microscopy (RCM). A total of 710 consecutive cutaneous lesions excised to exclude malignancy (216 MMs, 266 nevi, 119 BCCs, 67 pigmented facial macules, and 42 other skin tumors) were imaged by RCM. RCM features were correlated with pathology diagnosis to develop diagnostic algorithms. It turned out that the diagnostic accuracy of the BCC algorithm defined on multivariate analysis of the training set (50%) and tested on the remaining cases was 100% sensitivity, 88.5% specificity. Positive features were polarized elongated features, telangiectasia and convoluted vessels, basaloid nodules, and epidermal shadowing corresponding to horizontal clefting. Negative features were nonvisible papillae, disarrangement of the epidermal layer, and cerebriform nests. Multivariate discriminant analysis on the training set (excluding the BCCs) identified seven independently significant features for MM diagnosis. The diagnostic accuracy of the MM algorithm on the test set was 87.6% sensitivity, 70.8% specificity. The four invasive MMs that were misdiagnosed by RCM were all of nevoid subtype. RCM is a highly accurate noninvasive technique for BCC diagnosis. Good diagnostic accuracy was achieved also for MM diagnosis, although rare variants of melanocytic tumors may limit the strict application of the algorithm [4].

As an example, multiple endocrine neoplasia type 2 (MEN2) may be considered. This is a rare autosomal dominant syndrome caused by mutations in the rearranged during transfection (RET) protooncogene and is characterized by a strong penetrance of medullary thyroid carcinoma (MTC) (all subtypes) and is often accompanied by pheochromocytoma (MEN2A/2B) and primary hyperparathyroidism (MEN2A). The evaluation and management of MEN2-related tumors is often different from that of sporadic counterparts. An overview of clinical manifestations, diagnosis, and surgical management of MEN2 patients, has been reported, along with applications of the most up-to-date imaging modalities to MEN2 patients that are tightly linked to the clinical management and aims to guide physicians toward a rationale for the use of imaging prior to prophylactic thyroidectomy, initial surgery, and reoperations for persistent/recurrent disease. It has also been concluded that, in the near future, it is expected that these patients will indeed benefit from newly developed positron emission tomography (PET) approaches which will target peptide receptors and protein kinases. Identification of MEN2-specific radiopharmaceuticals will also soon arise from molecular profiling studies. Furthermore, subtotal (cortical-sparing) adrenalectomy, which is a valid option in MEN2 for avoiding long-term steroid replacement, will benefit from an accurate estimate through imaging of differential adrenocortical function [5] (Fig. 7.1).

Another important case is that of malignant pleural mesothelioma (MPM), an asbestos-related neoplasm that originates in pleural mesothelial cells and progresses locally along the pleura until it encases the lungs and mediastinum, ultimately causing death. Imaging plays a crucial role in diagnosis and optimal management. Computed tomography (CT) continues to be the primary and initial imaging modality. Magnetic resonance imaging (MRI) complements CT scan and is superior in determining chest wall and diaphragmatic invasion. FDG18-PET/CT provides anatomometabolic information and is superior to both CT and MRI in overall staging and monitoring response to therapy. This chapter will detail the imaging finding of MPM and role of imaging in guiding management [7]. Newly identified rearrangements during transfection point mutations have helped with MTC prognosis and have resulted in the establishment of new treatment guidelines. Screening for MTC in the United States with basal serum calcitonin for patients with thyroid nodules would cost

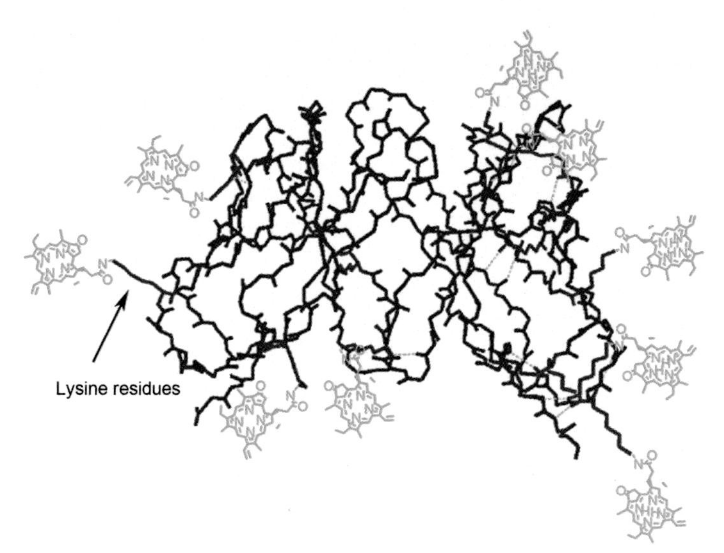

FIGURE 7.1 Monoclonal antibody fragment with covalently linked pheophorbide, a sensitizer [6].

11,793$ per life-year saved, compared with colonoscopy and mammography screening. For metastatic or recurrent disease, neck ultrasound, chest CT scan, liver MRI, bone scintigraphy, and axial skeleton MRI have been proven superior to FDG18-PET/CT. For patients with nonoperable metastatic disease, novel chemotherapeutic agents, such as vandetanib, targeting rearranged during transfection, vascular endothelial growth factor receptor, and epidermal growth factor receptor, are showing promise. Such agents are currently in phase II trials. By potentially downstaging of disease, and treating metastatic disease more effectively, overall survival and outcomes of patients may improve [8].

Carcinoid tumors account for less than 1% of all malignancies and the majority arise in the gastrointestinal system. These tumors are slow growing compared with adenocarcinomas and they differ from the other neuroendocrine malignancies by their protean clinical presentation. Carcinoid tumors were previously considered indolent, but they can manifest malignant characteristics with metastatic spread, which often results in a poor prognosis. Although there have been advances in diagnostic and treatment modalities, carcinoid tumors are still often diagnosed late, often when the tumor has metastasized and patients develop carcinoid syndrome. A high concentration of urinary 5-hydroxyindoleacetic acid and elevated plasma serotonin and chromogranin. A levels help to establish the initial diagnosis of carcinoid tumors. In addition to the CT and MRI, molecular imaging modalities, such as a type of scintigraphy used to find carcinoid, pancreatic neuroendocrine tumors, and to localize sarcoidosis (OctreoScan), as neuroblastomas and pheochromocytomas. A small amount of a substance called radioactive ^{131}I-metaiodobenzylguanidine (MIBG) is used in scintigraphy imaging and more recently PET imagings are vital in detection of primary malignancy and metastatic involvement. Surgery is the mainstay of treatment of nonmetastatic carcinoid tumors. Cytotoxic chemotherapy is not beneficial due to the chemoresistant nature of these tumors. Because carcinoid tumors express somatostatin receptors, somatostatin analogs, which inhibit the release of serotonin and other neuroendocrine peptides, are often used, but their use is limited to symptom control. Treatment using high doses of radionuclides such as radiolabeled somatostatin analogs and MIBG is a more recent option which offers a definite advantage in management. Typical features of the carcinoid tumors and contemporary methods of detecting and assessing carcinoid tumors the role of various diagnostic and therapeutic options are discussed [9].

Chronic fatigue syndrome (CFS) is an illness currently defined entirely by a combination of nonspecific symptoms. Despite this subjective definition, CFS is associated with objective underlying biological abnormalities, particularly involving the nervous system and the immune system. Most studies have found that active infection with human herpesvirus-6 (HHV-6)--a neurotropic, gliotropic, and immunotropic virus--is present more often in patients with CFS than in healthy control and disease comparison subjects, yet it is not found in all patients at the time of testing. Moreover, HHV-6 has been associated with many of the neurological and immunological findings in patients with CFS. Finally, CFS, multiple sclerosis, and seizure disorders share some clinical and laboratory features and, like CFS, the latter two disorders are also being associated increasingly with active HHV-6 infection. Therefore it is plausible that active infection with HHV-6 may trigger and perpetuate CFS in a subset of patients [10].

Cyclooxygenase (COX) is an enzyme that catalyzes the first two steps in the biosynthesis of prostanoids. The constitutively expressed isoform COX-1 is regarded as a housekeeping enzyme that is responsible for the normal production of prostanoids. The inducible isoform COX-2, on the other hand, is transiently induced during inflammation by various stimuli. Increasing evidence has shown that COX-2 is not only implicated in inflammation but also in oncogenesis. Overexpression of COX-2 was observed in a variety of tumors. Prostaglandins produced by COX-2 affect important processes in carcinogenesis, including angiogenesis, tissue invasion, metastasis, and apoptosis. Several studies indicate that COX-2 is also involved in neurological disorders, like Alzheimer's disease, Parkinson's disease, and ischemia, where COX-2 overexpression leads to neurotoxicity. Many aspects of the role of COX-2 in (patho)physiological paths, however, remain unclear. At present, COX-2 expression is determined by ex vivo laboratory analysis, but the results could be greatly affected by the instability of COX-2 mRNA and protein and by sampling errors. A noninvasive imaging method to monitor COX-2 expression, like PET or single photon emission CT, could overcome this complication and may provide novel insights in the role of COX-2, especially in neurological disorders where repetitive sampling is not possible. Such a technique could also be applied to the in vivo evaluation of novel selective COX-2 inhibitors and in dose-escalation studies [11].

Acute and chronic lung diseases are almost invariably associated with some degree of inflammation. Cells that evolved as an effective mechanism to counter infection and heal lung tissue may, in some circumstances, lead themselves to be partially responsible for the pathogenesis of chronic lung disease that leads to irreversible lung damage and loss of lung function. Although standard measurements of lung function can document the progression of disease, the contributions of the numerous interacting elements to the process are difficult to measure in life. The use of molecular imaging techniques allows the different components of the inflammatory response to be monitored in situ in humans. In particular, PET of selected markers targeted to specific cells and biochemical pathways can provide accurate measurements of disease activity, enabling a better understanding of inflammatory processes at all stages of disease. The practicability of sequential measurements allows one to monitor the natural history of different lung diseases. More importantly, imaging provides a unique tool for quantification of the modulation of discrete and specific aspects of inflammatory lung disease by targeted interventions. This should facilitate the development of new treatment strategies with better specificity for key elements of each disease [12].

Over the past decades, laser use in medicine has expanded from its initial application as a light-based scalpel to a plethora of clinical uses, ranging from surgical treatment through composite polymerization, dental ablation, vision correction, and skin resurfacing to diverse diagnostic modalities. Recently, the concept of light-based diagnostics and therapy has come under investigation. Low light intensities are used to excite endogenous or exogenous fluorophores, some of which have characteristic fluorescence emissions in pathological tissues. Thus premalignancy and malignancy potentially can be detected and diagnosed. Photosensitized superficial lesions can subsequently be destroyed selectively by using higher intensities of laser light. The application of fluorescence emission-based detection and diagnosis of precancer and cancer is reviewed, based on its application to the oral cavity--the author's primary anatomical area of expertise. This approach is justified as the same principles apply throughout the human body; to any area accessible to the clinician either directly or by some sort of fiberoptic probe [13].

Many important emitters have been prepared by interactions between biomolecules and antibodies. Immunochemistry offers simple, rapid, robust yet sensitive, and easily automated methods for routine analyses in clinical laboratories. Immunoassays are based on highly specific binding between an antigen and an antibody. An epitope (immunodeterminant region) on the antigen surface is recognized by the antibody's binding site. The type of antibody and its affinity and avidity for the antigen determine assay sensitivity and specificity. Depending on the assay format, immunoassays can be qualitative or quantitative. They can be used for the detection of antibodies or antigens specific for bacterial, viral, and parasitic diseases as well as for the diagnosis of autoimmune diseases. Immunoassays can measure low levels of disease biomarkers and therapeutic or illicit drugs in patient's blood, serum, plasma, urine, or saliva. Immunostaining is an example of an immunochemical technique, which combined with fluorescent labels allows direct visualization of target cells and cell structures (Figs. 7.2–7.4).

However, the recent literature point on emission of rare earth salts, which are considerably advantageous. Mechanoluminescence (ML) is the emission of light consecutive to a mechanical force or stress imposed to a crystalline material. Many inorganic and organic compounds present this phenomenon that is known for over 400

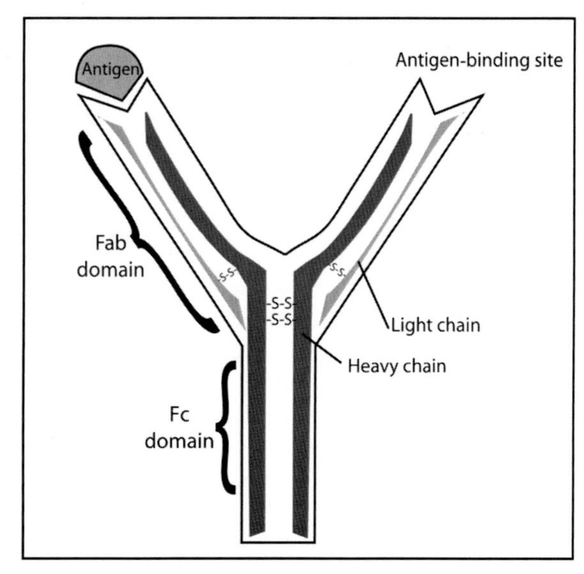

FIGURE 7.2 Schematic of antibody structure [14].

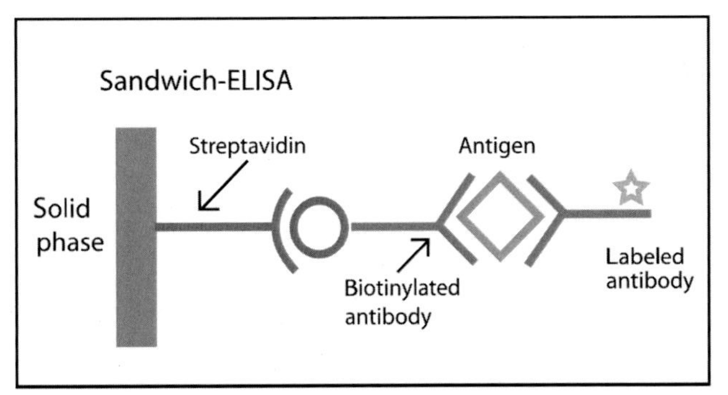

FIGURE 7.3 Biotinylated antibodies can be bound to the solid support via streptavidin. The four biotin-binding sites on one streptavidin molecule facilitate the increase in assay sensitivity through amplification [14].

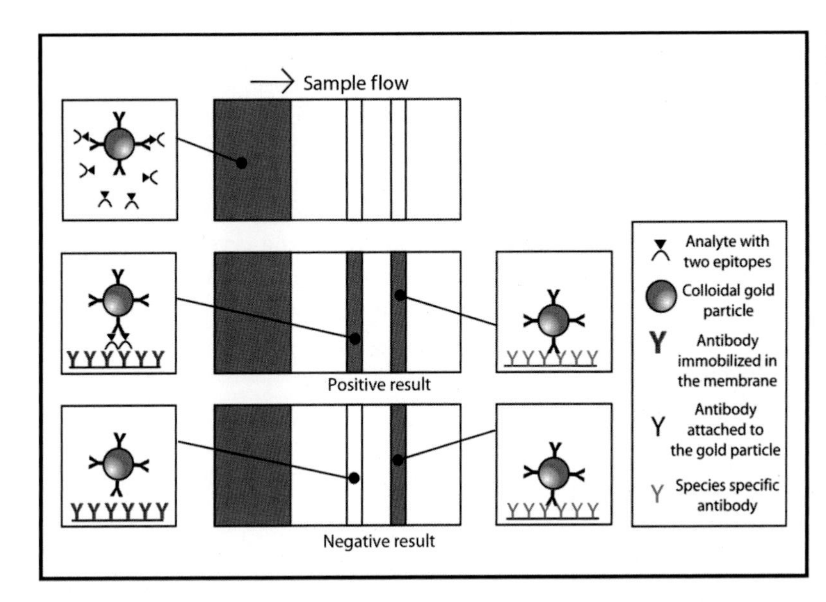

FIGURE 7.4 A schematic of a lateral flow capture (sandwich) assay used for detection of analytes (like hormones) with multiple epitopes [14].

years. Lanthanide and uranyl salts were among the first substances investigated for this property. ML, also referred to as triboluminescence, is often considered as being a badly understood phenomenon. Different mechanical stresses, from simple rubbing to applied pressure, crushing, impact of a weight, ultrasound, laser-generated shock wave, crystallization, dissolution of crystals, or even wind can trigger it [15].

Highly efficient photoemitting materials have been prepared by using lanthanides. The electronic and coordination properties of these cations, the prominent issues related to the design and synthesis of efficient luminescent antenna complexes, and their photophysical properties have become of great interest in the last decades. The basic principles of ligands design to yield systems featuring a coordination site for the metal cation with appended suitable chromophores as sensitizers (leading two-component approach). When properly designed, these ligands are capable of forming

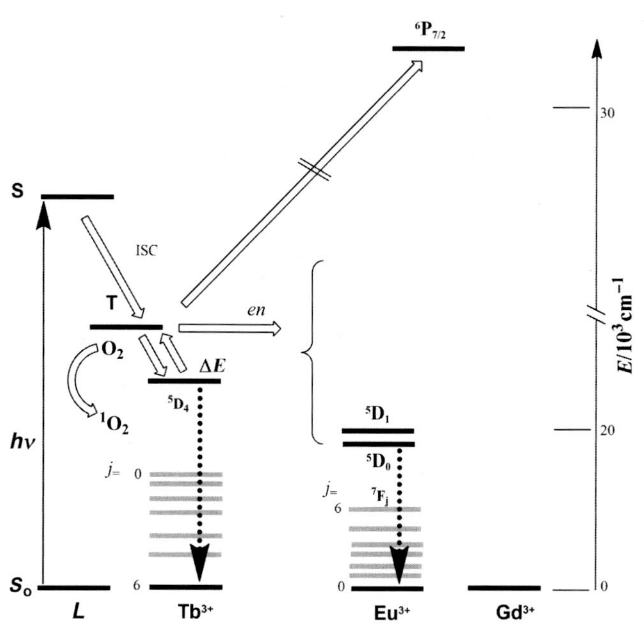

FIGURE 7.5 The antenna effect for sensitization of the luminescence in some lanthanide cations (selected levels are displayed); empty arrows indicate nonradiative processes, dot arrows indicate radiative processes. As exemplified here for the Tb^{3+}, Eu^{3+}, and Gd^{3+}centers, three cases are encountered with regard to energy transfer (en) as regulated by the energy gap, ΔE, between the triplet level (*T*) of the chromophore (*L*) and the emitting level of the cation: (A) when $\Delta E \leq 1500\ cm^{-1}$ back energy transfer takes place and, as a consequence, O_2 effects (particularly intense in solution at room temperature given the long lifetime of the ligand triplet level) are observed, see text; (B) when $\Delta E \geq 1500\ cm^{-1}$ energy transfer is complete; and (C) energy transfer is exothermic and does not take place [16].

highly luminescent complexes (overall sensitization yield, $\phi_{se} > 0.05$ in aqueous medium). Highly luminescent color tunable films for applications in lighting and light conversion technologies have been prepared by combining the peculiar luminescence properties of Eu^{3+} and Tb^{3+} antenna complexes with optically transparent inorganic matrices [16] (Fig. 7.5).

Owing to their unmatched optical properties, lanthanide ions have been used as commercial probes in bioanalysis since the early 1980s and are now starting to be explored as probes for optical imaging and substituting antibodies. For example, Eu^{3+} chelates by nanoparticles (NPs) in immunoassays results in an improvement of the limit of detection by 2–3 orders of magnitude [17]. Most of spectra reported involve "fingerprint region," but since the last decade, the advances in theoretical chemistry allow to predict them. What remains for the experimentalist is "only" to prepare the sample in a suited way (Scheme 7.1).

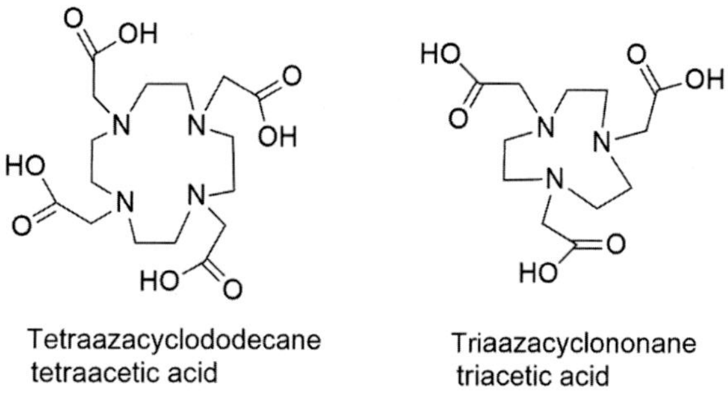

Tetraazacyclododecane tetraacetic acid

Triaazacyclononane triacetic acid

SCHEME 7.1 Ligands used for preparing lanthanides complexes.

7.2 Two-photon conversions in absorbance-emission

Emission studies are quite important in biology and medicine because of their gaining in vivo target-specific information is of obvious significance for understanding how metabolism occurs and signals are interchanged among cells. Furthermore, diagnostic agents delivered along with a drug, it would be possible to understand which is the best dose and ultimately arrive at a personalized medicine, where healing effects are maximized, while toxic ones are minimized. All of the techniques invented by chemists have been adapted to the study of biomolecules in vitro or in vivo [18].

Certainly the compounds chosen for this job have to comply with the characteristics above, as well as being photostable (notably in the presence of oxygen and water). All of the methods evolved by chemistry in the last decades and have been adapted to biological studies, in vitro or in vivo and are nowadays covering all of the wavelength ranges achievable and the progress of reliable calculation methods have allowed the attribution of structure and energy of reasonable intermediates as well as the application of some convenient methods of after analysis with considerable improvement of the signal range [19]. Intelligent polymeric micelles provide great potential for accurate cancer theranostics. For example, gemcitabine (GEM)-conjugated redox-responsive prodrug micelles based on pH-responsive charge-conventional dual-responsive micelles with aggregation-induced emission (AIE), PMPC-b-P (DEMA-*co*-SS-GEM-*co*-TPMA) copolymer [20] (Fig. 7.6).

During the last few years, several clinical studies have demonstrated that circulating

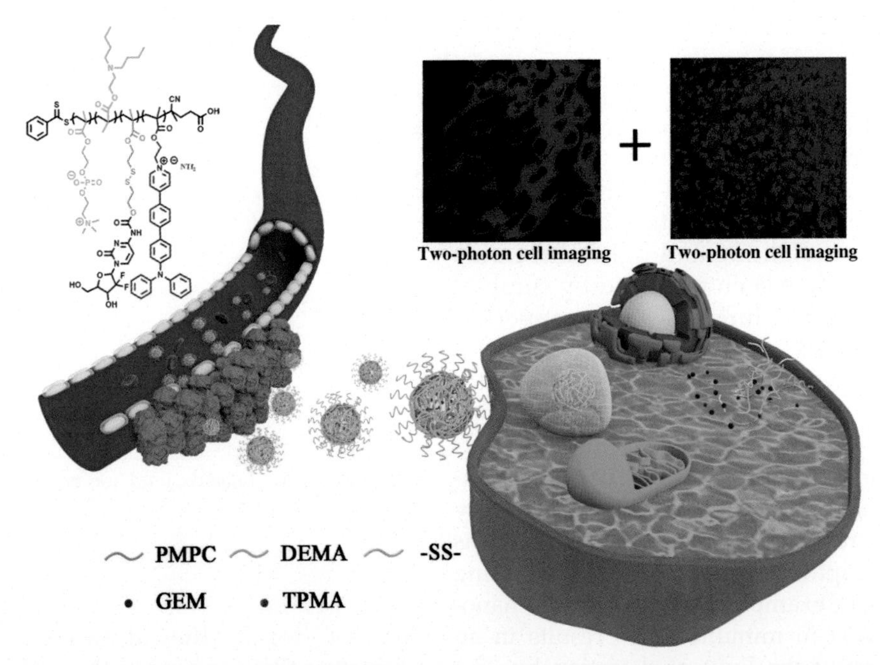

FIGURE 7.6 Schematic diagram of gemcitabine (GEM)-conjugated polymeric micelles based on PMPC-b-P (DEMA-*co*-SS-GEM-*co*-TPMA) copolymer for pH- and redox-triggered drug delivery and AIE-active two-photon bioimaging [20,21].

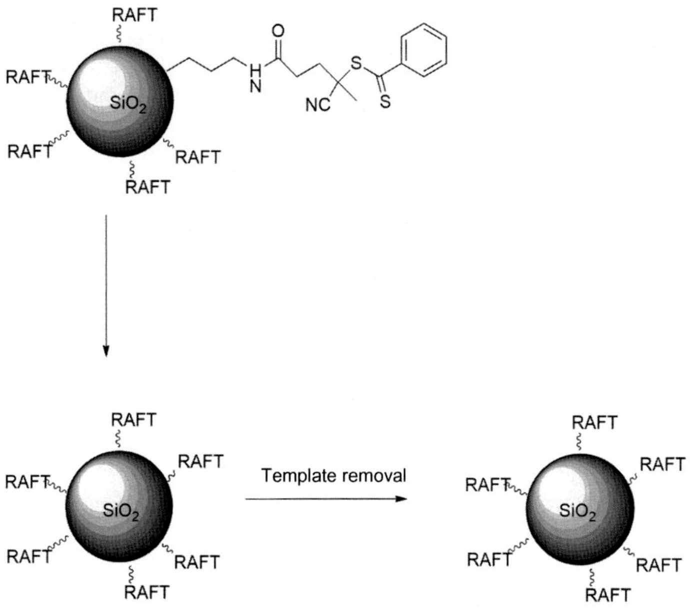

SCHEME 7.2 Synthesis of MIP for NLLGLIEAK via RAFT-modified mesopoeous silica [23]. *MIP*, molecular imprinting polymers; *RAFT*, reversible addition fragmentation chain transfer polymerization.

tumor cells can be used as a marker for understanding metastatic development, a key factor in increasing the overall survival rate for liver cancer [22] (Scheme 7.2).

Photonic crystals have been synthesized by a reversible addition fragmentation chain transfer polymerization (RAFT) reaction and used, to advantage, for sensing biomolecules [24].

S-values of nine positron-emitting radionuclides (^{11}C, ^{13}N, ^{15}O, ^{18}F, ^{64}Cu, ^{68}Ga, ^{82}Rb, ^{86}Y and ^{124}I) in 48 source regions for 10 anthropomorphic pediatric hybrid models, including the reference newborn, 1-, 5-, 10-, and 15-year-old male and female models, using the Monte Carlo N-particle extended general purpose [25].

A hydrophobic two-photon absorbing (2PA) red emitter (R) was successfully incorporated into micelles formed from two block copolymers, poly(ε-caprolactone)-*block*-poly(ethylene glycol), for imaging and toxicity studies. In micelles, the chromophore R exhibits a 2PA cross section of 400 GM (1 GM $= 1 \times 10^{-50}$ cm^4 s/photon/molecule, GM is the international accepted coefficient for two photons absorption, named in the honor of Prof. Goeppert Mayer) at 820 nm, which is among the highest values reported for red 2PA emitters. The micelles with a cationic amino moiety-containing poly(ethylene glycol) corona showed an enhancement of cell internalization and delivered the dye into the cytoplasmic regions of the mouse macrophage RAW 264.7 cells [21].

The structure of the red 2PA emitter (R) is based on a D-π-A-π-D motif as shown in Fig. 7.7. The two amino groups act as the electron-donating units while the aqueous bilirubin ditaurate (BTD) functions as the electron-accepting moiety in the center. Almost no shift of its absorbance maximum in the above listed organic solvents was observed, indicating that the solvent polarity did not

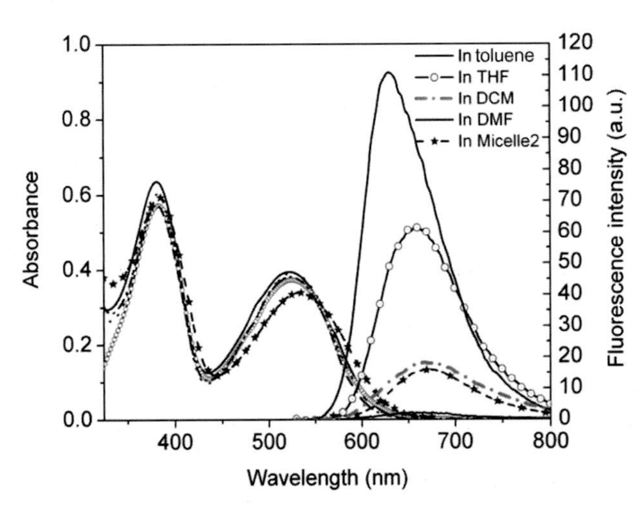

FIGURE 7.7 UV-Vis and fluorescence emission spectra of R in toluene, THF, DCM, and diluted micelle 2. For these measurements, the dye concentration was 5 μM [26]. *THF*, tetrhydroflavin.

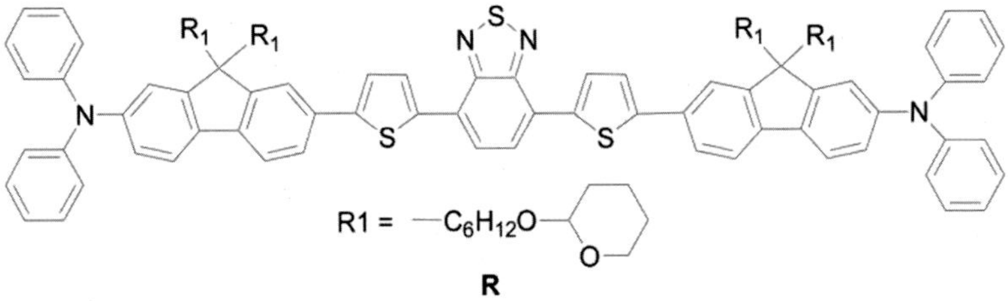

R1 = —$C_6H_{12}O$—

R

SCHEME 7.3 Molecular structure of the red emitter R [21].

affect its electronic states. The absorption spectrum of the compound has two bands around 330–425 and 425–625 nm arising from π–π* and charge-transfer transitions, respectively. The material emits in the red spectral window at the wavelength region of 550–800 nm, and the emission spectrum in a certain solvent is static regardless of excitation wavelengths [26] (Scheme 7.3).

Observation of the activation and inhibition of angiogenesis processes is important in the progression of cancer. Application of targeting peptides, such as a small peptide that contains adjacent L-arginine (R), glycine (G), and L-aspartic acid (D) residues can afford high selectivity and deep penetration in vessel imaging [26]. To facilitate deep tissue vasculature imaging, probes that can be excited via 2PA in the near-infrared (NIR) and subsequently emit in the NIR are essential [27].

Fluorescence imaging of tissues offer an essential means for studying biological systems. Autofluorescence becomes a serious issue in tissue imaging under excitation at UV–Vis wavelengths where biological molecules compete with the fluorophore. To address this critical issue, a novel class of fluorophores that can be excited at ~900 nm under two-photon excitation conditions and emits in the red wavelength region (≥600 nm) has been disclosed. The new π-extended dipolar dye system shows several advantageous features including minimal autofluorescence in tissue imaging and pronounced solvent-sensitive emission behavior, compared with a widely used 2PA dye, acedan. As an important application of the new dye system, one of the dyes was developed into a fluorescent probe for amyloid-β plaques, a key biomarker of Alzheimer's disease [28].

Simulations were performed using the Penetration and Energy Loss of Positrons and Electrons (radiation matter interaction simulation software) Penelope and an anthropomorphic mathematical phantom, which was built specifically for these calculations and that mimics the Medical Internal Radiation Dose-type mathematic phantoms. In the examples studied, the source organ was the thyroid gland and the target organs were the urinary bladder, the testicles, the ovaries, and the uterus. Photons of energies between 30 keV and 2 MeV were assumed to be emitted isotropically from initial positions distributed uniformly over the volume of the source organ. The results of a set of simulations performed without applying any virus removal tool were compared with those obtained by applying interaction forcing and Russian roulette and splitting driven by an ant colony algorithm [29].

Colorectal cancer is a disease that can be prevented if is diagnosed and treated at preinvasive stages. Thus the monitoring of colonic cancer progression can improve the early diagnosis and detection of malignant lesions in the colon. This monitoring should be performed with appropriate image techniques and be accompanied by proper quantification to minimize subjectivity [30].

7.3 Theoretical background

FRET is the acronym of the Förster (Fluorescence) Resonance Energy Transfer. This is a physical phenomenon where an excited donor transfers energy (not an electron) to an acceptor group through a nonradiative process. This process is very fast, but strongly dependent on the distance and this characteristic, allows to finally prove biological structures. In particular, a simple and large used approach involves attaching appropriate donor-acceptor groups to the biological polymer and directly measure the distance between two positions of interest (Figs. 7.8 and 7.9; Scheme 7.4).

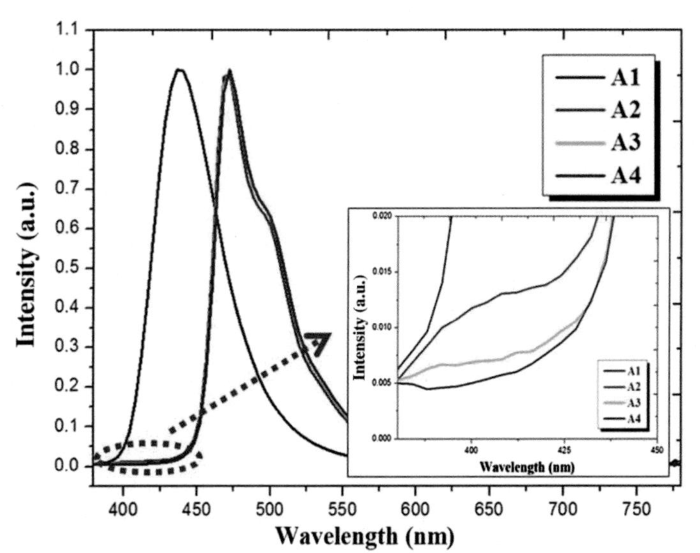

FIGURE 7.8 Electroluminescence spectra measured for a series of devices with different distances (A1-A4). It provides evidence of energy transfer from 1,3-di(9H-carbazol-9-yl) benzene (mCP) host to the phosphorescent dopant following the doping concentration from 0% to 12%. This result shows a complete energy transfer from host to phosphorescent dopant, except in the case of device A2 (with lower doping concentration of 4%), as mCP's emission peak at 410 nm has almost disappeared. A slight emission remained at around 410 nm (see the inset) in the spectra of device A2 and is caused by incomplete energy transfer from mCP [31].

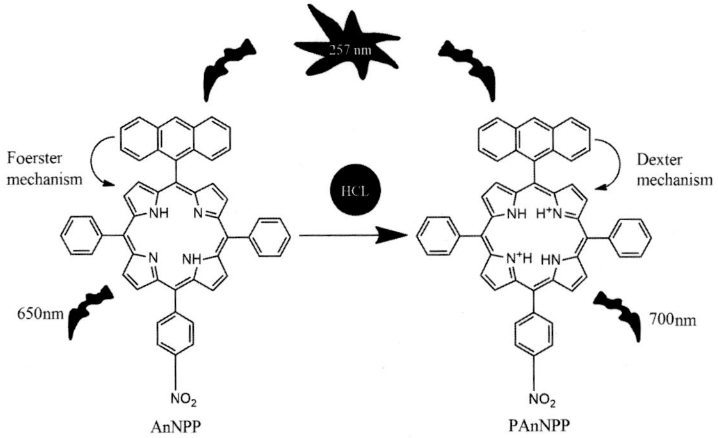

FIGURE 7.9 Bichromophoric anthracene-porphyrin molecule namely 5-(4-nitrophenyl)-10,20-bisphenyl-15-(9-anthryl) porphyrin (AnNPP) was synthesized as a model to determine the short-range energy transfer between anthracene and porphyrin moieties. Switching of Förster to Dexter mechanism was studied in the free base (AnNPP) and protonated form of AnNPP (PAnNPP). The complete protonation of AnNPP was carried out by using hydrogen chloride (HCl) and was confirmed by UV – Vis spectroscopy. The steady-state fluorescence spectroscopy shows that anthracene quantum yield is decreased in both AnNPP and PAnNPP. This was further proved by measuring fluorescence lifetime using time correlated single photon counting technique. A remarkable decrease in fluorescence lifetime of anthracene is observed for both AnNPP and PAnNPP. Quantum chemical calculations were performed for AnNPP and PAnNPP to support the short distance energy transfer with switching of excited state energy transfer (EET) mechanism. The comparison of the EET rates reveals that the Förster mechanism is followed in AnNPP, whereas the Dexter mechanism is predominant in PAnNPP [32].

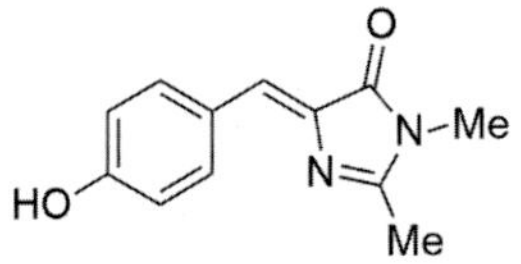

SCHEME 7.4 Prosthetic group of green fluorescent protein.

Enhanced cyan and yellow fluorescent proteins are widely used for dual color imaging, but are thermal sensitive and show a tendency for aggregation. A site-directed mutagenesis approach was successfully used for improving these fluorescent proteins, which fold faster and more efficiently at 37°C and have superior solubility and brightness. Bacteria expressing SCFP3A were ninefold brighter than those expressing ECFP and 1.2-fold brighter than bacteria expressing Cerulean. SCFP3A has high quantum yield (0.56) and fluorescence lifetime. Bacteria expressing SYFP2 were 12 times brighter than those expressing EYFP(Q69K) and almost twofold brighter than bacteria expressing Venus. In HeLa cells, the improvements were less pronounced; nonetheless, cells expressing SCFP3A and SYFP2 were both 1.5-fold brighter than cells expressing ECFP and EYFP(Q69K), respectively. The enhancements of SCFP3A and SYFP2 are most probably due to an increased intrinsic brightness (1.7-fold and 1.3-fold for purified recombinant proteins, compared with ECFP and EYFP(Q69K),

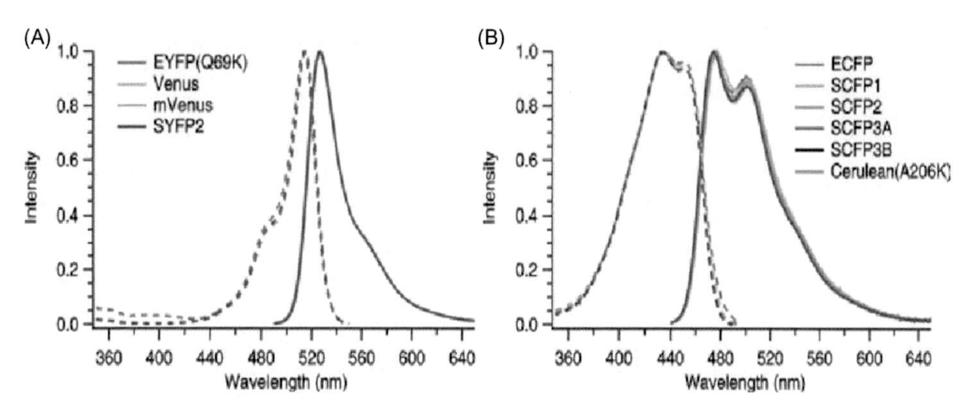

FIGURE 7.10 Absorbance and emission spectra. Comparison of absorbance (dotted lines) and emission spectra (solid lines) between YFP (A) and CFP (B) variants. Excitation wavelengths were 480 and 430 nm for YFPs and CFPs, respectively. The spectra represent the average of at least three measurements from three independent protein isolations [33].

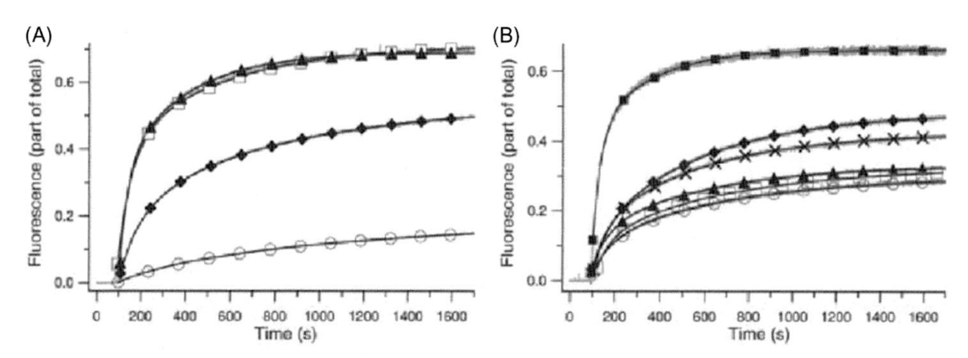

FIGURE 7.11 Refolding of fluorescent proteins after denaturation. Representative refolding curves with curve fits for the YFP variants (A) EYFP(Q69K) (O), Venus (□), mVenus (▲), and SYFP2 (♦), and for the CFP variants (B) ECFP (O), SCFP1 (■), SCFP2 (▲), SCFP3A (♦), SCFP3B (×), and Cerulean (A206K) (□). Fluorescence intensities are normalized to the fluorescence of an equal amount of native protein [33].

respectively) and due to enhanced protein folding and maturation. The latter enhancements most significantly contribute to the increased fluorescent yield in bacteria whereas they appear less significant for mammalian cell systems. SCFP3A and SYFP2 make a superior donor-acceptor pair for fluorescence resonance energy transfer, because of the high quantum yield and increased lifetime of SCFP3A and the high extinction coefficient of SYFP2. Furthermore, SCFP1, a CFP variant with a short fluorescence lifetime but identical spectra compared to ECFP and SCFP3A, was characterized. Using the large lifetime difference between SCFP1 and SCFP3A enabled us to perform for the first time dual-lifetime imaging of spectrally identical fluorescent species in living cells [33] (Figs. 7.10–7.12).

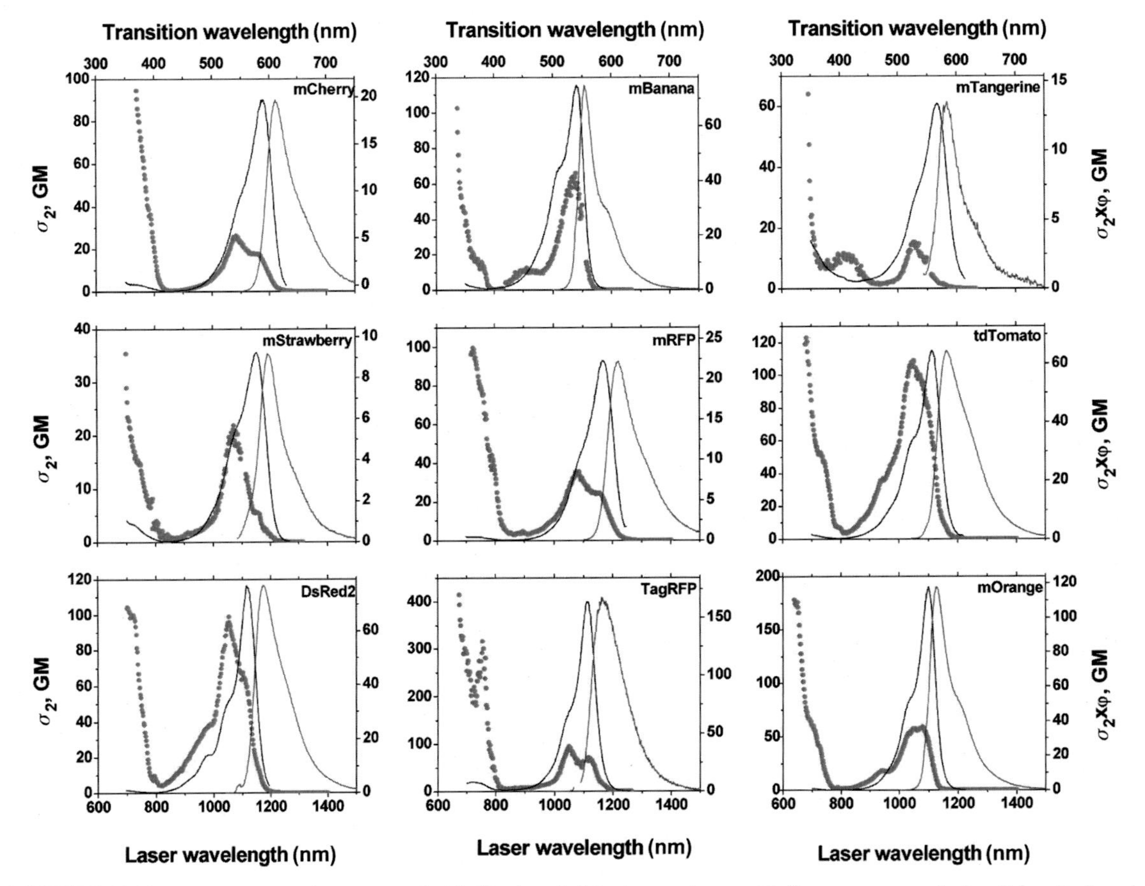

FIGURE 7.12 2PA spectra of orange and red FPs (symbols) shown along with fluorescence emission (blue line) and one-photon fluorescence excitation (black line) spectra. The left vertical scale shows the 2PA cross section. The scale on the right represents two-photon brightness. One-photon excitation and emission spectra are shown in arbitrary units [34]. *FP*, Fluorent Proteins.

References

[1] R. Ferraro, A. Agarwal, E.L. Martin-Macintosh, P.J. Peller, R.M. Subramaniam, MR imaging and PET/CT in diagnosis and management of multiple myeloma, Radiographics. 35 (2015) 438–454.

[2] T. Yoshida, H. Inoue, S. Usui, H. Satodate, N. Fukami, S.-E. Kudo, Narrow-band imaging system with magnifying endoscopy for superficial esophageal lesions, Gastrointest Endoscopy. 59 (2004) 288–295.

[3] A. Sieroń, K. Sieroń-Stołtny, A. Kawczyk-Krupka, W. Latos, S. Kwiatek, D. Straszak, et al., The role of fluorescence diagnosis in clinical practice, Onco Targets Ther 6 (2013) 977–982.

[4] P. Guitera, S.W. Menzies, C. Longo, A.M. Cesinaro, R. A. Scolyer, G. Pellacani, *In Vivo* Confocal Microscopy for Diagnosis of Melanoma and Basal Cell Carcinoma Using a Two-Step Method: Analysis of 710 Consecutive Clinically Equivocal Cases, J Investigative Dermatology 132 (2013) 2386–2394.

[5] D. Taieb, E. Kebebev, F. Castinetti, C.C. Chen, J.F. Henri, K. Pacak, Diagnosis and preoperative imaging of multiple endocrine neoplasia type 2: current status and future directions, Clin. Endocrinol. (Oxf. UK) 81 (2014) 317–328.

[6] D. Phillips, Light relief: photochemistry and medicine, Photochem. Photobiol. Sci. 9 (2010) 1589–1596.

[7] R.R. Gill, Imaging of mesothelioma, Recent Results Cancer Res. 189 (2011) 27–43.

[8] S. Roma, P. Mehta, J.A. Sosa, Medullary thyroid cancer early detection and novel treatments, Curr. Opin. Oncol. 21 (2008) 5–10.

[9] M.U. Khan, R.E. Coleman, Diagnosis and therapy of carcinoid tumors-current state of the art and future directions, Nucl. Med. Biol. 35 (Suppl. 1) (2008) S77–S91.
[10] A.L. Komaroff, Is human herpesvirus-6 a trigger for chronic fatigue syndrome? J. Clin. Virol. 37 (Suppl. 1) (2006) S39–S46.
[11] E.F.J. de Vries, Imaging of cyclooxygenase-2 (COX-2) expression: potential use in diagnosis and drug evaluation, Curr. Pharm. Des. 12 (2006) 3847–3856.
[12] H.A. Jones, Inflammation imaging, Proc. Am. Thorac. Soc. 2 (545–548) (2005) 513–514.
[13] P.W. Smith, Fluorescence emission-based detection and diagnosis of malignancy, J. Cell. Biochem. 2002 (Suppl. 39) (2002) 54–59.
[14] M.E. Koivunen, R.L. Krogsrud, Principles of immunochemical techniques used in clinical laboratories, Labmedicine 37 (8) (2006) 490–497.
[15] J.C.G. Bunzli, K.L. Wong, Lanthanide mechanoluminescence, J. Rare Earths 36 (1) (2018) 1–41.
[16] L. Armelao, S. Quici, F. Barigelletti, G. Accorsi, G. Bottaro, M. Cavazzini, et al., Design of luminescent lanthanide complexes: from molecules to highly efficient photo-emitting materials, Coord. Chem. Rev. 254 (2009) 487–505.
[17] J.C.G. Bunzli, Rising stars in science and technology: luminescent lanthanide materials, Eur. J. Inorg. Chem. 2017 (2017) 5058–5063.
[18] G.H. Beastall, I.D. Watson, Clinical chemistry and laboratory medicine: an appreciation, Clin. Chem. Lab. Med. 51 (3) (2013) 4.
[19] Y. Tao, W. Zhuang, X. Su, B. Ma, H. Hu, G. Li, et al., Dual-responsive micelles with aggregation-induced emission feature and two photon aborsption for accurate drug delivery and bioimaging, Bioconjugate Chem. 30 (2019) 2075–2077.
[20] J. Kapeleris, A. Kulasinghe, M.E. Warkiani, I. Vela, L. Kenny, K. O'Byrne, et al., The prognostic role of circulating tumor cells (CTCs) in lung cancer, Front. Oncol. 8 (2018) 311.
[21] T. Yu, W. Zhuang, X. Su, B. Ma, J. Hu, H. He, et al., Dual-responsive micelles with aggregation-induced emission feature and two-photon absorption for accurate drug delivery and bioimaging, Bioconjug Chem. 30 (2019) 2075–2087.
[22] C. Rossetti, A. Abdel Qader, T.G. Halvorsen, B.R. Sellergren, L. Reubsaet, Antibody-free biomarker determination: exploring molecularly imprinted polymers for pro-gastrin releasing peptide, Anal. Chem. 86 (2014) 12291–12298.
[23] W. Chen, Z. Meng, M. Xue, K.J. Shea, Molecular imprinted photonic crystal for sensing of biomolecules, Mol. Impr 4 (2016) 1–12.
[24] T. Xie, W.E. Bolch, C. Lee, H. Zaidi, Pediatric radiation dosimetry for positron-emitting radionuclides using anthropomorphic phantoms, Med. Phys. 40 (2013) 102502.
[25] Y. Tian, W.C. Wu, C.Y. Chen, S.H. Jang, M. Zhang, T. Strovas, et al., Utilization of micelles formed from poly(ethylene glycol)-block-poly(epsilon-caprolactone) block copolymers as nanocarriers to enable hydrophobic red two-photon absorbing emitters for cells imaging, J. Biomed. Mater. Res. A 1 (93) (2010) 1068–1079.
[26] X. Yue, A.R. Morales, G.W. Githaiga, A.W. Woodward, S. Tang, J. Sawada, et al., RGD-conjugated two-photon absorbing near-IR emitting fluorescent probes for tumor vasculature imaging, Org. Biomol. Chem. 13 (2015) 10716.
[27] D. Kim, H. Moon, S.H. Baik, S. Sungha, Y.W. Jun, T. Eang, et al., Two-photon absorbing dyes with minimal autofluorescence in tissue imaging: application to in vivo imaging of amyloid-β plaques with a negligible background signal, J. Am. Chem. 137 (2015) 6781–6789.
[28] G. Diaz-Londono, S. Garcia-Pareja, F. Salvat, A.M. Lallena, Monte Carlo calculation of specific absorbed fractions: variance reduction techniques, Phys. Med. Biol. 60 (2015) 2625–2644.
[29] J. Ardur, L. Erbes, M. Bianchu, S. Ruff, A. Zeitoune, M.F. Izaiguirre, et al., Quantitative assessment of colorectal cancer progression: a comparative study of linear and nonlinear microscopy, Techniques BioRxiv 2018 (2018) 1–26.
[30] A. Sadowska-Rociek, M. Surma, E. Cieslak, Comparison of different modifications on QuEChERS sample preparation method for PAHs determination in black, green, red and white tea, Environ. Sci. Pollut. Res. Int. 21 (2013) 1326–1328.
[31] J.W. Kim, S.I. You, N.H. Kim, J.A. Yoon, K.W. Cheah, F.R. Zhu, et al., Study of sequential Dexter energy transfer in high efficient phosphorescent white organic light-emitting diodes with single emissive layer, Sci. Rep. 4 (2014) 7009.
[32] K. Sudha, S. Sundharamurthi, S. Karthikaikumar, K. Abinaya, P. Kalimuth, Switching of Förster to Dexter mechanism of short-range energy transfer in meso -anthrylporphyrin, J. Phys. Chem. C 121 (2017) 5941–5948.
[33] G.J. Kremers, J. Goedhart, E.B. van Munster, T.W.J. Gadella Jr, Cyan and yellow super fluorescent proteins with improved brightness, protein folding, and FRET Förster radius, Biochemistry 45 (2006) 6570–6580.
[34] M. Drobizhev, S. Tillo, N.S. Makarov, T.E. Hughes, A. Reban, Absolute two-photon absorption spectra and two-photon brightness of orange and red fluorescent proteins, J. Phys. Chem. B 113 (2009) 855–859.

CHAPTER

8

Conclusion and outlook

The few examples presented above should give at least a first idea of what excited state is able to do. The contribution that photochemistry and allied sciences have given to the development of human physiology is really large. What next? New instruments are now available and should be the basis of new development aiming at answering further key questions. From the synthetic point of view, the peculiar ability of photochemistry is transforming weak interactions into covalent bonds. As a simple example, *n* donors form solid complexes with halogens and other strong electrophiles [1,2], but they independently absorb light and react when irradiated (and in different ways when different modes of complexation exist) [3] On the contrary, the mode of formation of such bonds can be guided to a degree. As shown in the cover of this book, the plunger looks for a single molecule and turns only that molecule reactive. This fact bewildered the first explorers of the field, in that the overall energy transferred to the solution irradiated seemed to be ridiculously lower than required for forming the resulting high-energy molecules. This, however, was easily explained when the selective formation of electronic excited states was recognized (the energy is taken up from the few molecules that absorb it, not from the part that overcomes the activation energy, and the rate does not increase with temperature).

As shown in the cover of this book, the plunger looks for a single molecule and turns only that molecule reactive. This characteristic bewildered the scientists who entered this realm, as it seemed that an unreasonably small amount of energy was transferred to the reacting solution. Once the role of electronic excited states was recognized, all became understandable. Many photochemists are convinced that there is still something to exploit in solar light, in particular, from the crowded region around 310 nm, close to ozone border, where many molecules absorb and exert key functions (a cyclohexene ring opening to previtamin D, erythema forming/tanning) (Fig. 8.1).

It must be possible to maximize pleasant and useful effects, while eliminating the unpleasant effects and arrive to a perfect light-healing hospital 120 years after the experimental establishment of the sun clinic in Copenhagen by Niels Fielden, the importance of which was recognized in 1903 after he won the Nobel Prize in medicine (for his contribution to the treatment of diseases, especially *lupus vulgaris*, with concentrated light radiation, whereby he opened a new avenue for medical science), but with a much more developed knowledge of (bio)chemistry to make the process of confronting difficult topics in a new way somewhat easier. It can be safely assumed that the most valuable applications will be in

Light, Molecules, Reaction and Health
DOI: https://doi.org/10.1016/B978-0-12-811659-3.00008-6

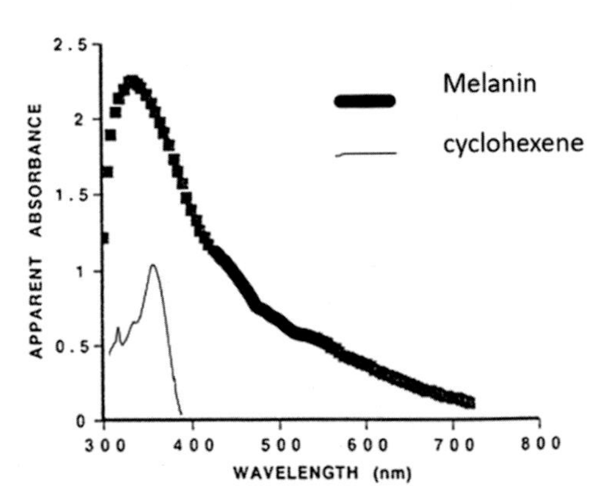

FIGURE 8.1 Absorption spectrum of melanin (thick line) and 7-dehydrocholesterol. No quantitative attribution is meant.

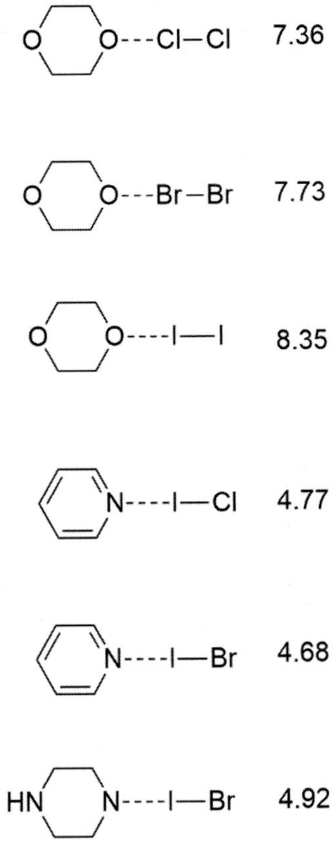

O---Cl—Cl 7.36

O---Br—Br 7.73

O----I—I 8.35

N----I—Cl 4.77

N----I—Br 4.68

HN N----I—Br 4.92

SCHEME 8.1 Distance of heteroatom–electron withdrawing species. Such complexes absorb light independently and react [1].

medicinal chemistry in a large variety of reactions, always taking advantage of the possibility to freely choose the "where" and "when" of photochemical reactions. As for "what," the most useful exploitation is when the photon energy is used for forming a covalent bond (see above). Thus drug D form complexes with a site in a biomolecule M (M ... D), or, in a simple example, n donor form complexes with halogens and related electrophiles, but they absorb and react when irradiated (independently when different modes of complexation exist). In particular, this may be referred to drugs since these (and particularly anticancer drugs) by definition involve complexation with a biochemical active site (see Scheme 8.1). As is well known, this is a question of balance between the fast establishing of a site-selective equilibrium (M ... D), in order to minimize damage to heal cells, and the fast reaction of the complex, to maximize destruction of the tumor cells. In practice, this results in the use of highly reactive drugs, and chemotherapy is often quite toxic. When shifting to photochemotherapy, however, one can resort to the use of a prodrug (PD) that is able to complex site M, but nonreactive (see Scheme 8.1A).

Photoactivating a molecule with spatial and temporal selectivity changes the very foundations of medicinal chemistry. Thus most of the methods elaborated as photoactivatable protecting groups are based on a strong increase in the electrophilicity of a group, which becomes susceptible to water hydrolysis (by the way, the same path followed by nature in the photochemical phenomenon par excellence for vision). Transforming a weak interaction into a strong covalent bond is the typical output of a photochemical reaction, for example, in the charge transfer examples shown in Schemes 8.1 and 8.2 [1].

As is well known, this is a question of balance between the fast establishing of a site-selective equilibrium (M ... D), in order to

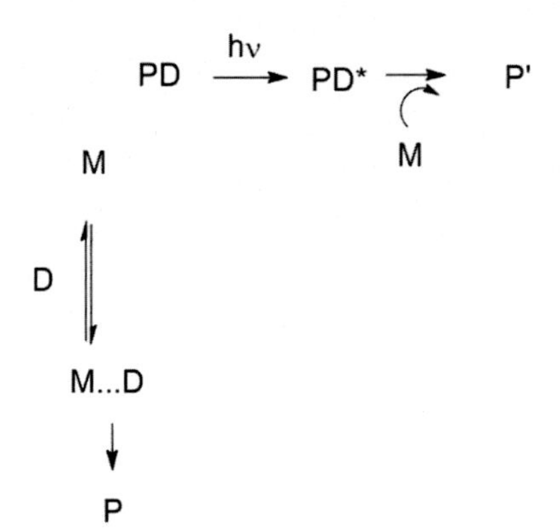

SCHEME 8.2 Easy cleavage of a molecule after irradiation and hydrolysis. All of these processes are well characterized and optimized, both of the fragments may be the one active as D.

minimize damage to heal cells and the fast reaction of the complex, to maximize destruction of the tumor cells. In practice, this results in the use of highly reactive drugs, and chemotherapy is often quite toxic. When shifting to photochemotherapy, however, one can resort to the use of a PD that is able to complex site M, but in a nonreactive manner (see Scheme 8.3). In this case one can take their time for letting the equilibrium establish and then irradiate in order to form actual drug D. Examples of molecules useful for this approach mostly contain substituents originally developed as photoremovable protecting groups, in which the electrophilic/nucleophilic character is strongly increased upon irradiation, so that they are smoothly hydrolyzed (Scheme 8.3) [4]. Another approach is based on maintaing the drug within a natural or man-made membrane (cyclodextrins, liposomes, etc.; see Fig. 8.2 [5]) and drilling a hole when deemed appropriate so as to discharge the cargo in the most useful location.

Considering the formation of covalent bonds is easy to contemplate the formation of chiral centers, not so much because of the interest in synthesis. It is important to remember that all of the reactions occurring in the cells are taking place under a chiral environment, and thus one has to take into account all of the possibilities. This is particularly interesting for biology and medicine practitioners and to simply open our eyes toward that perspective. Notice that light itself is chiral (it travels at ~300,000 km/vibration) perpendicularly to the axe of propagation, either with a right or left handedness. This characteristic is difficult to exploit and arrive to an absolute chemical synthesis, because of the quite limited differences between the absorptivity of two enantiomers with the polarized light, but the other possibilities are certainly working (chiral transfer from an existing molecule, enantioselective catalysis, and atropoisomeric). This ability has always been a speciality of photochemistry since Pasteur's time and from them the "croce e delizia" of chemistry [6] (Scheme 8.3 and Fig. 8.3).

A different application again involves the separation of the two steps and refers to the retaining of drug within a synthetic or natural membrane (micelles, liposomes, etc.), and involves irradiation and cleavage of the bond, when it is felt appropriate. Suitable PD should contain at least an antenna group and a cleaving group; the active drug D may result, in general, from either of the two parts at the extremes of the cleaving bond. In contrast, such selectivity also manifests in several subsequent reactions. Important is the case of crystals where the transformation may occur over several chiral selective steps, as in well-known case of santonin that undergoes four selective rearrangement reactions, the atom displacements in the crystal become excessive and the crystal bursts. This property was noticed when this molecule was first reported in 1834 as the first light-caused reaction, although the actually occurring reactions in the solid state were fully different from those occurring in the solution state, which were understood only 180 years later [7]. Reacting molecules when associated with another chiral, or simply the

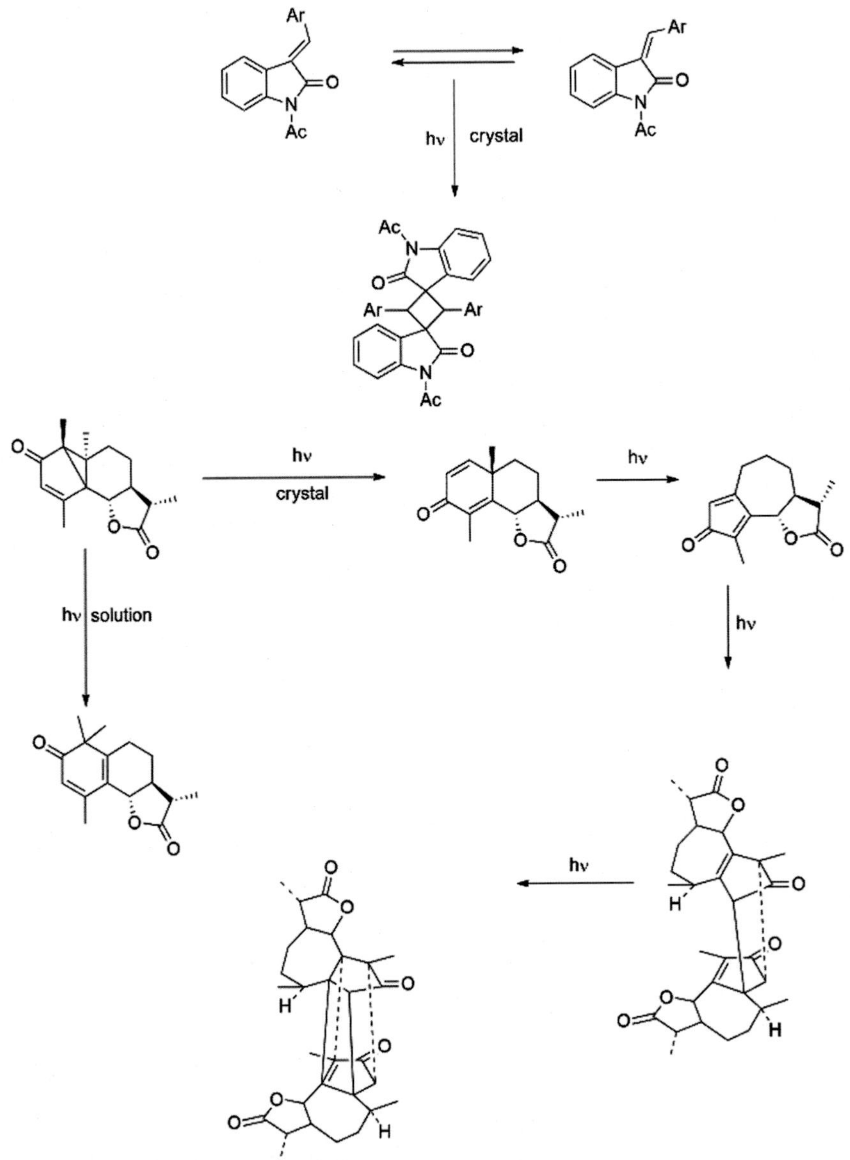

SCHEME 8.3 Photochemical rearrangement in crystal-to-crystal reactions of oxindole and santonin (four times) [7,8].

cocrystal in a chiral nonsymmetric form, may give excellent results in chiral synthesis.

The impression a chemistry practitioner receives by hearing all of the above is that of being assigned too much homework to the class, which leads to; many aspects being retained only on the paper. In fact, the principle that anything that chemistry can do, photochemistry may do better, may be accepted by now, and in particular, photochemistry is an excellent method for discussing reverse asymmetric synthesis in a group meeting, as

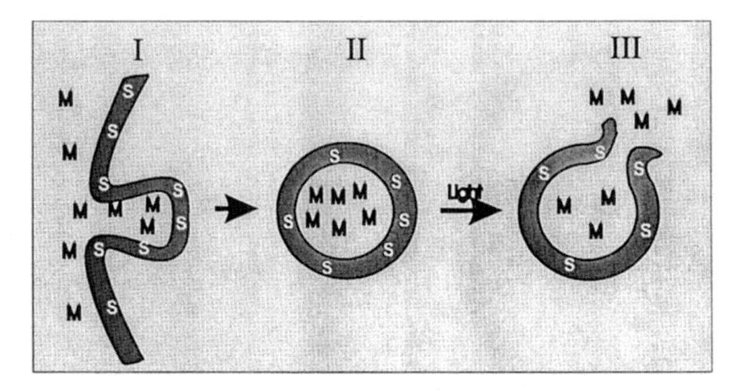

FIGURE 8.2 Photochemical internalization, the carrier vesicle takes up a cargo of a drug and carries it on the desired target, where photoreaction drills a hole in the membrane and discharges the cargo. Both passive (no interaction) and active (specific) interactions may occur [5].

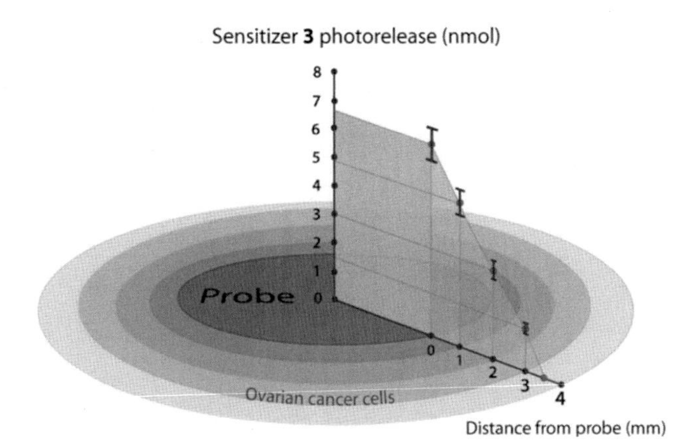

FIGURE 8.3 Sensitizer released (nM after 1-h irradiation, equated to singlet oxygen formation), and thus oxidative cleavage occurring into the OVCAR-5 cell films (mm) as a function of treatment with the fiber optic-based sensitizer delivery device equipped with a Vycor probe tip. Amount of sensitizer released directly beneath the probe tip and the amount that diffused away from the periphery of the probe tip [9].

cleaving and making bonds by a pencil has a reasonable probability to work. Consider a typical example that any of the (~200) photo-activated enzymes that operate on reversing the 2 + 2 cycloaddition reaction of thymine. It can be safely anticipated that such enzymes are ready for new medicinal applications [7]. Indeed, it has been applied in a large range of reactions.

Furthermore, the delicate function of light takes in the circadian system also suggests that light-activated drugs will have a key role in governing the mood of people [10]. Optogenerapy [11] stands for combined optogenetic and gene therapy, and holds the promise of a new modern syringe era. The strategy combines the engineering of synthetic near-infrared (NIR) optogenetic pathway with a macroencapsulation device equipped with a wireless powered optoelectronic circuit. In the electronic implant, the NIR light is used as a cell–machine interface.

The strategy offers the possibility to use genetically modified cell line in confinement that first of all protects the therapeutic cells from the circulating immune cells and also protects the patient from the common risk associated with the cell therapy. As the device is hermetically sealed, it is also possible to explant it. The proof of the optogenerapy concept was performed by connecting a brain–computer interface (BCI) to a powering antenna wirelessly powering the bioelectronic cell-based implant placed subcutaneously in the back of a mouse. The user connected to a BCI could trigger the activation of the optogenerapy device following a specific mental task and thus could control the secretion of a biomarker in the blood circulation in a rodent animal model. The wireless powered cell-based implant method is gaining attention as researchers have tested its efficacy for the release of insulin in the context of diabetic disorder in rodent animal model. Light-sensitive therapeutic delivery cells can be genetically programmed to produce any therapeutic recombinant proteins. NIR light bioelectronic cell-based implant potentially represents a disruptive innovation to support light-controlled drug delivery.

Most applications are expected to be successful in medicine, in particular, on the basis of the well-known fact that tumor cells differ from the normal ones due to the enhanced permeability and retention effect with liposomes and nanoparticles. Thus liposomes cross the blood vessels and either accumulate in the tumor area through a passive effect (passive targeting) or are taken up by the cells (active targeting in Fig. 8.2). If, at this point, the membrane is destroyed, the drug contained in the liposome is liberated in the vicinity of, or within, tumor cells. For this job, a number of liposomes have been elaborated by inserting photoactivatable groups in the photophatides. Actually, this idea remains on the papers rather than reacting as expected, liposomes are preferentially captured by the reticulondothelial system [12], but different approaches may be considered, as an example, in-depth studies of photoremovable groups.

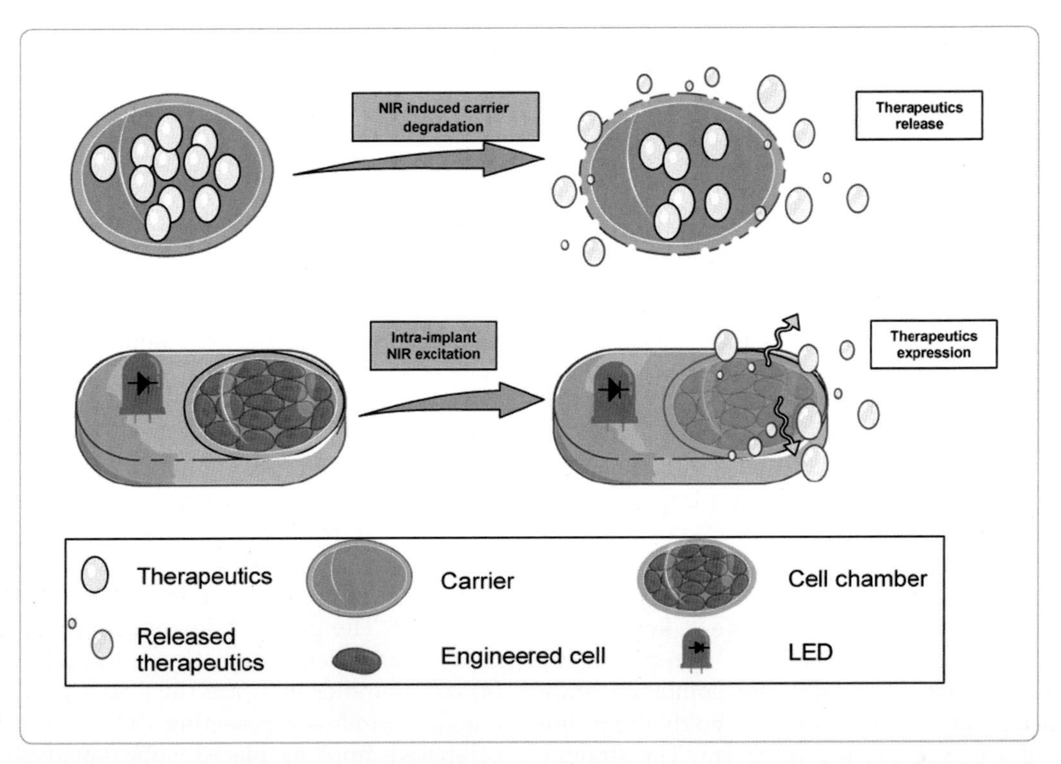

FIGURE 8.4 Near-infrared (NIR)-controlled drug-delivery devices: smart vesicles loaded with NIR photothermal agent. Wireless powered optogenetic cell-based implant; a NIR light–emitting device (LED) controls the gene expression of engineered light-sensitive therapeutic cells residing in the device cell chamber [10,15,16].

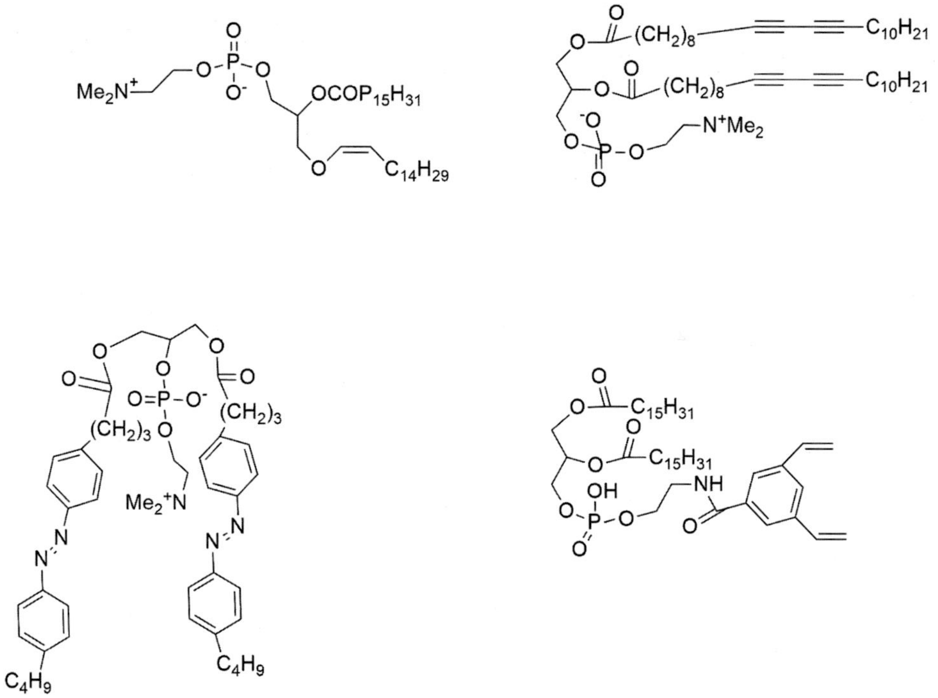

SCHEME 8.4 Light-activated phosphatides used in drugs delivery.

Another topic that demands attention is the fate of drugs. As it is well known, drugs act through complexation, and are not consumed. Thus drugs are excreted while their structure is conserved, or after only minor changes, and generally maintains the photoreactivity [13]. The general stability of such compounds favor accumulation (mostly in the case of veterinary drugs in part of the countries where husbandry is important or in the vicinity of hospitals). Thus a "green" lifetime programmed drug can not be put on the market it would be too weak a medicament. In turn, this leads to a progressive bacterial resistance, as soon as new aggressive strain develops. Here again photochemistry comes handy, as photocalysis in oxygen equilibrated water bodies has proved to be one of the best advanced oxidation methods. Surely, there is no ground to think that this topic will not encounter any trouble and disappointmens, indeed it has already happened, but then not much, and after all, this is what gains (photo)chemists a living (and much fun). [14]

On the nonmedicinal topics, an importance should be reserved to disinfection, and in particular to the usual principle of the iron hand in a velvet glove, to the disinfection of food, as selectivity and efficiency have to play between them in order to have a savory, yet save food [15,16] (Fig. 8.4; Scheme 8.4).

References

[1] O. Hassel, Structural aspect of interatomic charge-transfer bonding, Science 170 (1970) 497–502.

[2] G. Cavallo, P. Metrangolo, R. Milani, T. Pilati, A. Priimagi, G. Resnati, et al., The halogen bond, Chem. Rev. 116 (2016) 2478–2601.

[3] P. Das, M. Bahoum, Y.P. Leem, Reactions between atomic chlorine and pyridine in solid para-hydrogen: infrared spectrum of the 1-chloropyridinyl (C_5H_5NCl) radical, J. Chem. Phys. 138 (2013) 054307.
[4] P. Klán, T. Šolomek, C.G. Bochet, A. Blanc, R. Givens, M. Rubina, et al., Typical photochemically detachable protecting groups, used here for generating drugs from prodrugs. Photoremovable protecting groups in chemistry and biology: reaction mechanisms and efficacy, Chem. Rev. 113 (2013) 119−191N.
[5] K. Berg, P.K. Selbo, L. Prasmickaite, T.E. Tjelle, K. Sandvig, J. Moan, et al., A novel technology for delivery of macromolecules into cytosol, Photochem. Internalization 59 (1999) 1180−1183.
[6] B.L. Feringa, R.A. van Delden, Absolute asymmetric synthesis: the origin, control, and amplification of chirality, Angew. Chem. Int. Ed. Engl. 38 (1999) 3418−3438.
[7] A. Natarajan, C.K. Tsai, S.I. Khan, P. McCarren, K.N. Houk, M.A. Garcia-Garibay, The photoarrangement of α-santonin is a single-crystal-to-single-crystal reaction: a long kept secret in solid-state organic chemistry revealed, J. Am. Chem. Soc. 129 (2007) 9846−9847.
[8] M. Milanesio, D. Viterbo, A. Albini, E. Fasani, R. Bianchi, M. Barzaghi, Structural study of the solid-state photoaddition reaction of arylidenoxindoles, J. Org. Chem 65 (2000) 3416−3425.
[9] D. Bartusik, D. Aebisher, A. Ghogare, G. Ghosh, I. Abramova, T. Hasan, et al., A fiberoptic (photodynamic therapy type) device with a photosensitizer and singlet oxygen delivery probe tip for ovarian cancer cell killing, Photochem. Photobiol. 89 (2013) 936−941.
[10] V. Pierroz, M. Folcher, From photobiolumination to optogenerapy, recent advances in NIR light photomedicine applications, J. Mol. Genet. Med. 2 (2018) 2.
[11] G.E. Lawson, Y. Lee, A. Singh, Formation of stable nanocapsules from polymerizable photopholipids, Langmuir 19 (2003) 6401−6407.
[12] X. Sun, X. Yan, O. Jacobson, W. Sun, Z. Wang, X. Tong, et al., Improved tumor uptake by optimizing liposome based RES blockade strategy, Theranostics 7 (2017) 319−328.
[13] A. Speltini, M. Sturini, F. Maraschi, A. Profumo, A. Albini, Analytical methods for the determination of fluoroquinolones in solid environmental matrices, TrAC 30 (2011) 1337−1350.
[14] A. Albini, E. Fasani, Photochemistry of Drugs, RSC, Cambridge, 2000.
[15] J.F. Diehl, C. Hasselmann, D. Kilcast, Regulation of food irradiation in the European Community: is nutrition an issue? Food Control 12 (1991) 212−219.
[16] J. Farkas, C. Mohacsi-Farkas, History and future of food irradiationTrends, Food Sci. Technol. 22 (2011) 121−126.

Index

Note: Page numbers followed by "*f*" and "*t*" refer to figures and tables, respectively.

D

E

Printed in the United States
By Bookmasters